Springer Proceedings in Mathematics & Statistics

Volume 359

This book series features volumes composed of selected contributions from workshops and conferences in all areas of current research in mathematics and statistics, including operation research and optimization. In addition to an overall evaluation of the interest, scientific quality, and timeliness of each proposal at the hands of the publisher, individual contributions are all refereed to the high quality standards of leading journals in the field. Thus, this series provides the research community with well-edited, authoritative reports on developments in the most exciting areas of mathematical and statistical research today.

More information about this series at http://www.springer.com/series/10533

Mohd Hafiz Mohd · Md Yushalify Misro ·
Syakila Ahmad · Doanh Nguyen Ngoc
Editors

Modelling, Simulation and Applications of Complex Systems

CoSMoS 2019, Penang, Malaysia, April 8–11, 2019

Springer

Editors
Mohd Hafiz Mohd
School of Mathematical Sciences
Universiti Sains Malaysia
Penang, Malaysia

Md Yushalify Misro
School of Mathematical Sciences
Universiti Sains Malaysia
Penang, Malaysia

Syakila Ahmad
School of Mathematical Sciences
Universiti Sains Malaysia
Penang, Malaysia

Doanh Nguyen Ngoc
Department of Science and Technology
Thuyloi University
Hanoi, Vietnam

ISSN 2194-1009 ISSN 2194-1017 (electronic)
Springer Proceedings in Mathematics & Statistics
ISBN 978-981-16-2628-9 ISBN 978-981-16-2629-6 (eBook)
https://doi.org/10.1007/978-981-16-2629-6

Mathematics Subject Classification: 92-10, 76-10, 93A16, 68T09, 90-10

This Springer imprint is published by the registered company Springer Nature Singapore Pte Ltd.
The registered company address is: 152 Beach Road, #21-01/04 Gateway East, Singapore 189721, Singapore

Preface

This volume collects research papers and survey articles of participants and speakers in the Workshop on Complex Systems Modelling and Simulation 2019 (CoSMoS 2019): IoT and Big Data Integration, which was held in Penang, Malaysia, from 8–11 April 2019. The event was jointly organized by the School of Mathematical Sciences, Universiti Sains Malaysia (USM), Water Resource and Disaster Management (WARM) Vietnam, South East Asia Centre of Unité de Modélisation Mathématique et Informatique des Systèmes Complexes (UMMISCO France-Vietnam), the French Research Institute for Development (IRD) and Thuyloi University, Vietnam. CoSMoS 2019 had also attracted international delegates and participants from across the globe such as Japan, French, Bangladesh, Nigeria, Egypt, China, Indonesia, the Philippines, Myanmar, Vietnam and Malaysia.

The main topics highlighted during this international event were complex systems and agent-based modelling (ABM) incorporating elements of big data analytics and also Internet of things (IoT). Big data methods are often employed to discover potentially interesting patterns in large data sets, while ABM is an approach to investigate complex systems that arise in different real-life situations such as biology, engineering and socio-economic problems. This modelling framework often concentrates on the interactions of heterogeneous agents such as people, animals, vehicles and other entities, and sometimes interesting insights can emerge from the behaviours and also interactions of these agents at an individual scale. During this workshop, the potential of combining big data methods and IoT with those of ABM techniques was discussed in order to make a prediction about complex (biological, physical or engineering) systems and to support the decision-making process. One of the main targets in CoSMoS 2019 was to employ complex systems and ABM techniques together with big data and IoT approaches in order to solve the distinct problems faced by the local community and other parts of the world. Among the issues discussed during this workshop were transportation, agriculture and other real-life problems.

Consistent with the themes of CoSMoS 2019, the Springer PROMS volume entitled "Modelling, Simulation and Applications of Complex Systems - CoSMoS 2019, Penang, Malaysia, 8–11 April 2019" is put together and specially designed to highlight some significant research results on broad disciplinary areas of complex systems. This edited volume takes into account a multidisciplinary approach in complex systems analysis that will encourage the transfer of ideas and methodology from modelling and simulation fields to the other areas of knowledge (and vice versa). In fact, one of the main highlights of this book is on the practical aspects of complex systems analysis in solving daily life problems faced by the local communities, government policy makers, industries and other relevant parties. Special attention is given to various applications on the techniques of complex systems in examining the transportation and mobility networks, social issues, epidemiology, ecological and conservation management, environmental problems, engineering and industrial applications. Many realistic biological and physical examples from recent research are employed in this book as illustrations. This book will be of interest to a broad readership including those interested in complex systems research and other related areas such as mathematical modelling, big data analytics, numerical simulation and agent-based modelling frameworks.

All authors contributing to the chapters in this book are in fact active researchers in the fields. The papers in this volume have also been through a rigorous refereeing process to ensure high scientific quality and standards. Through the publication of this edited volume, it will provide a platform for researchers from various scientific backgrounds the opportunity of disseminating recent research and knowledge in complex systems modelling and analysis. While other existing books in the fields often give more attention to a particular topic, this edited volume concentrates on the state-of-the-art research on complex systems in a broader sense. This is crucial to ensure that readers can grasp different concepts and ideas in complex systems research and also can demonstrate how the techniques from modelling and simulation and computational methods being employed to investigate the complexity of real-life systems. The development of new computational tools can improve our understanding of complex systems under consideration and highlight the emergent observations of complex biological and physical phenomena of interest.

We would like to thank all participants, course lecturers and invited speakers of the workshop for making the event a great success. Thanks to the organizing committee as well for their great efforts in conducting a well-run event. We are grateful to all the authors for their contributions to this volume and to all the reviewers for their timely and detailed feedbacks on the manuscripts. Additionally, we would like to extend our deepest gratitude to the School of Mathematical Sciences USM, WARM, UMMISCO, IRD France, Thuyloi University, Divison of Academic and International USM, Divison of Research and Innovation USM, Malaysian Mathematical Sciences Society (PERSAMA), Penang Convention and Exhibition Bureau (PCEB), Uni Paper Products Sdn. Bhd., Penang State Museum and Universiti Sains Malaysia Press for their valuable support and sponsorship to make the event successful.

April 2019

Organization

Workshop on Complex Systems Modelling and Simulation 2019 (CoSMoS 2019): IoT and Big Data Integration was jointly organized by the School of Mathematical Sciences, Universiti Sains Malaysia (USM), Water Resource and Disaster Management (WARM) Vietnam, South East Asia Centre of Unité de Modélisation Mathématique et Informatique des Systèmes Complexes (UMMISCO France-Vietnam), the French Research Institute for Development (IRD) and Thuyloi University, Vietnam.

USM Organizing Committee

Advisors

Hailiza Kamarulhaili
Ahmad Izani Md. Ismail

Chairperson

Mohd Hafiz Mohd

Deputy Chairperson

Majid Khan Majahar Ali

Secretary

Noor Saifurina Nana Khurizan

Assistant Secretary

Norshafira Ramli

Treasurer

Md Yushalify Misro

Assistant Treasurer

Yusnita Yusop

Programme Committee

Yazariah Mohd Yatim
Syakila Ahmad
Mohd Nadhir Ab Wahab
Ahmad Sufril Azlan Mohamed

Food and Accommodation Committee

Fam Pei Shan
Siti Zulaikha Mohd Jamaludin

Technical, Logistic and Transportation Committee

Norazrizal Aswad Abdul Rahman
Syed Mohamed Hussain Syed Osman

Publicity and Promotion Committee

Ong Wen Eng
Hartini Ahmad

Secretariat

Siti Amirah Abd Rahman
Ahmad Wafi Sahedan
Shamani Supramaniam
Ng Zhen Chuan
Nur Atiqah Jamaluddin

International Scientific Committee

Advisor

Alexis Drogoul

Chairperson

Nguyen Ngoc Doanh

Secretary

Le Huu Ton

Scientific Programme Committee (Course Speakers)

Benoit Gaudou
Kevin Chapuis
Tran Giang Son
Le Huu Ton

Contents

Deterministic Modelling and Simulation of Complex Systems

Data Science and Optimization of Complex Systems

Editorial Notes

Mohd Hafiz Mohd, Md Yushalify Misro, Syakila Ahmad, and Doanh Nguyen-Ngoc

Abstract This is an introductory chapter to the edited volume entitled "Modelling, Simulation and Applications of Complex Systems, Penang, Malaysia, April 8–11, 2019" that is published in conjunction with the organisation of Workshop on Complex Systems Modelling & Simulation 2019 (CoSMoS 2019): IoT & Big Data Integration. This event was held in the School of Mathematical Sciences, Universiti Sains Malaysia (USM), from April 8–11, 2019.

Keywords Agent-Based Modelling · Deterministic Modelling and Simulation · Data Science and Optimization · Complex systems

1 Introduction

This is an introductory chapter to the edited volume entitled "Modelling, Simulation and Applications of Complex Systems, Penang, Malaysia, April 8–11, 2019" that is published in conjunction with the organisation of Workshop on Complex Systems Modelling & Simulation 2019 (CoSMoS 2019): IoT & Big Data Integration. This event was held in the School of Mathematical Sciences, Universiti Sains Malaysia (USM), from April 8–11, 2019. This volume discusses the latest progresses and developments on complex systems research. It intends to give an exposure to

M. H. Mohd (✉) · M. Y. Misro · S. Ahmad
School of Mathematical Sciences, Universiti Sains Malaysia, 11800 Penang, Gelugor, Malaysia
e-mail: mohdhafizmohd@usm.my

M. Y. Misro
e-mail: yushalify@usm.my

S. Ahmad
e-mail: syakilaahmad@usm.my

D. Nguyen-Ngoc
Department of Science and Technology, Thuyloi University, 175 Tay Son Street, Dong Da District, Hanoi, Vietnam
e-mail: doanhnn@tlu.edu.vn

M. H. Mohd et al. (eds.), *Modelling, Simulation and Applications of Complex Systems*, Springer Proceedings in Mathematics & Statistics 359,
https://doi.org/10.1007/978-981-16-2629-6_1

prospective readers about the theoretical and practical aspects of mathematical modelling, numerical simulation, data science, optimization techniques and agent-based modelling frameworks in complex systems. The main purpose of this book is to emphasise a unified approach to complex systems analysis, which goes beyond to examine complicated phenomena of numerous real-life systems; this is done by investigating a huge number of components that interact with each other at different (microscopic and macroscopic) scales; new insights and emergent collective behaviours can evolve from the interactions between individual components and also with their environments. These tools and concepts will permit us to better understand the patterns of various real-life systems and help us to comprehend the mechanisms behind which distinct factors shaping some complex systems phenomena being influenced.

This volume compiles contributions on the analysis of complex systems and related applications. The collection of research work presented here contributes to a better understanding on how the techniques from modelling and simulation, data science and optimization being employed to investigate the complexity of real-life systems. These articles provide important insights into a multidisciplinary approach in the analysis of complex systems, which will encourage the transfer of ideas and methodology from modelling and simulation fields to the other areas of knowledge (and vice versa), as well as providing new perspectives in the understanding of the emergent observations of complex phenomena of interest. A total of 21 research papers [1–21] were accepted after a single-blind review process by at least two regional and international experts using the Springer PROMS review guidelines.

Among the computational tools and modelling frameworks that have been employed in this volume are Agent-Based Modelling (ABM) in Complex Systems (Section I), Deterministic Modelling and Simulation of Complex Systems (Section II) and Data Science and Optimization of Complex Systems (Section III). In ABM section, different real-life problems such a social election system [1], epidemiology [2, 5], traffic-related air pollution issues [3], disaster emergency and evacuation planning [4] have been considered. It is shown how this bottom-up approach can be used to help in understanding relationships and thus possible causal mechanisms that can shape behaviours of distinct complex systems of interest.

Additionally, in Section II, differential equations-based systems are employed in order to model distinct complex systems observations under consideration. Special attention is also given to the applications of mathematical modelling techniques in biology (e.g., epidemiology, ecology, and cell biology) and physics. Several biological and physical examples from recent research are utilised as illustrations in this volume. For instance, the techniques from differential equations, dynamical systems, optimal control and numerical simulation are used to examine the transmission dynamics of diseases such as a Fasciola epidemic with treatment and quarantine [6], a cardiovascular-respiratory system model [7], a cancer cell model with alternative therapies [8] and a tuberculosis model with the risk of re-infection [9]; similar approaches can also be used to analyse different key issues in biology and physics e.g., lake eutrophication [10] and nanofluids problems [11].

To emphasise parallel with the theme of CoSMoS 2019 i.e., Internet of Things (IoT) & Big Data Integration, a dedicated section on Data Science and Optimization of Complex Systems is compiled in Section III. By considering an epidemiological problem such as dengue infections [12, 13], which emerge as a public health burden in Southeast Asia, this infectious disease is studied in greater depth using network analysis and statistical techniques. This epidemic issue is crucial to be addressed with the advancement of data science techniques as the Southeast Asian countries are bracing for possible outbreaks of infectious diseases that show symptoms similar to those of the novel coronavirus (COVID-19) ahead of the rainy season, complicating the work of front-line medical workers [22–26]. Using spatio-temporal statistics, this volume also examines distinct infectious diseases e.g., measles [14] and typhoid [15] and study the persistence of epidemiological patterns over time and space. Other current research trends in data science are also demonstrated in this edited volume through the chapters on the application of machine learning in chaotic systems [16] and optimization techniques of production planning problems [17–19]. Additionally, recent applications of optimization methods in traffic and transportation problems [20, 21] are also discussed and these are in fact some of the research outputs from CoSMoS 2019 working groups.

Different works corresponding to distinct chapters in this volume were carried out by considering some geographical areas (and/or real datasets) from: Japan [1], Vietnam [3, 4], Senegal [6], Austria [7], China [10], Malaysia [2, 5, 10], the Philippines [8, 12–15] and the United States [20, 21]. This is possible through scientific collaboration formed under CoSMoS 2019 initiatives between numerous research institutions such as the School of Mathematical Sciences USM, Water Resource and Disaster Management (WARM) Vietnam, South East Asia Center of Unité de Modélisation Mathématique et Informatique des Systèmes Complexes (UMMISCO France-Vietnam), The French Research Institute for Development (IRD) and Thuyloi University, Vietnam.

While the contributions collected in this edited volume are encouraging, it is evident that more discussions and research are needed to improve our understanding of the complexity science and this subject warrants further studies. These promising research directions can help to shed some light in better comprehending the differences between complex systems in different fields and understanding the causes of emergent behaviour in terms of system components, their attributes, and their interactions. We hope that this edited volume will stimulate and promote new research work by theoreticians, modellers, and data scientists on the intriguing open problems in complex systems.

References

1. Haga, H., Takatani, T.: Simulation of Japanese national diet members election system using agent-based modeling. In: Mohd, M.H., Misro, M.Y., Ahmad, S., Ngoc, D.N. (eds.) Modelling, Simulation and Applications of Complex Systems. CoSMoS 2019. Springer Proceedings in Mathematics & Statistics (2021)
2. Feng, L.L.W., Mohd, M.H.: Modelling the dilution and amplification effects on Sin Nombre Virus (SNV) in deer mouse. In: Mohd, M.H., Misro, M.Y., Ahmad, S., Ngoc, D.N. (eds.) Mod-

elling, Simulation and Applications of Complex Systems. CoSMoS 2019. Springer Proceedings in Mathematics & Statistics (2021)

3. Brugière, A., Pham, M.D., Chapuis, K., Drogoul, A., Gaudou, B., Grignard, A., Marilleau, N, Tri, N.H.: Experimenting the impact of pedestrianisation on urban pollution using tangible agent-based simulations. In: Mohd, M.H., Misro, M.Y., Ahmad, S., Ngoc, D.N. (eds.) Modelling, Simulation and Applications of Complex Systems. CoSMoS 2019. Springer Proceedings in Mathematics & Statistics (2021)
4. Chapuis, K., Elwaqoudi, T., Brugiere, A., Daude, E., Drogoul, A., Gaudou, B., Ngoc, D.N., Nghi, H.Q., Zucker, J.D.: An agent-based co-modeling approach to simulate the evacuation of a population in the context of a realistic flooding event. In: Mohd, M.H., Misro, M.Y., Ahmad, S., Ngoc, D.N. (eds.) Modelling, Simulation and Applications of Complex Systems. CoSMoS 2019. Springer Proceedings in Mathematics & Statistics (2021)
5. Feng, L.L.W., Mohd, M.H.: Agent-based modelling using GAMA 1.8 with applications to biological system in epidemiology. In: Mohd, M.H., Misro, M.Y., Ahmad, S., Ngoc, D.N. (eds.) Modelling, Simulation and Applications of Complex Systems. CoSMoS 2019. Springer Proceedings in Mathematics & Statistics (2021)
6. Diaby, M., Diop, O., Nassouri, E., Sene, A., Sene, M.: Mathematical analysis of Fasciola epidemic model with treatment and quarantine. In: Mohd, M.H., Misro, M.Y., Ahmad, S., Ngoc, D.N. (eds.) Modelling, Simulation and Applications of Complex Systems. CoSMoS 2019. Springer Proceedings in Mathematics & Statistics (2021)
7. Calderon, P.G.B., Palma, L.V., Kappel, F., Aurelio, A.: Control, sensitivity and identification of a cardiovascular-respiratory system model. In: Mohd, M.H., Misro, M.Y., Ahmad, S., Ngoc, D.N. (eds.) Modelling, Simulation and Applications of Complex Systems. CoSMoS 2019. Springer Proceedings in Mathematics & Statistics (2021)
8. Lao, A., Edra, J.M.C., Talag, K.D., Tan, D.S., Oyong, G., Noel, M., Enriquez, M.L., Tan, M.C.: Cytotoxic activity of Raphanus sativus Linn. On selected cancer cell lines and mechanistic pathways predicted through mathematical modeling. In: Mohd, M.H., Misro, M.Y., Ahmad, S., Ngoc, D.N. (eds.) Modelling, Simulation and Applications of Complex Systems. CoSMoS 2019. Springer Proceedings in Mathematics & Statistics (2021)
9. Sulayman, F., Mohd, M.H., Abdullah, F.A.: Bifurcation analysis of a tuberculosis model with the risk of re-infection. In: Mohd, M.H., Misro, M.Y., Ahmad, S., Ngoc, D.N. (eds.) Modelling, Simulation and Applications of Complex Systems. CoSMoS 2019. Springer Proceedings in Mathematics & Statistics (2021)
10. Tay, C.J., Teh, S.Y., Koh, H.L, Mohd, M.H., Zhang, Z.: Managing regime shift in lake systems by modelling and simulation. In: Mohd, M.H., Misro, M.Y., Ahmad, S., Ngoc, D.N. (eds.) Modelling, Simulation and Applications of Complex Systems. CoSMoS 2019. Springer Proceedings in Mathematics & Statistics (2021)
11. Ramli, N., Ahmad, S.: Stability analysis of magnetohydrodynamic mixed convection flow and heat transfer over a moving flat plate in ferrofluids with suction and slip effects. In: Mohd, M.H., Misro, M.Y., Ahmad, S., Ngoc, D.N. (eds.) Modelling, Simulation and Applications of Complex Systems. CoSMoS 2019. Springer Proceedings in Mathematics & Statistics (2021)
12. Oryan, R.R., Addawe, J.M., Panes, D.T.: Modeling and analysis of the dengue activity in Baguio city using two-mode and one-mode networks. In: Mohd, M.H., Misro,, M.Y., Ahmad, S., Ngoc, D.N. (eds.) Modelling, Simulation and Applications of Complex Systems. CoSMoS 2019. Springer Proceedings in Mathematics & Statistics (2021)
13. Balino, L.V.A., Caasi, K.S., Addawe, R.C.: Spatio-temporal distribution of dengue infections in Baguio City, Philippines. In: Mohd, M.H., Misro, M.Y., Ahmad, S., Ngoc, D.N. (eds.) Modelling, Simulation and Applications of Complex Systems. CoSMoS 2019. Springer Proceedings in Mathematics & Statistics (2021)
14. Addawe, R.C., Angeles, G.M., Balolong, M.: Spatio-temporal analysis of measles cases in Baguio City, Philippines from 2010–2018. In: Mohd, M.H., Misro, M.Y., Ahmad, S., Ngoc, D.N. (eds.) Modelling, Simulation and Applications of Complex Systems. CoSMoS 2019. Springer Proceedings in Mathematics & Statistics (2021)

15. Ignacio, N., Liwag, R., Addawe, R.C.: Spatio-temporal analysis of typhoid cases in Baguio City, Philippines. In: Mohd, M.H., Misro, M.Y., Ahmad, S., Ngoc, D.N. (eds.) Modelling, Simulation and Applications of Complex Systems. CoSMoS 2019. Springer Proceedings in Mathematics & Statistics (2021)
16. Ahmad, N.A., Javed, S.: Chaotic time series prediction using random Fourier feature kernel least mean square algorithm with adaptive kernel size. In: Mohd, M.H., Misro, M.Y., Ahmad, S., Ngoc, D.N. (eds.) Modelling, Simulation and Applications of Complex Systems. CoSMoS 2019. Springer Proceedings in Mathematics & Statistics (2021)
17. Zehi, R.Y., Khurizan, N.S.N.: A robust DEA model to handle the uncertainty in production trade-offs. In: Mohd, M.H., Misro,, M.Y., Ahmad, S., Ngoc, D.N. (eds.) Modelling, Simulation and Applications of Complex Systems. CoSMoS 2019. Springer Proceedings in Mathematics & Statistics (2021)
18. Malumfashi, M.L., Ismail, M.T., Bature, B., Sani, D., Ali, M.K.M.: An EPQ model for delayed deteriorating items with two-phase production periods, variable demand rates and linear holding cost. In: Mohd, M.H., Misro, M.Y., Ahmad, S., Ngoc, D.N. (eds.) Modelling, Simulation and Applications of Complex Systems. CoSMoS 2019. Springer Proceedings in Mathematics & Statistics (2021)
19. Malumfashi, M.L, Ismail, M.T., Rahman, A., Sani, D., Ali, M.K.M.: An EPQ model for delayed deteriorating items with variable production rates, two-phase demand rates and shortages. In: Mohd, M.H., Misro, M.Y., Ahmad, S., Ngoc, D.N. (eds.) Modelling, Simulation and Applications of Complex Systems. CoSMoS 2019. Springer Proceedings in Mathematics & Statistics (2021)
20. Rahman, M.M., Ismail, M.T., Sadullah,, A.F.M., Khurizan, N.S.N., Ali, M.K.M.: Binary decision for discretionary lane changing by time-effects factors. In: Mohd, M.H., Misro, M.Y., Ahmad, S., Ngoc, D.N. (eds.) Modelling, Simulation and Applications of Complex Systems. CoSMoS 2019. Springer Proceedings in Mathematics & Statistics (2021)
21. Rahman, M.M., Ismail, M.T., Ali, M.K.M.: Lane changing effects on surrounding vehicles by incorporating the risk factors. In: Mohd, M.H., Misro, M.Y., Ahmad, S., Ngoc, D.N. (eds.) Modelling, Simulation and Applications of Complex Systems. CoSMoS 2019. Springer Proceedings in Mathematics & Statistics (2021)
22. Rampal, L., et al.: Battling COVID-19 pandemic waves in six South-East Asian countries: a real-time consensus review. Med. J. Malaysia **75**(6), 613 (2020)
23. Salman, A.M., et al.: Scenario analysis of COVID-19 transmission dynamics in Malaysia with the possibility of reinfection and limited medical resources scenarios. Comput. Biol. Med. **133**, 104372 (2021)
24. Drogoul, A., et al.: Designing social simulation to (seriously) support decision-making: COMOKIT, an agent-based modelling toolkit to analyse and compare the impacts of public health interventions against COVID-19. Revi. Artif. Soc. Soc. Simul. (2020)
25. Mohd, M.H., Fatima, S.: Unravelling the myths of R0 in controlling the dynamics of COVID-19 outbreak: a modelling perspective. Chaos, Solitons Fractals **138**, 109943 (2020)
26. Gaudou, B., et al.: COMOKIT: a modeling kit to understand, analyze, and compare the impacts of mitigation policies against the COVID-19 epidemic at the scale of a city. Front. Pub. Health **8** (2020)

Agent-Based Modelling (ABM) in Complex Systems

Simulation of Japanese National Diet Members Election System Using Agent-Based Modeling

Satoshi Takatani and Hirohide Haga

Abstract This article reports the investigation of the influence to the result of election with the change of electoral system. The latest election in Japan was held in 2017. In this election, the leading party, Liberal Democratic Party (LDP), got 48% of total votes and got 76% of seats. Some people considered this result as inadequate. In other words, current electoral system does not or cannot reflect the voters' willing appropriately. What is the appropriate election system? Current Japanese election system uses the combination of single-seat system and proportional representation system. Is this the optimal system? This is our research question.

In this article, we assume that there are distinct n political issues. Each people (voter), political party, and candidate has its own opinion. This opinion is represented by the value from -1 to $+1$. We also represent each people, political party, and candidate as agents. Each agent has its own political position represented as a n-dimensional vector. All voters' willing is represented as a composition vector of all voters' political vectors. We compare the composition vector of all voters' political vectors and that of all successful candidates. Difference of voters' composite vector and successful candidates' composite vector is observed when we change the electoral system. Our simulation results show that multiple-seat selection system is less reflects the people's opinion and there is no significant difference in other electoral systems.

Keywords Agent-based modeling · Election system · Vector space modeling

1 Introduction

In this article, we will report how the changes of the electoral system will influence the difference between public opinion and the will of successful candidates. We assumed that an electoral system that minimizes such differences between the opinions of the

S. Takatani
Meitec Corporation, Nagoya, Japan

H. Haga (✉)
Graduate School of Science and Engineering, Doshisha University, Kyotanabe 610-0321, Japan
e-mail: hhaga@main.doshisha.ac.jp

M. H. Mohd et al. (eds.), *Modelling, Simulation and Applications of Complex Systems*, Springer Proceedings in Mathematics & Statistics 359,
https://doi.org/10.1007/978-981-16-2629-6_2

public and those of the successful candidates would better. Based on this assumption, we investigated what sort of electoral system would minimize such differences by using a multi-agent model to build an electoral system simulator.

Agent-based Modeling (ABM) or Multi-agent Modeling (MAM) becomes popular in various fields such as natural, engineering and social sciences [1]. This method is often considered as an opposition to equation-based modeling [2]. There are several works about the application of ABM to social science [3, 4]. This is mainly because it is virtually impossible to describe the equation models in social science. However, in ABM no specific equations are needed. Instead, some laws which describe the local interaction between agent-agent and environment-agent are needed. Comparing the equation modeling, it is relatively easier to describe local interaction laws embedded in agents.

The electoral system is one of the most important factors for democracy. Citizens select representatives through elections, and the representatives are entrusted with governing. Therefore, the electoral system must be designed to reflect the intentions of the citizens as accurately as possible. Current electoral system used for general elections in Japan is a combination of single-seat constituencies and proportional representation [5]. Recently especially in single-seat constituency elections, there has been a problem with the "gap between the proportion of votes and the proportion of parliamentary seats gained," raising the question of whether this system really reflects public opinion. In the results of the 2017 election of the House of Representatives in Japan, Liberal Democratic Party received 47.8% of the votes for single-seat constituencies but gained 75.4% of the parliamentary seats. On the other hand, for example, Party of Hope received 20.6% of the votes and gained only 6.2% of the seats. This party received over 20% of the votes, it gained only about 6% of the seats in the House of Representatives [6]. This result shows that the current electoral system may not reflect public opinion [7].

It is virtually impossible to find the optimal electoral system that better reflects public opinion by actually changing the electoral system in the real society. Therefore, in this article, we investigate how electoral systems reflect public opinion by simulating them using a multi-agent model. Multi-agent models now receive much attention recently as a method for simulating complex phenomena such as social and natural phenomena. There are not so many works about the simulation of election by ABM. Some of them are [8, 9]. However, their model does not deal with the change of electoral system. These works mainly consider the difference of the results by changing the details of models. Therefore, these works do not find the appropriate electoral system. Furthermore, these works deal with simple electoral systems. However, for example, Japanese electoral system is relatively complex compared to the models of [8, 9]. Therefore, we started to construct new model of election for simulation.

We built a multi-agent-model electoral system simulator and used it to examine the following types of electoral system.

- Proportional representation system only
- Single-seat constituency system only
- Multiple-seat constituency system only

- Combination of single-seat constituency and proportional representation in a ratio of 1:0.610 (close to current electoral system)
- Combination of single-seat constituency and proportional representation in a ratio of 1:4
- Combination of single-seat constituency and proportional representation in a ratio of 4:1.

Furthermore, conditions varied in simulations include the following:

- Vote decision factors
- Number of candidates per electoral region
- Regional differences in distribution by electoral region.

For each of these conditions, we set two alternatives. Therefor a total of $2^3 = 8$ combinations are possible. We then conducted simulations for all combinations of the six electoral systems and eight sets of conditions, totaling 48 cases. As a result, we can say that the multiple-seat constituency system does not reflect public opinion well for any other simulation cases, but we did conclude that there were no significant differences among the other five electoral systems.

When we try to design an electoral system which reflects the public opinion more appropriately, we first need to define the term "public opinion." Generally, public opinion is defined somewhat vaguely as "the total will of the citizens," [10] but this is too ambiguous to use for our simulation. In this article, we define the public opinion as follows. In current society, there are several political issues, and individual voters and the political parties each have their own stances with respect to each issue. Political parties express these stances to voters in the form of political manifest. Thus, in this article, we express the will of each voter with respect to n current political issues by an n-dimensional vector of values ranging from +1 (completely agree) to -1 (completely disagree) and define public opinion as a composite vector of these vectors for all voters. Similarly, we define the will of the House by the composite vector of the agree/disagree vectors representing the will of the successful candidates with respect to the n political issues. With the respect to these definitions, minimizing gap between public opinion and the will of House is equal to minimize the difference of the composite vectors for public opinion and for the House. The difference between public opinion and the will of the successful candidates can be expressed by an angle between two composite vectors representing public opinion and the will of the House of Representatives respectively. Taking the cosine of this angle, the closer the value is to 1, the smaller is the difference between public opinion and that of the House, and as it gets farther from 1, the difference between public opinion and the will of the House becomes large. Thus, we can evaluate the difference between these two vectors expressing the gap between public opinion and the election results by the cosine value of two composite vectors.

2 Preparing Electoral System Simulation

2.1 Agents and Their Attributes

Our model includes three types of agents.

- Political parties: p_i $(1 \leq i \leq x)$
- Voters: v_j $(1 \leq j \leq y)$
- Candidates: c_k $(1 \leq k \leq z)$.

x, y, and z are arbitrary natural numbers. Normally, x is 10^1 order, y is 10^5–10^6 order per electoral district and z is also 10^1 order. Each political party, voter and candidate has its own opinion on each political issue. For this research, a real value is used to represent the opinion of each agent on each political issue. We will refer to the opinion of an agent regarding a political issue as their political issue weighting. Weightings are real values in the range from -1 to $+1$, with $+1$ indicating "completely agree" and -1 indicating "completely disagree" on the political issue. Let us assume that there are n political issues, and each agent possesses a political issue weighting for each of the n issues. Thus, the weightings for the n political issues can be represented by an n-dimensional vector. The political issues are common to all agents, so all agents have political issue weightings and are represented by n-dimensional vectors. As such, we will refer to the n-dimensional political-issue weightings for each agent as their **political vector**. The political issue vectors for political parties, voters, and candidates are represented as $\boldsymbol{p_i}$, $\boldsymbol{v_j}$ and $\boldsymbol{c_k}$, respectively. Furthermore we will use following notations:

- p_i^m, v_j^m, c_k^m: weighting for political issue m for each agent, $\boldsymbol{p_i}$, $\boldsymbol{v_j}$, $\boldsymbol{c_k}$.

2.2 Generating Political Issue Weightings

The political issue weightings for each political parties, voters and candidates cannot be uniquely defined because they involve complex and uncertain factors that vary for each individual, including social attributes such as regional and economic conditions, and personal psychological attributes such as psychological state. The results of each election also differ, so results must differ in each simulation as well. As such, in this simulation, the political issue weightings for each agent were generated based on distributions with specific tendencies. All agents are generated by giving them political issue weightings that are random values generated based on pre-defined distributions. The following seven distributions were used to generate political issue weightings and will be referred to as distribution 1 to 7 respectively.

1. Opinions are uniformly distributed
2. Opinions are polarized
3. Most opinions are intermediate (normal distribution)

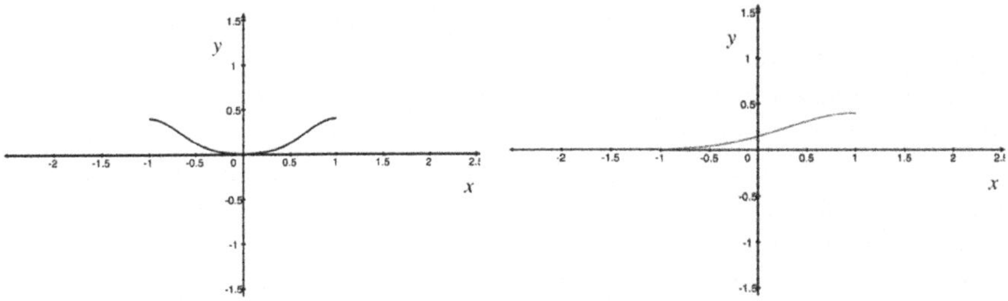

Fig. 1 Graph of distribution function 2 (left) and 4 (right)

4. More opinions agree than disagree
5. More opinions disagree than agree
6. Everyone agrees
7. Everyone disagrees.

For example, a graph of distribution functions 2 and 4 are shown in Fig. 1.

2.3 Political Party Agents

Political parties are formed by the collection of politicians with similar political objectives. There are currently several political parties in Japan, including the Liberal Democratic Party [11], the Constitutional Democratic Party [12], and the Communist Party [13] and so on, each with different opinions on the various political issues. Sometimes multiple political parties work together toward the same objectives, but they are different parties, so while their directions on certain political issues may be the same, they will differ in the weightings they place on each issue. Also, since political parties in Japan have opinions on all kinds of political issues, it is unlikely that all political parties will have the same direction on a given political issue. As such, when generating political party agents, their political issue opinions were generated using distributions 1 to 5 as described in Sect. 2.2, excluding distributions 6 and 7. Thus, the political party attributes consist of n political issue weightings (an n-dimensional political issue vector) generated using these distribution functions.

2.4 Voter Agents

The opinion of each voter on an individual political issue i is expressed as the degree to which they agree or disagree with the issue. Thus, a voter's opinion on the n political issues is represented by an n-dimensional vector, same as that of party agents. The n-dimensional vectors representing individual voters are set randomly based on distribution functions 1 to 7 in Sect. 2.2 for each political issue with introducing a concept of "regional differences."

In an election, voters will vote in their own electoral district, and will basically vote for the candidate with opinions on political issues that match voters' own opinions. However, there is generally no candidate that matches a voter's opinions exactly, so voters select and vote for the candidate that best matches the voter's opinions according to some kind of rule.

We expect voters and candidates to hold different opinions on various political issues according to their electoral district. For example, on the issue of the Trans-Pacific Partnership (TPP) agreement, most people in rural areas are opposed because it will have a greater negative effect on them, but most people in urban areas support it because it will improve business profitability in those areas. On the issue of increasing consumption tax, almost everyone is opposed, regardless of electoral district. In this way, there are political issues with arguments on both sides, varying according to region. To express such differences in this research, we have assigned political issue weightings for each electoral district, generating these weightings using distributions 1 to 7.

2.5 Candidate Agents

Political parties are composed of members with similar political objectives. Thus, we can basically assume that the political issue vectors for candidate agents will be similar to those of the party to which they belong. However, each candidate also has his/her own differing opinions for objectives, so it is unlikely that the values in a candidate agent's political issue vector will exactly match those in the political issue vectors of their political party or other candidates. Thus, political parties will include both candidate agents that are compliant to the party opinions and also those that are somewhat more distant. As such, the n-dimensional vectors for candidate agents were assigned according to normal distributions whose mean values were the values for the n political issue opinions of the agent for their political party. For example, in the n-dimensional political issue vector for political party p_1, if the weighting for political issue 1, p_1^1, is 0.4, a random number on a normal distribution with mean of 0.4 is generated and assigned in the n-dimensional political issue vector for candidate agent c_1, as the weighting for political issue 1, or c_1^1. Weightings for political issues are assigned in this way for candidate agents. According to this definition, weightings for political issues are real values in the range from -1 to 1.

2.6 Differences (gaps) Between Public Opinion and the Will of Successful Candidates

Currently in Japan, voters select a candidate or political party in an election, and candidates that gain the most votes are elected, winning a seat in the House of Representatives. Within the House, the majority opinion is respected. Political issues are decided and acted upon by the members selected in elections (successful candidates). If the electoral system is perfect, the opinions of the successful candidates should closely match those of the voters (public opinion). In this research, we define such difference between public opinion and that of a successful candidate quantitatively as follows.

1. **Representation of public opinion**: Public opinion is the total sum of all opinions on political issues for all voters. Thus, it can be expressed by a *composite vector* combining the political issue vectors of all voters. If $\boldsymbol{V}$ is public opinion, then,

$$\boldsymbol{V} = \sum \boldsymbol{v}_i$$

   Here, $\sum$ is vector composition, and each $\boldsymbol{v}_i$ is the political issue vectors of all voters.
2. **Opinion of the House**: The opinion of the House, $\boldsymbol{S}$, similar to public opinion, is a composite vector of the political issue vectors of all who hold seats in the House; that is, defined as follow;

$$\boldsymbol{S} = \sum \boldsymbol{s}_i$$

   Here $\sum$ is vector composition, and each $\boldsymbol{s}_i$ is the political issue vectors of all successful candidates.

The difference Θ between the public opinion vector $\boldsymbol{V}$ and the House opinion vector $\boldsymbol{S}$ then is computed by following expression (1) shown below.

$$\Theta = \frac{\boldsymbol{V} \cdot \boldsymbol{S}}{\|\boldsymbol{V}\| \, \|\boldsymbol{S}\|} \tag{1}$$

Here $\boldsymbol{V} \cdot \boldsymbol{S}$ is the inner product of vectors $\boldsymbol{V}$ and $\boldsymbol{S}$, and $\|\boldsymbol{V}\|$ is the norm of vector $\boldsymbol{V}$. Θ ranges between -1 and 1, with values closer to 1 indicating that the difference is small. On the other hand, values closer to -1 indicate the large difference and that public opinion is not reflected. In this research, we simulate how the difference changes by changing the electoral system.

3 Simulation Conditions

The agents and their attributes in our model built for this research were discussed in Sect. 2. These elements did not change while the simulations conducted. Section 3 discusses factors that were changed in the simulations. We observed how various conditions that involve uncertainty, such as drivers of voting behavior, could affect the difference between public opinion and those of successful candidates in elections, by performing simulations while changing such elements. In our research, we changed following three factors:

- Voting behavior
- Regionality
- Changes in number of candidates in one electoral district.

We used two patterns for each of these three conditions, yielding a total of eight ($2^3 = 8$) pattern combinations.

3.1 Voting Behavior

The primary idea regarding voting behavior is that voters will vote for the candidate whose views on political issues are most similar to voters' own. We used two measures of similarity as follows.

1. Voters will vote for the political party or candidate in their electoral district whose political issue vector has the smallest angle (difference) with their own. In Eq. (2) below, the political issue vectors of individual voters are represented by $\boldsymbol{v}_i$, those of candidates are $\boldsymbol{c}_j$, and the angles between them are θ_{ij}. This assumes that the voters understand the political issue vectors of the candidates correctly.

$$\theta_{ij} = \cos^{-1} \frac{\boldsymbol{v}_i \cdot \boldsymbol{c}_j}{\|\boldsymbol{v}_i\| \, \|\boldsymbol{c}_j\|} \tag{2}$$

2. In a real election, it is often not the case that voters understand the political issue vectors of candidates or political parties completely. For this reason, we considered another similarity measure, which may be more appropriate when voters do not understand the political issue vectors of the candidates or political parties with sufficient accuracy, which we call voting by direction. This similarity measure only considers the sign of the weightings in individual elements of the political issue vectors of candidates and voters, converting them to vectors of just the signs. Voters then vote for the candidate whose sign vector matches their own for the greatest number of elements. If there are multiple such candidates, one is selected at random.

We will observe how the difference of public opinion and that of successful candidates changes in accordance with each of these two types of voting behavior.

3.2 Regionality

The political issues embraced by voters often differ by region. There are political issues for which regional differences are relatively small, such as national security and social welfare, but for other issues such as tax system or foreign trade, support or opposition can differ significantly by region, particularly for urban vs. rural areas. As such, for our simulations, we divided the entire country into multiple electoral districts and assigned biases in each district, supporting or opposing each political issue. To represent these biases, we used the very simple approach of moving the origin of the distribution functions left or right in parallel. In doing so, values exceeding ± 1 were set to ± 1 respectively.

3.3 Number of Candidates per Electoral District

We consider two patterns for the number of candidates per electoral district in the single-seat constituency system.

1. **Number of candidates is not fixed:** In the current electoral system, there are two or more candidates per electoral district. Political parties have different numbers of candidates in different electoral districts, and we assume some intention in fielding candidates, since it is unlikely that candidates are fielded randomly. In this research, we set the condition that political parties always field candidates in electoral districts that they have prospects of winning. This depends on the number of supporters they have in each region, and parties will always support candidates in regions where they have many supporters. On the other hand, parties may also have candidates in electoral districts where they have little prospect of winning. They may field candidates in hopes of gaining votes from unaffiliated voters (the group of voters that do not support a particular party). In particular, the number of candidates is computed by computing the angles, θ_{ij}, between the i-th party's political issue vector and all voters, j, in an electoral district. If the value is positive, we determine that the party has prospects of winning, and a candidate is fielded. Parties also field candidates randomly in districts with negative values, to gain votes from voters not affiliated with a party.
2. **Number of candidates is fixed:** In this case, all political parties field candidates in all electoral districts. This involves more candidates than we could expect in a real election, but having many candidates should enable voters to vote for a candidate who better represents their own views. Thus, we can observe what effect such conditions have on the difference between public opinion and that of the successful candidate.

3.4 Overview of Electoral Systems

We will simulate the following types of electoral system.

1. **Single-seat constituency system:** Only one successful candidate is selected from each electoral district.
2. **Multiple-seat constituency system:** Multiple successful candidates are selected from each electoral district.
3. **Proportional representation system:** Numbers of successful candidates from each party are assigned based on the proportion of votes gained by the political parties in each electoral district.

In 1 and 1 voters vote for candidates, while in 1 they vote for political parties. The current system in Japan is a combination of system 1 and system 3. Thus, in addition to electoral systems 1, 2, and 3 individually, we consider systems that use both 1. and 3 in some fixed proportion. We examined the following combined patterns.

4. Single-seat constituency:Proportional representation ratio of 1:0.6.[1]
5. Single-seat:Proportional ratio of 1:4
6. Single-seat:Proportional ratio of 4:1.

Simulations will be conducted using the above six types of electoral system.

3.5 Summary of Simulation Conditions

A summary of simulation conditions used in this research is given below.

1. **Voter decision rule:** Two types as described in Sect. 3.1.
2. **Regionality:** With and without regionality as discussed in Sect. 3.2.
3. **Candidates per electoral district:** Not-fixed and fixed, as discussed in Sect. 3.3.
4. **Electoral system:** Six types, as discussed in Sect. 3.4.

This totals $2 \times 2 \times 2 \times 6 = 48$ cases. Random numbers based on certain distribution functions are used when generating political party, candidate, and voter agents, so for each of these cases, 100 sets of conditions will be generated, producing a total of 4,800 cases that were simulated.

4 Overview of Electoral System Simulator

A screen shot of the electoral system simulator developed for this research is shown in Fig. 2. In Fig. 2, the 100×100 grid of cells is the election space, representing

[1]This ratio is very close to that used in the current electoral system in Japan.

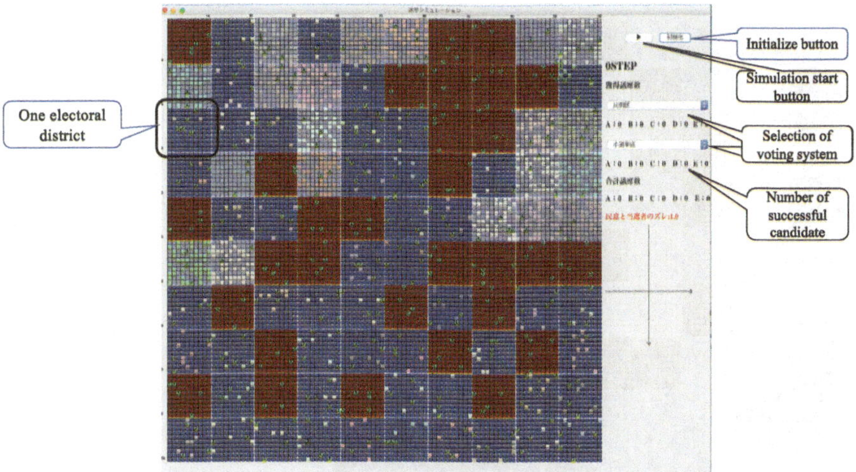

Fig. 2 Screenshot of Electoral system simulator

what would be a country in real space. This space is divided into 10 by 10 cell sections, which are the electoral districts. Thus, electoral districts are 10×10 sections of 100 cells. Each electoral district includes voters and candidates. The number of voters varies by electoral district, and districts are colored according to number of voters for clarity. Candidates are shown with a green circle (○). The various electoral systems can be simulated by selecting them in pull-down menus and other controls in the panel on the right. The simulator was built using the Java language and the Swing GUI toolkit with `jfreechart` library [14]. All source code and complied class files are downloadable from google drive [15]. Note that this simulator uses the `jfreechart` library, this library must be installed before compiling. Furthermore, as this simulator is designed for Japanese users, some Japanese characters are included in this code. Figure 2 shows the initial state of the simulator. In Fig. 2 each small regular rectangle located at the left-hand side represents the electoral district. The difference of the color means the difference of the attitude of political issues. In each electoral district, there are several voters and candidates. Voters are represented by the small rectangles of each district. Candidates are represented by green-colored circle. Right-hand side of the simulator includes the control buttons and menus. They include "initializing the simulator button," "start and resume button," and "selection menu of electoral system." By pushing the "start" button, simulation will start. Users can resume the simulation by pushing the "resume" button. The electoral system can be changed using the pull-down menus. When the selected electoral system is changed, a new election space is generated, suitable for the selected electoral system.

Figure 3 is an example simulation result snapshot. In Fig. 3, the candidates shown in red were elected in the electoral district where they were fielded. On the right side in the simulator, the total number of seats won by the political parties of the elected candidates are shown. The proportional representation system was selected in this

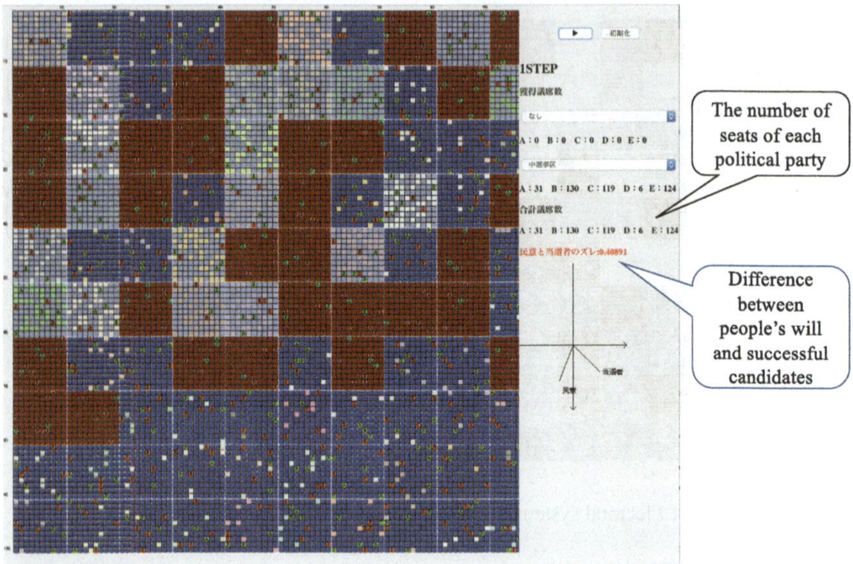

Fig. 3 Simulation result screenshot of Electoral system simulator

simulation, so the total number of seats and the number of seats gained by multiple-seat constituencies is the same. The difference between public opinion and that of the successful candidate is 0.40891, and the conditions that produced that difference value are represented in 2D space.

5 Simulation

5.1 Overview of Simulation

An overview of the experiments is given below.

1. For the six electoral systems and eight simulation conditions, totaling 48 cases, measure the difference between public opinion and the successful candidates. For each case, perform 100 simulations and take the average result.
2. The political issue vectors for voters, political parties, and candidates have five dimensions and there are five political parties, named A to E.
3. Electoral districts have 100, 300, or 1000 voters, and there are 20, 30, and 50 districts of each size, respectively. Positions on political issues are assigned to each electoral district using probabilistic distributions, and these are used to assign political issue vectors to voters.

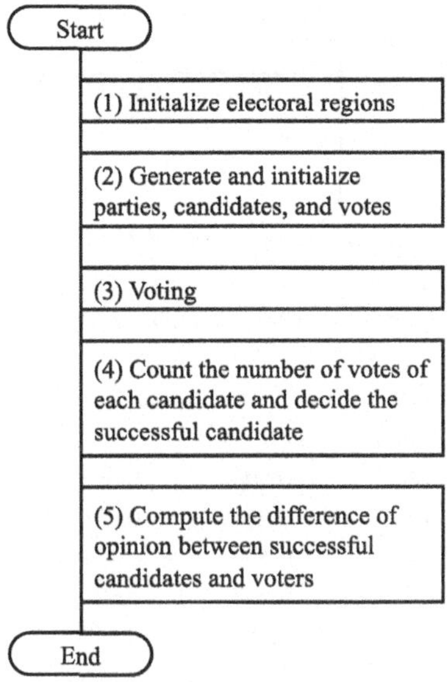

Fig. 4 Flowchart of simulation (1)

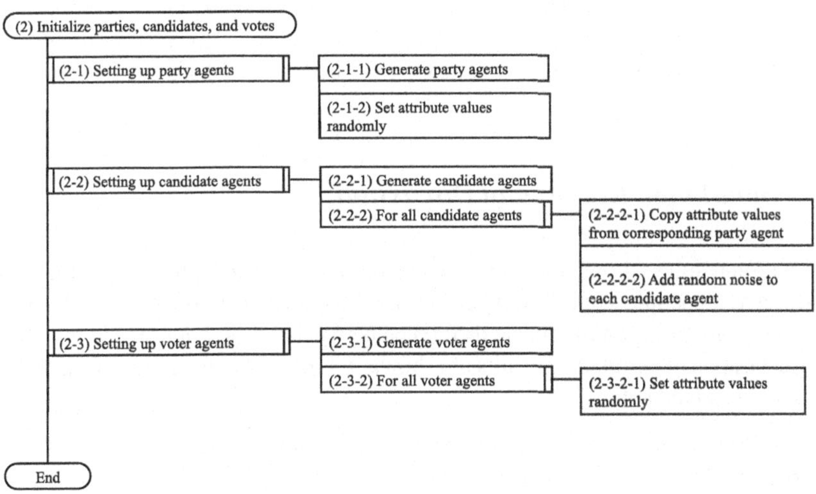

Fig. 5 Flowchart of simulation (2)

General procedure of simulation is shown in Fig. 4 and some details of procedures are shown in Figs. 5 and 6. Note that, these procedures are represented by PAD (Problem Analysis Diagram) designed by Y. Futamura [16].

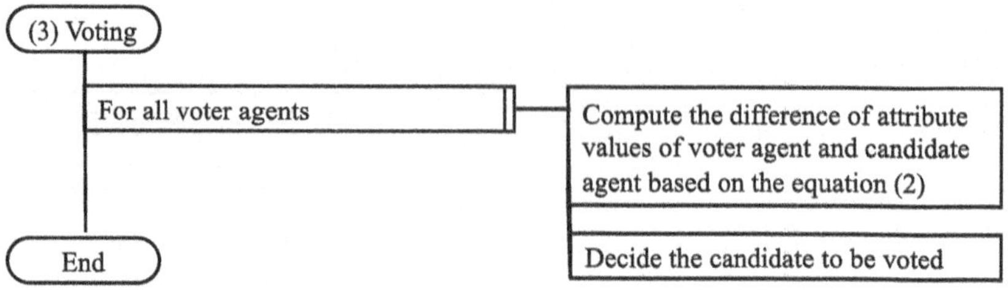

Fig. 6 Flowchart of simulation (3)

Table 1 Changes in difference from people's opinion and that of the House by electoral system

Pattern	PS	SS	MS	SS1+PS1	SS1+PS4	SS4+PS1
000	0.508	0.655	0.368	0.775	0.719	0.686
001	0.644	0.651	0.389	0.694	0.75	0.66
010	0.502	0.451	0.297	0.444	0.523	0.505
011	0.602	0.52	0.375	0.627	0.597	0.505
100	0.456	0.653	0.322	0.676	0.663	0.738
101	0.493	0.556	0.271	0.646	0.663	0.626
110	0.42	0.422	0.374	0.563	0.513	0.595
111	0.513	0.402	0.351	0.513	0.439	0.559
Average	0.517	0.539	0.343	0.617	0.608	0.609

5.2 *Simulation Results and Discussion*

Simulation result is shown in Table 1. All values in this table represent the difference (gap) between people's opinion and that of the House. The values mean that the closer the value is to 1, the smaller is the difference between public opinion and that of the House. Graphs of simulation results are shown in Fig. 7. The meanings of each of the symbols are given below.

- PS: Proportional Seats (proportional representation)
- SS: Single Seat (single-seat constituency system)
- MS: Multiple Seat (multiple-seat constituency system)
- SSm+PSn: Combined system with SS and PS in m:n ratio.

The three-digit binary numbers in each chart have the following meanings.

- 1st digit: Voting decision rule (0: minimum opinion gap, 1: direction only)
- 2nd digit: Number of candidates (0: Not-fixed, 1: Fixed)
- 3rd digit: Regionality (0: No regionality, 1: Regionality).

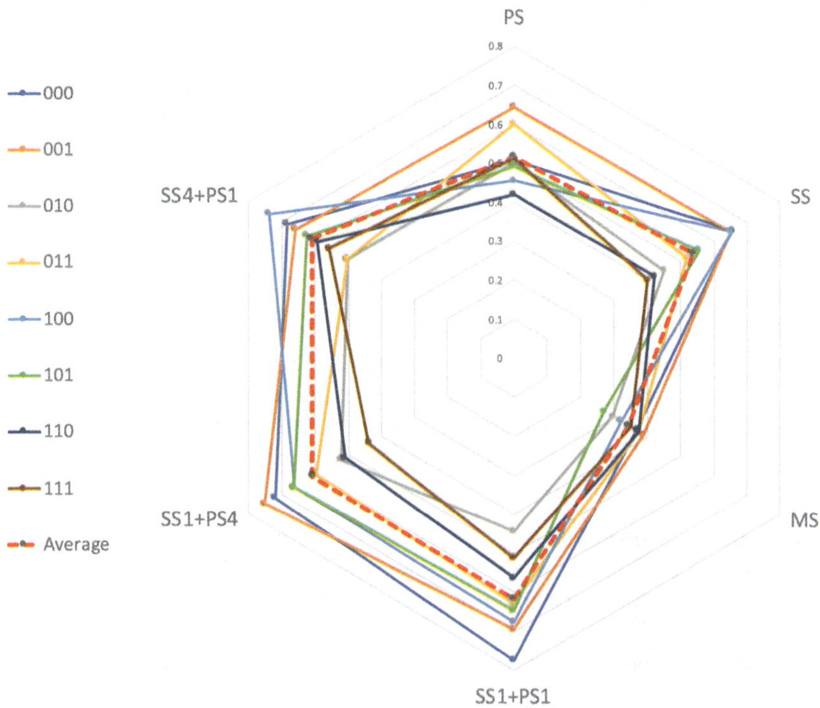

Fig. 7 Visualising changes in difference from people's opinion and that of the House by electoral system

The following conclusions can be derived from this simulation result.

(1) Except for the multiple-seat constituency system, there are no large differences among the systems. For the multiple-seat constituency system, the gap with public opinion was clearly larger than the other systems. With multiple-seat constituencies, many candidates can be fielded by all political parties and two or more candidates are elected in each electoral district, so there can be more conflict of opinion between political parties than with single-seat constituencies, which could be the reason that opinions of the House diverged from that of the public.
(2) After the multiple-seat constituency system, the proportional representation system, which is generally thought to best match public opinion, has large gap. This may be due to the fact that we introduced regionality in our simulations. In particular, we did not incorporate consideration for actual conditions in different regions, so we may not have adequately supported regional differences in public opinion for individual political issues. This is also demonstrated by the fact that differences from public opinion were smaller for systems combining single-seat constituencies and proportional representation than for the other individual systems.

(3) The results of the simulations indicate that in systems combining single-seat constituencies and proportional representation, the proportion with which the two systems are combined did not have a great influence. The system with the smallest difference was the one with SS and PS in equal proportion (SS1+PS1) with an average value of 0.617, but this value is not significantly different from the values for SS1+PS4 (0.608) or SS4+PS1 (0.609). Thus, from the results of our simulations, we conclude that the proportion used in combining single-seat constituency and proportional representation systems does not make a significant difference.

We conducted simulations while varying the electoral system and various simulation conditions, but except for the multiple-seat constituency system, no significant differences were observed. Thus, whichever electoral system is used other than multiple-seat constituencies, no significant differences are produced. Of the systems used in this research, the multiple-seat constituency system reflected public opinion the least. However, we think that the models used are not yet perfect, and this could affect the validity of the result. Three possible reasons for this are as follows.

(a) It is impossible to implement an electoral system that is the same as a real-world system. For example, there is no consideration for other elements that could affect results, such as voter abstention, or dual candidacy, and this could affect validity of the results.
(b) We have not considered how political parties perceive public opinion. This is because awareness of public opinion by political parties comes from surveys done by the media, and the political parties adjust their own positions on political issues based on this understanding. These surveys are done on many non-specific voters, so results may differ from public opinion obtained in an election, with abstaining voters. As such, it may also be necessary to account for how political parties consider opinions gained from such anonymous surveys of public opinion.
(c) Finally, we have not incorporated the concept of "supported party" factor in our model. Most people who support a particular political party will vote based on the name of the political party rather than the various political issues. This is particularly noticeable for the single-seat constituency system, and differences regarding political issues may not have a significant effect on voting behavior in that case. In our simulations, all voters voted according to differences regarding political issues, and this may be different to behavior in a real election.

For these reasons, there is still room to improve our model, and our results may not be valid as such. We did conduct simulations of different electoral systems under various conditions, however, and results suggested that the electoral system tends not to produce significant changes in the difference between public opinion and those of the successful candidates. This can be taken as a significant conclusion.

6 Conclusion and Future Works

In this article, we presented that we conducted simulations of how changes in the electoral system can affect the gap between public opinion and the opinions held by members of the House of Representatives. We proposed a method that expresses the will of individual actors in the simulations in terms of their attitudes with respect to n different political issues, and represents them as an n-dimensional political issue vector. We defined public opinion as a composite vector of the political issue vectors of all voters, and the will of the House of Representatives as a composite vector of the political issue vectors of the members of the House. Thus, the difference between public opinion and the will of the House is represented by the difference between these two vectors, which can be expressed in terms of the inner product and norms of the vectors. Using this representation, we simulated the gap between public opinion and the will of the House for multiple electoral systems, yielding the result that there was no significant change among the electoral systems, except for the multiple-seat constituency system. Even for the multiple-seat constituency system, our model may still be inadequate, so it will be necessary to add further conditions and conduct more detailed and real-world like simulations in the future.

Acknowledgment This work was supported by JSPS KAKENHI Grant Number JP18K04623.

References

1. Wilensky, U., Rand, W.: An Introduction to Agent-based Modeling - Modeling Natural, Social, and Engineered Complex Systems With Netlogo. MIT Press, Cambridge (2015)
2. Cecconi, F., Campenni, M., Andrighetto, G., Conte, R.: What do agent-based and equation-based modelling tell us about social conventions: the clash between ABM and EBM in a congestion game framework. J. Artif. Soc. Soc. Simul. **13**(1), 6 (2010)
3. Gilber, N., Troitzsch, K.G.: Simulation for the Social Scientist. Open University Press, Maidenhead (2005)
4. Epstein, J.: Generative SOCIAL SCIENCE - Studies in Agent-Based Computational Modeling. Princeton University Press, Princeton (2006)
5. Er, L.P., Stockwin, J.A.: Governing Japan: Divided Politics in a Resurgent Economy (Modern Governments), 4th edn. Wiley, Pondicherry (2008)
6. Ministry of Internal Affairs and Communications: Result of 2017 General Election (2017). http://www.soumu.go.jp/senkyo/48sansokuhou/index.html
7. Wada, J.: The Japanese Election System – Three Analytical Perspectives. Routledge, Oxford (2016)
8. Lou, J., et al.: A social diffusion model with an application on election simulation. Sci. World J. (2014). https://doi.org/10.1155/2014/180590
9. De Ita, G., Contreras, M, and Chantes, I.: A multiagent system for modeling democratic elections. In: Proceedings of the Twenty-Fourth International Florida Artificial Intelligence Research Society Conference, pp. 91–92 (2011). https://aaai.org/ocs/index.php/FLAIRS/FLAIRS11/paper/view/2576/2998
10. Shapiro, R.T., and Jacobs, L. R.: The Oxford Handbook of American Public Opinion and the Media. Oxford University Press, Oxford (2013)
11. Libral Democratic Party. https://www.jimin.jp/english/

12. Constitutional Democratic Party. https://cdp-japan.jp/english/
13. Japanese Communist Party. https://www.jcp.or.jp/english/
14. Java chart library. http://www.jfree.org/jfreechart/
15. https://drive.google.com/file/d/1-RiWp-eHFbHhcPablIBpTS4iqOCUoXCb/view?usp=sharing
16. Futamura, Y., Kawai, T., Horikoshi, H., Tsutsumi, M.: Development of computer programs by problem analysis Diagram (PAD). In: Proceedings of the 5th International Conference on Software Engineering (ICSE 1981), pp. 325–332 (1981)

Modelling the Dilution and Amplification Effects on Sin Nombre Virus (SNV) in Deer Mouse in GAMA 1.8

Lloyd W. F. Lee and Mohd Hafiz Mohd

Abstract Sin Nombre Virus (SNV) is a species of hantavirus that can cause hantavirus pulmonary syndrome in humans. To investigate the biodiversity effect on the SNV transmission in deer mouse, we formulated a stochastic agent-based model (ABM) to compare the impact between the presence of a dilution agent and an amplification agent in the deer mouse population. The ABM simulations were done in GAMA 1.8 and the results were then compared with the deterministic counterpart of the model. The deterministic results showed the dilution agent has better effectiveness in reducing the infected density compared to the amplification agent. However, this was not observed for the stochastic results with small populations. Instead, the infected densities were at a similar level for both dilution and amplification agent in the ABM results. This suggests that the investigation on the role of the community assemblage may not be relevant in reducing SNV transmission when the population density is small, and further research is needed to better understand the discrepancy between the stochastic and deterministic result and its implications. Our study highlights the importance of ABM in eco-epidemiological studies, and has established a methodological discussion regarding the usability of different simulation approaches e.g., deterministic and stochastic ABM in order to produce robust observations of eco-epidemiological phenomenon under consideration.

Keywords Sin Nombre Virus (SNV) · Biodiversity · GAMA

L. W. F. Lee (✉) · M. H. Mohd
School of Mathematical Sciences, Universiti Sains Malaysia, 11800 USM Penang, Malaysia

M. H. Mohd
e-mail: mohdhafizmohd@usm.my

M. H. Mohd
School of Mathematical Sciences, Faculty of Science and Technology, Universiti Kebangsaan Malaysia, 43600 Bangi, Selangor, Malaysia

M. H. Mohd et al. (eds.), *Modelling, Simulation and Applications of Complex Systems*, Springer Proceedings in Mathematics & Statistics 359,
https://doi.org/10.1007/978-981-16-2629-6_3

1 Introduction

In 1993, the Four Corners region in the United States had a mysterious disease outbreak which killed many humans. The disease was later identified as the hantavirus pulmonary syndrome (HPS), which was caused by a hantavirus species called the Sin Nombre Virus (SNV) [1]. The SNV is primarily hosted by the deer mouse, *Peromyscus maniculatus*, and humans can be infected through contacts with the saliva, urine and excreta of the infected rodents [2]. To better understand the dynamics of SNV among the deer mouse population, Abramson and Kenkre [3] were one of the pioneers to mathematically model this eco-epidemiological problem. They proposed a susceptible-infected (SI) spatio-temporal model to investigate the dynamics of the SNV. Peixoto and Abramson [4] later extended the model to include the biodiversity effect on the SNV transmission in deer mouse. Based on their theoretical model, they observed that the presence of a non-host alien species was able to reduce the SNV prevalence. Empirical studies such as [5, 6] have further supported such hypothesis.

However, Randolph and Dobson [7] warned that the biodiversity effect may not necessarily reduce a disease prevalence as amplification effect may occur instead. Studies such as [8, 9] investigated the amplification effect (increase in disease prevalence with increase species diversity) and dilution effect (decrease in disease prevalence with increase species diversity) for the case of Lyme disease. Authors from [8] concluded that the occurrence of amplification or dilution effect is dependent on the mechanism of competition, the host contact rates with ticks and acquired host resistant to ticks. Furthermore, different work such as [10] has investigated the mechanisms which cause the dilution or amplification effect for the endemic case of a disease. They highlighted that factors such as the type of disease transmission, relationship between the host competence and community assembly, and identity of hosts contributing to disease transmission, should be investigated to uncover whether a dilution or amplification effect can occur. For the case of SNV in deer mouse, Luis et al. [11] observed the occurrence of both amplification and dilution effects from their empirical data. Both effects occurred concurrently, and a net dilution was observed due to the dilution effect being greater between the two effects. Hence, we were interested in formulating a mathematical model to better understand the dynamics of the amplification and dilution effects of a non-host species and the impacts of stochasticity on SNV transmission in deer mouse.

In recent years, stochastic agent-based model (ABM) has received much utilization in eco-epidemiology studies. Unlike deterministic model, an ABM manages to incorporate the noise feature, which allows it to better mimic the reality. Allen [12] showed the importance of stochastic modelling of epidemics especially when the number of infectious individuals is small, or when there occurs a variability in transmission, recovery, births, deaths, or the environment which impacts the epidemic outcome. Authors from [13] were able to observe the extinction of disease in a general multi-host epidemic model given that the level of prevalence in the spillover species is relatively low and the reproduction number in the reservoir host is less than one. However, such occurrence was not observed in their deterministic model;

thus, showcasing the capability of stochastic modelling in understanding mechanisms underlying natural phenomenon. Eco-epidemiology studies such as [14, 15] have utilized the stochastic approach in their studies for modelling dengue disease and disease transmission among Tilapia with Pelican respectively. Guzzeta et al. [16] highlighted the incorporation of stochastic effect into modelling the dynamics of zoonotic pathogen has allowed them to gauge the probability and severity of potential future outbreaks. Besides that, several studies on the dynamics of hantavirus [17–19] also utilized the stochastic approach. Therefore, we would like to opt for a stochastic ABM for this study as well. It is the goal of this paper to show the importance of ABM in eco-epidemiological studies, and to establish a methodological discussion regarding the usability of different simulation approaches e.g., deterministic and stochastic ABM in order to produce robust observations of eco-epidemiological phenomenon under consideration.

In the next section, the deterministic model would first be introduced; then, the stochastic ABM counterpart would be introduced along with its implementation in GAMA version 1.8, a software which supports agent-based modelling. While some studies, e.g. Mohd [20] and Mohd [21], have simulated agent-based models (ABM) using Matlab package, we opt to study the dynamics of ABM by employing GAMA 1.8 to give an alternative approach on simulating the biological system using the techniques of stochastic process and differential equations. To the best of our knowledge, this approach has not been employed before (upon checking the literature review on GAMA), which is one of the main novelties of this study. This has contributed to the methodological discussions on the use of different modelling techniques and computer packages to examine the biological phenomena of interest. Similar to other platforms, GAMA provides similar flexibility to code the agent's characteristics and behavior according to the researcher's choice of techniques and assumptions. As mentioned above, what we would like to show is that GAMA can serve as an alternative towards the other platforms and it is up to the researcher's discretion to choose whichever platform they are comfortable with.

2 Model Formulation

2.1 Deterministic Model

The deterministic model we would like to introduce is based on the proposed model in our previous research [22]. It is a "single host, single non-host" endemic model that accounts for density-dependent restricted logistic growth, with the non-host having a certain amount of influence on the SNV transmission rate depending on its amplification or dilution role in a closed system. The biodiversity effect is accounted for through the inclusion of the non-host into the deer mouse community. The deterministic model is given as below:

$$\begin{aligned}\frac{dS}{dt} &= N\left[b_1 - ar_1\left(\frac{N+q_1Z}{K_1}\right)\right] - S\left[d_1 + r_1(1-a)\left(\frac{N+q_1Z}{K_1}\right)\right] - \gamma(1+\delta Z)SI \\ \frac{dI}{dt} &= \gamma(1+\delta Z)SI - I\left[\mu + d_1 + r_1(1-a)\left(\frac{N+q_1Z}{K_1}\right)\right] \\ \frac{dZ}{dt} &= r_2\left[1 - \frac{Z+q_2N}{K_2}\right]Z\end{aligned} \tag{1}$$

where $N = S+I$ is the total average population density of the deer mouse per hectare, S is the average population density of the susceptible deer mouse per hectare, I is the average population density of the infected deer mouse per hectare, and Z is the average population density of the non-host individuals per hectare. The descriptions for the rest of the parameters can be found in Table 1. Most of the parameter values were based on [11] while the rest were based on the work of [22]. The parameter values obtained from [11] were based on their observational study at several sites, while the parameter values in [22] were based on a modelling study, which was motivated by the ecological studies of Luis et al. [11]. Interested readers can refer to [11, 22] for the assumptions and derivation.

Similar to our previous study, we shall present 2 case studies to account for the dilution and amplification role of the non-host in a small population community. Our interest in modelling for a small population was to observe for potential differences between the deterministic and stochastic results at this level. For case study 1, the non-host (dilution agent) has a relatively weaker interspecific competition strength compared to the deer mouse ($q_1 = 0.2$) but it does not contribute any positive influence towards the SNV transmission rate ($\delta = 0$). For case study 2, the non-host (amplification agent) has a relatively stronger interspecific competition strength compared to the deer mouse ($q_1 = 0.4$) and has a positive influence on the SNV transmission rate ($\delta = 0.0543$). These parameter values were chosen to best reflect the reality whereby a dilution agent, e.g. desert pocket mouse, is timid towards the deer mouse; while the amplification agent, e.g. Merriam's kangaroo rat, is aggressive towards the deer mouse and thus, the deer mouse would avoid encountering them [23]. In the presence of the Merriam's kangaroo rat (amplification agent), the activity area of the deer mouse becomes smaller and this may increase the stress in deer mouse, which indirectly influence its susceptibility towards the SNV infection [24].

2.2 *Stochastic ABM*

To model the stochastic ABM counterpart of model (1), we utilized the discrete-time Markov chain approach. We assumed that each individual has a probability in executing one of the three events, namely "reproduce", "die" or "do nothing", in every small time step, Δt. When an individual executes the "reproduce" event in the $[t, t + \Delta t)$ interval, a new individual of the same category will be created in the system. For the "die" event, the individual would be permanently deleted from the system. Finally, nothing will happen to the individual if it executes the "do

Table 1 Descriptions and parameter values for model (1)

Parameters	Descriptions	Parameter Value	
		Case Study 1	**Case Study 2**
K_1	The carrying capacity of the deer mouse (per hectare)	20[b]	
K_2	The carrying capacity of the non-host. (per hectare)	15[b]	
r_1	The net density dependent growth rate for the deer mouse, $b_1 - d_1$. (per month)	3.1496×10^{-1} [a]	
b_1	The density dependent birth rate for the deer mouse. (per month)	0.315[a]	
μ	The disease induced mortality rate (per month)	0.085[a]	
d_1	The death rate of the deer mouse. (per month)	3.66×10^{-5} [a]	
q_1	The interspecific pressure exerted by the non-host onto the deer mouse	0.2[b]	0.4[b]
a	The proportion of density dependence due to density dependence of the deer mouse in birth rates	0.614[a]	
r_2	The net density dependent growth rate for the non-host, $b_2 - d_2$. (per month)	3.9996×10^{-1} [b]	
b_2	The density dependent birth rate for the non-host. (per month)	0.4[b]	
d_2	The death rate of the non-host. (per month)	4.0×10^{-5} [b]	
q_2	The interspecific pressure exerted by the deer mouse onto the non-host	0.3[b]	
γ	Initial disease transmission rate of the deer mouse without the influence of additional species. (hectare per month)	0.0130[b]	
δ	Proportional constant of the disease transmission rate with the non-host density. (hectare)	0[b]	0.0543[b]
Δt	Small time step (per month)	0.001	

[a]The values were based on [11]
[b]The values were based on [22].

nothing" event in the $[t, t + \Delta t)$ interval. Figure 1 shows the schematic diagram of the possible events of an individual.

Following the approach depicted in Fig. 1, we could then do the same for the susceptible and infected deer mouse, and the non-host. By expressing the equations in model (1) in per capita form, we could then equate the probabilities of the "die", "reproduce" and "do nothing" events for each susceptible, infected and non-host individuals. The per capita form of model (1) is as follow:

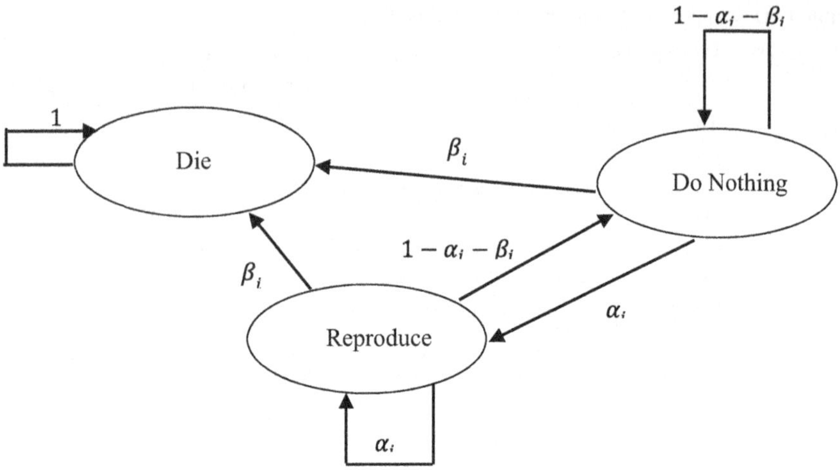

Fig. 1 Schematic diagram of the possible event path for an individual with α_i = probability of executing "reproduce" event and β_i = probability of executing "die" event in the $[t, t + \Delta t)$ interval, where $i = 1, 2, 3, \ldots$

$$\begin{aligned} \frac{1}{S}\frac{\mathrm{d}S}{\mathrm{d}t} &= r_1 + \frac{I}{S}(b_1) - r_1\left(\frac{N+q_1Z}{K_1}\right)\left[1 - a\left(\frac{I}{S}\right)\right] - \gamma(1+\delta Z)I \\ \frac{1}{I}\frac{\mathrm{d}I}{\mathrm{d}t} &= \gamma(1+\delta Z)S - \left[\mu + d_1 + r_1(1-a)\left(\frac{N+q_1Z}{K_1}\right)\right] \\ \frac{1}{Z}\frac{\mathrm{d}Z}{\mathrm{d}t} &= r_2\left[1 - \frac{Z+q_2N}{K_2}\right] \end{aligned} \tag{2}$$

For abbreviation purpose, let us express the following transition probabilities:

1. $\alpha_i(S) = Pr\{\text{a } S \text{ reproduces a new } S \text{ in } [t, t + \Delta t)\}$
2. $\beta_i(S) = Pr\{\text{a single } S \text{ dies in } [t, t + \Delta t)\}$
3. $1 - \alpha_i(S) - \beta_i(S) = Pr\{\text{a single } S \text{ does nothing in } [t, t + \Delta t)\}$
4. $\alpha_i(I) = Pr\{\text{a } I \text{ reproduces a new } I \text{ in } [t, t + \Delta t)\}$
5. $\beta_i(I) = Pr\{\text{a single } I \text{ dies in } [t, t + \Delta t)\}$
6. $1 - \alpha_i(I) - \beta_i(I) = Pr\{\text{a single } I \text{ does nothing in } [t, t + \Delta t)\}$
7. $\alpha_i(Z) = Pr\{\text{a } Z \text{ reproduces a new } Z \text{ in } [t, t + \Delta t)\}$
8. $\beta_i(Z) = Pr\{\text{a single } Z \text{ dies in } [t, t + \Delta t)\}$
9. $1 - \alpha_i(Z) - \beta_i(Z) = Pr\{\text{a single } Z \text{ does nothing in } [t, t + \Delta t)\}$

Thus, we can now equate the transition probabilities as below:

$$\alpha_i(S) = \left[r_1 + \frac{I}{S}(b_1)\right]\Delta t$$

$$\beta_i(S) = \left\{r_1\left(\frac{N + q_1Z}{K_1}\right)\left[1 - a\left(\frac{I}{S}\right)\right] + \gamma(1 + \delta Z)I\right\}\Delta t$$

$$1 - \alpha_i(S) - \beta_i(S) = 1 - \left[r_1 + \frac{I}{S}(b_1)\right]\Delta t - \left\{r_1\left(\frac{N + q_1Z}{K_1}\right)\left[1 - a\left(\frac{I}{S}\right)\right] + \gamma(1 + \delta Z)I\right\}\Delta t$$

$$\alpha_i(I) = \left[\gamma(1 + \delta Z)S\right]\Delta t$$

$$\beta_i(I) = \left[\mu + d_1 + r_1(1 - a)\left(\frac{N + q_1 Z}{K_1}\right)\right]\Delta t$$

$$1 - \alpha_i(I) - \beta_i(I) = 1 - \left[\gamma(1 + \delta Z)S\right]\Delta t - \left[\mu + d_1 + r_1(1 - a)\left(\frac{N + q_1 Z}{K_1}\right)\right]\Delta t$$

$$\alpha_i(Z) = r_2 \Delta t$$

$$\beta_i(Z) = r_2\left(\frac{Z + q_2 N}{K_2}\right)\Delta t$$

$$1 - \alpha_i(Z) - \beta_i(Z) = 1 - r_2\Delta t - r_2\left(\frac{Z + q_2 N}{K_2}\right)\Delta t$$

With these probabilities, we implemented the stochastic simulations for both case studies in GAMA version 1.8. It should be reminded that any software which supports individual-based modelling such as MATLAB can also be used for the simulations.

2.3 *Implementation in GAMA 1.8*

GAMA 1.8 is a free modelling and simulation development software [25]. It specializes in spatially explicit agent-based simulations. It is developed by several teams from France and Vietnam under the IRD/SU international research unit UMMISCO. It has been widely used by researchers to study problems related to epidemiology, urban planning, transportation, etc. The stochastic ABM was implemented through the steps depicted by the flowchart in Fig. 2.

The simulations were first done with the parameter values from Table 1. Then, we reran the simulations with different values for K_2 and q_1 to observe for any potential differences between case study 1 and 2. Due to our computer limitation, we only ran 100 simulations for each scenario.

3 Results

3.1 *Simulations Based on Parameter Values from Table 1*

The deterministic and stochastic results for both case study 1 and 2 are depicted in Fig. 3 respectively. By comparing the deterministic results from both cases, we could see that the non-host population (Z) stabilized at similar levels; whereas, the

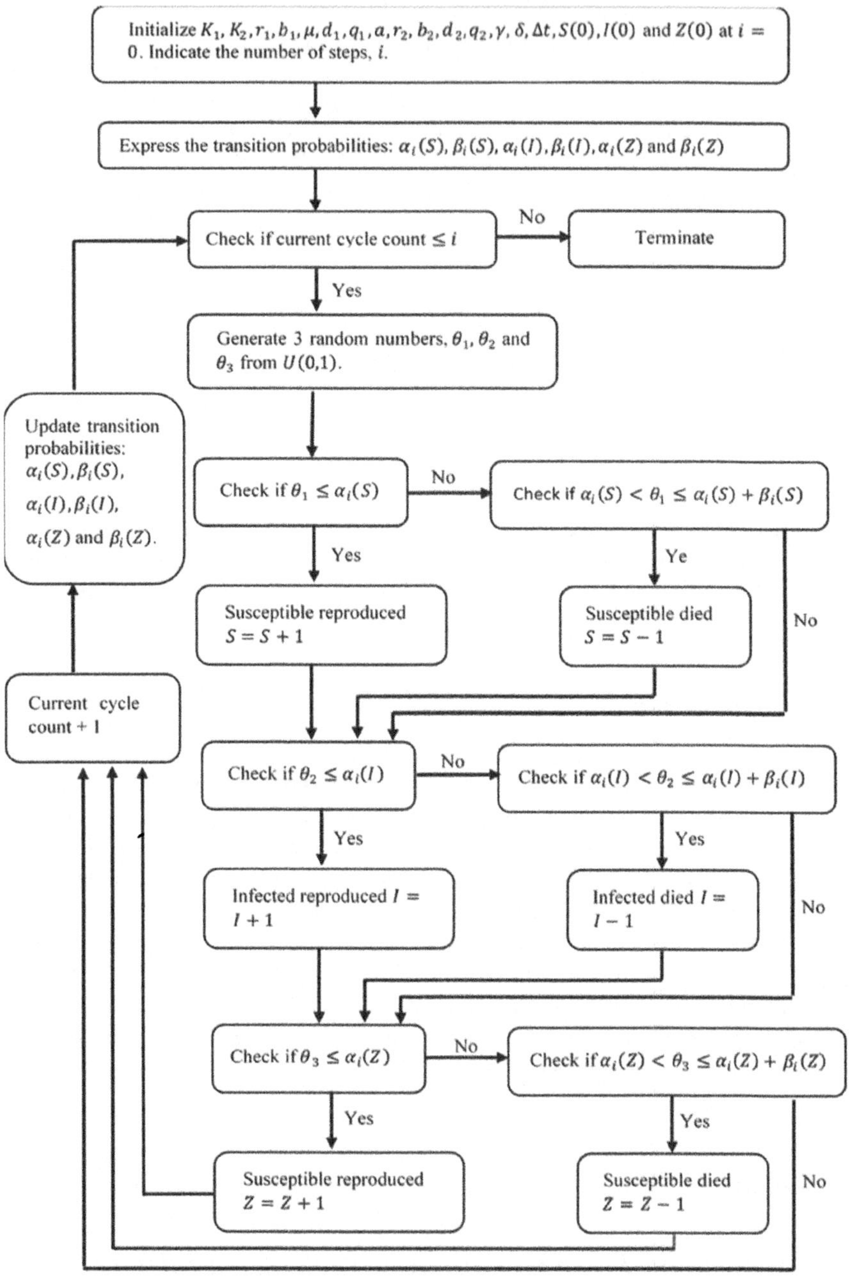

Fig. 2 Flowchart for simulating the stochastic ABM in GAMA

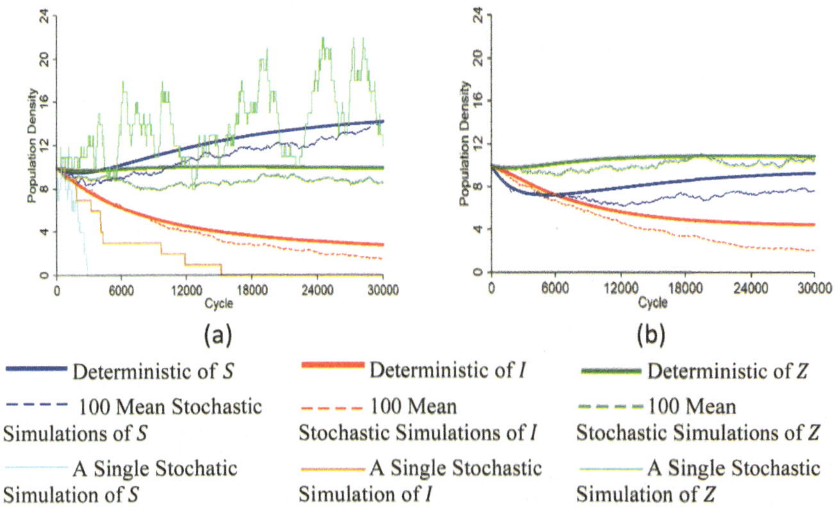

Fig. 3 Population density versus time for (a) case study 1 and (b) case study 2 with initial population $(S(0), I(0), Z(0)) = (10, 10, 10)$ and parameter values from Table 1. The (S^*, I^*, Z^*) densities at the end of the 30,000 cycles simulation were (a) Deterministic: (14.2, 2.8, 9.9); Stochastic: (14.4, 1.5, 8.6) (b) Deterministic: (9.3, 4.5, 10.9); Stochastic: (7.6, 2.1, 10.5)

susceptible deer mouse density was higher when the non-host acted as a dilution agent compared to it being an amplification agent. Despite the lower density in the susceptible, the population density for the infected deer mouse was higher for the case of amplification agent; and it is also interesting to note that the infected density was higher than the susceptible density before the simulation reached 6000 cycles. This showed that a dilution agent not only reduces the SNV transmission but also preserve the healthy deer mouse population; whereas, the inclusion of an amplification agent in the system was rather ineffective in reducing SNV transmission and the reduction of the SNV was at the cost of further reduction of the healthy deer mouse population.

However, the stochastic simulations showed a slightly different result in terms of the comparison between case study 1 (dilution agent) and 2 (amplification agent) for the infected deer mouse population. By comparing the stochastic results of the infected density in Fig. 3a and 3b, we could see the densities between both dilution agent and amplification agent cases were quite similar (stochastic infected deer mouse density at 30000th cycle for: study case 1 = 1.5; study case 2 = 2.1). This indicated that the role of the non-host might not be significant in reducing the SNV prevalence in small deer mouse population compared to the deterministic results. Furthermore, small gaps were observed between the deterministic and stochastic results. This phenomenon was due.

to the inherent characteristic of the extinction probability in the stochastic ABM model which matches the observation in [26].

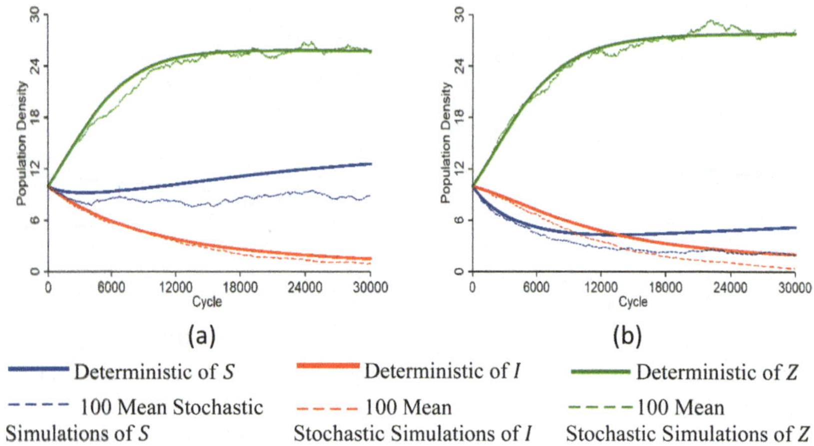

Fig. 4 Population density versus time for (a) case study 1 and (b) case study 2 with initial population $(S(0), I(0), z(0)) = (10, 10, 10)$ and parameter values from Table 1 except for $K_2 = 30$. The (S^*, I^*, Z^*) densities at the end of the 30,000 cycles simulation were (a) Deterministic: (12.6, 1.5, 25.8); Stochastic: (9.0, 1.0, 25.6) (b) Deterministic: (5.2, 2.1, 27.8); Stochastic: (2.0, 0.4, 28.2)

3.2 *Varying K_2 and q_1*

To investigate the impact of the carrying capacity and the interspecific strength of the non-host, we reran the simulations with an initial population of $(S(0), I(0), Z(0)) = (10, 10, 10)$ with $K_2 = 30$ (Fig. 4), and $(S(0), I(0), Z(0)) = (10, 10, 10)$ with $q_1 = 0.6$ (Fig. 5), while keeping the other parameter values as of Table 1. $K_2 = 30$ represents a much more favorable environment for the non-host while $q_1 = 0.6$ indicates the increase in aggressiveness of the non-host towards the deer mouse.

Based on Fig. 4 and 5, we could see that the intensity of the susceptible and infected densities for both cases in terms for both deterministic and stochastic results were generally lower compared to $K_2 = 15$, or $q_1 = 0.2$ (for case study 1) and $q_1 = 0.4$ (for case study 2). This showed that the carrying capacity and the interspecific strength of a non-host has a positive relationship in reducing the SNV prevalence irrespective of its role being an amplification or dilution agent. By comparing the total density of the deer mouse (N) between the varied K_2 and q_1 cases with the original parameter values in Table 1 at the end of the simulations, we could see there was a significant reduction in the deer mouse density; e.g. the N in Fig. 3(a) was at 17 (deterministic) and at 15.8 (stochastic) while the N in Fig. 4(a) was at 14.1 (deterministic) and at 10.0 (stochastic). Hence, we hypothesized that the reduction of the infected population was through the mechanism of decreasing the host density. Interestingly, the infected stochastic results for both dilution and amplification cases appeared to decrease to a similar level as opposed to the deterministic counterpart. The infected density at the end of the simulations for varied K_2 was 1.0 (Fig. 4(a)) and 0.4 (Fig. 4(b)); whereas it was 1.6 (Fig. 5(a)) and 2.2 (Fig. 5(b)). These observations were similar to the results

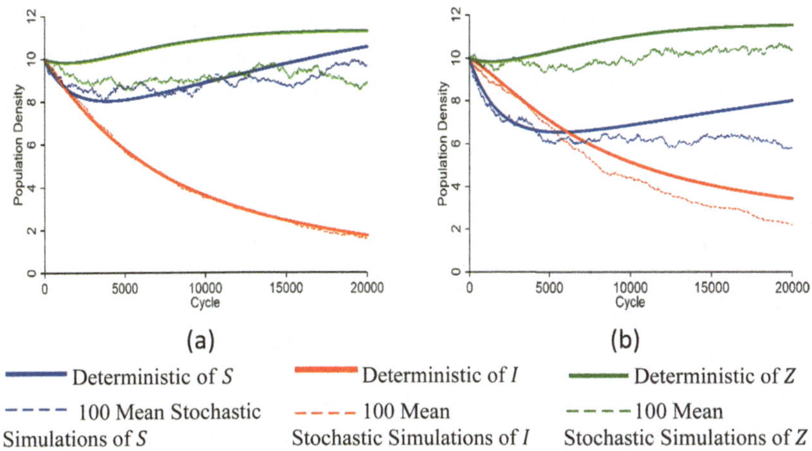

Fig. 5 Population density versus time for (a) case study 1 and (b) case study 2 with initial population $(S(0), I(0), z(0)) = (10, 10, 10)$ and parameter values from Table 1 except for $q_1 = 0.6$. The (S^*, I^*, Z^*) densities at the end of the 20,000 cycles simulation were (a) Deterministic: (10.6, 1.7, 11.3); Stochastic: (9.7, 1.6, 8.9) (b) Deterministic: (8.0, 3.4, 11.5); Stochastic: (5.8, 2.2, 10.5)

in Fig. 3, which led us to question the importance of investigating the community assemblage in terms of the biodiversity effect on SNV transmission in deer mouse especially when the population densities are small.

4 Discussion

From the above results, we managed to show and compare the impact of a non-host being a dilution agent and amplification agent towards the SNV transmission in deer mouse from deterministic and stochastic perspectives. Our deterministic model has shown that there were inherent differences in the infected density in the presence between a dilution agent and an amplification agent. The results showcased that a dilution agent performed much better in reducing the infected deer mouse density as well as preserving a much larger susceptible deer mouse population compared to an amplification agent. This finding aligns with the work of Milholland et al. [27] which highlights the identification of a species' role within the assemblage is as crucial as other factors (e.g. environmental conditions and species competition strength) to identify the biodiversity effect on disease transmission in an eco-epidemiological problem. However, such results were not observed in the case of stochastic ABM simulations. Our stochastic simulations revealed that the infected density levels were similar in the presence between a dilution agent and an amplification agent in small population. This suggests that the role of a non-host species may not have much of a difference between being a dilution or amplification agent when the population density is small, as both manage to reduce the infected deer mouse to a similar density from a stochastic perspective. It may not be surprising to observe such discrepancies between our deterministic and stochastic model as Mohd et al. [28] managed to observe contrasting results on alternative stable states between their stochastic model and deterministic multiple species models. As such, our study managed to highlight the importance of ABM in eco-epidemiological studies as well as providing a methodological discussion regarding the usability of different simulation approaches, e.g. deterministic and stochastic ABM in order to produce robust observations of eco-epidemiological phenomenon under consideration.

Nonetheless, further investigation, especially comparison with experimental or field data, is required to confirm the discrepancies between the deterministic and stochastic results. If possible, we would suggest field researchers to conduct experimental studies at both small and large scales to investigate the transmission of SNV among deer mouse in the presence of a dilution agent, e.g. desert pocket mouse, in a close community. The results should then be compared to the experimental study conducted with an amplification agent, e.g. Merriam's kangaroo rat. The small and large scales studies are intended to represent the stochastic and deterministic simulations respectively. However, such experimental studies may be extremely difficult to perform. Alternatively, observational studies, which investigate the comparison of SNV transmission through role identification within a species assemblage at low and high densities population level, may be a more feasible approach. The challenges posed for such observational studies would then lie in the role identification played by the species on disease transmission, and the possible indirect effects caused by other species interactions within the assemblage. It should be noted the experimental and observational studies mentioned were just proposals based on our ideas. Opinions from experts and researchers on other viable study designs are much welcome to validate our findings.

Our study has successfully employed GAMA 1.8 as an alternative approach on simulating the biological system using the techniques of stochastic process and differential equations. This has contributed to the methodological discussions on the use of different modelling techniques and computer packages to examine the biological phenomena of interest. Similar to other platforms, GAMA provides similar flexibility to code the agent's characteristics and behavior according to the researcher's choice of techniques and assumptions. Some beginners may find GAMA a bit intimidating as it requires hard coding to simulate the ABM of their choice. They will need to browse through the tutorials to understand the available GAMA functions and their logic to construct their models of choice. Understanding these struggles, the GAMA developers and its community are actively providing skeleton codes along with examples in several fields to help with the learning process. All in all, GAMA is a good alternative platform for modelling biological systems and it is up to the researcher's discretion to choose whichever platform they are comfortable with.

There are a few limitations to this study. For starters, we only considered temporal modelling; but in reality, the rodents move around the environment. The lack of spatial considerations might cause us to miss some important observations, e.g. [3]

observed the presence of "refugia" for the SNV infected deer mouse in a spatio-temporal model. When the overall environmental condition is less favorable, the SNV infected deer mouse will find "refuge" in an area with better environment, which harbors the SNV and will then act as a source of transmission when the environmental condition has improved. We probably could follow the footstep of [17] in utilizing the agent-based modelling to incorporate not only the spatio-temporal feature but also to include detailed rodents' characteristics in a simulation. We also limited ourselves by only considering a non-host species. Other studies observed that other *Peromyscus* species [29] and desert woodrat [30] can serve as secondary reservoirs for the SNV. As pointed out by Ostfeld and Keesing [31], the quality in disease transmission by the secondary reservoir needs to be investigated as the presence of the secondary reservoir may not necessarily amplified the infection but may dilute it instead. Hence, it would be interesting to model for such scenario to better understand its dynamics and mechanisms.

5 Conclusion

This study managed to simulate and compare the effects of an amplification and dilution agent in regulating the SNV transmission in small deer mouse population from deterministic and stochastic perspectives. We implemented our stochastic ABM model in GAMA 1.8. Our deterministic results showed the effectiveness of a dilution agent comparatively to an amplification agent in reducing SNV. However, our stochastic results showed rather indifferent results between the dilution and amplification agent in small population density. Based on these contrasting results between the deterministic model and stochastic ABM, further investigations are required to better understand this discrepancy. As such, we would like to highlight the importance of utilizing ABM, especially in the eco-epidemiological field, as its usage might produce additional information which the deterministic models might fail to capture.

Acknowledgements The authors would like to thank the School of Mathematical Sciences and the Universiti Sains Malaysia (USM) for the support. Mohd Hafiz Mohd is supported by the USM Fundamental Research Grant Scheme (FRGS) No. 203/PMATHS/6711645.

References

1. MacNeil, A., Nichol, S.T., Spiropoulou, C.F.: Hantavirus pulmonary syndrome. Virus Res. **162**, 138–147 (2011). https://doi.org/10.1016/j.virusres.2011.09.017
2. Jiang, H., Zheng, X., Wang, L., Du, H., Wang, P., Bai, X.: Hantavirus infection: a global zoonotic challenge. Virol. Sin. **32**, 32–43 (2017). https://doi.org/10.1007/s12250-016-3899-x
3. Abramson, G., Kenkre, V.M.: Spatiotemporal patterns in the Hantavirus infection. Phys. Rev. E **66**, 011912 (2002). https://doi.org/10.1103/PhysRevE.66.011912

4. Peixoto, I.D., Abramson, G.: The effect of biodiversity on the hantavirus epizootic. Ecology **87**(4), 873–879 (2006). https://doi.org/10.1890/0012-9658(2006)87[873:teobot]2.0.co;2
5. Dizney, L.J., Ruedas, L.A.: Increased host species diversity and decreased prevalence of sin nombre virus. Emerg. Infect. Dis. **15**(7), 1012–1018 (2009). https://doi.org/10.3201/eid1507.081083
6. Suzán, G., Marcé, E., Giermakowski, J.T., Mills, J.N., Ceballos, G., Ostfeld, R.S., Armién, B., Pascale, J.M., Yates, T.L.: Experimental evidence for reduced rodent diversity causing increased hantavirus prevalence. PLoS ONE **4**(5), e5461 (2009). https://doi.org/10.1371/journal.pone.0005461
7. Randolph, S.E., Dobson, A.D.M.: Pangloss revisited: a critique of the dilution effect and the biodiversity-buffers-disease paradigm. Parasitology **139**, 847–863 (2012). https://doi.org/10.1017/S0031182012000200
8. Ogden, N.H., Tsao, J.I.: Biodiversity and Lyme disease: dilution or amplification? Epidemics **1**(3), 196–206 (2009). https://doi.org/10.1016/j.epidem.2009.06.002
9. Levi, T., Keesing, F., Holt, R.D., Barfield, M., Ostfeld, R.S.: Quantifying dilution and amplification in a community of hosts for tick-borne pathogens. Ecol. Appl. **26**(2), 484–498 (2016). https://doi.org/10.1890/15-0122
10. Faust, C.L., Dobson, A.P., Gottdenker, N., Bloomfield, L.S. P., McCallum, H.I., Gillespie, T.R., Diuk-Wasser, M., Plowright, R.K.: Null expectations for disease dynamics in shrinking habitat: dilution or amplification? Philos. Trans. R. Soc. Lond. B. Biol. Sci. **372**(1722) (2017). https://doi.org/10.1098/rstb.2016.0173
11. Luis, A.D., Kuenzi, A.J., Mills, J.N.: Species diversity concurrently dilutes and amplifies transmission in a zoonotic host–pathogen system through competing mechanisms. Proc. Natl. Acad. Sci. **115**(31), 7979–7984 (2018). https://doi.org/10.1073/pnas.1807106115
12. Allen, L.J.S.: A primer on stochastic epidemic models: formulation, numerical simulation, and analysis. Infect. Dis. Model. **2**(2), 128–142 (2017). https://doi.org/10.1016/j.idm.2017.03.001
13. McCormack, R.K., Allen, L.J.S.: Stochastic SIS and SIR multihost epidemic models. In: Proceedings of the Conference on Differential and Difference Equations and Applications, pp. 775–785 (2006)
14. Otero, M., Solari, H.G.: Stochastic eco-epidemiological model of dengue disease transmission by *Aedes aegypti* mosquito. Math. Biosci. **223**(1), 32–46 (2010). https://doi.org/10.1016/j.mbs.2009.10.005
15. Mandal, P.S., Banerjee, M.: Deterministic chaos vs. stochastic fluctuation in an eco-epidemic model. Epidemiology **7(3)**, 99–116 (2012). https://doi.org/10.1051/mmnp/20127308
16. Guzzetta, G., Tagliapietra, V., Perkins, S.E., Hauffe, H.C., Poletti, P., Merler, S., Rizzoli, A.: Population dynamics of wild rodents induce stochastic fadeouts of a zoonotic pathogen. J. Anim. Ecol. **86**, 451–459 (2017). https://doi.org/10.1111/1365-2656.12653
17. Kaplan, M., Manore, C.A., Bagamian, K.H.: Agent-based hantavirus transmission model incorporating host behavior and viral shedding heterogeneities derived from field transmission experiments. Lett. Biomath. **3**(1), 209–228 (2016). https://doi.org/10.1080/23737867.2016.1248507
18. Wesley, C.L., Allen, L.J.S., Jonsson, C.B., Chu, Y., Owen, R.D.: A discrete-time rodent-hantavirus model structured by infection and developmental stages. In: Advances in Dicrete Dynamical Systems, 387–398, Mathematical Society of Japan, Tokyo, Japan (2009). https://doi.org/10.2969/aspm/05310387
19. Escudero, C., Buceta, J., de la Rubia, F.J., Lindenberg, K.: Effects of internal fluctuations on the spreading of Hantavirus. Phys. Rev. E Stat. Nonlin. Soft. Matter. Phys. **70**, 061907 (2004). https://doi.org/10.1103/PhysRevE.70.061907
20. Mohd, M.H.B.: Modelling the presence-absence of multiple species. Ph.D. Thesis (2016)
21. Mohd, M.H.: How can modelling tools inform environmental and conservation policies? Int. J. Eng. Technol. **7**(4.28), 333–337 (2018). https://doi.org/10.14419/ijet.v7i4.28.22610
22. Lee, L.W.F., Mohd, H.M.: The effects of amplification and dilution agent in transmitting sin nombre virus (SNV) in deer mouse population. MATEMATIKA: MJIAM, **36**(2), 85–98 (2020). https://doi.org/10.11113/matematika.v36.n2.1244

23. Rubio, A.V., Castro-Arellano, I., Mills, J.N., List, R., Avila-Flores, R., Suzan, G.: Is species richness driving intra- and interspecific interactions and temporal activity overlap of a hantavirus host? An experimental test. PLoS ONE **12**(11), e0188060 (2017). https://doi.org/10.1371/journal.pone.0188060
24. Brenner, G.J., Moynihan, J.A.: Stressor-induced alterations in immune response and viral clearance following infection with Herpes Simplex Virus-Type 1 in BALB/c and C57Bl/6 Mice. Brain Behav. Immun. **11**, 9–23 (1997). https://doi.org/10.1006/brbi.1997.0480
25. Taillandier, P., Gaudou, B., Grignard, A., Huynh, Q.-N., Marilleau, N., Caillou, P., Philippon, D., Drogoul, A.: Building, composing and experimenting complex spatial models with the GAMA platform. GeoInformatica **23**(2), 299–322 (2019). https://doi.org/10.1007/s10707-018-00339-6
26. Allen, L.J., McCormack, R.K., Jonsson, C.B.: Mathematical models for hantavirus infection in rodents. Bull. Math. Biol. **68**(3), 511–524 (2006). https://doi.org/10.1007/s11538-005-9034-4
27. Milholland, M.T., Castro-Arellano, I., Arellano, E., Nava-Gracia, E., Rangel-Altamirano, G., Gonzalez-Cozatl, F.X., Suzan, G., Schountz, T., Gonzalez-Padron, S., Vigueras, A., Rubio, A.V., Maikis, T.J., Westrich, B.J., Martinez, III., J.A., Esteve-Gassent, M.D., Torres, M., Rodriguez-Ruiz, E.R., Hahn, D., Lacher, Jr., T.E.: Species identity supersedes the dilution effect concerning hantavirus prevalence at sites across Texas and Mexico. ILAR J. **58**(3), 401–412 (2017). https://doi.org/10.1093/ilar/ily001
28. Mohd, M.H., Murray, R., Plank, M.J., Godsoe, W.: Effects of dispersal and stochasticity on the presence-absence of multiple species. Ecol. Model. **342**, 49–59 (2016). https://doi.org/10.1016/j.ecolmodel.2016.09.026
29. Abbott, K.D., Ksiazek, T.G., Mills, J.N.: Long-term hantavirus persistence in rodent populations in central arizona. Emerg. Infect. Dis. **5**(1), 102–112 (1999). https://doi.org/10.3201/eid0501.990112
30. Dearing, M.D., Mangione, A.M., Karasov, W.H., Morzunov, S., Otteson, E., St. Jeor, S.: Prevalence of hantavirus in four species of *neotoma* from arizona and utah. J. Mammal. **79**(4), 1254–1259 (1998). https://doi.org/10.2307/1383016
31. Ostfeld, R.S., Keesing, F.: Effects of host diversity on infectious disease. Annu. Rev. Ecol. Evol. Syst. **43**, 157–182 (2012). https://doi.org/10.1146/annurev-ecolsys-102710-145022

Experimenting the Impact of Pedestrianisation on Urban Pollution Using Tangible Agent-Based Simulations: Application to Hoan Kiem District, Hanoi, Vietnam

Arthur Brugière, Minh Duc Pham, Kevin Chapuis, Alexis Drogoul, Benoit Gaudou, Arnaud Grignard, Nicolas Marilleau, and Nguyen-Huu Tri

Abstract The development of permanent or temporary pedestrian areas, whether for leisure or to decrease air pollution, has become an integral part of urban planning in numerous cities around the world. Hanoi, the capital of Vietnam, began to implement its first area, around the iconic Hoan Kiem lake, a few years ago. In most of cases, however, road closure is likely to deport traffic to nearby neighbourhoods with the consequences of intensifying congestion and, possibly, increasing air pollution in these areas. Because this outcome might appear counter-intuitive to most stakeholders, it is becoming more and more necessary to analyse, assess and share the impacts of these developments in terms of traffic and pollution shifts before implementing them. In this project, we used the GAMA platform to build an agent-based model that simulates the traffic, its emissions of air pollutants, and the diffusion of these pollutants in the district of Hoan Kiem. This simulation has been designed so as to serve either as a decision support tool for local authorities or as an awareness-raising tool for the general public: thanks to its display on a physical 3D model of the district,

A. Brugière · M. D. Pham · K. Chapuis · A. Drogoul · B. Gaudou (✉) · N. Marilleau · N.-H. Tri
UMI 209 UMMISCO, Sorbonne Université, IRD, Bondy, France

A. Brugière · M. D. Pham · A. Drogoul · B. Gaudou
ICTLab, USTH, Vietnam Academy of Science and Technology, Hanoi, Vietnam

K. Chapuis
UMR 228, ESPACE-DEV, IRD, Montpellier, France

B. Gaudou
UMR 5505, IRIT, Université Toulouse 1 Capitole, Toulouse, France

A. Grignard
City Science, MIT Media Lab, Cambridge, USA

N.-H. Tri
IXXI, ENS Lyon, Lyon, France

M. H. Mohd et al. (eds.), *Modelling, Simulation and Applications of Complex Systems*, Springer Proceedings in Mathematics & Statistics 359,
https://doi.org/10.1007/978-981-16-2629-6_4

people can effectively and naturally interact with it at public venues. Although more accurate data and more realistic diffusion models are necessary and will need further research in the future, the simulation is already able to reflect traffic and air pollution peaks during rush hours, allowing residents and developers to understand the impact of pedestrianization on air quality in different scenarios.

Keywords Agent-based model · Air pollution · Urban traffic · GAMA platform · Interactive simulation

1 Introduction

According to the World Health Organization (WHO), air pollution, particularly in urban areas, is the leading environmental health risk worldwide, causing about one in nine deaths [26]. Awareness of the extent of this phenomenon is recent and still confused, particularly in developing countries where there is an accelerated, and sometimes uncontrolled urban development. Even if urban air pollution sources are multiple (local factories and agriculture burning, or regional atmospheric streams [8]...), the urban traffic and congestion have definitely (even if it is not clearly quantified) an impact on the amount of pollutant in air [39]. Adapted traffic and more generally urban management strategies are thus necessary to reduce traffic-related emissions and inhabitant exposure [5]. The difficulty encountered by urban planners lies not only in the diversity of possible solutions (*e.g.* development of public transport, infrastructure development, changes in regulations) but also in their acceptance by the inhabitants concerned by the risk and often poorly, or not at all, informed about their effects. Better consultation between planners and residents is therefore a necessary step to avoid finding solutions that are rejected, misunderstood, or even counterproductive.

However, this stage comes up against uncertainty concerning the possible evolutions of the city and the great complexity of the studied phenomena: their understanding, analysis and projections into possible futures are currently a real scientific challenge, particularly because it is difficult to measure and predict the impact of individual and collective adaptation behaviours, whether to pollution or regulations. The complex nature of a city dynamics (in the sense of the complexity science [23]) with the importance of spatial and individual heterogeneities lead us to choose the agent-based modelling [22] approach to tackle this issue. But we argue here that we need to go further to enhance the understanding of the simulated phenomena (both by decision-makers and by random people) and the discussions about result patterns. We chose to make the simulation tangible (by providing a physical substrate) [27] and interactive, so that the user can formulate questions on the simulation it observes and test them in runtime.

Following these principles, we propose in this article a decision-making and consultation support tool based on the design and implementation of an interactive and tangible agent-based simulator. In the frame of the HoanKiemAir project, we couple a 3D printed map of the studied area (the Hoan Kiem district of Hanoi city, Vietnam) used as a screen to show simulations of an agent-based model (implemented using the GAMA platform software [33]) coupling a traffic sub-model and an air pollution sub-model (air pollution is limited to pollutants emitted by vehicles). Users can interact with the simulation using an Android application to modify the traffic and the urban environment and observe, in real time, the impact of these choices on urban air quality. The agent-based model has been designed to be able to take into account individual mobility and adaptation behaviours in order to reproduce Vietnamese urban traffic and its evolution, particularly in the case of new urban planning (*e.g.* various pedestrian areas implementations).

The model has previously been presented in details in [28]. We focus this article on the analysis and exploration of the model, with the key objective of providing insights to the question of the impact of alternative pedestrian areas scenarios on the emitted pollution.

The article is organized as follows. Section 2 details the case study: the area of Hoan Kiem district and the problematic linked to the implementation and extension of the pedestrian area around its central lake. Section 3 presents, first, a quick state of the art related to modelling of urban traffic and air pollution, and secondly, an overview of tangible and interactive simulations and in particular the ones used for urban planning. Section 4 presents a summary of the HoanKiemAir agent-based model. The two following sections describe the two possible uses of the model: Sect. 5 describes its interactive use through the 3D map and the Android application, while Sect. 6 presents the first results from its sensitivity analysis and exploration. Finally, Sect. 7 concludes the paper and discusses perspectives and future extensions.

2 Case Study

For several years, the Hanoi People's Committee drew up a plan to organize some pedestrian zones in Hanoi, where vehicles would be banned on weekends. This plan was tested and implemented in two different areas of the city, one of which is located around the iconic Hoan Kiem Lake, which is the centre of the city old quarter, and the key touristic spot of the city. From 7PM every Friday evening until 2PM on Monday morning, several streets are closed to any vehicle and thus invaded by thousands of pedestrians (see Figs. 1a and 1b for the map of the closed roads in the current pedestrian area). As time goes by, this area is becoming a popular tourist attraction as well as a gathering point for residents.

However, in a city as dense as Hanoi, the closure of roads, especially in the city centre, is not without consequences on traffic flows and, albeit indirectly, on the air pollution induced by this temporarily reorganized traffic. While local air quality is clearly improving around the lake and the area have become much more pleasant for pedestrians, the effects of the induced reorganization of the traffic do not have been yet evaluated on the rest of the district and even at the scale of the city. We can

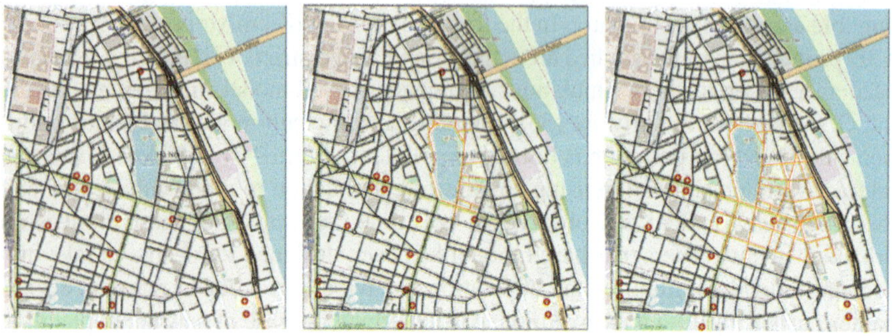

(a) No closed road (b) Current area (c) Possible extension plan

Fig. 1 Pedestrian area scenarios: roads in black are opened for vehicles, while those in orange are closed.

suppose that this will lead to a degradation elsewhere, due to the report of the traffic in other streets of the district, but no simulation or visualisation tools are available in order to observe this impact.

Thanks to the popular success of the implementation of the weekend pedestrian area, authorities are thinking about extending the pedestrian area and thus to close more roads during the weekend (one of the possible extension plans is displayed on Fig. 1c). Whereas the current plan is focused mainly on the roads around the lake and some small ones close to it, such a new plan could close many streets in the South and the East of the district and have a much more important impact on the traffic.

One of the objective of the HoanKiemAir model is to be able to take into account such alternative closure plans and compute their impact on the traffic and pollution emitted.

3 Related Works

3.1 Traffic Flow Models

Modelling of traffic (in particular in an urban context) is now a widely spread approach to optimize traffic flow, reduce traffic jams or improve infrastructures, and help urban planning. [16] or [6] identified three main modelling approaches, mainly related to the modelling scale.[1]

Macroscopic models represent traffic flow in the streets as a fluid flow in a pipe. Mathematical models can thus be used to describe the evolution of some macroscopic values of the flow: density, average speed, volume... The paradigmatic model is LWR [21, 29]. Although they can produce realistic outputs with low computational cost and

[1] [37] illustrates a genealogy of traffic flow models.

in a deterministic way, there are not expressive enough to represent heterogeneities in terms of vehicles and drivers' behaviours.

On the other hand, **microscopic models** describe vehicles individually, with their individual attributes (location, speed, acceleration...) and behaviours. As an example the classical NaSch model [24] has first been implemented as a Cellular Automaton and later as an agent-based model [3]: each car is represented individually with its own speed. Its behaviour is to go straight, accelerating as much as it can (given an acceleration parameter) and decreasing its speed when another vehicle is before it. Such models are highly expressive, able to represent any kind of vehicles, drivers' behaviours and traffic situations. However, computation load can be very high as the number of vehicles in the network becomes larger.

Finally, **mesoscopic models** are hybrid models, combining features of the two previous approaches. A hybrid model can for example model vehicles not as individuals but as small groups. They allow vehicles to make certain decisions, but at the same time limit the level of detail on their behaviours. An example of mesoscopic model is the classical Underwood model [35] which considers individual vehicles, but their speed is constrained by the total number of vehicles on a road edge. Another interesting approach relies on the coupling between several models from different approaches in order to build a model that can adapt the simulated depending on the traffic and on the situation [4].

In the HoanKiemAir, we chose the mesoscopic approach by developing an agent-based model (ABM) [22], *i.e.* a model representing the studied system at an individual level: each vehicle is represented, with its individuals features and behaviours (perception, decision-making process, learning...).[2] The overall dynamics is expected to emerge at a macroscopic level from individual agents' behaviours and interactions. ABMs have the advantage to be easily spatialized (in a continuous 2D or 3D environment, or on a discrete grid of graph). ABMs have now a wide variety of applications, from physical and life sciences to economics and social sciences. Such agent-based models has been developed for example for evacuation using vehicles [7] or to investigate various mobility means [12]. Other advantages of ABMs are their flexibility and expressivity, *e.g.* to easily integrate urban management policies or to be coupled with various other modelling approaches, such as the ones used to model air pollution.

3.2 Pollution Model

To model air pollution, two main dynamics have to be taken into account: **emission** and **dispersion**. An emission model calculates the amount of pollutants released by any source (factories, vehicles...), whereas dispersion models aim to describe how air pollutants disperse in the atmosphere.

---

[2]The model is mesoscopic because we rely on the Underwood model to compute vehicle speeds.

When considering air pollution in an urban context, a widely used dispersion model is the Operational Street Pollution Model (OSPM) [15], suitable for areas with street canyons (*i.e.* streets surrounded by tall buildings). It assumes that two factors contribute to the concentration of pollutant in a street canyon: traffic flow (with direct emission from vehicles and air recirculation in the street canyon) and other sources (pollution from other streets, factories, power plants...). The European emission model COPERT IV [25] is integrated into OSPM as an emission module. This model (mainly funded by the European Union) is now recognized as a standard in both the academic and industrial communities. The emission values are computed based on fuel type, emission factor as well as vehicle type and diurnal traffic frequency. An example of integration of such a model with an agent-based traffic model can be found in the MarrakAir project [10] aiming at investigating the impact of the number, type and age of vehicles on air pollution in the case-study of Marrakesh (Morocco).

Whereas OSPM is mainly focused on European context, [34] applied it on the Vietnamese context (in particular on the case study of Hanoi) and provided data about Vietnamese vehicles emissions. Using a simplified version of OSPM and data from the previous study, [30] have proposed a GAMA simulation to study traffic-induced air pollution in Hanoi, Vietnam.

3.3 *Tangible User Interface and Simulation*

The idea of tangible and interactive simulations is to use a physical support (often named Tangible User Interface (TUI), see [11] for an attempt of classifications) such as a 3D printed map, sandbox, laser cut map or any other physical objects [31] to display digital information coming from a simulation. The benefits of such tangible objects to help the user to think spatial information and dynamics has now been shown (see Chapter 1 of [27] for more details about the various cognitive benefits). Similar benefits have been demonstrated in an educational context [38].

TUI can take many shapes [11] and often the setup is specific to the case study. When considering simulations supported by a TUI, a few main components can be identified in most of the projects:

(i) the **simulation**, representing the virtual world and its (spatial) evolution; it should be able to manage inputs from the user, modify the virtual world in accordance and take into account these modifications in future computation.
(ii) **controllers**, that are devices controlled by the user to modify the simulations or sensors perceiving some modifications made in the environment by the user; examples include 3D Kinect camera (to detect modifications in a physical environment, such as a sandbox) [31, 38], physical objects with RFID tags [17, 20], Lego blocks [2, 13] or application on tablet [10, 19].

(iii) **physical objects**, that can be a simple 3D map of the virtual environment used as a screen to display the simulation [2, 10, 13, 27],[3] a fully modifiable environment (as a Sandbox [38], that works also as a controller, through the camera), or a set of physical objects representing elements in the simulation and that the user can manipulate (*e.g.* traffic lights in [20] or buildings in [31]);
(iv) **visualisation supports**, in addition to a display of the simulation on the 3D map [2, 10, 13, 27] or Sandbox [38], or on an interactive table (such as the TangiSense table [18]), it is common to add screens to display additional information (3D rending of the environment in Tangible Landscape [27] or additional indicators [17]).

Such an approach is now used in most of the topics that rely on a spatial representation of the case studies and that can be tackled by simulation.

One of the first implementation of such an approach is perhaps the Augmented Reality sand boxes, such as the one developed in the Lake visualisation 3D project.[4] Given a sandbox and a Kinect camera (a low-cost camera able to detect movements in 3D), users can play with the sand to modify the ground topology, by digging rivers or "building" mountains or volcanoes. The camera is able to identify in real-time modifications of the elevation and given its values, fill the gaps with water. [38] built such a setup and evaluate its positive impact in geology classes on students to make them understand some key concepts and interactions between elements (*e.g.* water with the landscape). Many other examples of Tangible User Interface in landscape study are presented in [27].

Concerning urban case studies, one of the first work we can identify [31] was dedicated to simulate wind streams in a city, depending on the buildings' location. It is composed on a 3D (laser cut map) map of the city, a Kinect camera to identify the position of buildings and a fluid simulator, taking as input these buildings and computing the wind flow. [20] used an interactive table (TangiSense table [18]) to simulate simple traffic situations; interactions with the simulation is made through some physical objects (traffic lights...) with an RFID tag. Later the CityScience research group of the MIT Media Lab developed several projects [2, 13] to give the possibility to visualize the link between urban planning and organization, mobility modes and traffic. [10] focuses on the link between traffic (in particular related to vehicle numbers and states) and air pollution.

Finally, [17] use a tangible table with RFID tag-based physical objects to define collaborative simulations to help crisis management.

[3][19] shown that displaying a flood simulation even on a flat table for participative simulations provides a much better immersion for participants and increase discussions than displaying it on a wall.

[4]LakeViz3D project: http://lakeviz.org.

4 Model

We present in this section the key elements of the HoanKiemAir model,[5] developed using the agent-based modelling and simulation GAMA platform [33]. An extensive description of the model is presented in [28] using the standard O.D.D. (Overview, Design concepts, Details) protocol [14].

4.1 Overview of the Model

The model aims to simulate the interrelation between traffic, air pollution, and road management (in terms of pedestrian areas scenarios) in the Hoan Kiem district (Hanoi). The main question tackled by the model is: how the various pedestrian area scenarios and the evolution of the number of vehicles influence congestion and air pollution in the area?

4.2 Description of the Model Entities

The model is composed of two main dynamics, the *traffic sub-model* and the *pollution sub-model*, influenced by pedestrianization scenarios.

The *traffic sub-model* generates the movement of vehicles on a road network. It is composed of two types of agents: `vehicle` agents move on `road` edges toward an intersection target.

The *pollution sub-model* is divided into an *emission model* and a *dispersion model*. The emission model is managed by a single agent `pollutant_manager`, which takes the output of the traffic model, calculates the amount of pollutants emitted by vehicles, and then passes them to the dispersion model. The dispersion model is based on a collection of agents `pollutant_cell`, which represent airborne packages. Each `pollutant_cell` has a volume (depending on its area) in order to compute a pollution concentration in each one. We consider here two kinds of dispersion models (that will be evaluated in the next section):

(i) `pollutant_cell` agents are cells of a grid covering all the space. Each cell will thus get the pollution from the `road` edges it overlaps and diffuses to its 8 neighbour cells.

(iii) `pollutant_cell` agents are buffers on top of each road edges. Each cell will thus get the pollution from the associated road, and will diffuse to the neighbour `pollutant_cell`. This model makes the assumption that pollution that pollutants cannot be dispersed over building rooftops, but only through streets.

[5]The model is open-source and available for its GitHub repository: https://github.com/WARMTeam/HoanKiemAir.

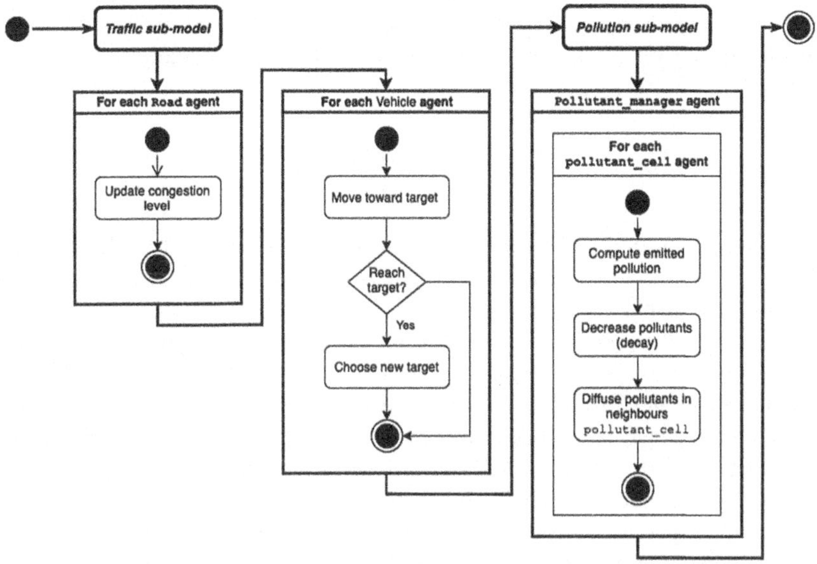

Fig. 2 Flowchart of the execution of one cycle of the HoanKiemAir model.

4.3 *Description of the Model Processes*

In a simulation cycle, the agents of the traffic model are scheduled first, followed by those of the emission and dispersion model (Fig. 2 illustrates the various steps of a simulation cycle). One simulation step represents by default 16 s, in order to be able to observe vehicles moving on the roads. When we want to observe the influence of the number of vehicles over time in the day on the pollution, we need to simulate on a period of several days. In order to keep the simulations interesting for users we choose in this case a simulation step representing 5 minutes.

In the traffic model, the `road` agents are executed first, updating their congestion levels. We choose indeed to use a traffic flow model close to the Underwood model [35]: roads compute a congestion factor, which will influence the speed of vehicles moving on it. Finally, the `vehicle` agents move to their target (a vertex of the road network, *i.e.* a road intersection) on the traffic graph. The vehicles move using built-in features implemented in the GAMA platform. When a `vehicle` reaches its target, it will choose a new intersection as next target.

Once all `vehicle` agents have completed their movement, the `pollutant_manager` agent computes the amount of traffic-induced pollutants and stores them in `pollutant_cell` agents. Then, it decreases the pollutant values in these cells according to a predefined decay rate[6] and disperses the pollutants in the neighbouring cells (the dispersion algorithm depends on the model implementation).

4.4 Input Data

The traffic and pollution sub-models are executed only on the Hoan Kiem district (an area of 2506 m × 2779 m). Nevertheless, to provide some landmarks for the users, we also display spatial elements on a wider area (a rectangle of 5458 m × 3052 m) that will not have any influence on the simulation. Figure 7 shows all the spatial data used: shapefiles impacting the simulation (roads and buildings) and the additional shapefiles loaded for contextual and visualisation purpose (lakes, rivers, special buildings and main roads shapefiles). These additional roads and the roads in the Hoan Kiem district are stored in 2 different shapefiles in order that only roads in the considered area are used as support of the traffic model.

Most of the spatial data (except buildings that have been manually sketched from satellite images) come from OpenStreetMap (OSM)[7] data source. In addition to geometric information, the OSM data also contain various attributes. For example, as far as roads are concerned, OSM can provide the number of lanes, maximum allowed speed, one-way or two-way, traffic light availability, etc. Although OSM provides some information for some roads, most of the values are missing, so we have defined a set of default values: the maximum permitted speed of 60 km/h.

Finally, the hourly distribution data of traffic (see Fig. 3) is provided by [1]. They estimated the number of trips per hour throughout the day and give us a realistic traffic flow over a full day.

4.5 Outputs

To make pollutant values easy to interpret, an Air Quality Index (AQI) value is calculated from the amount of each pollutant. We use the method given by [36]:

$$AQI^h = max_x(AQI_x^h) \text{ with } AQI_x^h = \frac{TS_x^h}{QC_x} * 100 \qquad (1)$$

where AQI_x^h is the AQI of a pollutant x at hour h, TS_x^h is the measured average amount of pollutant x in h, and QC_x is the allowed average amount of pollutant x in an hour (Table 1). Ranges of AQI are associated to colours, as illustrated in Fig. 4b.

---

[6] To be more precise, the decay factor has to be understood as the remaining ratio of pollutant after decay. If it values 0.99, this means that after decay, 99% of pollutants are kept or in other words, that they have been reduced by 1%.

[7] OpenStreetMap: www.openstreetmap.org.

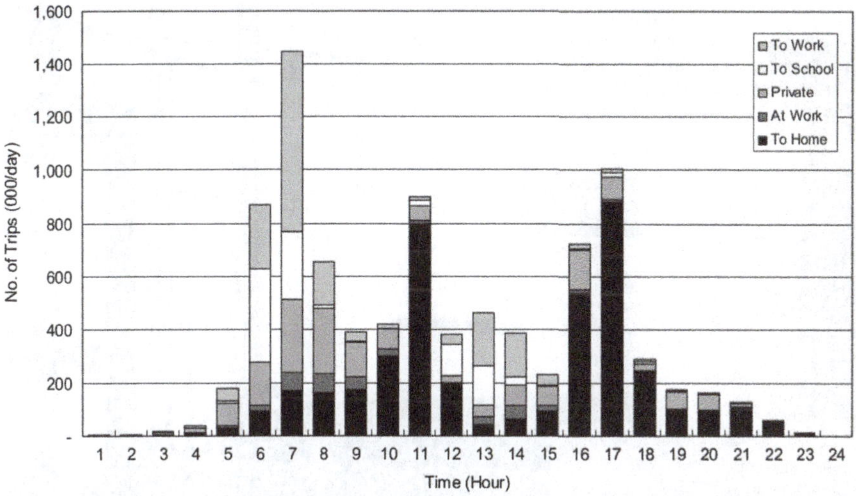

Fig. 3 Example of hourly traffic distribution (from [1]).

Table 1 Allowed hourly average amount for different pollutants (measured in $\mu g/m^3$) [36]

CO	NO_x	SO_2	PM10
30000	200	350	300

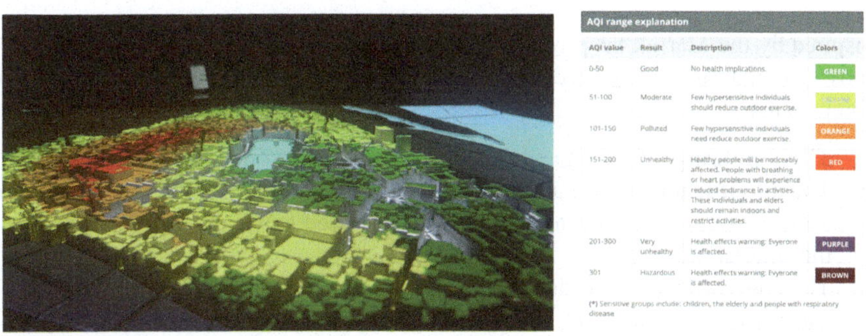

(a) 3D map (b) AQI color scale

Fig. 4 (a) Example of display on the 3D map (using colors defined in the AQI scale (b). Source: https://moitruongthudo.vn/).

The pollutant_cell agents are not made visible and the pollution is displayed on buildings (see Fig. 4a for an example on the 3D map).

In addition, in order to observe the overall evolution over time of the AQI, we display a chart of the maximum value of AQI in the district over time. Figure 6 shows the full simulation display.

Fig. 5 The full setup for tangible and interactive simulations (including the 3D map, the tablet with the Android application, the projector).

5 Interactive Simulations

Inspired by the MarrakAir project [10], we built a full setup for tangible and interactive simulations that is accessible by any random user. It consists of three main components:

- The **simulation** of traffic and pollution implemented on GAMA platform;
- The **tangible part** is managed by a 3D-printed map of a part of Hanoi city, with a focus on the Hoan Kiem district, which is used as **visualisation support** of the simulation;
- The **controller** is an Android application, installed on a tablet, that communicates with the simulation via networking, allowing users to modify in runtime simulation parameters and the visualisation mode.

Figure 5 illustrates the whole setup when it was located at the French Embassy in Hanoi during the summer 2019. It shows in particular the video projector located on top of the 3D map.

Fig. 6 Screenshot of the full simulation, including the pollution map, the pollution charts and the value of the pedestrian scenario, number of cars and of motorbikes parameters.

5.1 3D Models

The simulation is running only on the Hoan Kiem district (cf. Fig. 1), nevertheless, as the 3D map is the only place to display information for the users, we made the choice to build the 3D map on an extended area. This brings two main advantages: (i) it allows us to display contextual elements (the Red River on the right, and some of the most well-known buildings of Hanoi on the left, some main roads and bridges of Hanoi...) in order to help users to locate in space; (ii) it provides empty space to provide more information, such as the state of parameters (number of vehicles, the kind of scenario...) and plots (in particular related to the evolution of the pollution). Figure 6 illustrates this combination of parameter values, (air pollution) plot and the display of the agents.[8]

To produce the 3D models that have been printed, we use some shapefiles used in the simulation (cf. Fig. 7 for an exhaustive description of the dataset used to produce the 3D map). The main challenge was to transform this set of shapefiles into a set of tiles in the STL format.[9] Here are the steps to create them[10]:

1. **Prepare the shapefile.** First, all the shapefiles are merged in a single one (roads, buildings, rivers and lakes). The roads, which are represented as a polylines are buffered in order to give them a surface. The empty holes are then filled

[8]A video is available to illustrate the dynamics of simulations at the address: https://youtu.be/U2w0GtLHACU.

[9]STL is a datafile format describing 3D objects and very common as an input for 3D printers.

[10]The full process and in particular the various softwares used are documented on the HoanKiemAir GitHub repository: https://github.com/WARMTeam/HoanKiemAir/tree/master/stl_convert.

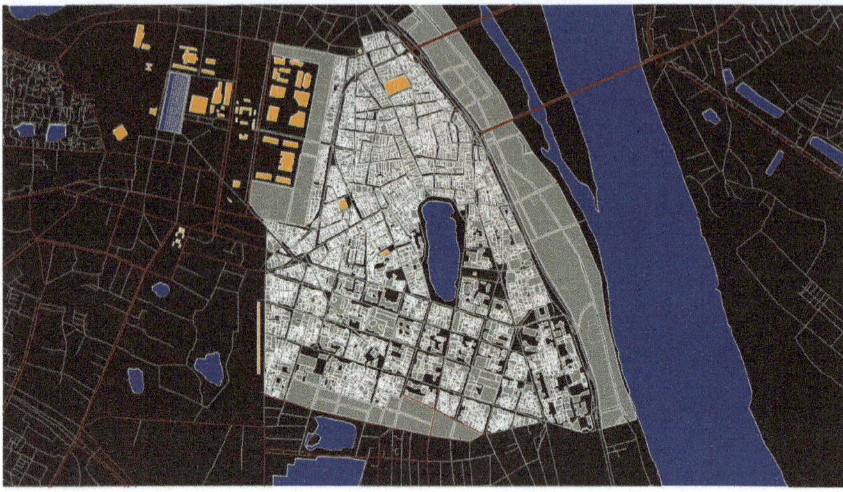

Fig. 7 Data used to build the 3D map: **blue** elements are rivers and lakes, **grey** elements are buildings, **orange** elements are special buildings (markets, Literature Temple, Ho Chi Minh Mausoleum, the National Assembly building, Presidential palace...), **white lines** represent roads (only the roads inside the Hoan Kiem District are used for traffic simulations), **red lines** are the main roads (roads outside Hoan Kiem district that are displayed in the simulation).

with polygons to represent the base of the 3D map, where there are no buildings, roads or lakes. All polygons are assigned a height value that follow this ordering: $(road,\ lake,\ river) < base < building$. We choose: 0.3 cm for roads, lakes and rives, 0.5 cm for the base, and 0.8 cm for special buildings. For the buildings, we do not have data about their elevation, but when looking at the Hoan Kiem district, it appears that buildings are highly heterogeneous in the area. We thus choose to give a random elevation value between 1 cm and 1.6 cm to reproduce this heterogeneity. Finally, we give the size of 2.5 cm to some specific buildings (the higher buildings in the area), in order to help users to recognize some key points on the map. It was important to keep a low elevation for the elements of the 3D map, to prevent from having huge shadows when the simulation will be displayed on the 3D map.

2. **Generate a single STL file.** From the shapefile produced in the previous step, we produce a 3D model file (STL format) with dimensions of approximately 122 cm by 64 cm.
3. **Split the STL file in tiles.** As the 3D map is too big to be printed in one go by standard 3D printers, the STL file generated in the previous step is split in square tiles of 16 cm per 16 cm following a 4×7 grid as presented in Fig. 8.

Fig. 8 Grid cut of the 3D map.

5.2 Remote Control Through an Android Application

As we expect random users to interact in an easy and intuitive way with the simulations, we choose not to let them manipulate directly the simulation interface and have developed an easy-to-use remote control application (running on an Android tablet). In order to keep the application intuitive, we limited the number of parameters accessible by users to the minimum ones: (i) the number of vehicles (motorbikes and cars), (ii) the pedestrian area scenarios and (iii) the visualisation mode (traffic or air pollution).

For communication between the application and GAMA, the Message Queuing Telemetry Transport (MQTT) protocol is used. It is a fast and lightweight publish/subscribe messaging protocol, commonly used in IoT projects. Basically, clients exchange messages through a server (often called a broker). A client can publish a message with a topic to the broker, then the broker will distribute that message to clients that are subscribed to the topic. In our setup, an Apache ActiveMQ[11] server is executed on the same computer as the GAMA simulator. Both the Android application and the GAMA simulator act as MQTT clients. An external plugin is provided by the GAMA community and implements communication through the MQTT protocol: it allows any agent to become an MQTT client, and thus to connect to an MQTT broker, publish messages and subscribe to certain topics.

Once the application is connected to the broker, it will publish messages with pre-defined topics when users modify any parameter value. A dedicated agent in the simulation receives these topics and modify the corresponding parameters (this can

[11]Apache ActiveMQ is an implementation of an MQTT broker, see http://activemq.apache.org/.

also induce more changes in the simulation, such as the creation or destruction of some agents).

6 Experiments

The HoanKiemAir simulator has been designed as an awareness-rising and comprehension-support system. As a consequence, simulations should be both scientifically grounded but also able to deliver its key messages to any users in an interactive way (and thus in a limited among of time). It should thus be explored to identify the parameter ranges fulfilling these two objectives. Finally, this exploration should be designed keeping in mind that the ultimate objective is to be able to provide insights about the impact of the pedestrian area scenarios on the air pollution quantity and distribution.

Given the high number of parameters (see next section for a detailed overview), we were not able to run a global sensitivity analysis. We thus performed some local analyses around the parameter values chosen empirically for the various exhibitions.

As the model is stochastic (mainly due to moving vehicles), we first investigate its sensibility to this randomness factor (Sect. 6.2). In particular, we need to check that simulations outputs are not qualitatively or quantitatively significantly different among replications. Finally, we will determine the number of replications needed to be launched in the following experiments. As stated in Sect. 4.3, several simulation step duration can be used in the different execution modes. In Sect. 6.3, we evaluate the influence of the step duration to check that the simulations keep close behaviours with different values.

Section 6.4 then investigates effects of pollution models and their parameters (decay and diffusion rates). Section 6.5 explores the impact of the number of vehicles on AQI indicators in order to identify whether there is some qualitative changes when the number of vehicles increase and if the AQI have an upper bound. The Sect. 6.6 then explores the impact of the evolution of the number of vehicles during the simulation (using the daytime traffic scenario). In particular, we aim at identifying whether there is a bootstrap period.

Finally, Sect. 6.7 explores the impact of the pedestrian area scenario by comparing the three scenarios, but also how the model behaves when the scenario changes during the simulation (to study what happens when the user switches between scenarios, or when we simulate a week the switch during the week-end).

All the experiments have been executed in headless mode on the following hardware configurations:

- Laptop MacBook Pro 2017
  - CPU—Intel(R) Core(TM) i7-7820HQ CPU @ 2.90 GHz (8 virtual cores)
  - RAM—16 Go
  - Operating System—MacOS 10.15.5 Catalina

- Laptop ASUSTeK GL553VE
  - CPU—Intel(R) Core(TM) i7-7700HQ CPU @ 2.80 GHz (8 virtual cores)
  - RAM—16 Go
  - Operating System—Linux Ubuntu
- Two nodes on universitary HPC
  - CPU—Intel(R) Xeon(R) CPU E5-2620 v4 @ 2.10 GHz (16 virtual cores)
  - RAM—24 Go
  - Operating System—Linux Debian

6.1 Parameters and Indicators

Parameters. The model has been designed to be exhibited during an undetermined time. As a consequence, a simulation **stop condition** cannot be defined as a specific situation; we thus define it as a final simulation step at which outputs are (expected to be) stable. In addition, given the model stochasticity, several replications (*i.e.* simulations launched with same parameter values) have to be launched to get significant results: this **number of replications** has thus to be determined before going further in the exploration. Finally, we explore the impact of the **simulation step duration**, in order to check that it can be adapted.

The evolution of the number of vehicles in the area is controlled by the parameter **traffic scenario**, that can take 2 values: *constant traffic*[12] or *daytime traffic*. The default value is set to *constant traffic*, which means that the number of vehicles is constant and determined by the three parameters: **number of cars** (by default equal to 700), **number of motorbikes** (by default equal to 2000), and the **traffic factor** (by default equal to 1) which multiplies the initial number of vehicle (car and motorbike) by a given factor. This factor will vary to explore the impact of the number of vehicles. When *daytime traffic* is selected, the number of vehicles evolves with the hour in the day (following data presented in Fig. 3).

As far as pollution diffusion is concerned, we will explore the two models implemented (depending on the **pollution diffusion model** parameter): *grid* and *roads*. By default, the model is *grid*. The two parameters influencing these pollution models are the **decay rate** (a real value between 0 and 1.0, with the default value set at 0.99) and the **diffusion rate** parameter (a real value between 0.0 and 0.12, to keep the uniform diffusion conservative, and set to 0.05 by default).

The set of closed roads is defined by the **pedestrian area scenario**. It can take 3 values: *no pedestrian area* (default value), *current scenario*, *extended area*.

Indicators. As it is the key quantity impacted by all the dynamics of the model, all the indicators are related to the air pollution, represented by the AQI value (cf.

[12]*constant traffic* is the mode also used when we want the user to change itself the number of vehicles through the application.

Sect. 4.5). First implemented indicators are thus the **mean** and **standard deviation** of AQI over all the buildings (as the AQI is computed on buildings in the simulation). Another interesting indicator would be the total amount of pollution (to assess whether closing roads increase total pollution or only move it). Nevertheless, as the number of buildings is constant, the AQI average and total values have the same evolution, and we only compute the former one. Finally, to observe the distribution of pollution values, we also compute the **maximum** and **minimum** value of AQI. As these values continue to change, even when the dynamics is stabilized, we compute the average value over last steps.

For each parameters values, we will run several replications. For each of them, we will save the value of each indicator at each simulation step. We will then plot, for each indicator and parameters configuration, the evolution over the simulation of the *average value* (over all the replications) as a line. It will be surrounded by an *extreme values interval* bounded by the maximum and minimum value of the indicator among all the replications.

6.2 Stochasticity Sensitivity Analysis

This experiment aims at exploring the impact of the stochasticity on the simulation outputs. The objective is twofold. First we want to identify some meta parameters that will be used in the following experiments: the number of steps that need to be executed before reaching a quite stable state (when the number of vehicles is constant) and the number of replications that will be necessary in the following experiments in order to find results that are statistically meaningful. Second, we want to determine the stability of the simulation results. In particular, if the results show a huge instability, it will be difficult for simulation users to understand what they observe.

First, to find the maximum step for the simulations, we run 4 replications (with a final step of 10000) and observe the average values of the mean and standard deviation of AQI over the 4 replications (cf. Fig. 9). We can observe that the values change a lot between steps 0 and 500. As the simulation is initially empty of pollutants and of vehicles, when the first vehicles emit pollution, the AQI increases quickly. After these first steps, the simulation reaches an equilibrium between pollutant emission (that tends to increase the AQI) and the pollution diffusion and decay (that tend to decrease the AQI) dynamics. After step 1000, we observe a continuous decrease of the mean AQI value, toward a stable value (around step 9000). Nevertheless, from step 1000 and 10000, we cannot observe any qualitative change in the simulation outputs. In a trade-off between the computation capabilities and the parameter space to explore, we will thus consider in most of the following experimentation a final step at 1500.

Even if we do not aim at quantitatively exploring or calibrating the model in this paper, we can notice that the effect of stochasticity increases with the number of steps (the extreme values interval amplitude increase with the number of steps).

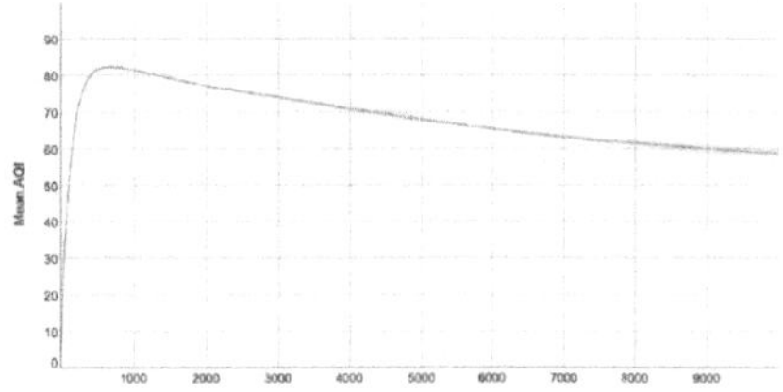

(a) Mean of AQI values.

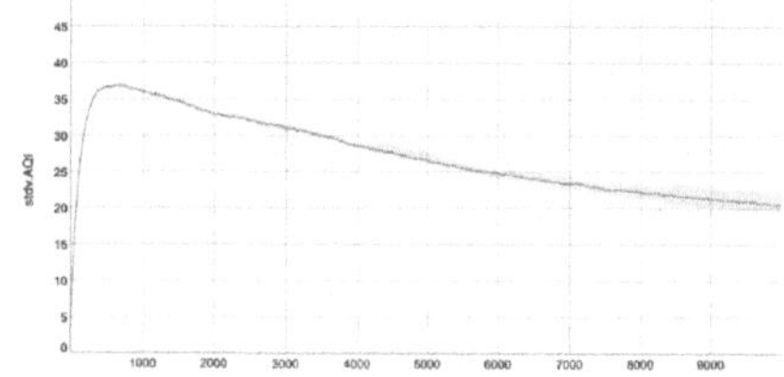

(b) Standard deviation of AQI values.

Fig. 9 Mean and standard deviation values of the AQI in buildings over 4 replications of 10000-step simulations (with a extreme values interval limited by the minimum and maximum values).

In order to investigate the number of needed replications, we adapt the method detailed in [9] using the standard error of the mean. The standard error of the mean $\sigma_{\overline{x}}$ is defined on a given indicator (*e.g.* mean, standard deviation of AQI) as:

$$\sigma_{\overline{x}} = \frac{\sigma}{\sqrt{N}}$$

where σ is the standard deviation of a given indicator and N the number of replications considered.

The main idea is to compute this standard error for an increasing number of replications and to find a threshold number where adding more replications does not bring more information nor a better stability.

So we first compute a big number of replications (here we ran 100 replications as a first attempt). Then incrementally we compute the standard error on an increasing number N of replications from 2 to 100. For each N, in order to prevent biases in the chosen replications, we compute the average of the standard errors over 10 picks of N replications. The results for the 4 indicators (mean, standard deviation, maximum and minimum of AQI) are displayed in Fig. 10. We can observe that on all the indicators, the main standard error's evolution are before 10 replications. After 20 replications, the standard errors seem to begin to stabilize and 60 replications seem to be the threshold after which adding more replications do not bring any new information. As a consequence, depending on the exploration purpose and on the required precision, one of these 3 thresholds could be chosen.

Again, following our objective of qualitative exploration and in order to be able to explore a bigger parameter space, we will limit to 10 replications for each parameter value. Further exploration could choose 20 or 60 replications for precision's sake.

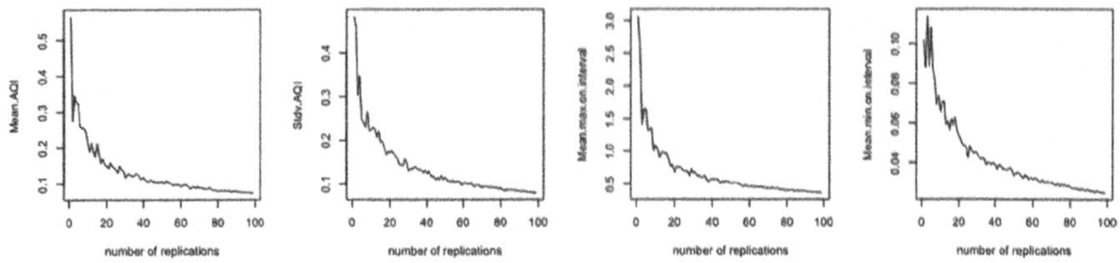

Fig. 10 Standard errors on mean, standard deviation, max, and min AQI, over 100 replications. Figures show a very fast decrease before 10 replications, a beginning of stabilization around 20 replications. After 60 replications, standard errors are stable and additional replications do not bring more information.

6.3 Impact of the Step Duration

This second experiment aims at exploring the impact of the simulation step duration over the results, but we went to 5 min to simulate the daytime traffic.

The objective of this experiment is to evaluate the impact of the simulation step duration on the air pollution index evolution: we aim at checking that the model behavior is still acceptable when changing the simulation step. To this purpose, we consider a single parameter (the step duration) and run 10 replications for each value of the step duration among {16 s, 30 s, 60 s, 120 s, 180 s, 300 s}. We plot the various indicators (mean, standard deviation and min and max of AQI) for each step duration (Figs. 11).

We can observe that the simulation step has an impact on the AQI values: the values for all the indicators increase with the simulation step duration. As in the previous experiments, the stochasticity has more impact on results when the simulation step is longer: this can be observed by the extreme values interval becoming bigger with the step duration. We can also observe a convergence of the indicators values: above 120 s, the step duration does not impact results anymore (and even above 60s for standard deviation). Finally, and this is the most important result of this experiment, even if the indicators values are altered by the simulation step duration, the qualitative evolution of the AQI is the same whatever the step duration value is. To sum up, this experiment shows that, in the following experiments, we can modify this duration step without changing the behavior of the simulation. Nevertheless, we need to keep in mind that the AQI values will be altered.

6.4 Effect of the Pollution Diffusion Model

As presented previously, two diffusion models have been implemented, each of them depending on 2 parameters: decay and diffusion rates. This experiment aims at evaluating the qualitative impact of these parameters on the simulation outputs. We thus explore the evolution of AQI indicators for all the combinations of **diffusion model**

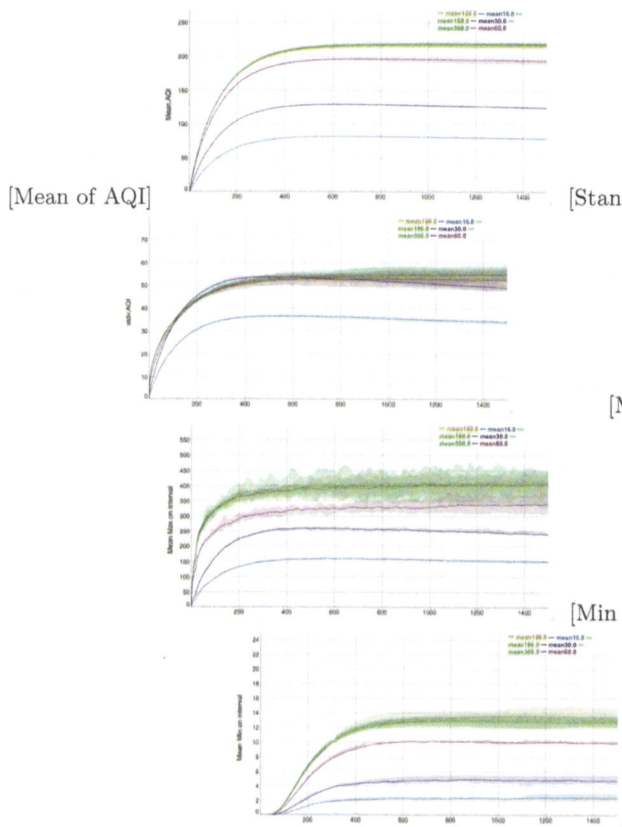

Fig. 11 Mean values of the mean, standard deviation, and min and max of AQI indicators for the step duration in {16 s, 30 s, 60 s, 120 s, 180 s, 300 s}. Figures illustrate the impact of the simulation step on results: when the step duration becomes longer, all indicators increase. The values converge when the step duration is greater then 120 s.

(among {*grid*, *roads*}), **decay rate** (among {0.01, 0.1, 0.3, 0.5, 0.7, 0.9, 0.99}) and **diffusion rate** (among {0.02, 0.04, 0.06, 0.08, 0.1, 0.12}). For each combination of parameter values, 10 replications are launched. Results are summarized in Fig. 12 for the *grid* model and Fig. 13 for the *roads* model.

Grid Diffusion Model. The Figs. 12 display the 42 time series produced by this experiment. First no quantitative change appears when exploring all the parameter values. Second, "groups" of series can be observed (in particular of the standard deviation and maximum), which could come from the fact that one of the 2 parameters has more influence than the other one.

To confirm this observation, and thus assess the impact of each parameter on the results, we conduct an ANalysis Of the VAriance (ANOVA) [32] at step = 500. To check the impact of this step value choice, we have computed the ANOVA for

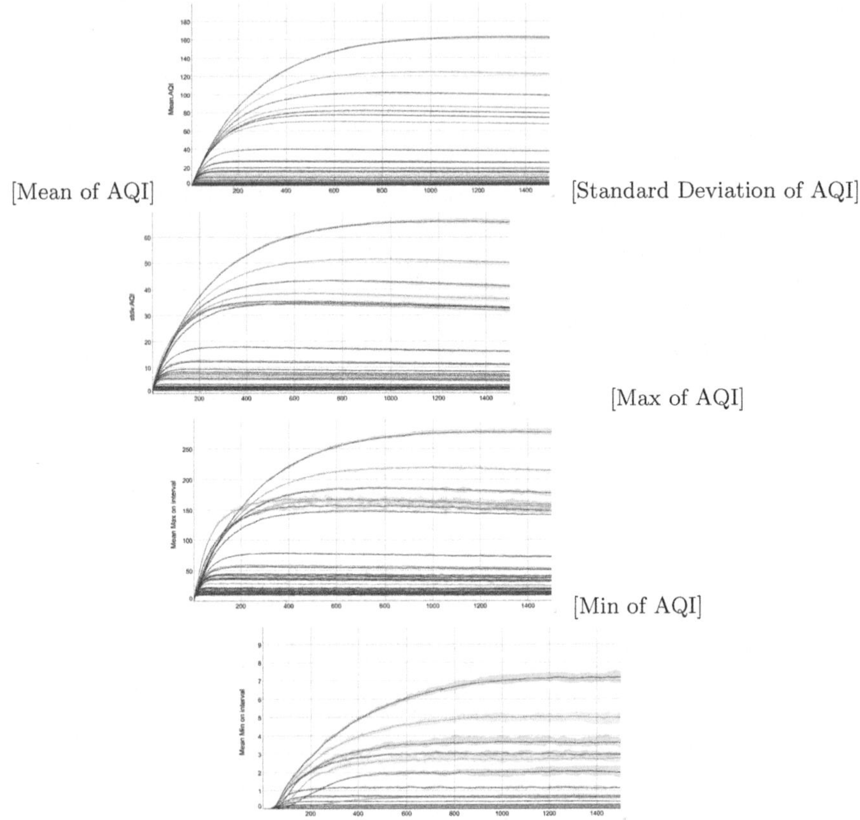

Fig. 12 Mean values of AQI indicators for the **grid diffusion model** with decay rate among $\{0.01, 0.1, 0.3, 0.5, 0.7, 0.9, 0.99\}$ and diffusion rate among $\{0.02, 0.04, 0.06, 0.08, 0.1, 0.12\}$. Some groups of time series appear (with the same decay value) appear for standard deviation and max of AQI in particular, giving the insight that the model is more sensitive to this parameter.

other step values and get very close results. Table 2 summarizes the relative impact of each factor (parameters *Decay* and *Diffusion*, the randomness (*Residuals*)) and combination of factors on the dispersion of values. The analysis shows clearly that the impact of the *Diffusion* parameter and the combination of the parameters *Decay-Diffusion* have a limited influence on all the indicators. They are mainly impacted by the *Decay* parameter and the randomness (which has even a more important effect). This means that it is really important to be careful in the choice of the *Decay* parameter as it will highly impact the values of the AQI.

Road Diffusion Model. The Figs. 13 display the 42 time series produced by the exploration of the roads diffusion model in the parameters decay rate and diffusion rate space. The observations are quite similar with the difference that the time series groups are much clearer: to confirm this observation, we conduct another ANOVA.

Table 2 Results of the ANOVA with the grid diffusion mode. Each cell gives the percentage impact of the parameter (row) on the indicator (column).

	mean	stdDev	max	min
Decay	41.16	43.37	46.00	30.41
Diffusion	9.04	7.40	5.89	11.91
Decay:Diffusion	3.91	2.74	1.64	12.17
Residuals	45.89	46.50	46.47	45.51

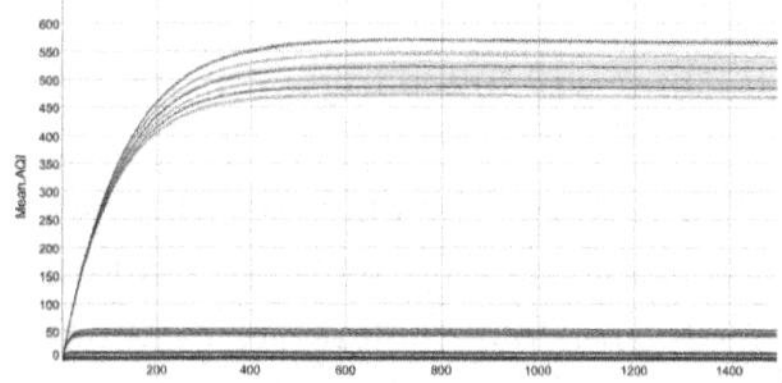

[Mean of AQI] [Standard Deviation of AQI]

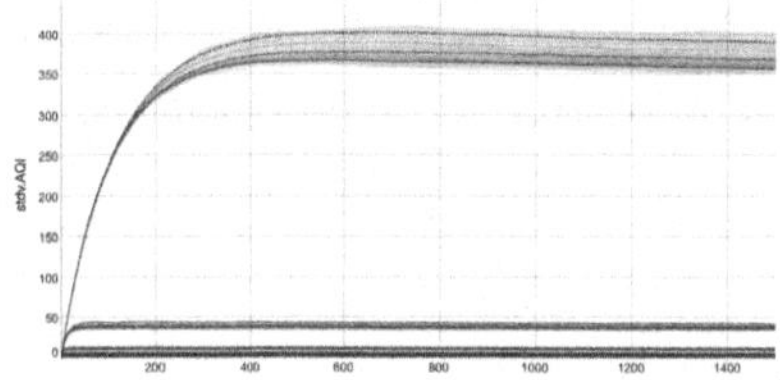

[Max of AQI]

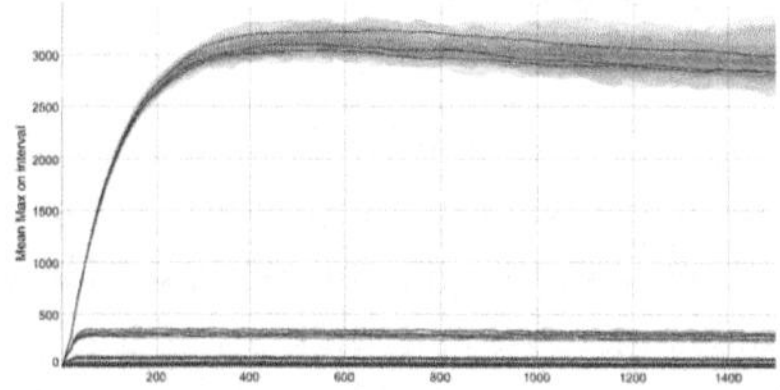

[Min of AQI]

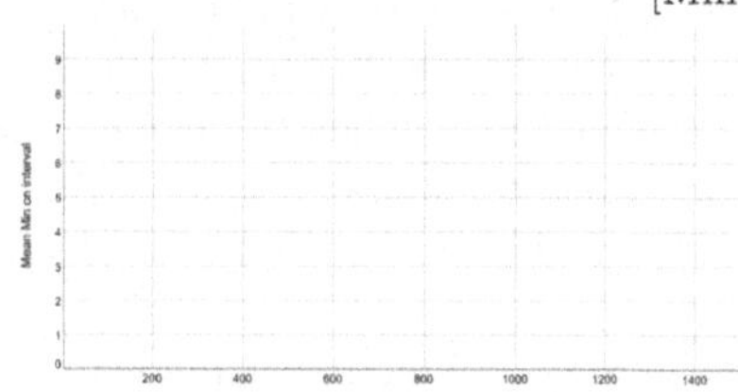

Fig. 13 Mean values of AQI indicators for the **roads diffusion model** with decay rate among {0.01, 0.1, 0.3, 0.5, 0.7, 0.9, 0.99} and diffusion rate among {0.02, 0.04, 0.06, 0.08, 0.1, 0.12}. Clusters of time series with the same value of decay illustrate the high sensibility of the model to this single parameter.

Table 3 Results of the ANOVA with the grid diffusion mode. Each cell gives the percentage impact of the parameter (row) on the indicator (column).

	mean	stdDev	max	min
Decay	38.61	39.46	39.23	NaN
Diffusion	0.07	0.02	0.01	NaN
Decay:Diffusion	0.16	0.05	0.02	NaN
Residuals	61.16	60.48	60.74	NaN

A second observation is that AQI values are much higher than with the grid diffusion model.

Table 3 summarizes the ANOVA results. The analysis shows once again that the parameter with the most influence is *Decay*. But with the road diffusion model the impact of *Diffusion* and *Decay-Diffusion* parameters are definitely negligible (this appears clearly on the Fig. 12 with clusters of series with the *Decay* value). No result can be computed for the minimum AQI indicator as results are constant (equal to 0.0). Finally, randomness appears to have even more impact in the road diffusion model than with the grid one.

6.5 *Effect of the Number of Vehicles*

The main objective of this experiment is to explore the impact of the number of vehicles. In particular, the key questions we want to address are: is there any qualitative change in the dynamics when the number of vehicles increase? How AQI indicators evolve when the number of vehicles increases and in particular is there a cap value?

To this purpose, we design an experiment with the parameter **traffic factor** taken values among {0.5, 1.0, 1.5, 2.0, 5.0}. This means that the number of vehicles varies between 350 cars and 1000 motorbikes (when *traffic factor* is equal to 0.5) to 3500 cars and 10000 motorbikes (when *traffic factor* = 5.0). The ratio between cars and motorbikes is thus preserved in all the simulations. Results (on 10 replications) are plotted in Figs. 14.

We observe a linear evolution of all the indicators with the multiplication factor *traffic factor*: as an example, when the mean AQI value at step = 400 is around 40 (for *traffic factor* = 0.5), it values 400 for *traffic factor* = 5.0. In particular, results do not show a cap value of the AQI when the number of vehicle varies (as we observed when the step duration increases). Nevertheless, the shape of the evolution of all the indicator is similar for all the values of *traffic factor*.

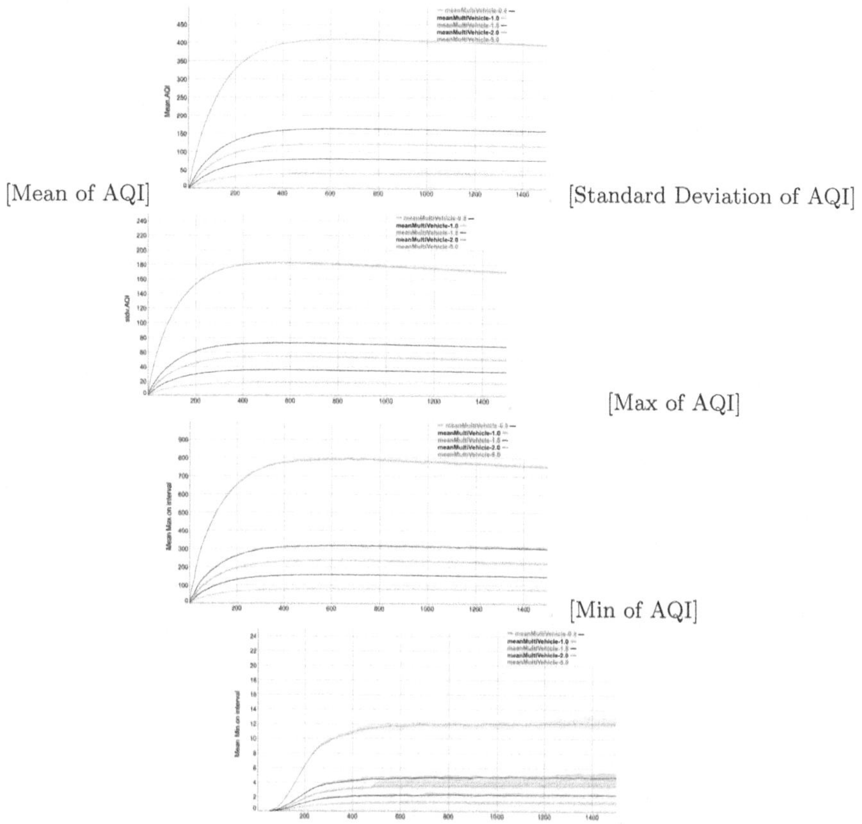

Fig. 14 Mean values of AQI indicators for *traffic factor* in {0.5, 1.0, 1.5, 2.0, 5.0}. Plots show a linear increase of the AQI indicators with the number of vehicles.

6.6 Effect of the Traffic Scenario

The evolution of the number of vehicles in the area can follow two simulation modes: *constant traffic* or *daytime traffic*. In this experiment, we explore the impact of the *daytime traffic* scenario on the behavior of the simulations. The numbers of cars and motorbikes will thus change each hour, following data illustrated in Fig. 3; but every day, the hourly numbers will be unchanged. We thus launched 10 replications for each value of the parameter *traffic scenario* (the simulations where the parameter value is set to *constant traffic* are used as a baseline). We launch simulations with a step of 5 min (1 day is thus simulated in 288 steps) and we stop the simulations after 3000 steps. Results are displayed in Figs. 15.

As expected, the AQI values follows the number of vehicles: data (presented Fig. 3) show 3 rush hours (around 7AM, 11AM and 5PM) and a low-traffic period during the night. The 3 local maximum and the minimum can clearly be observed on

[Mean of AQI] [Standard Deviation of AQI] [Max of AQI] [Min of AQI]

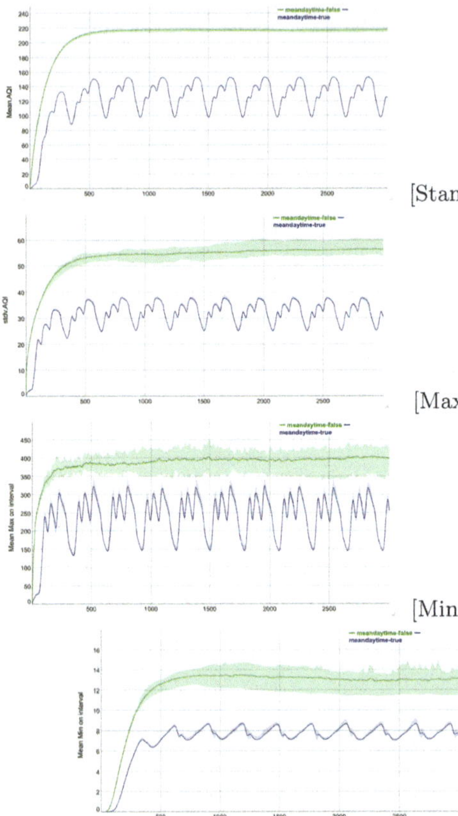

Fig. 15 AQI indicator results for each traffic scenario: green line for a number of vehicles *constant traffic*, purple for a number of vehicles following *daytime traffic*. Plots show that pollutant emissions become stationary after one day of simulation and correlated to the evolution of the number of vehicles.

all the indicators (except the minimum of AQI). The extreme values intervals are very small showing the low impact of the stochasticity on the results when the number of vehicles evolves over time (the main exception is for the optimums of the max and min of AQI). The most interesting observation is the fact that indicator values are periodic (after the first day). The simulation bootstrap period is thus limited to the first day and after the second day, results are stabilized.

6.7 Effect of Pedestrian Area Scenario

The ultimate objective of this section is to explore the impact of the various pedestrian area scenarios on the AQI. We thus launched simulations with the **pedestrian**

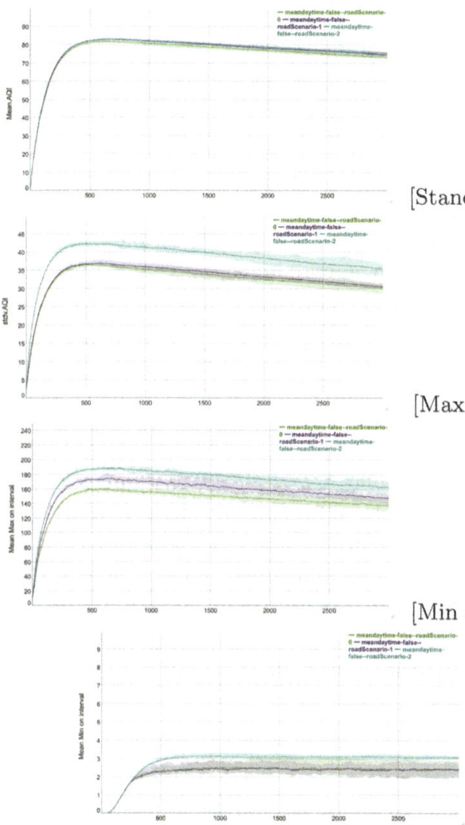

[Mean of AQI] [Standard Deviation of AQI]

[Max of AQI]

[Min of AQI]

Fig. 16 Simulation results for the three scenarios of pedestrian area: green line for *no pedestrian area*, purple for *current scenario*, and indigo blue for *extended scenario*, with a constant number of vehicles. Figures show that the application of any of the pedestrian scenario does not change qualitatively the pollutant emissions, but only increase the maximum value and the spatial distribution heterogeneity.

area scenario parameter taken values among {*no pedestrian area*, *current scenario*, *extended area*}, for both traffic scenarios (*daytime* or *constant traffic*). In addition, as the user is able to change the pedestrian area in the interactive mode, it appears necessary to check the impact of this change on indicators to prevent unstable states; this will also be conducted on both traffic scenarios.

Simulations of the Pedestrian Area Scenario in Constant Traffic Mode. In this experiment, we set the traffic mode to *constant traffic*, and thus set the simulation step duration to its default value (16s). We launched 10 replications for each pedestrian area scenario. Results are summarized in Figs. 16.

A first notable observation is that the mean values of AQI are not significantly different among the three pedestrian scenarios. More generally, the 2 scenarios *no pedestrian area* and *current scenario* produce very similar results on all the indicators; the main difference is on the max AQI, which is higher for the *current scenario*. But as the extreme values intervals overlap, this difference is not really significant. The *extended area* scenario produces some differences in its results: the main one being a standard deviation value that is higher than for the 2 other scenarios and a slightly higher maximum of AQI. As a consequence, the *extended area* scenario is the one producing the most change, with higher maximum of AQI and a distribution of AQI among the buildings that is more heterogeneous than for the other scenarios.

Simulations of the Pedestrian Area Scenario in Daytime Traffic Mode. In this experiment, we set the traffic mode to *daytime* traffic, and thus the simulation step duration to 5 min. We launched 10 replications for each pedestrian area scenario. Results are summarized in Figs. 17.

Once again, we can observe that *no pedestrian area* and *current scenario* produce very similar results on all the indicators. The only slight difference is that during rush hours, the mean AQI is a little bit higher when pedestrian area is activated. Contrarily to the previous experiment, when the daytime traffic is activated, all the indicators increase when the *extended scenario* is activated: the AQI is higher in average, and so are the max and standard deviation. In this specific configuration, we can conclude that the extended pedestrian area increases not only the peak of AQI but also the overall pollutant quantities.

Effects of Pedestrian Area Scenarios Changes. During simulations, through the remote control, users can decide to change the pedestrian area scenario to observe the impacts of alternatives. To assess the impact of this switch between scenarios, we implement a new experiment in which the scenario is initialized to *no pedestrian area*. During the simulation, every 864 steps (in daytime mode, *i.e.* 3 days) or 800 steps (in constant traffic mode), the scenario is changed: if the current scenario is *no pedestrian area* then it becomes *current scenario* (resp. *extended area*), otherwise it becomes *no pedestrian area*. As we run 3000 simulation steps, we will observe the alternation: *no pedestrian area-current scenario-no pedestrian area-current scenario*. Results are summarized in Figs. 18 for the current traffic mode and Figs. 19 for the daytime traffic mode.

When the traffic mode is set to **constant traffic**, we observe, as previously, that the activation of any pedestrian scenario does not modify significantly the mean of AQI (even if a very small increase can be observed on the time series in Fig. 18a). The change is more significant for the standard deviation (when activating the *extended area* scenario) and the max value of AQI (with a higher max value for the *extended area* scenario). Finally, we can observe that, when the pedestrian area scenario is switched from *current* or *extended area* scenario to *no pedestrian area*, the indicator values does go back to the baseline value immediately: on average it will take around 100 step to reach the step without scenario switch.

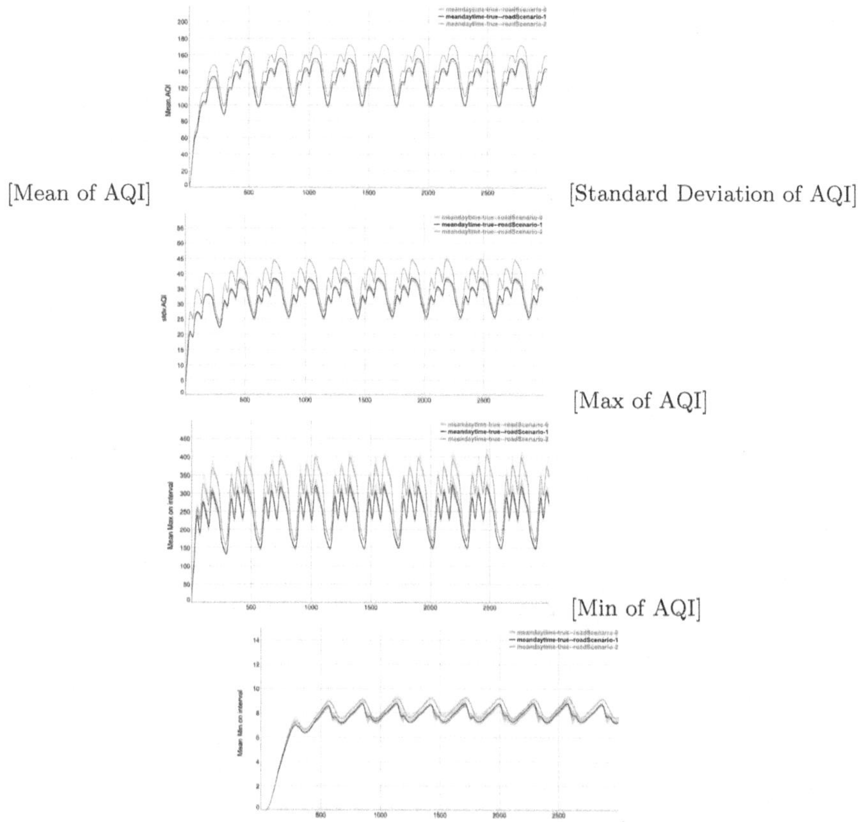

Fig. 17 Simulation results for the three scenarios of pedestrian area: green line for *no pedestrian area*, purple for *current scenario*, and indigo blue for *extended scenario*. In a situation of varying traffic, the application of each of the scenario from the beginning of the simulation highlight the fact that the situations without any pedestrian area and with the current scenario show both quantitatively and qualitatively similar results. In contrary, the extended scenario show an increase of the amount of the air pollutants and of their spatial distribution heterogeneity.

When the *daytime traffic* mode is activated, we observe similar results to the previous experiment concerning the impact of the pedestrian area scenario on the indicators. The mean AQI is not significantly impacted by the scenario as Fig. 19a does not show clear difference between the period without pedestrian area (3 first days, *i.e.* before step 850) and period with a pedestrian area (next 3 days, roughly between steps 850 and 1700). Conversely, when the *extended area* scenario is activated, an increase in all the indicators is visible.

More interestingly, we can observe that, if we omit the initial bootstrap phase of the first day (that has been identified in the previous experiment), no transition phase appears later when switching between *no pedestrian area* and *current scenario*. In particular, it would have been possible to have an unstable period after each transition

[Mean of AQI]

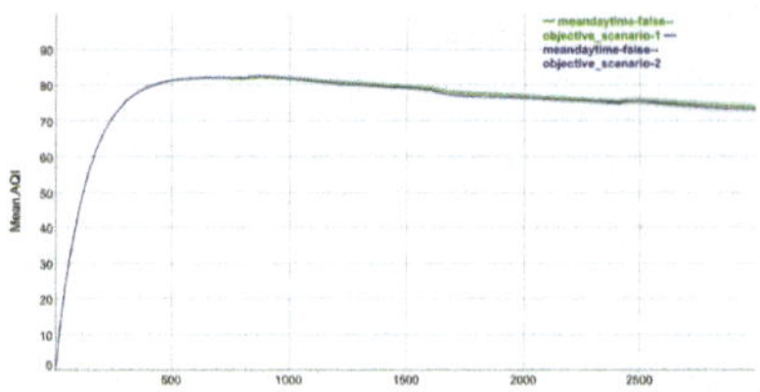

[Standard Deviation of AQI]

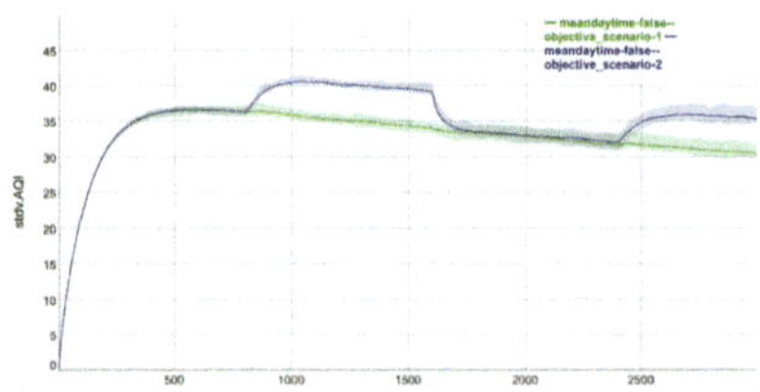

[Max of AQI]

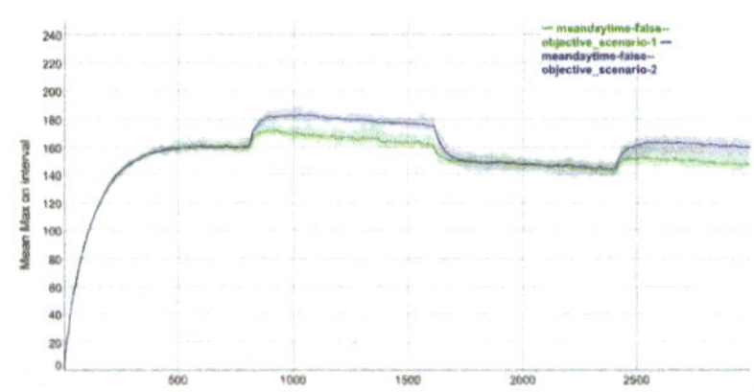

[Min of AQI]

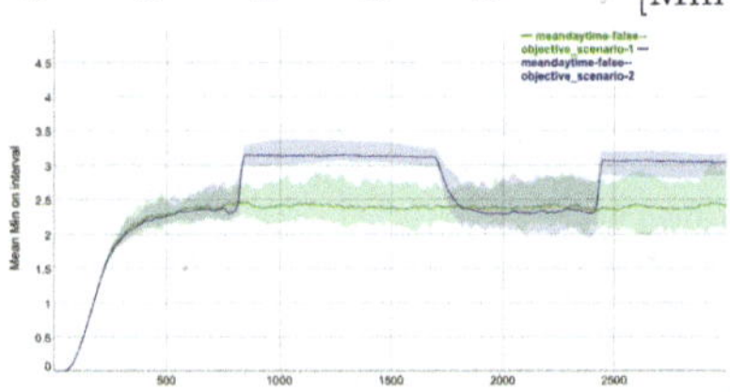

Fig. 18 Simulation (in constant traffic mode) results when the scenario of pedestrian area changes during the simulation (green line represents results with for *current scenario* and purple lines with *extended scenario*). In a hypothetical situation of constant traffic, these figures show that the application of the two alternative scenarios does not change the overall quantity of air pollutant but induces an increase of the maximum peak value. The application of the extended scenario is also characterized by a huge increase of the spatial distribution heterogeneity.

between pedestrian area scenarios, but the results of the first day of a period appears very similar to the other days of the period. Nevertheless, when the *extended area* scenario is deactivated, we can observe that the mean AQI values of the two series are slightly different during the first day after the switch, meaning that simulation requires one day to fully recovered from the pedestrian area scenario change (even if the difference between both series is not huge).

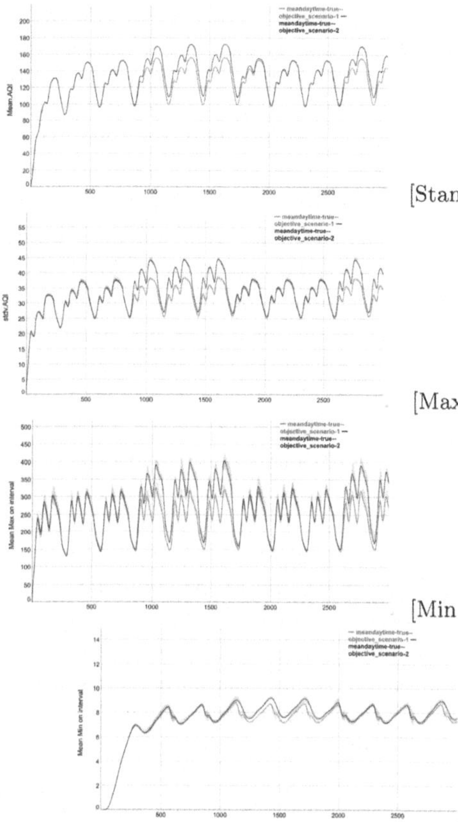

[Mean of AQI]
[Standard Deviation of AQI]
[Max of AQI]
[Min of AQI]

Fig. 19 Simulation (in daytime traffic mode) results when the scenario of pedestrian area changes during the simulation: green line represents results with for *current scenario* and green lines with *extended scenario*. In a situation of traffic varying over the day, these figures show that the extended scenario application induces an increase of the amount of air pollutants and of their spatial distribution heterogeneity. Conversely, the application of the current scenario does not show significant modification on these indicators.

7 Conclusion

This article presents the HoanKiemAir project that aims at providing a tangible and interactive simulation to represent the impact of various pedestrian area scenarios around the Hoan Kiem lake (Hanoi, Vietnam), on traffic and emitted air pollutants. It is expected to help authorities and people audiences to understand the interrelations and feedback loops between these dynamics using a 3D map as a screen for the simulation and a tablet application to control it. Whereas the model is described in details in [28], this paper focuses on the analysis of the simulator and in particular on the sensibility of the results when exploring the parameter space.

The results of experiments presented here tend to show that the step duration can be changed without changing qualitatively the results, and that there is a threshold above which, it does not have any quantitative impact either. Conversely, not such a threshold can be identified on the number of vehicles: the increase of the number of vehicles induces a linear increase on the AQI indicators. As far as the traffic scenario is concerned, the evolution of AQI is coherent with the traffic evolution in the *daytime traffic* and a bootstrap period of 1 day as been identified. Finally, the activation of the current pedestrian area does not significantly impact the AQI (even if we can notice a small increase of the maximum of AQI) in both constant and daytime traffic configurations. Concerning the extended area scenario, its activation definitely increases the maximum of AQI and the standard deviation (*i.e.* increases the heterogeneity of the spatial AQI distribution). In the case of the daytime traffic, it also increases the mean AQI (and thus the overall quantity of pollutants in the system). When a switch is made between the *extended area* and the *no pedestrian area*, we can observe that a period of time is necessary to come back to state before the activation of the pedestrian area (around 100 steps in *constant traffic* and 1 day for *daytime traffic*).

The analysis of the simulation behaviour when exploring the parameter values reinforce the trust we can have on the use of the simulator in an interactive way with various audience. Obviously the model still need to be improved both with additional data about the real traffic in the area and the other pollution sources (factory, agriculture, etc.), and additional dynamics, and in particular a weather model and how it interacts with the pollution diffusion model. Finally, we still need to calibrate the model more accurately with real data, mainly coming from sensors localized in the area.

In the end, apart from this theoretical analysis of the model, the full setup has been demonstrated several times to diverse audiences. Preliminary observations show that the use of a 3D printed map definitely attracts people and arouses their interest and curiosity. Feedback were very positive. Nevertheless, we have observed that the full setup in its current state is not fully self-content: when users are alone (*i.e.* without any facilitator around) in front of the map and simulation, they are not able, in general, to fully understand what they observe. There is still a need to improve contextual information (either on the simulation or around the table). A second point is that the tablet application is not naturally used by the users. An interesting alternative could be to build a physical remote control with actual sliders and buttons to evaluate if users are more likely to use it rather than the Android application.

The simulation (in an extended version) could be used by both decision-makers and urban planners to test several scenarios of pedestrian areas and compare their results in an interactive way. Extensions of the interactions with the model could allow the user to define interactively the pedestrian area to test, at the scale of the single road. This would give insights about the amount of roads that can be closed before the appearance of critical consequences in the other areas of the district and even in other districts. This could also allow decision-makers to identify key roads, roads which will highly and even qualitatively impact the overall traffic and air pollution emission. As far as the public is concerned, the simulation results would help it to understand the complex feedback loop between the urban-planning decisions and traffic and air pollutants emission. This could help it to accept a debatable or provide it elements to factually criticize it. We could also imagine using the setup to let people propose new pedestrian plans in a crowd sourcing approach.

Acknowledgement The HoanKiemAir project was funded by the French embassy in Vietnam, in collaboration with the Hoan Kiem district People's Committee and PRX Vietnam.

References

1. ALMEC Corp, Nippon Koei Co Ltd., Yachiyo Engineering Co Ltd.: The comprehensive urban development programme in Hanoi Capital City of the Socialist Republic of Vietnam (HAIDEP) (2007)
2. Alonso, L., Zhang, Y.R., Grignard, A., Noyman, A., Sakai, Y., ElKatsha, M., Doorley, R., Larson, K.: CityScope: a data-driven interactive simulation tool for urban design. use case Volpe. In: International Conference on Complex Systems, pp. 253–261. Springer (2018)
3. Arnaud, B.: Simulating pedestrian behavior in complex and dynamic environments: An agent-based perspective. In: F., B., Mager, C. (eds.) European Handbook of Theoretical and Quantitative Geography, pp. 1–27. Faculty of Geosciences and Environment of the University of Lausanne (2009)
4. Banos, A., Corson, N., Lang, C., Marilleau, N., Taillandier, P.: Multiscale modeling: application to traffic flow. In: Agent-Based Spatial Simulation with NetLogo, vol. 2, pp. 37–62. Elsevier (2017)
5. Bigazzi, A.Y., Rouleau, M.: Can traffic management strategies improve urban air quality? A review of the evidence. J. Transp. Health **7**, 111–124 (2017)
6. Chao, Q., Bi, H., Li, W., Mao, T., Wang, Z., Lin, M.C., Deng, Z.: A survey on visual traffic simulation: models, evaluations, and applications in autonomous driving. Comput. Graph. Forum **39**(1), 287–308 (2020)
7. Chapuis, K., Taillandier, P., Gaudou, B., Drogoul, A., Daudé, E.: A multi-modal urban traffic agent-based framework to study individual response to catastrophic events. In: International Conference on Principles and Practice of Multi-Agent Systems, pp. 440–448. Springer (2018)
8. Chifflet, S., Amouroux, D., Bérail, S., Barre, J., Van, T.C., Baltrons, O., Brune, J., Dufour, A., Guinot, B., Mari, X.: Origins and discrimination between local and regional atmospheric pollution in Haiphong (Vietnam), based on metal(loid) concentrations and lead isotopic ratios in PM 10. Environ. Sci. Pollut. Res. **25**(26), 26653–26668 (2018)
9. Cohen, P.R.: Empirical Methods for Artificial Intelligence, vol. 139. MIT Press, Cambridge (1995)
10. Emery, J., Marilleau, N., Martiny, N., Thévenin, T., Nguyen-Huu, T., Badram, M., Grignard, A., Hbdid, H., Laatabi, A.M., Toubhi, S.: Marrakair: une simulation participative pour observer les émissions atmosphériques du trafic routier en milieu urbain. In: Treizièmes Rencontres de Théo Quant (2017)
11. Fleck, S., Rivière, G., Ticona-Herrera, R., Couture, N.: Classifying tangible user interfaces with cladistics: criteria and forms for a collaborative inventory. In: Proceedings of the 30th Conference on l'Interaction Homme-Machine, pp. 209–218 (2018)
12. Grignard, A., Alonso, L., Taillandier, P., Gaudou, B., Nguyen-Huu, T., Gruel, W., Larson, K.: The impact of new mobility modes on a city: a generic approach using ABM. In: International Conference on Complex Systems, pp. 272–280. Springer (2018)
13. Grignard, A., Macià, N., Alonso Pastor, L., Noyman, A., Zhang, Y., Larson, K.: Cityscope Andorra: a multi-level interactive and tangible agent-based visualization. In: Proceedings of AAMAS, pp. 1939–1940 (2018)

14. Grimm, V., Berger, U., Deangelis, D., Polhill, J., Giske, J., F. Railsback, S.: The ODD protocol: a review and first update. Ecol. Model. **221**, 2760–2768 (2010)
15. Hertel, O., Berkowicz, R.: Operational Street Pollution Model (OSPM). Evaluation of the model on data from St. Olavs street in Oslo. Technical report, National Environmental Research Institute, Danish Ministry of the Environment (01 1989)
16. Hoogendoorn, S.P., Bovy, P.H.L.: State-of-the-art of vehicular traffic flow modelling. J. Syst. Control Eng. **215**(4), 283–303 (2001)
17. Kobayashi, K., Narita, A., Hirano, M., Kase, I., Tsuchida, S., Omi, T., Kakizaki, T., Hosokawa, T.: Collaborative simulation interface for planning disaster measures. In: CHI 2006 Extended Abstracts on Human Factors in Computing Systems, pp. 977–982 (2006)
18. Kubicki, S., Lepreux, S., Kolski, C.: RFID-driven situation awareness on Tangisense, a table interacting with tangible objects. Pers. Ubiquitous Comput. **16**(8), 1079–1094 (2012)
19. Laatabi, A., Becu, N., Marilleau, N., Pignon-Mussaud, C., Amalric, M., Bertin, X., Anselme, B., Beck, E.: Mapping and describing geospatial data to generalize complex models: the case of LittoSIM-GEN. Int. J. Geospatial Environ. Res. **7**(1), 6 (2020)
20. Lebrun, Y., Adam, E., Mandiau, R., Kolski, C.: A model for managing interactions between tangible and virtual agents on an RFID interactive tabletop: case study in traffic simulation. J. Comput. Syst. Sci. **81**(3), 585–598 (2015)
21. Lighthill, M.J., Whitham, G.B.: On kinematic waves II. A theory of traffic flow on long crowded roads. Proc. Roy. Soc. Lond. Ser. A Math. Phys. Sci. **229**(1178), 317–345 (1955)
22. Macal, C., North, M.: Tutorial on agent-based modelling and simulation. J. Simul. **4**, 151–162 (2010)
23. Mitchell, M.: Complexity: A Guided Tour. Oxford University Press, Oxford (2009)
24. Nagel, K., Schreckenberg, M.: A cellular automaton model for freeway traffic. Journal de physique I **2**(12), 2221–2229 (1992)
25. Ntziachristos, L., Gkatzoflias, D., Kouridis, C., Samaras, Z.: COPERT: a European road transport emission inventory model. In: Information Technologies in Environmental Engineering, pp. 491–504, January 2009
26. Organization, W.H.: WHO releases country estimates on air pollution exposure and health impact (2016). https://www.who.int/news-room/detail/27-09-2016-who-releases-country-estimates-on-air-pollution-exposure-and-health-impact
27. Petrasova, A., Harmon, B., Petras, V., Mitasova, H.: Tangible Modeling with Open Source GIS. Springer (2015)
28. Pham, M.D., Chapuis, K., Drogoul, A., Gaudou, B., Grignard, A., Marilleau, N., Tri, N.H.: HoanKiemAir: simulating impacts of urban management practices on traffic and air pollution using a tangible agent-based model. In: The 2020 RIVF International Conference on Computing and Communication Technologies, pp. 1–7. IEEE (2020)
29. Richards, P.I.: Shock waves on the highway. Oper. Res. **4**(1), 42–51 (1956)
30. Rodrique, K., Ho, T., Nguyen, M.H.: An agent-based simulation for studying air pollution from traffic in urban areas: the case of Hanoi city. IJACSA **10**, 596–604 (2019)
31. Salim, F.: Tangible 3D urban simulation table. In: Proceedings of the Symposium on Simulation for Architecture & Urban Design, vol. 23. Society for Computer Simulation International (2014)
32. Saltelli, A., Chan, K., Scott, E.M.: Sensitivity Analysis, vol. 1. Wiley, New York (2000)
33. Taillandier, P., Gaudou, B., Grignard, A., Huynh, Q.N., Marilleau, N., Caillou, P., Philippon, D., Drogoul, A.: Building, composing and experimenting complex spatial models with the GAMA platform. GeoInformatica **23**(2), 299–322 (2019)
34. Tho Hung, N., Ketzel, M., Jensen, S., Oanh, N.T.: Air pollution modeling at road sides using the operational street pollution model-a case study in Hanoi. Vietnam. J. Air Waste Manage. Assoc. **1995**(60), 1315–1326 (2010)
35. Underwood, R.T.: Speed, volume and density relationships. Quality and theory of traffic flow (1961)
36. Vietnam Env. Admin.: Quyet dinh 878/QD-TCMT ve viec ban hanh so tay huong dan tinh toan chi so chat luong khong khi (AQI) (2011)

37. van Wageningen-Kessels, F., Van Lint, H., Vuik, K., Hoogendoorn, S.: Genealogy of traffic flow models. EURO J. Transp. Logist. **4**(4), 445–473 (2015)
38. Woods, T.L., Reed, S., Hsi, S., Woods, J.A., Woods, M.R.: Pilot study using the augmented reality sandbox to teach topographic maps and surficial processes in introductory geology labs. J. Geosci. Educ. **64**(3), 199–214 (2016)
39. Zhang, K., Batterman, S.: Air pollution and health risks due to vehicle traffic. Sci. Total Environ. **450**, 307–316 (2013)

An Agent-Based Co-modeling Approach to Simulate the Evacuation of a Population in the Context of a Realistic Flooding Event: A Case Study in Hanoi (Vietnam)

Kevin Chapuis, Taha Amine Elwaqoudi, Arthur Brugière, Eric Daudé, Alexis Drogoul, Benoit Gaudou, Doanh Nguyen-Ngoc, Huynh Quang Nghi, and Jean-Daniel Zucker

Abstract According to recent studies, Vietnam is one of the twenty countries most affected by natural disasters in the world, and particularly by floods either on the low elevation coastal zones (risk of submersion) or along the Red River and the Mekong River (risk of flooding). In this context, dams are both means of mitigation but also threats given the possible failures and ruptures. The authorities must, therefore, prepare warning systems and evacuation plans for the downstream population to avoid loss of life. Agent-based models are now the approach of choice to support such preparedness by considering the system as a whole and integrating dynamics of different natures: hydrology, population behavior, evacuation, crisis management, etc. To design such a decision-support tool, modelers generally need to couple different formalisms, such as diffusion equations when considering the hydrodynamic part, and agent-based modeling when considering inhabitants' behaviors. This is the goal of the ESCAPE project, which uses agent-based simulations to explore evacuation

K. Chapuis (✉) · T. A. Elwaqoudi · A. Brugière · A. Drogoul · B. Gaudou · D. Nguyen-Ngoc · H. Q. Nghi · J.-D. Zucker
UMI 209 UMMISCO, Sorbonne Université, IRD, Bondy, France

T. A. Elwaqoudi · A. Brugière · B. Gaudou
ICTLab, Vietnam Academy of Science and Technology, USTH, Hanoi, Vietnam

K. Chapuis
UMR 228, ESPACE-DEV, IRD, Montpellier, France

B. Gaudou
UMR 5505, IRIT, Université Toulouse 1 Capitole, Toulouse, France

H. Q. Nghi
Can Tho University, Can Tho, Vietnam

D. Nguyen-Ngoc
Thuyloi University, Hanoi, Vietnam

E. Daudé
UMR 6266 IDEES, CNRS, Normandie University, Rouen, France

M. H. Mohd et al. (eds.), *Modelling, Simulation and Applications of Complex Systems*, Springer Proceedings in Mathematics & Statistics 359,
https://doi.org/10.1007/978-981-16-2629-6_5

strategies and contribute to the development and evaluation of evacuation plans. In this study, to improve the ESCAPE framework, we propose to combine a hydraulic dam failure model with an agent-based evacuation model using the GAMA platform. We focus on the evacuation of a Hanoi city (Vietnam) district, threatened by flooding due to the failure of the Hoa Binh dam located more than 80 km upstream of the city. We demonstrate how to methodologically and operationally couple a hydrodynamic water diffusion model (implemented using the HEC-RAS software) and a multi-paradigm evacuation model (using the ESCAPE framework). Our goal is to extend and enrich this population evacuation model by coupling it with flood simulation.

Keywords Agent-based model · Dam break · Co-modeling · GAMA platform · Population evacuation

1 Introduction

The high concentrations of people and critical infrastructures in increasingly urbanized spaces make risk and disaster management highly complex [32, 51] and amplify population vulnerability to hazards. Considering the poorly controlled urban development, the increasing intensity of monsoon rains, the sea level rise and the ground level subsidence in a large part of the mega cities of South East Asia located along the coasts and rivers, the population evacuation is often the only effective response to face imminent hazards. According to the world risk index [26], Vietnam is one of the twenty countries around the world with the highest vulnerability to natural hazard disasters. The country is recurrently hurt by storms, mass floods, and typhoons. As an example, in November 2017, the typhoon Damrey has killed more than 100 persons [15] in the South part of the Central Region of Vietnam. It has blown the roofs or destroyed more than 25 000 houses and forced authorities to evacuate more than 30 000 persons.[1] In response to the potential lost and cost due to disasters, Vietnamese authorities have launched in 2018 a large project to mitigate natural and industrial risks.[2] The failure of the Hoa Binh dam is among the envisaged risks. Located 80 km upstream of Hanoi city on a tributary of the Red River, the dam is regularly forced to discharge overloaded water during the rainy season (*e.g.* see [62]) and its age (it has been built in the early 90') raises concerns about potential future failures.

Among the various measures envisaged to reduce the exposure of populations to this hazard, the planning of evacuation strategies becomes a crucial concern. But evaluation of these plans coherence and of their benefits outside a crisis context is difficult. Although evacuation drills and the improvement of risk culture among the

[1] "Typhoon Damrey hits Vietnam with deadly force", The Guardian. Link: https://www.theguardian.com/world/2017/nov/04/typhoon-damrey-hits-vietnam-with-deadly-force. [Last access: August 2nd 2020].

[2] Vietnam Disaster Management Authority: http://phongchongthientai.mard.gov.vn/en/Pages/priority-programs-for-disaster-risk-reduction-in-vietnam.aspx [Latest access: October 2nd 2020].

population can mitigate losses, they are generally difficult to implement [23]. The use of computer simulations to help authorities to understand the key outcomes and to anticipate the main issues related to mass population evacuation is therefore of primary interest [16]. Agent-based models (ABMs) [41] are a class of computational models able to support such a preparedness by considering the system as a whole, integrating dynamics of various natures: hazards (*e.g.* floods), people's behaviors, crisis management... Nevertheless, to describe each of these dynamics, modelers usually have to rely on the most appropriate and efficient formalism, such as diffusion equations when considering the hydrodynamic part, or agent-based modeling when considering inhabitants' behaviors. Coupling these various models is often a challenge.

The objective of this paper is thus to describe a promising co-modeling approach to couple an evacuation model of a population from a micro perspective using an ABM, and the dynamics of the disaster (an inundation) using a hydrological model. The proposed implementation relies on the GAMA platform [56], a generic agent-based modeling platform.

The paper is organized as follows. Section 2 presents the case study of a catastrophic flood caused by a dam failure in northern Vietnam. Section 3 briefly describes the state of the art concerning agent-based evacuation models and the use of hydrological models in such approaches. Section 4 details the agent-based evacuation model and the hydrological flood model separately, while Sect. 5 focuses on the co-model implementation and presents preliminary results on both toy and real case study coupled models. Finally Sect. 6 summarizes the proposed approach and opens discussions about futures perspectives.

2 Case Study

The Red River basin contains two large reservoirs (created by two dams): Thac Ba and Hoa Binh. Recently, two other dams have been built upstream of the Red River basin: Tuyen Quang and Son La Reservoirs [39]. The Hoa Binh dam, built in 1984 on the Black River, has been the largest hydropower dam in Vietnam (until 2012) and the second largest in South-East Asia with a height of 128 m and a length of 970 m. Its reservoir has an area of 208 km^2 with a storage capacity of 9.5 km^3 [64]. Therefore, it has a very important strategic interest for North Vietnam and also constitutes a threat that the authorities have clearly identified.[3] In addition to all the people living directly downstream of the Hoa Binh dam, it is assumed that Hanoi, the capital of Vietnam, may also be at risk of flooding.

[3]Tran Quang Hoai, permanent member of the Central Steering Committee for the Prevention and Control of Natural Disasters, stated in 2018: "According to the plan, we are not allowed to let the Hoa Binh Hydropower Plant's dam break. It is necessary to actively divide the water and keep the dams safe and ensure the security of the people." [62].

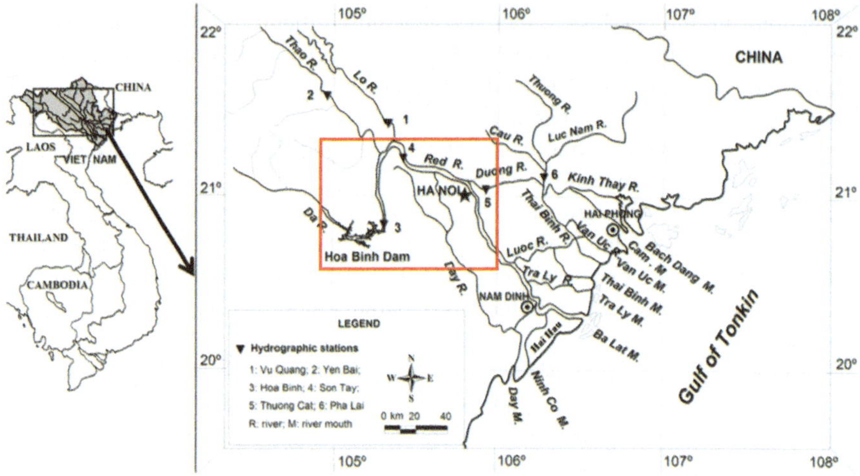

Fig. 1 Map of the Red River basin with its three main upstream tributaries and its delta area [63]. The red rectangle is the simulated area.

The area of study (indicated by the red rectangle in Fig. 1) is thus the portion of the Red River basin located between the Hoa Binh reservoir, the dam (point 3 on Fig. 1), and Hanoi city along the Black River (Da Song) and the Red River. The dam is about 60 km from Hanoi as the crow flies and more than 120 km along the river.

The main watercourses that we consider in our study are: the Red River (Hong Song) and its upstream part called Thao River (Thao Song), its confluence Clear River (Lo Song), the Black River which flows out of the Hoa Binh Reservoir, and the Duong River (Duong Song) which is a tributary of the Red River just near Hanoi.

From a topographic point of view, our case study can be divided into two parts. The Black River flows in an area surrounded by hills, but when it joins the Red River, the delta area begins and the terrain becomes flat: in particular, the entire area to Hanoi (10 m above sea level) is mainly a flat anthropised landscape of small villages, rice fields, and dike infrastructures. The Red River delta is indeed an area that has been heavily diked for several centuries. Facing a huge amount of floods all over its history [57, 58], Vietnam authorities have started to build dikes in the Red River delta from 1099. P. Gourou estimates in [30] that the embankment of the whole delta has been achieved in 1865 and maintained and repaired from that time.

As far as the agent-based evacuation model is concerned, we focus on the Phuc Xa district (Fig. 2). This district of Hanoi city is located between the dike and the Red River, in the north of the old quarter of Hanoi. It is quite a small urban district (around 0.92 km^2) but highly densely populated (15767 inhabitants in the census of 2009).

Located in a risk-prone area, it has been affected by many flooding events. As an example, it was overflowed by the huge inundations that occurred in the provinces

Fig. 2 Map of the Red River basin highlighting the Phuc Xa district location (Latitude: 21.04574—Longitude: 105.848663) in the flooding area. (Source: Google Maps; access: August 1st 2020)

of the Red River Delta in mid-August 1971,[4] killing a huge number of people[5] and causing property damages evaluated to 455 million USD. More recently, the 2008 off-season heavy rain in the North and North Central Coast surpassed all expectations and caused a historic flood in Hanoi [40]. These heavy rains are considered a record rainfall in the last 100 years (in 2018) and killed 17 people in the capital of Vietnam.[6] The damages have impacted nearly 13,000 households along the dike, flooding most houses around the Red River and tributaries. The overall amount of damage in Hanoi is estimated at least to 1260 million USD. Up to now, there is no mass evacuation plan of Hanoi identified by local authorities.

[4]Le Hien, "A report on the big flood in 1971." People's Police Newspaper, 9, 2015. Link: http://antg.cand.com.vn/Kinh-te-Van-hoa-The-Thao/Nhin-tu-con-dai-hong-thuy-nam-1971-289808/ [Last access: August 1st, 2020. In Vietnamese].

[5]Statistics indicate that 594 people have been killed, whereas other sources state the number of 100,000 people.

[6]"Vietnam floods kill 14, Hanoi streets under water", Reuters. Link: https://in.reuters.com/article/idINIndia-36242120081031 [Last access: August 1st 2020].

3 State of the Art

In this section, we review the necessity and possibilities of coupling ABM and hydrological models, in particular for crisis management. In Sect. 3.1, we briefly introduce the main approaches and tools to model mass evacuation management using ABM. Section 3.2 discusses the various methodological and theoretical approaches to couple ABMs with dedicated domain-specific hazard models, while in Sect. 3.3 we consider more closely the coupling of ABMs and hydrological models (with a focus on natural catastrophic events). Finally, Sect. 3.4 introduces hydrodynamic modeling of flooding due to a dam failure.

3.1 Agent-Based Modeling of City Evacuation and Mass Forced Displacement Management

The agent-based modeling paradigm [41] aims at representing the behavior of a system as an emerging property of computational entities (agents) interacting with each other in a simulated environment. ABM has become an approach of choice to represent complex socio-environmental systems, especially when they require the inputs of different disciplines.

The scope of ABMs applied to the problem of evacuation, and the massive forced displacement of the urban population in particular, can be defined along one major point of departure: the general perspective considering the regime of agent mobility (*e.g.* horizontal or vertical) and the collective management of the evacuation (*e.g.* dry or during hazard, synchronous or asynchronous). When it is possible to forecast the path or extension of hazards from several hours to several days in advance, like inundations or typhoons, the focus is usually put on horizontal evacuation with the main issue being the displacement of the population to exit the risky zones [20]. When the considered catastrophic event is very localized or short standing, like flash floods or tsunami, the model usually focuses on vertical evacuation [1]. In this case, the safety zone may not require massive displacement but rather the proper protective behavior [61]. While the first set of models heavily rely on mobility modeling (including pedestrians, motorised traffic or multi-modal mobility) [18], social and individual responses to the alert system [54], or a global infrastructure/resource management [22], the other part of the ABM effort pays more attention to *in-situ* individual and collective behaviors including socio-cognitive and emotional aspects [2, 6, 8, 60]. In recent years, several models explore the outcome of combined vertical and horizontal evacuations [44, 66] even if research efforts to support this trend remain limited.

Many agent-based traffic modeling frameworks are now available to design evacuation models, the main ones include MATSim [36, 65], SUMO [25, 38], or SimMobility [3] (see [20] for a synthetic summary). Nevertheless, when it is required to integrate individual behaviors and attitudes related to the evacuation, these platforms have limited modeling capabilities. Hence, it is often more appropriate to rely on more generic, flexible, and expressive tools such as generic agent-based modeling and simulation platforms, *e.g.* Netlogo [67], Repast [49] or GAMA [56]. They can indeed integrate spatial data and define heterogeneous individual agents with no *a priori* on the architecture of agents' decisions and behaviors.

When we look at disaster management and security, the required accuracy of the hazard model will mostly depend on its suddenness and the impact of actors on its evolution. When studying the evacuation of a population in anticipation of floods that will occur several hours later, the hazard does not necessarily need to be simulated: a simpler dry evacuation plan simulation model can be considered sufficient [20]. The hazard would have been simulated independently and the time available to evacuate is the main information of interest for modelers and evacuation managers. Furthermore, if the hazard is expected to interfere with the evacuation (*e.g.* by blocking roads), pre-calculated data can be integrated into the simulation. However, when human beings' actions may alter the hazard or when the evacuation occurs during the hazard, then its dynamic needs to be coupled with the individual behavior model. For example, this is the case in a model of bushfires in the state of Victoria (Australia), where inhabitants could either evacuate in advance or stay and protect their homes [1], or during the flood of 1926 in Hanoi (Vietnam), where the inhabitants tried to protect their city from flooding by repairing or increasing the dikes [27].

In these specific cases, human behaviors models and hazard models need to be coupled in order to account for the potential interactions: human beings' actions can have an impact on the course of the hazard, while it will push people to fight or to flee the threat.

3.2 Model Coupling

Necessity of Model Coupling in Socio-environmental Studies. Model coupling has nowadays become increasingly popular for answering research questions arising from complex systems science. Coupling can be seen as the integration or combination of different models, referred to as strong coupling. Alternatively, weak coupling can be found in a kind of data transmission between different tools and/or platforms [24].

When it comes to model socio-environmental systems, taking into account their various dimensions, interdisciplinary collaboration becomes necessary. But each research field comes with its language, vision of the systems, modeling paradigm, and even existing models. Being able to couple these models without altering them is thus a necessity to ensure the expressivity and the quality of the whole model. Besides, relying on a flexible coupling approach also improves the model scalability, providing the possibility to switch between a precise and time-consuming model and a lighter, faster, but less accurate one when upscaled or downscaled.

Several frameworks have proposed technical and operational solutions to the model coupling. Most famous ones include HLA [19], DEVS [70] or FMI [9]. As an example, DEVS proposes an event-based formalism to describe the simulation execution, considering each sub-model as a black-box only characterized by its inputs and outputs; the whole model is thus defined as a set of interconnected sub-models.

Agent-Based Co-modeling Approaches. The question of coupling models has also been tackled in the agent-based modeling and simulation community, mainly using ad hoc approaches (such as when it comes to couple ABM and equation-based models [7, 43]), while generic approaches such as HLA or DEVS are rarely applied when it comes to couple socio-environmental models [50].

In the following, we will rely on the **co-modeling approach**, introduced in [21, 33]. *Co-models* have been defined as "a particular sort of agent-based models, where agents wrap one or several instances of the models to be coupled, with their life-cycle, operations, collaborations and conflict resolution mechanisms [...]". In this sense, they allow a recursive description of the system (such as in the holonic approaches[7] [53], or the MADKIT platform [31]). A key point of the co-modeling approach is its ability to describe in a homogeneous way *atomic agents* (whose behavior is described through any given formalism) and *micro-model agents*, *i.e.* agents wrapping another model and executing it or agents calling an external simulator and reading its result data files. The main objective of the approach is to increase the expressivity of the model (by associating several modeling approaches) and reusability of its components. A key challenge of this modeling approach is the need to be able to provide a way to synchronize every agents and models in space and time.

From a general point of view, the modeling of urban evacuations relies on many dynamics such as physical models (representing the hazard), mobility models, warning systems, human behaviors, and decision-making models. Using the co-modeling approach to implement such a model could definitely take advantage of the existing literature in each of these domains and in particular from hydrodynamic modeling as far as floods are concerned.

3.3 *Coupling Agent-Based and Hydrological Models*

Coupling an Agent-Based Evacuation Model and a Hydrological Inundation Model. Despite active researches in various domains like economics, social science, biology, military, public policy, ecology, and traffic, the literature review shows however that the agent-based modeling approach applied to stream-flow or flood forecasting problems is relatively limited [55] as it is not appropriate. Therefore this prevents a straight-forward coupling of hydrological models with agent-based models.

[7]Holonic approaches consider agents as "holons". "Holons are self-similar entities that represent whole-part constructs and can be viewed either as higher-level system components or as wholes composed of other holons as substructures." [53].

Conversely, the HEC-RAS software suite has recently added a module called HEC-LifeSim to simulate evacuation within the context of a realistic inundation. The main purpose is to analyse the impact of the catastrophic event on the human loss and the cost of degradation. The tool makes use of pre-simulated inundation scenarios and extends them in two main ways: first, users can define an emergency plan made of warning diffusion to alert population and protective actions initiation to define population response. The second aspect of the evacuation model is the use of an agent-based model that simulates cars or pedestrians' flight using roads, destination points, and considering damaged environment (*e.g.* closed road due to the inundation).

Hydrology in ABMs Using the GAMA Platform. The GAMA agent-based modeling and simulation platform [56] has been designed, in particular, to implement large-scale models of socio-environmental systems. As a consequence, many case studies have integrated a water-related model, either by coupling the GAMA model with another existing model or by implementing an *ad hoc* model. The choice mainly depends on the studied area extension, but also on the simulation step duration and available data. As an example, the ARCHIVES model [27] simulates the flooding of Hanoi city in 1926 and the management of this crisis by authorities. No precise data are available related to the flood state or to the topology of the river, so we built a very simple Cellular Automaton model based on the few available data (mainly a hand-made hydrograph on a single point of the river). The crisis management model and the hydraulic model are linked through the dike agents: the water presses on the dikes until they break, while authorities try to repair and reinforce them fast enough to prevent breaches. This hydrological model choice fits well with the simulation of actors' actions (which step is set to 2 h) localized on dike as it can adapt to this time and space precision.

The MAELIA project [28] aims at assessing the social, economic, and environmental impacts of various water withdrawal policies on the Adour-Garonne (France) draining basin. Given the large area of interest and a simulation step sets to 1 day, we choose the semi-distributed watershed hydrological transport model SWAT [4]. As only a subset of all the dynamics included in the SWAT model was relevant for the project, the needed equations have been implemented in the GAMA model, other dynamics being replaced by more relevant ones (*e.g.* the plant growth and the farmer decision-making models).

Another project [21, 59] explored the land use and land cover evolution in the Mekong delta due to the impact of saltwater intrusion, that reduces yields of the classical rice crops of the delta. The land-use change is controlled by yearly decisions of farmers (in terms of crops) made depending on the previous years' yields. Given the studied process, the simulation step has been set to 1 month. As a consequence, in this case, the water evolution is not computed using a hydrology model, but data are simply imported as time-series data about water height in the rivers and in the sea, to compute the intrusion of salty water.

Finally, the LittoSim [35] project built a simulation-based participative game aiming at letting decision-makers think about alternative urban planning approaches to face sea submersion and to improve the risk culture of stakeholders. To this purpose, players can manage their city with the possibility to build dikes, improve houses, or to preserve dunes... Every year a submersion is computed, given the current land use (that impacts the elevation model). To this purpose, the Littosim model is coupled with the 2D hydrodynamic model LISFLOOD-FL [48], which computes the submersion given the updated city infrastructures. In terms of coupling, at each step, a new Digital Elevation Model is sent from GAMA model to LISFLOOD-FL. This latter computes the submersion and returns to GAMA the casualties induced by the water.

In this paper, we focus on floods due to dam failures; the next section presents thus a short overview of the modeling approaches regarding this kind of hazard.

3.4 *Modeling of Dam Breaks*

Dam Break Causes. Dams are under the threat of failures due to various types of events. However, the vast majority of failures have occurred on earthen embankment dams caused by extreme rainfall events [12]. Other events that can induce a dam break include natural hazards (flooding, landslide, earthquake...), failures (equipment, structure, foundation), or upstream dam failures [12].

Effects of Dam Breaks. Compared to other types of floods, floods caused by a dam failure are characterized by their suddenness, high flow and strong impulsive force. During a dam failure, the spread of flooding is influenced by a number of factors, such as the mode of failure, the presence and the shape of dikes along the downstream river, the elevation of the environment, the sediment transport and associated downstream hydrological and hydraulic parameters [37]. Although there are many examples of dam failures in history, only a few of them are well documented and preserved. Existing records are extremely insufficient and most of them have been documented by visual observation, which implies that data related to these events are highly uncertain [42].

Modeling Bam Failure and Induced Floods. It exists a large literature about the modeling of floods due to dam failures, using a wide variety of modeling approaches. As an example, the FLDWAV hydraulic model (a generalized flood routing model) has been used in various studies to determine the hydraulic characteristics of a flood, including its discharge, velocity, elevation, and depth at various times and distances downstream of the dam [47]. Other approaches using a 2D Cellular Automaton spatial distribution model have simulated the spatio-temporal process of a dam-break flood routing [42, 69].

The two main approaches used to model flooding are the hydrological approach (*e.g.* Muskingum method [17]) and the hydrodynamic approach (*e.g.* solving the St. Venant equations [29]) [68]. Several works have compared the 2 approaches on several case studies (*e.g.* dam failure-induced waves in alpine regions [52]) or in a controlled academic environment [5]. During dam failure, the saturation of the catchment area has little effect on the overall damage, as this type of flood is characterized by its suddenness and high destructiveness. Consequently, the hydrological approach is not suitable to compute the parameters in question.

In this study, we use the HEC-RAS software [11] which allows us to do 1D modeling, 2D modeling and 1D-2D coupled modeling at the same time. The latest versions of this software also have the "RAS Mapper", a tool that greatly facilitates the modeling process, the visualization of simulation results, and even more importantly, offers more data export options, which will help us later for the model coupling.

4 Models

In this section, we introduce the two models (the ESCAPE evacuation ABM in Sect. 4.1 and the Hoa Binh dam break HEC-RAS model in Sect. 4.2) that will be coupled in Sect. 5.

4.1 ESCAPE Model in a Nutshell

The ESCAPE agent-based framework [20] is dedicated to the study of urban evacuation strategies in the context of natural or technical hazard threats. It is based on four modules that describe entities and mechanisms of (i) the spatial and social environment, (ii) individual and collective decision-making processes, (iii) evacuation plans and their management, and (iv) the catastrophic event (an in-depth description has been provided in [20]). The key entity of the model is nevertheless the agents representing human beings evacuating the area, with their heterogeneity in terms of individual behavior and social characteristics. The model has been built upon a previous model of mobility in Rouen (France) [18] and extended to several case studies of urban evacuation including the Vietnamese case study of Phuc Xa (Hanoi) [13].

Social and Spatial Features of the Model. Figure 3 depicts the GIS data used to set up the spatial environment of the evacuation model in the quarter of Phuc Xa, Hanoi, Vietnam. The road network (yellow lines) and building footprint (grey polygons) have been gathered from OpenStreetMap (OSM).[8] In a second step we updated the data by digitalizing missing buildings from latest available Google satellite images. Because of data scarcity, we consider all roads to be two ways roads and practicable either for two wheels and car vehicles. The Phuc Xa quarter is surrounded by the Red River on the East and a dike (protecting the city) on the West; the evacuation points (green circles) are thus the gates in the dike accessible by roads. The height of the dike walls constitutes a major obstacle for pedestrians to cross it, they thus have to make

[8]OpenStreetMap is a collaborative online GIS opendata repository: https://www.openstreetmap.org/.

Fig. 3 Spatial data used to initialise the ESCAPE agent-based model on the Phuc Xa case study, including the ward boundary (red line), buildings (grey polygons), roads (yellow polylines), evacuation points (green circles), and the closest extends of the Red River (blue).

their way towards the nearest exit gate. Considering the social environment, values of inhabitant agents' attributes have been generated using the available census data[9]: age, gender, and occupation have been assigned to reflect known aggregated demographic marginals. The location of activities (*i.e.* workplaces and schools) and homes have been elicited uniformly using the synthetic population generator Gen* [14].

Mobility Behavior and Decision During Evacuation. In the ESCAPE framework, every inhabitant agent has its own set of activities during the day, *i.e.* its agenda. This means that when the hazard occurs, agents might already be in motion or in the course of an activity (*e.g.* staying at home or working). The agents' evacuation behavior can either be triggered by the event itself or an alert broadcast (*e.g.* siren or mobile phone messages) defined in the evacuation plan. For the Phuc Xa case study, a simplistic alert system has been designed to cope with the lack of information about the disaster as well as the absence of evacuation plan: all the agents are alerted at the same time and have to go to the nearest exit point. Most of the agents' decisions pertain to the choice of a mobility model, either using a motorbike or walking: if they are in motion, they will keep their mobility mode, whereas if they are in the middle of an activity they will choose to run directly to the exit point if the distance to their vehicle is, on average, higher than half of the distance to the exit point. Otherwise, agents go to their vehicle and follow the road to the evacuation point. The mobility

[9]Demographic data where taken from 2009 census and available at the General Statistics Office Of Vietnam https://www.gso.gov.vn/Default_en.aspx [Last access: October 11th 2020].

Fig. 4 ESCAPE simulation snapshot: the green triangles represent people agents and the blue rectangles the motorbikes [13].

speed will be altered by the congestion (correlated to the number of agents on the mobility road or pedestrian network) which thus impacts the evacuation time.

Evacuation Plan. As mentioned previously, no evacuation plans are defined in advance to prepare the management of crises in Vietnam. Research questions regarding evacuation management thus move from the assessment of evacuation plans to the improvement and ease of self-evacuation, *i.e.* when people evacuate by themselves. In [13], we thus investigate dry pedestrian and two wheels evacuation, with a simplistic alert system (everyone is alerted at once) to study the expected time to empty the area.

Model Catastrophic Event. In the ESCAPE framework, the hazard may not be implemented at all when the model focuses mainly on dry evacuation. When it is implemented in the model, it can either come as a static hazard, a set of scenarios (defining the extend areas of the hazard) or can be dynamically simulated. In all the cases, it will mainly impact the mobility of people by blocking roads and other mobility networks. In the model presented in this paper, we integrate the hazard in the evacuation simulation as results of a hydrodynamic model (see Sect. 4.2). It will impact the evacuation in a very simple way: when the water depth is lower than 0.5 m, it has no impact, otherwise, it makes the evacuation impossible for the overflowed agents. Once again, we choose a simple but realistic threshold value.[10]

Simulations. Figure 4 shows a snapshot of the evacuation in progress. We can observe some crowded areas for pedestrians (green triangles) and congestion on the main intersections for vehicles (blue rectangles). Because of the topology of the quarter

[10] A deeper investigation shows that this threshold also depends on the flow speed and human being profile, and not only on the water height. At low speed (less than 0.25 m/s), 0.5 m is the limit for a child and 1 m for a trained adult. At high speed (0.75 to 1 m/s), this threshold of 0.5 m is for the trained adult, the children cannot walk above 0.2 [10]. The chosen value of 0.5 has thus been chosen as an average value between the height, the speed and the human profile.

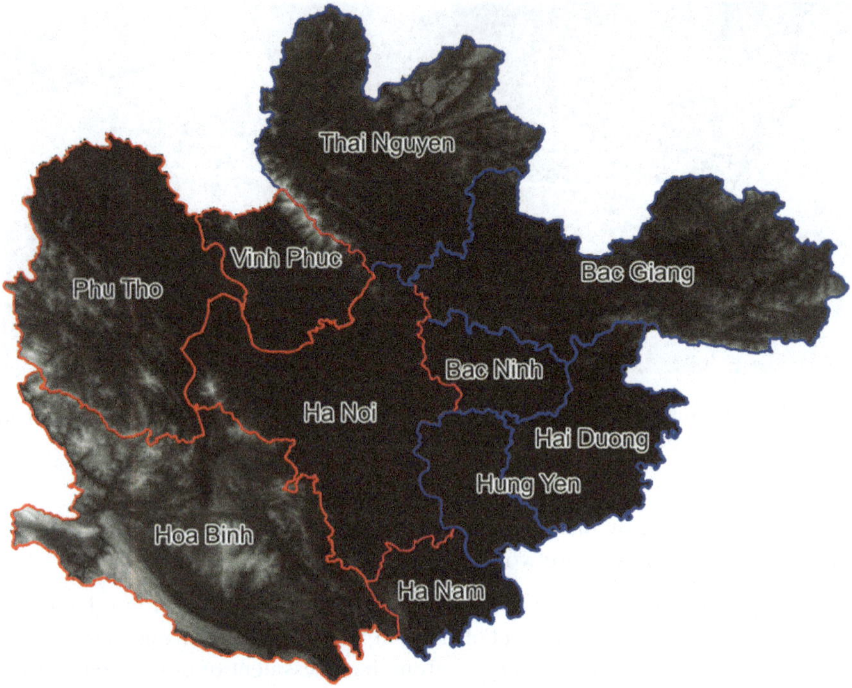

Fig. 5 Digital Elevation Model of the 10 provinces involved in the hydrodynamic model (provinces with red boundaries are the ones overlapped by the Red River, the ones with blue boundaries are the additional provinces impacted by the flooding).

(with only few main roads and many small pedestrian passages) and the induced congestion, the efficiency of massive unstaged evacuations (evacuations where every agent is alerted at the same time) might be highly reduced.

4.2 Hydrodynamic Model of the Dam Break

In order to model the hydrology of the studied area after the Hoa Binh dam failure, we choose to rely on the 2D modeling approach using the HEC-RAS software. To build the model, two kinds of input data are necessary: (i) topography data, including elevation and bathymetry, and (ii) water-related boundary conditions.

Input Data: Elevation Data. The most important data to gather for hydrological modeling is the studied area elevation and topography. In this subsection we introduce the most important source, *i.e.* the Digital Elevation Model (DEM), and all sources to further enhance knowledge about topography related to water flows, *i.e.* dikes and bathymetric data among other sources.

Digital Elevation Model (DEM). For the Digital Elevation Model (DEM), we used the NASA Shuttle Radar Topography Mission (SRTM) Version 3.0 with a resolution of 30 m (Fig. 5).[11] The studied area covers the 4 provinces of Hanoi, Hoa Binh, Phu Tho and Vinh Phuc. But some preliminary simulations showed that the water front of the flood tends to go beyond the boundaries of these provinces. So we need to add 6 more neighboring provinces: Thai Nguyen, Bac Giang, Bac Ninh, Hai Duong, Hung Yen, and Ha Nam.

Dikes. Preliminary simulations have shown that, even with daily flows lower than the average flows, some water "leaks" appear in the system: the water disperses and leaves the minor river bed before the dam break. The most likely cause is that the DEM, with a resolution of 30m, does not take into account the dikes, which are generally 5m wide. Given the importance of the dikes in the studied area, it has been necessary to integrate them more finely in the model: on the one hand to limit these "leaks", and on the other hand to have a more realistic simulation after dam failure. Using the data provided by [34], we modified the DEM to take into account the dikes.

Bathymetry. In addition to the DEM, we need bathymetric data (elevation below the water surface) along the rivers and the Hoa Binh Reservoir. However, complete bathymetric data for the entire area (Red, Black, and Clear rivers, Hoa Binh reservoir...) were not available. Therefore, we used the available data from [34] which provide bathymetry for 12 profiles across the Red River, all of them located close to the city of Hanoi. For the other sections of the Red River and other rivers, bathymetric data have been completed from various informal literature sources.[12] The main criterion used to select these data was the similarity of the geodetic system used. The possibility to use a small number of profiles to run the simulations is supported by related works: [45] has shown that the performance of the model is not significantly degraded when using a small number of sections (compared to the same model with a large number of cross-sectional profiles). Final dataset is shown on Fig. 6.

Input Data: Boundary Conditions. In addition to the water quantity in the reservoir and the water flow coming from the dam, the simulation requires knowledge about the flows from the various tributaries and at the system outlet. For this purpose, we used flow data from a daily flow database of 5 stations over a period from 1975 to 2016 [46].

From these data, an inflow transient flow hydrograph has been created for the Black River (*i.e.* in the upstream of the Hoa Binh dam), Red River, and Clear River to serve as a condition for the inflow boundaries. Similarly, hydrographs have been created for outflows of the system in the Red River and Duong River.

Description of the Simulation. Before simulating the dam failure itself, it is necessary to have a hydrological state of the system close to the real state. To this purpose, we have first simulated over a period of 1 month the water flow in the system using

[11] Source: https://earthexplorer.usgs.gov/.

[12] Other sections come from the following source: https://www.geosci-model-dev-discuss.net/gmd-2019-40/.

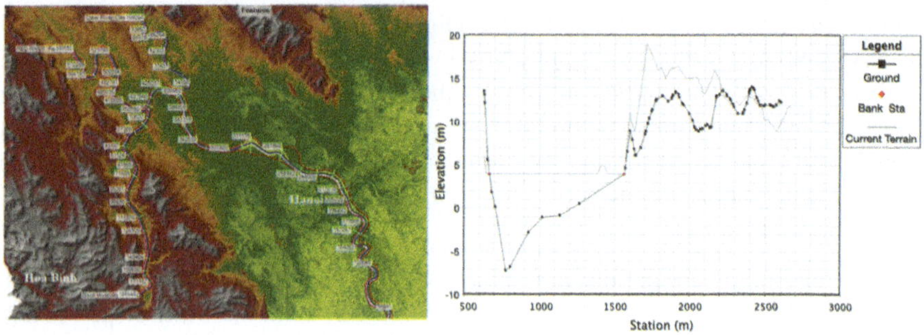

Fig. 6 Map of cross sections (left), the cross profile N 196893 after the modification of its bathymetry (right).

the current boundary conditions. This allowed us to fill the rivers and the reservoir and to reach a stationary state as correct as possible.

Once the stationary state reached, the dam failure can start. Based on the characteristics of the Hoa Binh dam (presented in Sect. 2), the most likely mode of failure is "Pipping", *i.e.* an infiltration erosion [12]. This is a type of failure for which, in situations where the population at risk and the study areas are sufficiently far downstream, the differences in peak flow and the shape of the hydrograph of the failure may not be significant when the flood wave reaches the downstream sections, and particularly for the Phuc Xa district in Hanoi.

This can be observed in the simulation results presented in the following section.

4.3 Preliminary Results and Discussions

The simulations show that a first wave affects the section of the Red River near Hanoi within 24 h. From then on, we can observe the increase in water speed and the flooding of the banks in the sections close to Hanoi. The second wave that floods a large section of the study area will not reach Hanoi before 48 more hours (72 h since the beginning of the simulation): as depicted in Figs. 7a and 7b, water velocity reach two time the regular water flow in the next 3 days at the intersection between the Black and Red Rivers.

This water front propagation time, which is longer than what was expected by some experts on the field, may suggest that we have to improve the data quality and refine the model by taking into account more parameters, and in particular hydrological ones such as water evaporation and infiltration. On a global perspective, Figs. 8a and 8b show respectively the initial and the final states (*i.e.* at the end of the simulation, 7 days after the dam failure) of the area in terms of water depth. We can notice that a large portion of the studied area have been over flooded because of the dam failure.

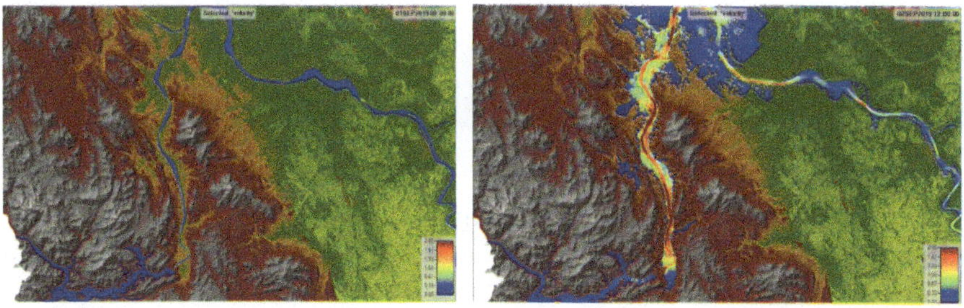

(a) Before the dam failure. (b) 72 hours after the dam failure.

Fig. 7 Water velocity in Black and Red Rivers watercourses before and after the dam failure. Blue color stands for initial regular velocity, while yellow to red mean 1.5 times to 2 times regular velocity.

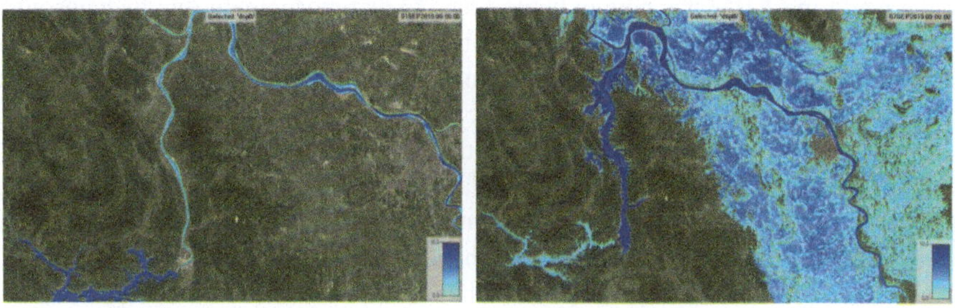

(a) Initial water depth. (b) Water depth 7 days after the dam break.

Fig. 8 Water depth in Black and Red Rivers watercourses before and after the dam failure. From Blue sea color stands for 10 m depth and above to sky blue color close but superior to 0 m depth.

It is important to notice that the city of Hanoi is not flooded after the simulated dam failure: the main reason is the quality of the DEM, which depicts building elevations rather than ground level. In order to obtain coherent flood results for the city of Hanoi, we need to improve the accuracy of data. This can be done reducing the elevation of the DEM using land registry and elevation data of buildings and roads. However, even if the model needs better data, more refinements, deeper analysis and a finer calibration procedure, the preliminary results we exposed, make it possible to use the flood model in conjunction with the agent-based model of evacuation as a proof of concept. The next section exposes the proposed coupling and its expected outcomes.

5 The Coupled Model

5.1 Coupling Principle

Main Principles. ESCAPE main purpose is to simulate the evacuation of an urban area and not the fight against flooding (as it is the case when inhabitants try to protect their city by repairing and elevating dikes, *e.g.* in [27]). As a consequence, we can limit the link between the hydrodynamic and the evacuation models to a weak coupling: first, the hydraulic simulation is executed, and then the results are integrated step by step in the ABM.

To this purpose, the agent-based and the hydraulic models have to share the same environment and thus the same DEM input data. Previously executed only at the scale of the ward, the ABM has now to be upscaled to an environment that matches with the DEM. The DEM file is thus used both to define the size of the environment, the dimensions of the grid of cells that will pave the whole environment, and initialise the elevation attribute of each cell. Conversely, the flood simulation will be downscaled.

With a given frequency, a result file produced by the hydraulic simulation is read and the water depth of each grid cell is updated accordingly. It is important to notice that the hydraulic and agent-based simulation steps can be different: either water depth will not be updated at each step (if the agent-based simulation step is shorter) or all the output files will not be used if the agent-based simulation step is longer. In our case study, the ESCAPE simulation step lasts 5 s (because the mobility model requires a short time step to be precise enough) whereas the hydraulic simulation step lasts 10 min. As a consequence, the water depth is updated once every 120 simulation steps in the evacuation model.

Implementation of the Co-model Coupling the HEC-RAS and GAMA Platforms. The co-model coupling the ESCAPE and HEC-RAS models is implemented as an extension of the ESCAPE model with a grid of cells containing a water depth and an agent dedicated to wrap the HEC-RAS simulator and to interact with the other agents of the agent-based model. The coupling is made in 2 steps: (i) at initialisation, the dedicated agent calls HEC-RAS to produce results (as tif files), (ii) at each step (or every N steps), the wrapper agent reads one of the result files and updates the water depth of the cells of the agent-based model.

Initialisation. The GAML (GAMA Modeling Language) code shown in Listing 1.1 illustrates the first step of the coupling.[13] The kind of agents `hydroManager` is in charge of the communication with the HEC-RAS platform. Compared to other kinds of agents, it is given additional capabilities, *i.e.* additional possible actions (encoded in the `hecrasSkill` skill). Once created, such an agent will execute at its initialisation:

[13]Source code is available in the Github repository of the GAMA extensions (https://github.com/gama-platform/gama.experimental) in the ummisco.gama.extension.hecras project.

1. **Establish a connection with the HEC-RAS software.** To this purpose, it first loads a new instance of the HEC-RAS engine.
2. **Generate the *rasmap* file specifying how to execute the plan.** The flooding simulation is driven by a plan that is configured through a so-called *rasmap* file. It specifies in particular the "terrain" and the start and end dates of the simulation.
3. **Ask HEC-RAS to open an existing project.** The project file contains a link to all the dataset necessary to run the simulations and to several possible plans defined by the user. A plan specifies the way a simulation is launched (which input data), if it is a steady or unsteady flow simulation in the case of a sediment simulation, and what is saved in output files every time step. The HEC-RAS project specifies the current plan to be used in step 4.
4. **Simulate the current plan specified by the project.** This will create all the result files containing the water depth as .tif files. One file is created for each save time step specified in the current plan. Each file is named with the simulation time. This simulation is executed in headless mode.
5. **Quit HEC-RAS software**, once all the computations have been done (note that the GAMA simulation waits for the HEC-RAS simulations to be over to continue its execution).

```
// The HEC-RAS related actions require the agent to have the
   hecrasSkill.
species hydroManager skills: [hecrasSkill] {

  init {
    // Load an HEC-RAS instance.
    do load_hecras();

    // Generate the Rasmap file specifying how the plan is
    executed.
    do Generate_RasMap(
        "../includes/Phuc Xa Simulation/PhucXaSimulation.rasmap"
    ,
        "Plan 01", "Phuc Xa DEM + 2 Bathymetries",
        "15AUG2019", 0, 24, 0, 59);

    // Specify the HEC-RAS project file that will be opened.
    file f <- file("../includes/HWC/HCW2.prj");
    do Project_Open(f);

    //Hide the GUI dialog of HEC-RAS
    do Compute_HideComputationWindow();

    // Compute the plan opened previously.
    // This produces the tif result files.
    do Compute_CurrentPlan();

    // Quit the Hec-RAS instance.
    do QuitRas();
  }

  // Update data
  // ...
}
```

Listing 1.1 GAML code of the initialisation of the agents wrapping HEC-RAS simulations.

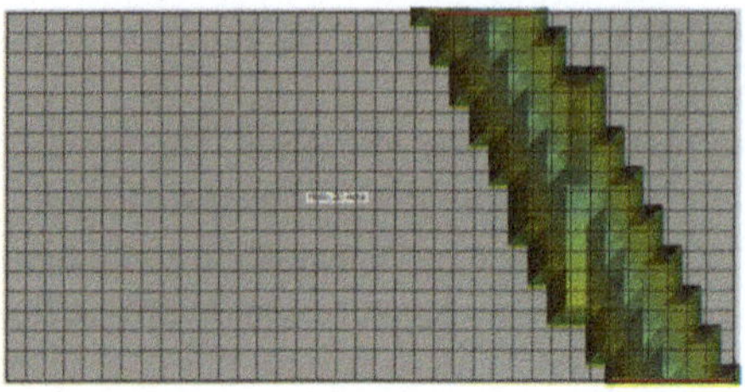

(a) Digital Elevation Model with a flat landscape and a simple river bed.

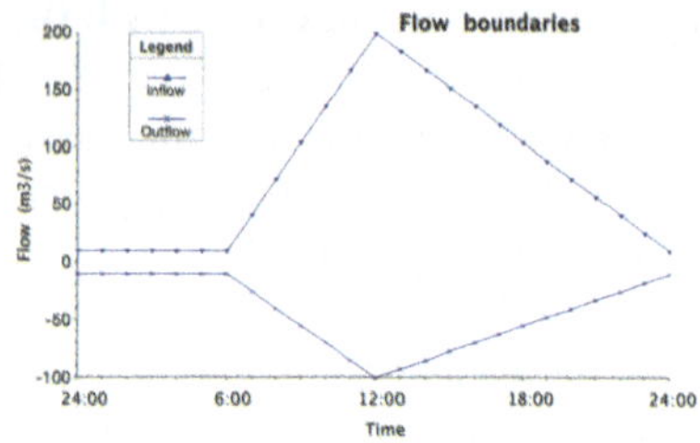

(b) Inflow and outflow hydrographs

Fig. 9 Input data of the minimal co-model.

Update Water Depth from the Hydrological Simulation. At every step of the ABM model, the main task for the `hydroManager` agent is:

1. **Identify the tif file to be opened.** Each result file can be identified by the simulated time it represents. As a consequence, the agent-based model has to compute a simulated time coherent with the ones of the HEC-RAS simulation. To this purpose, we rely on GAMA capabilities to manage dates: given a starting date and a duration of the simulation step, GAMA automatically computes the simulated time of the current step.
2. **Read water depth of each grid cell in the opened file.** The model reads the tif file and stores the results as a list of ordered float values. Given the encoding of tif files, this order is the same in all the files and it also corresponds to the order of cell agents created in the agent-based simulation.
3. **Update cells.** In this last step, each cell water depth value is updated. As any other agents of the simulation, cells can now impact the course of the simulation with the up-to-date flood value from the hydrological simulation.

Illustration on a Minimal Co-model. As an illustration of the co-model, we built a minimal co-model coupling a simple HEC-RAS model with a simple GAMA model limited to the hydrology part. The HEC-RAS model is made from data presented in Fig. 9: (i) a DEM of a flat landscape with only a river bed (*i.e.* the greeny area), (ii) unsteady inflow and outflow hydrographs. The hydrogaphs represent the evolution of flows over a single day: from 00:00 to 06:00, flows are constant (this will fill the river and keep a stable water volume in the bed), then from 06:00 to 12:00, the inflow increases, while the outflow decreases. The river will thus soon reached its capability and the flood will occur. During the second half of the day, the inflow decreases while the outflow increases, which will reduce the flood.

Figure 10 plots the evolution of the water volume over the whole day (in the GAMA simulation), showing the constant water depth before 6AM, and after 6AM, the water depth increase all over the day. The reduction of the inflow volume (after 12:00) only limits the increase of the water level, a maximum value seems to be reached at 0:00. The plot presents a discontinuity few minutes after 6AM: when

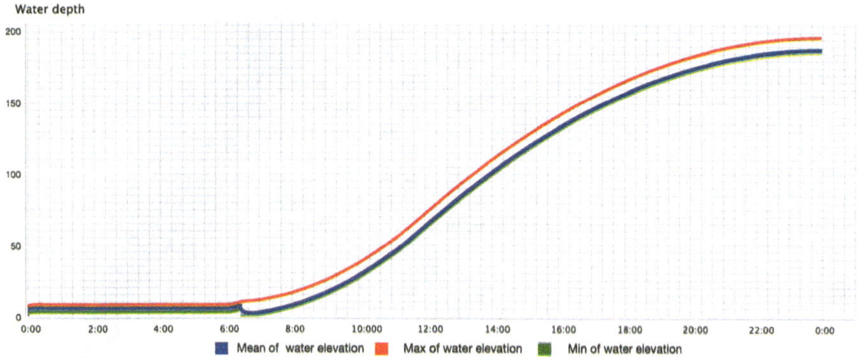

Fig. 10 Evolution of the water depth over the full day: series represent the average (blue), minimum (green) and maximum (red) water depth values in cells with water. The discontinuity at 6:20 reflects the time of flooding start (*i.e.* when cells outside of the river bed start to be covered by a low depth of water).

water starts to reaches cells outside of the riverbed, they first have a very low water depth, which makes the minimum water depth decreases suddenly.

5.2 *Minimal Evacuation Co-model*

To go one step further and illustrate the coupling between HEC-RAS flooding simulations and a GAMA agent-based evacuation model, we implement a minimal evacuation co-model, based on the dataset used for the previous flooding co-model. It thus contains a grid of cell agents created from the DEM data file. Each cell contains an elevation (coming from the DEM file) and a water depth (updated by the wrapper agent). We add to this model (cf. Fig. 11):

1. an *evacuation_point* agent, located on the left of the environment (on a random cell, located on the column $y = 0$).
2. a set of *house* agents: houses will contain initially *people* agents. Each house covers a full cell. Initially, the simulation creates 50 houses, located in the band of cells close to the river bed; which means they are prone to be quickly impacted by overflows.
3. *people* agents: one people agent is created in each house. As soon as it is alerted, it will try to escape toward the evacuation point. We force the people's move to be on the cells, and that only one agent can be located on a cell at the same time. As soon as a people agent reaches an evacuation point, it is removed from the simulation and counted as an evacuee. When a people agent is located on a cell with a water depth higher than 2 m, we consider that it will not be able to evacuate, it is thus removed from the simulation and counted as a casualty.

Fig. 11 Screenshot of the minimal evacuation co-model: houses (in grey) contain initially one people agent (black circle). The evacuation point is displayed with a green circle.

People agents will evacuate only when they are alerted by a higher-level authority. This authority (not represented explicitly in the current model) is simply characterized by an alert time parameter: when this time is reached in the simulation, all the agents are alerted and start their evacuation. To illustrate the whole workflow of coupling, we assess the evacuation process given various alert times. The explored parameter is the alert time (between 6:20 and 7:00, with a step of 5 min). As for indicators, we compute the number of casualties and evacuees at the end of the simulation. Figure 12 shows the result of this exploration. We can notice that, on this toy case study, the alert can be triggered even after the increase of the inflow (and the decrease of the outflow, conditions that induce the flooding) without jeopardising the chance for everybody to be saved. Similarly, we can notice a threshold after which nobody can reach exits of the area (which might highlight the usefulness of a vertical evacuation strategy given the situation).

5.3 Co-model Coupling the ESCAPE Model on the Phuc Xa Area with the HEC-RAS Flooding Model

The objective of this section is to demonstrate the practical capability of implementing a co-model coupling a real evacuation model (the ESCAPE model) with an HEC-RAS flooding simulation (flooding of Hanoi following the Hoa Binh dam break). A key challenge is to couple them without altering the behavior of each of them and with the minimum of modifications. This justifies the use of the co-modeling approach, wrapping the HEC-RAS simulator in an agent.

A first adaptation is related to the dimensions of the simulation environment of both models because their spatial extensions are very different: the Phuc Xa ward is

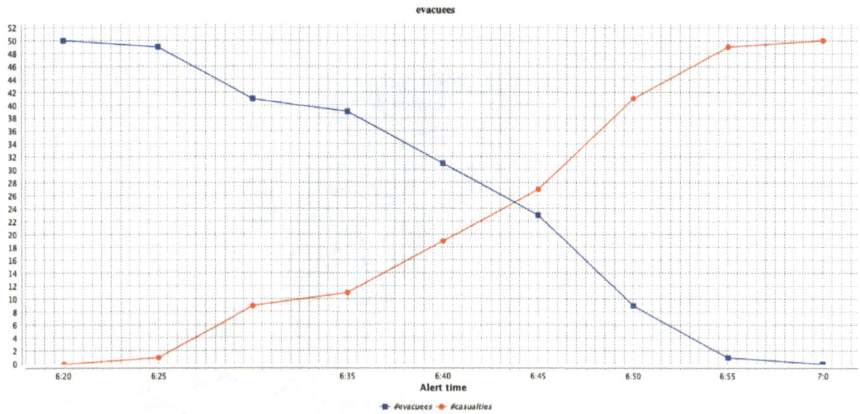

Fig. 12 Evolution of the casualties (red) and evacuees (blue) depending on the alert time.

a space of less than 1 km^2 whereas the hydrology is simulated on a space of around 4000 km^2. We choose to run both simulations in the same environment: we thus create a new DEM file covering the Phuc Xa ward and a part of the Red River (see the central picture in Fig. 13). The in- and outflow boundary conditions for this new HEC-RAS simulation come for the simulation run on the whole flooding area. The evacuation simulation is executed on a wider area. But, as the evacuation is constrained by the road and pedestrian networks, extending the overall environment dimensions will not have any impact on the evacuation simulation.

As mentioned previously, the DEM file used for the flooding simulation integrates the building height in the elevation computation, making the city not flooded as it should be. To be able to integrate the impact of the flooding on the evacuation process, the area needs to be flooded. We thus modified the elevation data in the Phuc Xa area: we apply to all the cells covering the ward the elevation of the cells without building located between the ward and the river. As depicted in Fig. 13, the overall simulated area of the Phuc Xa quarter can now eventually be submerged: the blue color intensity represents the water depth in the cell and white cells are cells without water.

The adaptation of the evacuation model has been limited to its minimum: it has simply been relocated in a bigger environment paved with a grid (defined by the DEM file used for the hydrological simulation). Each cell agent contains the water depth, updated by the wrapper agent reading the files produced by HEC-RAS.

The interaction between water level and evacuating agents will be managed by the cell agents: at each step, if a cell water depth is higher than a given threshold (chosen to 0.5 m in the following simulations), all the agents located on it will be removed from the simulation (and counted as non-evacuees). Finally, the alert time (the time when all the agents time are ordered to evacuate) is set as a parameter of the simulation to be explored in the following.

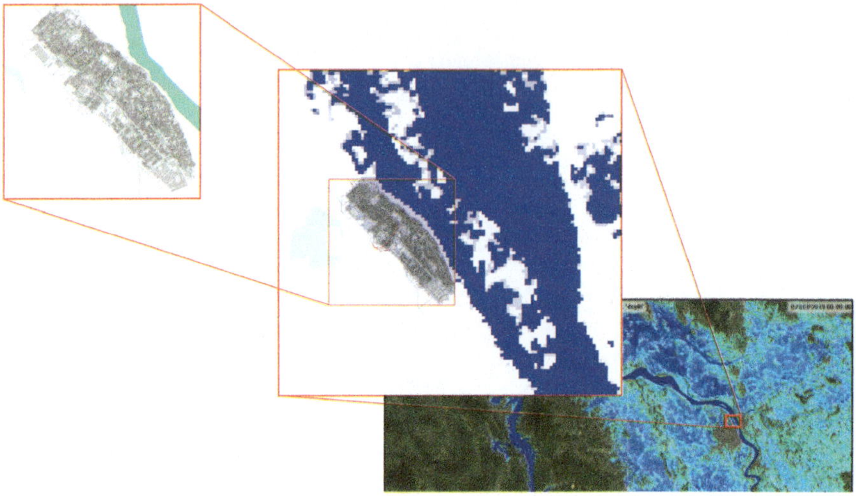

Fig. 13 ESCAPE simulation on Phuc Xa area (on the left), HEC-RAS simulation of the Hoa Binh dam break (on the right), and the coupled simulation (in the center).

As a proof of concept, we explore the co-model given 2 parameters: the **alert time** (taken values between 6:40 and 8:00, every 10 min) and the **number of inhabitants** (among the possible values {157, 785, 4710, 7850, 10990, 15700}). As an indicator that can be common to all the population sizes, we chose to compute the **evacuee rate**: it is computed as the number of evacuee agents (at the end of the simulation) over the total number of inhabitants. An evacuation rate of 1 means that everybody was able to evacuate. Results are summarized in Fig. 14a.[14]

First of all, we observe again that there is a range of alert time in which the model is highly sensitive: when we observe the time series for the smallest numbers of agents (175 and 785), a difference of 30 min in the alert time can change the output of the evacuation from a state where all the inhabitants have successfully evacuated to a state where no one was able to evacuate. After 7:40 PM, we can also notice that nobody can evacuate. Looking at Fig. 14b, we can notice that the period between 7:20PM and 7:40PM corresponds to the period where water depths reach 0.5 m (at 7:40PM all the cells have reached the threshold value).

Second, we can definitively observe the effect of the number of agents on the evacuation efficiency: as soon as there is a number of agents high enough (4710 in the experiment), the rate of evacuees is sharply reduced. In addition, it appears that the steepness of the curve decreases with the number of agents, which could let us imagine that the alert time (and thus the evacuation duration) needed to evacuate everyone will not be linear with the number of agents. And this can only be due to

[14]We ran only 1 simulation for each value of the parameters. As the ABM is stochastic, many more replications should be necessary to get accurate results. But this is out of the scope of this paper, as we focus here on the coupling from a methodological point of view.

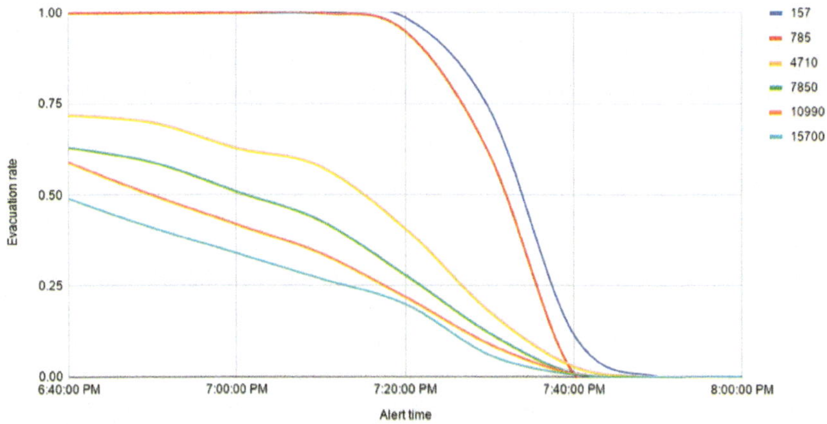

(a) Evolution of the rate of evacuees depending on the alert time and the number of inhabitants.

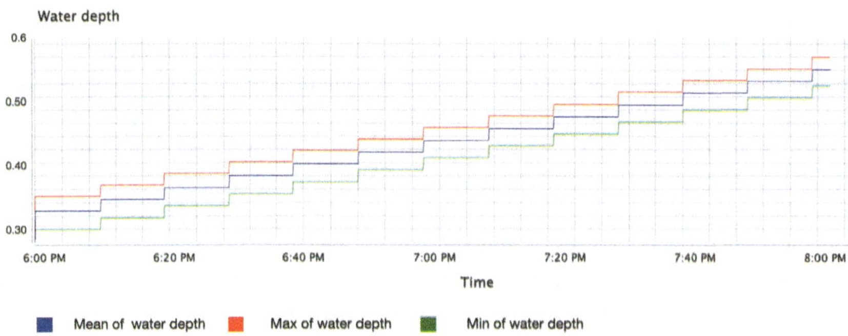

(b) Water depth evolution in the Phuc Xa ward.

Fig. 14 Input data of the minimal co-model.

the congestion (*i.e.* traffic jam and crowded area) induced by a bigger number of agents evacuating at the same time.

6 Conclusion

This paper focuses on the extension of the existing ESCAPE agent-based evacuation model, and more specifically its implementation on the Phuc Xa ward in Hanoi, threaten by flooding in case of the Hoa Binh dam failure. To this purpose we propose to combine a hydraulic dam failure model with the agent-based evacuation model using the GAMA platform. We demonstrate how to methodologically and operationally couple a hydrodynamic water diffusion model (implemented using the HEC-RAS software) and the multi-paradigm evacuation model (using the ESCAPE framework). The objective here is to integrate the evolution of the hazard, in order to be more precise in the assessment of the evacuation and in particular during the period of the flooding occurrence.

The main contribution of the paper is methodological about the coupling of two different models (different in their purpose and in their formalism), but also operational, with the extension of the GAMA platform to manage this coupling. To be more precise, contributions of this paper are (i) the design, implementation, and simulation of the inundation model on the wide area between the Hoa Binh dam and Hanoi city, (ii) the extension of the GAMA platform to interact with the HEC-RAS platform, (iii) the implementation of a minimal co-model example illustrating the coupling principles in a generic way, and (iv) the integration of the flood simulation in the agent-based evacuation model relying on the co-modeling paradigm.

This first attempt to integrate the hydrodynamic model into the ESCAPE agent-based framework makes it possible from now on to further study individual or collective responses and evacuation management strategies taking into account the evolution of the catastrophic event in a realistic way. A promising way to explore such complex interactions between the hazard and human behaviors will be to tackle thematic questions such as the ones related to synchronous versus asynchronous evacuation plans: within the context of a realistic flooding, not only congestion but fine grain water level impacts on mobility can be included as a major determinant of the evacuation strategy feasibility, drawbacks and advantages.

On a methodological aspect, even if the coupling between GAMA and HEC-RAS platforms is operational, it is up-to-now only in one direction: GAMA asks HEC-RAS to execute the simulation and return results that are integrated into the agent-based model during the simulation. Future works will focus on retro-action between both models in order that actions of agents can have an effect on the environment (*e.g.* on the elevation) which will impact back the HEC-RAS simulation.

A second interesting future work will be to propose an automatic workflow allowing the modeler to downscale the hydrodynamic simulations to a scale appropriate for the agent-based model and extract in- and outflow boundary conditions to be applied to the lower-scale simulation.

Even if it still needs to be improved, this methodological and operational coupling is really promising and it now makes it possible to tackle the question of improving crisis planning and evacuation plan strategies, in particular when dry evacuation is not an option.

Acknowledgement This work is supported by the ANR ESCAPE project, grant ANR-16-CE39-0011-01 of the French Agence Nationale de la Recherche.

References

1. Adam, C., Gaudou, B.: Modelling human behaviours in disasters from interviews: application to Melbourne bushfires. J. Artif. Soc. Soc. Simul. **20**(3), 12 (2017). http://jasss.soc.surrey.ac.uk/20/3/12.html
2. Adam, C., Taillandier, P., Dugdale, J., Gaudou, B.: BDI vs FSM agents in social simulations for raising awareness in disasters: a case study in Melbourne bushfires. Int. J. Inf. Syst. Crisis Response Manag. (IJISCRAM) **9**(1), 27–44 (2017)
3. Adnan, M., Pereira, F.C., Azevedo, C.M.L., Basak, K., Lovric, M., Raveau, S., Zhu, Y., Ferreira, J., Zegras, C., Ben-Akiva, M.: SimMobility: a multi-scale integrated agent-based simulation platform. In: 95th Annual Meeting of the Transportation Research Board Forthcoming in Transportation Research Record (2016)
4. Arnold, J.G., Srinivasan, R., Muttiah, R.S., Williams, J.R.: Large area hydrologic modeling and assessment part I: model development 1. JAWRA J. Am. Water Resour. Assoc. **34**(1), 73–89 (1998)
5. Bahrami, M., Nasab, S.B., Naseri, A.A., Albaji, M., et al.: Comparison of Muskingum-Cunge model and irrigation hydraulic models in estimation of furrow irrigation advance phase. Res. Crops **11**(2), 541–544 (2010)
6. Bañgate, J., Dugdale, J., Beck, E., Adam, C.: A multi-agent system approach in evaluating human spatio-temporal vulnerability to seismic risk using social attachment. WIT Trans. Eng. Sci. **121**, 47–58 (2018)
7. Banos, A., Corson, N., Gaudou, B., Laperrière, V., Coyrehourcq, S.R.: The importance of being hybrid for spatial epidemic models: a multi-scale approach. Systems **3**(4), 309–329 (2015)
8. Beck, E., Dugdale, J., Van Truong, H., Adam, C., Colbeau-Justin, L.: Crisis mobility of pedestrians: from survey to modelling, lessons from Lebanon and Argentina. In: International Conference on Information Systems for Crisis Response and Management in Mediterranean Countries, pp. 57–70. Springer (2014)
9. Bertsch, C., Ahle, E., Schulmeister, U.: The functional mockup interface-seen from an industrial perspective. In: Proceedings of the 10th International Modelica Conference, Lund, Sweden, 10–12 March 2014, pp. 27–33. Linköping University Electronic Press (2014)
10. Boissier, L.: La mortalité liée aux crues torrentielles dans le Sud de la France: une approche de la vulnérabilité humaine face à l'inondation. Ph.D. thesis, Université Paul Valéry - Montpellier III, December 2013
11. Bruner, G.: HEC-RAS, river analysis system users manual, versi6n 4.0. Technical report, US Army Corps of Engineers (2008). http://www.hec.usace.army.mil
12. Brunner, G.: Using HEC-RAS for dam break studies. Technical report, USACE HEC Technical (2014)
13. Chapuis, K., Taillandier, P., Gaudou, B., Drogoul, A., Daudé, E.: A multi-modal urban traffic agent-based framework to study individual response to catastrophic events. In: The 21st International Conference on Principles and Practice of Multi-Agent Systems (PRIMA 2018), p. 8 (2018)
14. Chapuis, K., Taillandier, P., Renaud, M., Drogoul, A.: Gen*: a generic toolkit to generate spatially explicit synthetic populations. Int. J. Geogr. Inf. Sci. **32**(6), 1194–1210 (2018). https://doi.org/10.1080/13658816.2018.1440563
15. Khanh Hoa Provincial Peoples' Committee: 2017 Vietnam post-typhoon damrey rapid damage and needs assessment. Technical report, WorldBank (2018). https://www.gfdrr.org/sites/default/files/publication/vietnam-damrey-rapid-assessment-report-en.pdf
16. Cuesta, A., Abreu, O., Alvear, D.: Future challenges in evacuation modelling. In: Cuesta, A., Abreu, O., Alvear, D. (eds.) Evacuation Modeling Trends, pp. 103–129. Springer, Cham (2016). https://doi.org/10.1007/978-3-319-20708-7_5
17. Cunge, J.: On the subject of a flood propagation computation method (Musklngum method). J. Hydraul. Res. **7**(2), 205–230 (1969)

18. Czura, G., Taillandier, P., Tranouez, P., Daudé, E.: MOSAIIC: city-level agent-based traffic simulation adapted to emergency situations. In: Takayasu, H., Ito, N., Noda, I., Takayasu, M. (eds.) Proceedings of the International Conference on Social Modeling and Simulation, Plus Econophysics Colloquium 2014, pp. 265–274. Springer, Cham (2015). http://link.springer.com/10.1007/978-3-319-20591-5_24
19. Dahmann, J.S., Morse, K.L.: High level architecture for simulation: an update. In: Proceedings of the 2nd International Workshop on Distributed Interactive Simulation and Real-Time Applications (Cat. No. 98EX191), pp. 32–40. IEEE (1998)
20. Daudé, E., Taillandier, P., Caron, C., Gaudou, B., Saval, A., Chapuis, K., Tranouez, P., Drogoul, A., Rey-Coyrehourq, S., Zucker, J.D.: ESCAPE: exploring by simulation cities awareness on population evacuation. In: ISCRAM 2019, pp. 76–93 (2019)
21. Drogoul, A., Huynh, N.Q., Truong, Q.C.: Coupling environmental, social and economic models to understand land-use change dynamics in the Mekong delta. Front. Environ. Sci. **4**, 19 (2016)
22. Durak, M., Durak, N., Goodman, E.D., Till, R.: Optimizing an agent-based traffic evacuation model using genetic algorithms. In: 2015 Winter Simulation Conference (WSC), pp. 288–299. IEEE (2015)
23. Fenet, J., Daudé, E.: Is population's prevention the great failure of territorial risk management: the case of the industrial risks, Rouen, France. Cybergeo **932**, 299–322 (2020)
24. Fianyo, Y.E.: Couplage de modèles à l'aide d'agents: le système OSIRIS. Ph.D. thesis, ANRT [diff.] (2001)
25. Flötteröd, Y.P., Erdmann, J.: Experiment study on the evacuation of bomb alert with sumo. In: SUMO 2016–Traffic, Mobility, and Logistics, pp. 39–50 (2016)
26. Garschagen, M., Hagenlocher, M., Comes, M., Dubbert, M., Sabelfeld, R., Lee, Y.J., Grunewald, L., Lanzendörfer, M., Mucke, P., Neuschäfer, O., et al.: World risk report 2016. Technical report, Bündnis Entwicklung Hilft and UNU-EHS (2016)
27. Gasmi, N., Grignard, A., Drogoul, A., Gaudou, B., Taillandier, P., Tessier, O., An, V.D.: Reproducing and exploring past events using agent-based geo-historical models. In: International Workshop on Multi-Agent Systems and Agent-Based Simulation, pp. 151–163. Springer (2014)
28. Gaudou, B., Sibertin-Blanc, C., Therond, O., Amblard, F., Auda, Y., Arcangeli, J.P., Balestrat, M., Charron-Moirez, M.H., Gondet, E., Hong, Y., et al.: The MAELIA multi-agent platform for integrated analysis of interactions between agricultural land-use and low-water management strategies. In: International Workshop on Multi-Agent Systems and Agent-Based Simulation, pp. 85–100. Springer (2013)
29. Gee, D.M., Brunner, G.W.: Dam break flood routing using HEC-RAS and NWS-FLDWAV. In: Impacts of Global Climate Change, pp. 1–9. American Society of Civil Engineers (2005)
30. Gourou, P.: Geographie du viet nam - les paysans du delta tonkinois. Publications de L'Ecole Francaise D'extreme-Orient, 661 p (1936)
31. Gutknecht, O., Ferber, J.: The MADKIT agent platform architecture. In: Workshop on Infrastructure for Scalable Multi-Agent Systems at the International Conference on Autonomous Agents, pp. 48–55. Springer (2000)
32. Hawe, G.I., Coates, G., Wilson, D.T., Crouch, R.S.: Agent-based simulation for large-scale emergency response: a survey of usage and implementation. ACM Comput. Surv. **45**(1), 1–51 (2012)
33. Huynh, Q.N.: CoModels, engineering dynamic compositions of coupled models to support the simulation of complex systems. Ph.D. thesis, Pierre and Marie Curie University (2016)
34. JICA: the study on the red river inland waterway transport system in the socialist republic of Vietnam: main report ; 1 - present situation: final. Technical report, Japan International Cooperation Agency. The Overseas Coastal Area Development Institute of Japan (OCDI). Japan Port Consultants, Ltd. (JPC) (2003)
35. Laatabi, A., Becu, N., Marilleau, N., Pignon-Mussaud, C., Amalric, M., Bertin, X., Anselme, B., Beck, E.: Mapping and describing geospatial data to generalize complex models: the case of LittoSIM-GEN. Int. J. Geospatial Environ. Res. **7**(1), 6 (2020)

36. Lämmel, G., Klüpfel, H., Nagel, K.: The MATSim network flow model for traffic simulation adapted to large-scale emergency egress and an application to the evacuation of the Indonesian city of Padang in case of a Tsunami warning. In: Pedestrian Behavior. Emerald Group Publishing Limited (2009)
37. Li, Y., Li, J.: Review of experimental study on dam-break. Adv. Water Sci. **20**(2), 304–310 (2009)
38. Lopez, P.A., Behrisch, M., Bieker-Walz, L., Erdmann, J., Flötteröd, Y.P., Hilbrich, R., Lücken, L., Rummel, J., Wagner, P., WieBner, E.: Microscopic traffic simulation using sumo. In: 2018 21st International Conference on Intelligent Transportation Systems (ITSC), pp. 2575–2582. IEEE (2018)
39. Lu, X.X., Oeurng, C., Le, T.P.Q., Thuy, D.T.: Sediment budget as affected by construction of a sequence of dams in the lower Red River, Viet Nam. Geomorphology **248**, 125–133 (2015)
40. Luo, P., Mu, D., Xue, H., Ngo-Duc, T., Dang-Dinh, K., Takara, K., Nover, D., Schladow, G.: Flood inundation assessment for the Hanoi central area, Vietnam under historical and extreme rainfall conditions. Sci. Rep. **8**(1), 1–11 (2018)
41. Macal, C., North, M.: Tutorial on agent-based modelling and simulation. J. Simul. **4**, 151–162 (2010)
42. Mao, J., Wang, S., Ni, J., Xi, C., Wang, J.: Management system for dam-break hazard mapping in a complex basin environment. ISPRS Int. J. Geo Inf. **6**(6), 162 (2017)
43. Marilleau, N., Lang, C., Giraudoux, P.: Coupling agent-based with equation-based models to study spatially explicit megapopulation dynamics. Ecol. Model. **384**, 34–42 (2018)
44. Mas, E., Adriano, B., Koshimura, S.: An integrated simulation of Tsunami hazard and human evacuation in La Punta, Peru. J. Disaster Res. **8**(2), 285–295 (2013). https://www.fujipress.jp/jdr/dr/dsstr000800020285
45. Md Ali, A., Di Baldassarre, G., Solomatine, D.P.: Testing different cross-section spacing in 1D hydraulic modelling: a case study on Johor River, Malaysia. Hydrol. Sci. J. **60**(2), 351–360 (2015)
46. Meteorological, V., Administration, H.: Data base of water discharge of 5 stations from 1975 to 2016. Technical report, LMI LOTUS - USTH (2016)
47. Michaud, J., Johnson, C., Iokepa, J., Marohnic, J.: Methods for estimating the impact of hypothetical dam break floods. In: Chemistry for the Protection of the Environment 4, pp. 195–199. Springer (2005)
48. Neal, J., Schumann, G., Fewtrell, T., Budimir, M., Bates, P., Mason, D.: Evaluating a new LISFLOOD-FP formulation with data from the summer 2007 floods in Tewkesbury, UK. J. Flood Risk Manag. **4**(2), 88–95 (2011)
49. North, M.J., Collier, N.T., Ozik, J., Tatara, E.R., Macal, C.M., Bragen, M., Sydelko, P.: Complex adaptive systems modeling with Repast Simphony. Complex Adapt. Syst. Model. **1**(1), 3 (2013)
50. Parrott, L.: Hybrid modelling of complex ecological systems for decision support: recent successes and future perspectives. Ecol. Inform. **6**(1), 44–49 (2011)
51. Perrow, C.: The Next Catastrophe: Reducing Our Vulnerabilities to Natural, Industrial, and Terrorist Disasters. Princeton University Press (2007)
52. Pilotti, M., Maranzoni, A., Milanesi, L., Tomirotti, M., Valerio, G.: Dam-break modeling in alpine valleys. J. Mt. Sci. **11**(6), 1429–1441 (2014)
53. Rodriguez, S., Hilaire, V., Gaud, N., Galland, S., Koukam, A.: Holonic multi-agent systems. In: Di Marzo Serugendo, G., Gleizes, M.P., Karageorgos, A. (eds.) Self-Organising Software: From Natural to Artificial Adaptation, pp. 251–279. Springer, Heidelberg (2011)
54. Ruin, I., Lutoff, C., Boudevillain, B., Creutin, J.D., Anquetin, S., Rojo, M.B., Boissier, L., Bonnifait, L., Borga, M., Colbeau-Justin, L., Creton-Cazanave, L., Delrieu, G., Douvinet, J., Gaume, E., Gruntfest, E., Naulin, J.P., Payrastre, O., Vannier, O.: Social and hydrological responses to extreme precipitations: an interdisciplinary strategy for postflood investigation. Weather Clim. Soc. **6**(1), 135–153 (2014)
55. Simmonds, J., Gómez, J.A., Ledezma, A.: The role of agent-based modeling and multi-agent systems in flood-based hydrological problems: a brief review. J. Water Clim. Change **11**(4), 1580–1602 (2019)

56. Taillandier, P., Gaudou, B., Grignard, A., Huynh, Q.N., Marilleau, N., Caillou, P., Philippon, D., Drogoul, A.: Building, composing and experimenting complex spatial models with the GAMA platform. GeoInformatica **23**(2), 299–322 (2019)
57. Tessier, O.: Outline of the process of Red River hydraulics development during the Nguyen dynasty (nineteenth century). In: Environmental Change and Agricultural Sustainability in the Mekong Delta, pp. 45–68. Springer (2011)
58. Tessier, O.: Hydrological development of the red river delta: a historical perspective of the role of the imperial then colonial state (from the XIIth century to the first half of the XXth century). Water and Its Many Issues. Methods and Cross-Cutting Analysis, Regional Social Sciences Summer University, pp. 130–154 (2012)
59. Truong, Q.C., Taillandier, P., Gaudou, B., Vo, M.Q., Nguyen, T.H., Drogoul, A.: Exploring agent architectures for farmer behavior in land-use change. A case study in coastal area of the Vietnamese Mekong Delta. In: International Workshop on Multi-Agent Systems and Agent-Based Simulation, pp. 146–158. Springer (2015)
60. Valette, M., Gaudou, B., Longin, D., Taillandier, P.: Modeling a real-case situation of egress using BDI agents with emotions and social skills. In: International Conference on Principles and Practice of Multi-Agent Systems, pp. 3–18. Springer (2018)
61. Velotti, L., Trainor, J.E., Engel, K., Torres, M., Myamoto, T.: Beyond vertical evacuation: research considerations for a comprehensive "vertical protection strategy". Int. J. Mass Emerg. Disasters **31**(1), 60–77 (2013)
62. VietnamNews (2018). https://vietnamnews.vn/society/462709/hoa-binh-son-la-hydroelectric-plants-ensure-safety.html. Accessed 29 June 2020
63. Vinh, V., Ouillon, S., Thanh, T., Chu, L.: Impact of the Hoa Binh dam (Vietnam) on water and sediment budgets in the Red River basin and delta. Hydrol. Earth Syst. Sci. **18**(10), 3987 (2014)
64. Vu, V.: The Hoa Binh reservoir and the general gestation of its basin area. General Centre for Meteorology and Hydrology of Vietnam 250 (2002)
65. Axhausen, K.W., Horni, A., Nagel, K.: The Multi-Agent Transport Simulation MATSim. Ubiquity Press, London (2016)
66. Wang, H., Mostafizi, A., Cramer, L.A., Cox, D., Park, H.: An agent-based model of a multimodal near-field tsunami evacuation: decision-making and life safety. Transp. Res. Part C Emerg. Technol. **64**, 86–100 (2016). http://www.sciencedirect.com/science/article/pii/S0968090X15004106
67. Wilensky, U.: NetLogo itself: NetLogo. Technical report, Center for Connected Learning and Computer-Based Modeling, Northwestern University, Evanston (1999). http://ccl.northwestern.edu/netlogo
68. Xiong, Y.F.: A dam break analysis using HEC-RAS. J. Water Resour. Prot. **3**(6), 370 (2011)
69. Yin, L., Zhu, J., Zhang, X., Li, Y., Wang, J., Zhang, H., Yang, X.: Visual analysis and simulation of dam-break flood spatiotemporal process in a network environment. Environ. Earth Sci. **74**(10), 7133–7146 (2015)
70. Zeigler, B.P., Moon, Y., Kim, D., Ball, G.: The DEVS environment for high-performance modeling and simulation. IEEE Comput. Sci. Eng. **4**(3), 61–71 (1997)

Agent Based Modelling Using GAMA 1.8 with Applications to Biological System in Epidemiology

Lloyd W. F. Lee and Mohd Hafiz Mohd

Abstract This paper is written as a guide for researchers on how to execute stochastic agent-based model (ABM) in GAMA 1.8. GAMA 1.8 is a free modelling software that supports both temporal and spatial ABMs. This paper will first briefly introduce GAMA 1.8, its installation and a quick overview of its performance compared to other platforms. As an example, this paper will provide a step-by-step approach on how to code a stochastic discrete time Markov chain ABM. A flowchart on the execution of ABM will be provided and graphical visualizations of the example ABM will be demonstrated.

Keywords Agent-based modelling (ABM) · GAMA 1.8 · Stochastic model

1 Introduction

In this paper, GAMA 1.8 was utilized to simulate a stochastic discrete time Markov chain agent-based model (ABM). This paper aims to serve as an introductory laboratory manual for GAMA 1.8 on stochastic modelling. GAMA 1.8 is modelling and simulation development environment for building spatially explicit agent-based simulations. It is developed by several teams under the umbrella of the IRD/SU international research unit UMMISCO [1]. The high-level and intuitive agent-based language of GAMA 1.8 allows users, especially for non-computer scientists, to build their models in several paradigms of modelling (including the incorporation of Geographic Information System (GIS)). Thus, GAMA 1.8 has been widely used by researchers for modelling transportation [2], urban planning [3], epidemiology [4],

L. W. F. Lee (✉) · M. H. Mohd
School of Mathematical Sciences, Universiti Sains Malaysia, USM, 11800 George Town, Penang, Malaysia

M. H. Mohd
e-mail: mohdhafizmohd@usm.my

M. H. Mohd
School of Mathematical Sciences, Faculty of Science and Technology, Universiti Kebangsaan Malaysia, 43600 Bangi, Selangor, Malaysia

M. H. Mohd et al. (eds.), *Modelling, Simulation and Applications of Complex Systems*, Springer Proceedings in Mathematics & Statistics 359,
https://doi.org/10.1007/978-981-16-2629-6_6

and environmental problems [5]. Not only GAMA 1.8 is a free software, it also has its own community in GitHub which users can contribute or troubleshoot for problems encountered in GAMA 1.8. New users can download GAMA 1.8 through this website: https://gama-platform.github.io/download, join its community at https://github.com/gama-platform, and refer to the tutorials and its documentation at https://gama-platform.github.io/wiki/Home. To showcase GAMA 1.8 capabilities in agent-based modelling, this paper will provide a step-by-step tutorial on simulating a stochastic discrete time Markov chain ABM as an example.

2 Installation of GAMA 1.8

To install GAMA 1.8, please go to this website: https://gama-platform.github.io/download. GAMA 1.8 is available in three operating environments, namely Windows, MacOS and Linux. It is highly recommended to download GAMA 1.8 with JDK (Java Development Kit) as it includes everything necessary to install and run the software. Users are advised to download the PDF documentation as well, which is located below the GAMA 1.8's Windows installation option on the installation website. The documentation covers almost everything in GAMA 1.8, from installation instructions to tutorials. Users are reminded that GAMA 1.8 requires an approximately 540 MB of disk space and a minimum of 4 GB of RAM. Users can launch GAMA 1.8 after extracting the zip installation file to their desired environment in their laptop or desktop. If there are any difficulties in getting the GAMA 1.8 installed, please refer to the PDF documentation or approach the GAMA community in the GAMA google group (https://groups.google.com/forum/#!forum/gama-platform) for assistance.

3 Why GAMA?

In recent years, there are different agent-based modelling systems that have been developed using various such as Agent Factory [6], AgentScape [7], GAMA [1], JADE [8], JASON [9], NetLogo [10], MATLAB [11–13] and etc. A survey conducted by Feraud and Galland [14] showed that GAMA scored the highest compared to NetLogo, JASON, JADE and Janus based on the opinions of 24 Universite de technologie de Belfort-Montbeliard (UTBM) students, whom had assessed the platforms through criteria such as the platform's inter-agent communication, code extensibility, graphic support for development and implementation, and support for agent environment. Similarly, GAMA was ranked quite well among 24 agent-based software in terms of its usability, operating ability, and pragmatics in a survey by Kravari and Bassiliades [15]. Since then, the GAMA development team has worked to further enhance GAMA; and now in GAMA 1.8, researchers can model with a powerful management of geographical data, flexible visualization tools and the capacity to carry out simulations with hundreds of thousands of agents in an easy-to-use agent-based language environment [1].

4 Simulating an Agent-Based Model in GAMA 1.8

4.1 Example Model

As an example, this paper will simulate a compartmental mathematical model in the context of eco-epidemiology. The model is based on [16], which studied the transmission of Sin Nombre Virus (SNV) among the deer mouse in the presence of an additional non-host species in small population level. SNV is a species of hantavirus which can cause the deadly hantavirus pulmonary syndrome (HPS) in humans [17]. In the Four Corners region, the SNV is primarily hosted by the deer mouse [18]. Studies such as [19–21] showed that the biodiversity (presence of species other than the deer mouse) has an effect on the SNV transmission in the deer mouse population. As observed by Luis et al. [22], the biodiversity effect does not necessarily guarantee a decrease in SNV transmission; and at times, it may worsen the SNV transmission. When the presence of a species other than the deer mouse increases the SNV infected population, that species is termed as an amplification agent. On the other hand, a species is termed as a dilution agent if its presence decreases the SNV infected population [23]. The work of [16] and [24] have compared the effect of a dilution agent with an amplification agent on SNV transmission among deer mouse from the stochastic and deterministic perspective. Interested readers can refer them to better understand the eco-epidemiological implications of the results as they will not be widely discussed in this paper. This paper only aims to provide a step-by-step manual on how to simulate the mathematical model presented in [16] and [24] in GAMA 1.8.

The mathematical model in [16] and [24] which describes the effect of an amplification agent or a dilution agent on SNV transmission among deer mouse population is given as such:

$$
\begin{aligned}
\frac{\mathrm{ds}}{\mathrm{dt}} &= N\left[b_1 - ar_1\left(\frac{N+q_1Z}{K_1}\right)\right] - S\left[d_1 + r_1(1-a)\left(\frac{N+q_1Z}{K_1}\right)\right] - \gamma(1+\delta Z)SI \\
\frac{dI}{dt} &= \gamma(1+\delta Z)SI - I\left[\mu + d_1 + r_1(1-a)\left(\frac{N+q_1Z}{K_1}\right)\right] \\
\frac{dZ}{dt} &= r_2\left[1 - \frac{Z+q_2Z}{K_2}\right]Z
\end{aligned}
\tag{1}
$$

where $N = S + I$ is the total population density of the deer mouse per hectare, S is the population density of the susceptible deer mouse per hectare, I is the population density of the infected deer mouse per hectare, and Z is the population density of a non-host individuals per hectare, which can either act as an amplification agent or a dilution agent. Basically, the first two terms in $\frac{dS}{dt}$ describes the birth and death of susceptible deer mouse per hectare per month; while the last term $\gamma(1+\delta Z)SI$ describes the conversion rate of susceptible deer mouse becoming infected. Since the SNV can only be transmitted horizontally (through aggressive encounters) [25], the SNV infected deer mouse population can only be increased through $\gamma(1+\delta Z)SI$. The second term in $\frac{\mathrm{d}I}{\mathrm{d}t}$ describes the removal rate (through death) of SNV infected deer mouse from the population. The $\frac{dZ}{dt}$ describes the logistic growth rate of the non-host individuals. The deer mouse and non-host species affects each other through the interspecific competition strength, q_1 and q_2. δ represents the effect of the non-host on the SNV transmission rate depending on whether the non-host is an amplification agent or dilution agent. In this example, we shall showcase the simulation whereby the non-host is an amplification agent. The descriptions for the rest of the parameters can be found in Table 1.

Discrete-time Markov chain approach was utilized to simulate model (1). It is assumed that each individual has a probability in executing one of the three events, namely "reproduce", "die" or "do nothing", in every small time step, Δt. When an individual executes the "reproduce" event in the $[t, t + \Delta t)$ interval, a new individual of the same category will be created in the system. For the "die" event, the individual would be permanently deleted from the system. Finally, nothing will happen to the individual if it executes the "do nothing" event in the $[t, t + \Delta t)$ interval. Thus, the

Table 1 Descriptions and parameter values for model (1)

Parameters	Descriptions	Parameter value
K_1	The carrying capacity of the deer mouse (per hectare)	20^*
K_2	The carrying capacity of the non-host (per hectare)	15^*
r_1	The net density dependent growth rate for the deer mouse, $b_1 - d_1$ (per month)	$3.1496 \times 10^{-1*}$
b_1	The density dependent birth rate for the deer mouse (per month)	0.315^*
μ	The disease induced mortality rate (per month)	0.085^*
d_1	The death rate of the deer mouse (per month)	$3.66 \times 10^{-5*}$
q_1	The interspecific pressure exerted by the non-host onto the deer mouse	0.4^*
a	The proportion of density dependence due to density dependence of the deer mouse in birth rates	0.614^*
r_2	The net density dependent growth rate for the non-host, $b_2 - d_2$ (per month)	$3.9996 \times 10^{-1*}$
b_2	The density dependent birth rate for the non-host (per month)	0.4^*
d_2	The death rate of the non-host (per month)	$4.0 \times 10^{-5*}$
q_2	The interspecific pressure exerted by the deer mouse onto the non-host	0.3^*
γ	Initial disease transmission rate of the deer mouse without the influence of additional species (hectare per month)	0.0130^*
δ	Proportional constant of the disease transmission rate with the non-host density (hectare)	0.0543^*
Δt	Small time step (per month**)	0.001

*The values were based on [24].
**The value was based on the observational data from [22].

transition probabilities of each susceptible, infected and non-host are depicted as below:

1. $Pr\{\text{a } S \text{ reproduces a new } S \text{ in } [t, t+\Delta t)\} = \left[r_1 + \frac{I}{S}(b_1)\right]\Delta t$
2. $Pr\{\text{a single } S \text{ dies in } [t, t+\Delta t)\} = \left\{r_1\left(\frac{N+q_1 Z}{K_1}\right)\left[1 - a\left(\frac{I}{S}\right)\right] + \gamma(1+\delta Z)I\right\}\Delta t$
3. $Pr\{\text{a single } S \text{ does nothing in } [t, t+\Delta t)\} = 1 - \left[r_1 + \frac{I}{S}(b_1)\right]\Delta t - \left\{r_1\left(\frac{N+q_1 Z}{K_1}\right)\left[1 - a\left(\frac{I}{S}\right)\right] + \gamma(1+\delta Z)I\right\}\Delta t$
4. $Pr\{\text{a } I \text{ reproduces a new } I \text{ in } [t, t+\Delta t)\} = \left[\gamma(1+\delta Z)S\right]\Delta t$
5. $Pr\{\text{a single } I \text{ dies in } [t, t+\Delta t)\} = \left[\mu + d_1 + r_1(1-a)\left(\frac{N+q_1 Z}{K_1}\right)\right]\Delta t$
6. $Pr\{\text{a single } I \text{ does nothing in } [t, t+\Delta t)\} = 1 - \left[\gamma(1+\delta Z)S\right]\Delta t - \left[\mu + d_1 + r_1(1-a)\left(\frac{N+q_1 Z}{K_1}\right)\right]\Delta t$
7. $Pr\{\text{a } Z \text{ reproduces a new } Z \text{ in } [t, t+\Delta t)\} = r_2\Delta t$
8. $Pr\{\text{a single } Z \text{ dies in } [t, t+\Delta t)\} = r_2\left(\frac{Z+q_2 N}{K_2}\right)\Delta t$
9. $Pr\{\text{a single } Z \text{ does nothing in } [t, t+\Delta t)\} = 1 - r_2\Delta t - r_2\left(\frac{Z+q_2 N}{K_2}\right)\Delta t$

Figure 2 displays the steps taken to implement the ABM in GAMA 1.8. For more information on model (1), readers can refer to [24].

4.2 *Simulating in GAMA 1.8*

Creating a New Project File for Simulation

To create the simulation in GAMA 1.8, it is recommended that the user create a new project file by following the steps below:

1. First, the user needs to navigate his or her cursor to "File". Then, select "New" from the drop-down option, and finally click on "Gama Project" as shown in Fig. 1.
2. A dialog box as shown in Fig. 3 will pop out and the user can name the project (in this example, the project was named as "SNV_Model"). Make sure the "Create a new model file" option is ticked before clicking on "Finish".
3. After that, a dialog box on the model file similar to Fig. 4 will prompt, which requires the user to input the file and model name or change the author under the "File name", "Model name" and "Author" input box respectively. The user has the option to give a description to the model file in the "Model description" input box if needed. In this example, the "File name" was named as "Modified_PA.gaml" (make sure the file name ends with the.gaml extension) and the "Model name" was given the similar name, which was "ModifiedPA". Click on "Finish" after everything has been done.
4. The user should get a similar result as Fig. 5. It can be seen that the "SNV Model" project is created in the "User Models" folder and the Modified_PA.gaml is

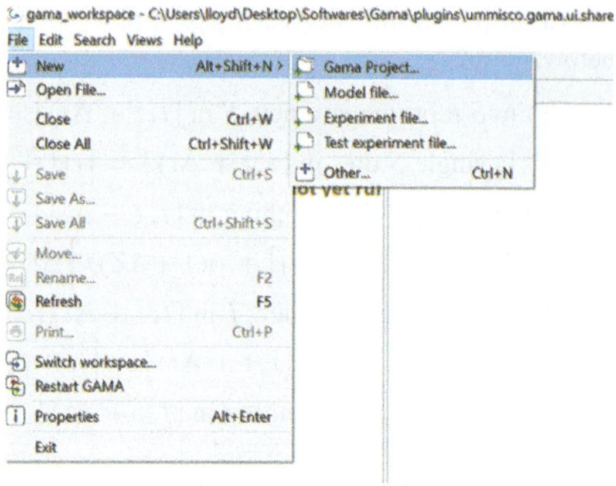

Fig. 1 Navigation to create a new project file in GAMA 1.8

created under the sub "models" folder in "SNV Model". On the right of the screen, the user can find the Modified_PA.gaml code sheet for model input.

Basics of Simulating an ABM in GAMA 1.8

The coding structure for simulating a model in GAMA 1.8 can be generally categorized into three parts, namely the "global", "species" and "experiment". A brief introduction for these categories is given below:

1. "global"—Refers to the world agent. Anything declared under "global" can be accessed anywhere in the code. There can only be one "global" for each model. Normally, common variables (to be accessed in other part of the code) are declared along with initial conditions for creating "species" agents. It creates an environment for the "species" agents to interact.
2. "species"—Refers to individual agent. User can specify the characteristics of their desired agent in this section. Multiple "species" agents can exist in a model. Variables declared within a "species" cannot be accessed by other "species". The "species" agents will act according to the characteristics specified after the simulation is played.
3. "experiment"—Refers to simulation to be executed. User can specify the output to be displayed after executing the simulation. A model can have multiple "experiment" but GAMA can only run one "experiment" at a time.

Additional noteworthy statement in GAMA 1.8 were also briefly discussed as below:

1. "init"—Refers to the initial condition statement of an agent. It can only be specified once inside each "global", "species" and "experiment". When the simulation is first executed, the "init" statement would first be executed and would not be executed in the next cycle.
2. "reflex"—Refers to the statement which will be executed at each cycle in a simulation. Multiple "reflex" can be specified within "global", "species" and "experiment".

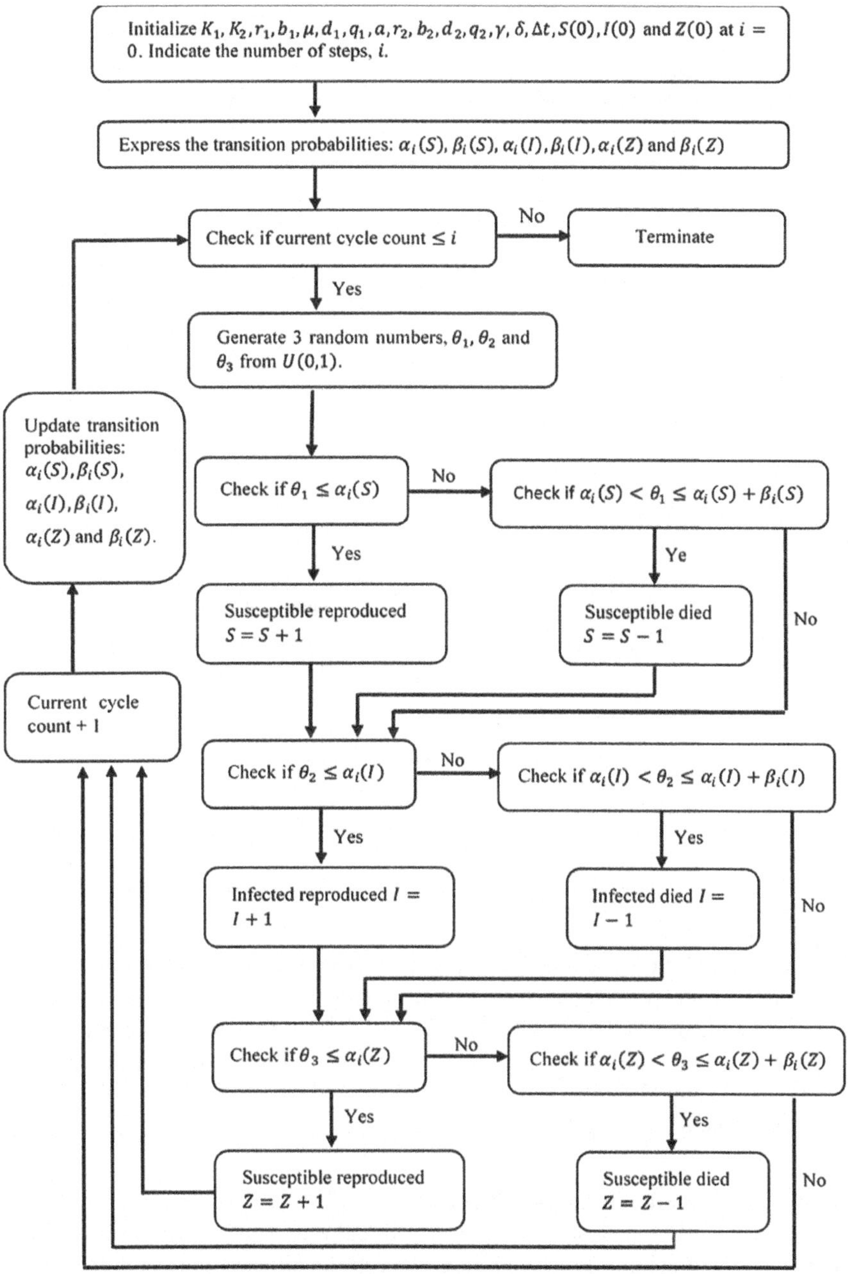

Fig. 2 Flowchart for simulating the stochastic ABM in GAMA with α_i = probability of executing "reproduce" event and β_i = probability of executing "die" event in the $[t, t + \Delta t)$ interval, where $i = 1$(Susceptible), 2(Infected), 3(Non-host)

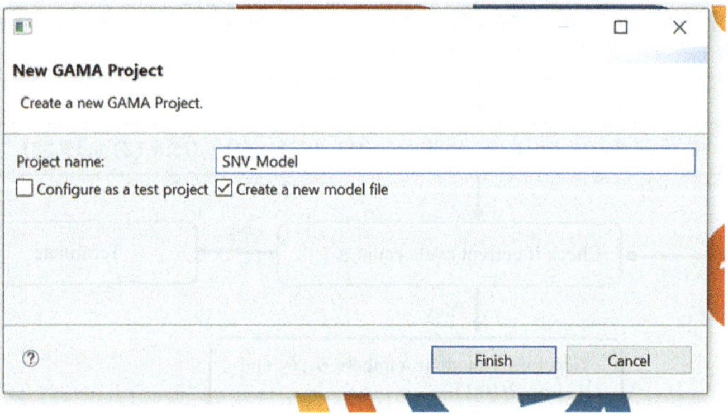

Fig. 3 Dialog box for new project file

Fig. 4 Dialog box for new model file

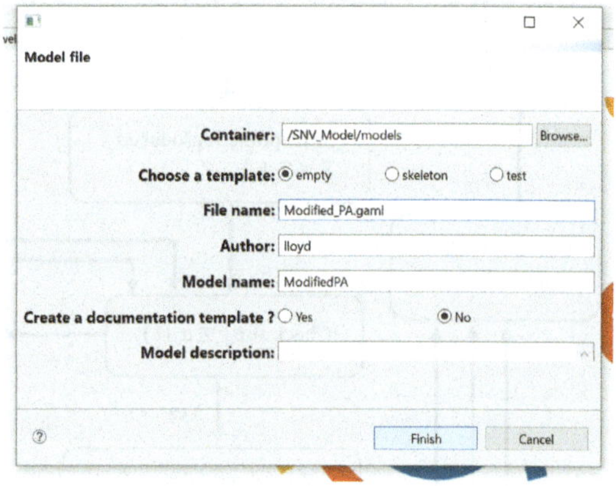

3. "action"—Refers to a function or procedure run by an instance of species. It can return a value provided the type of return value is specified. Unlike the "reflex" statement, the "action" function will only execute when it is called through the statement "do" or by assigning a variable to the "action" function which returns value. Multiple "action" can be specified within "global", "species" and "experiment".

"Global" Section for Example Model

The variables listed in Table 1 were declared and initialized. Species tracker were created to track the number of stochastic susceptible, infected and non-host individuals in the model. Under the "init" statement, the stochastic susceptible, infected and non-host species were declared and created with the specified initial number along with the declaration of the deterministic model (in this case, the deterministic model is considered as a "species"). A "reflex" statement was created to save the number of individuals at each cycle in the specified file path directory. An additional "reflex'

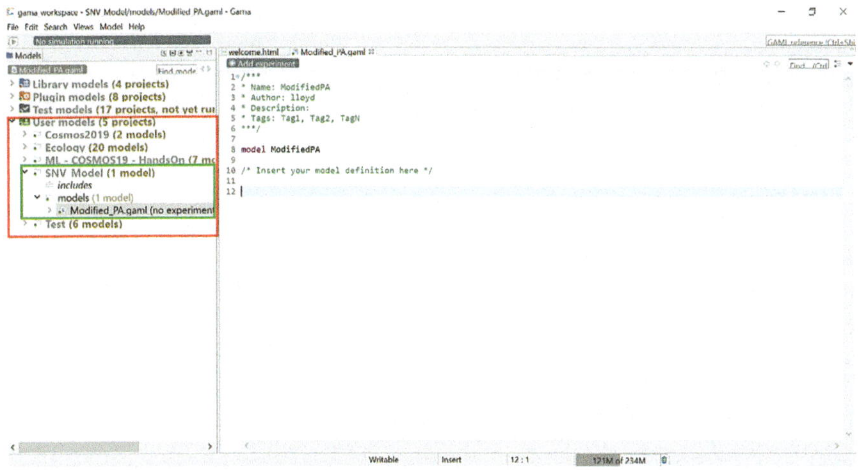

Fig. 5 The display after a new project file and model file are created. The red box depicts the "User models" folder and under this folder, the new project file was created (shown in the green box)

statement was created to automatically stop the simulation at a specified cycle/time. The code for "global" is depicted as below:

```
global{
  int init_S <- 10;//initial S population
  int init_I <- 10;//initial I population
  int init_Z <- 10;//initial Z population
  float K <- 20.0; //carrying capacity of deer mouse
  float b <- 0.315;//birth rate of deer mouse
  float d <- 0.0000366;  //death rate of deer mouse
  float a <- 0.614;  //proportion of density dependence
  float theta <- 0.0543; //influence of amplification
agent
  float mu <- 0.085; //disease induced mortality rate
  float gamma <- 0.0130; //initial disease transmission
rate
  float K_Z <- 15.0; //carrying capacity of non-host
  float b_Z <- 0.4;//birth rate of non-host
  float d_Z <- 0.00004;  //death rate of non-host
  float q_Z_onto_SI <- 0.4;  //interspecific competition
strength of non-host
  float q_SI_onto_Z <- 0.3;  //interspecific competition
strength of deer mouse
  float step <- 0.001; //time step
  float desired_stop_time <- 30.0;
```

```
  int loop_num <- 100; //number of simulations

  int nb_S -> length(S); //number of S in the system
  int nb_I -> length(I); //number of I in the system
  int nb_Z -> length(Z); //number of Z in the system

  init{
    create S number: init_S; //create a number of init_S
stochastic species S in the model
    create I number: init_I; //create a number of init_I
stochastic species I in the model
    create Z number: init_Z; //create a number of init_Z
stochastic species Z in the model
    create SIZ_ODE number: 1; //create a deterministic
model agent
  }

  //save the simulations into specific datafile. The file
is changed for every new experiment.
  reflex updateSave {
    save (""+cycle+" "+first(SIZ_ODE).S_ODE+"
"+first(SIZ_ODE).I_ODE+" "+first(SIZ_ODE).Z_ODE) to:
"C:/file_path_directory/filename.txt" rewrite: false;
    save (""+cycle+" "+nb_S+" "+nb_I+" "+nb_Z) to: "C:/
file_path_directory/different_filename.txt" rewrite:
false;
  }

  //stop the simulations at the desired time
  reflex pausing when: cycle=desired_stop_time/step{
    do pause;
  }
}
```

"Species" Section for Example Model
For this example, four "species" agents were created which were the stochastic susceptible agent, stochastic infected agent, stochastic non-host agent and deterministic model agent. In each of the stochastic agents, two "action" statements were created. The "update_prob" action statement is to generate a number between 0 and 1, update the transition probabilities presented at the end of Sect. 4.1 and return these values to the call statement. The "update_actions" action statement is to compare the randomly generated number between 0 and 1 to the transition probabilities of the agent. The agent will then execute the specified event. A "reflex" statement was created to execute and store the returned values of "update_prob". Then, these stored values are passed to "update_actions" which then helps the agent to decide the event to execute. For the deterministic model agent, the deterministic equations were specified and were solved with Runge-Kutta 4th Order method [26]. The following depicts the code:

```
species S{
  float prob_birth;//transition probability of "Repro-
duce" event for S
  float prob_death;//transition probability of "Die"
event for S
  float prob;  //variable to store generated random num-
ber from U(0,1)
  list temp_prob <- [prob,prob_birth,prob_death];//Store
prob, prob_birth and prob_death in a list

  //The function that updates the transition probability
  action update_prob (list passed_prob){
    passed_prob[0] <- rnd(1.0);  //generate a random num-
ber for S at every cycle
    passed_prob[1] <- (b-d+b*nb_I/nb_S)*step;
    passed_prob[2] <- ((b-
d)*(1+a*nb_I/nb_S)*(float(nb_S+nb_I)+q_Z_onto_SI*nb_Z)/K
+ gamma*(1+theta*nb_Z)*nb_I)*step;
    return passed_prob;  //return the list which contain
random number and transition probabilities
  }
```

```
  //The function that compares the random number to the
transition probabilities
  action update_actions (list passed_prob){
    float temp_probS <- passed_prob[0];
    float temp_birthS <- passed_prob[1];
    float temp_deathS <- passed_prob[2];

    //"Reproduce" event
    if (temp_probS <= temp_birthS){
      create species(self) number: 1;
    }
    //"Die" event
    else if ((temp_probS>temp_birthS) and
(temp_probS<=(temp_birthS+temp_deathS))) {
      do die;
    }
    //"Do nothing" event
    else {
      /*do nothing*/
    }
    }
    //Execute the update_prob action to update the transi-
  tion probabilities and pass them to the update_actions
  action to decide which event to execute at every time
  step
    reflex exeAction{
      temp_prob <- update_prob(temp_prob);  //executes the
  update_prob action and store the return list
      do update_actions(temp_prob); //executes the up-
  date_actions with the returned list values
    }
  }

  species I{
    float prob_birth;//transition probability of "Repro-
  duce" event for I
    float prob_death;//transition probability of "Die"
  event for I
    float prob;  //variable to store generated random num-
  ber from U(0,1)
```

```
  list temp_prob <- [prob,prob_birth,prob_death]; //Store
prob, prob_birth and prob_death in a list

  //The function that updates the transition probability
  action update_prob (list passed_prob){
    passed_prob[0] <- rnd(1.0); //generate a random num-
ber for I at every cycle
    passed_prob[1] <- gamma*(1+theta*nb_Z)*nb_S*step;
    passed_prob[2] <- (mu+d+(b-d)*(1-
a)*(float(nb_S+nb_I)+q_Z_onto_SI*nb_Z)/K)*step;
  }

  //The function that compares the random number to the
transition probabilities
  action update_actions (list passed_prob){
    float temp_probI <- passed_prob[0];
    float temp_birthI <- passed_prob[1];
    float temp_deathI <- passed_prob[2];

    //"Reproduce" event
    if (temp_probI <= temp_birthI){
      create species(self) number: 1;
    }
    //"Die" event
    else if ((temp_probI>temp_birthI) and
(temp_probI<=(temp_birthI+temp_deathI))) {
      do die;
    }
    //"Do nothing" event
    else {
      /*do nothing*/
    }
  }

  //Execute the update_prob action to update the transi-
tion probabilities and pass them to the update_actions
action to decide which event to execute at every time
step
  reflex exeAction{
    temp_prob <- update_prob(temp_prob); //executes the
update_prob action and store the return list
    do update_actions(temp_prob); //executes the up-
date_actions with the returned list values
  }
}
```

```
species Z{
  float prob_birth;//transition probability of "Repro-
duce" event for Z
  float prob_death;//transition probability of "Die"
event for Z
  float prob;  //variable to store generated random num-
ber from U(0,1)
  list temp_prob <- [prob,prob_birth,prob_death];//Store
prob, prob_birth and prob_death in a list

  //The function that updates the transition probability
  action update_prob (list passed_prob){
    passed_prob[0] <- rnd(1.0); //generate a random num-
ber for Z at every cycle
    passed_prob[1] <- (b_Z-d_Z)*step;
    passed_prob[2] <- ((b_Z-
d_Z)*(q_SI_onto_Z*float(nb_S+nb_I)+nb_Z)/K_Z)*step;
  }

  //The function that compares the random number to the
transition probabilities
  action update_actions (list passed_prob){
    float temp_probZ <- passed_prob[0];
    float temp_birthZ <- passed_prob[1];
    float temp_deathZ <- passed_prob[2];

    //"Reproduce" event
    if (temp_probZ <= temp_birthZ){
      create species(self) number: 1;
    }
    //"Die" event
    else if ((temp_probZ>temp_birthZ) and
(temp_probZ<=(temp_birthZ+temp_deathZ))) {
      do die;
    }
    //"Do nothing" event
    else {
      /*do nothing*/
    }
  }
```

```
  //Execute the update_prob action to update the transi-
tion probabilities and pass them to the update_actions
action to decide which event to execute at every time  .
step
  reflex exeAction{
    temp_prob <- update_prob(temp_prob);  //executes the
update_prob action and store the return list
    do update_actions(temp_prob); //executes the up-
date_actions with the returned list values
  }
}
//Generate the deterministic model
species SIZ_ODE{
  float S_ODE <- float(init_S);
  float I_ODE <- float(init_I);
  float Z_ODE <- float(init_Z);
  float t;
  float h <- 0.01;

  equation ODE{
    diff(S_ODE,t)=(S_ODE+I_ODE)*(b-a*(b-
d)*(S_ODE+I_ODE+q_Z_onto_SI*Z_ODE)/K)-S_ODE*(d+(b-d)*(1-
a)*(S_ODE+I_ODE+q_Z_onto_SI*Z_ODE)/K)-
gamma*(1+theta*Z_ODE)*S_ODE*I_ODE;
    diff(I_ODE,t)=gamma*(1+theta*Z_ODE)*S_ODE*I_ODE-
I_ODE*(mu+d+(b-d)*(1-
a)*(S_ODE+I_ODE+q_Z_onto_SI*Z_ODE)/K);
   diff(Z_ODE,t)=(b_Z-d_Z)*Z_ODE*(1-
(Z_ODE+q_SI_onto_Z*(S_ODE+I_ODE))/K_Z);
  }
  reflex solving{
    solve ODE method: rk4 step: h ;
  }
}
```

"Experiment" Section of the Example Model

To run a simulation, the user needs to create an "experiment" statement. Within this "experiment" statement, the user can opt to generate graphs and/or track the number of individual agents present at the current cycle. Multiple "experiment" statements can exist in a model but only one "experiment" can be executed at a time. As an example, the following display the code without any graphical outputs and the code with graphical outputs in GAMA 1.8.

1. The code without graphical output which simulates for 100 times with random seed number at each simulation.

```
experiment Base_value_amp type: gui{
  init{
    loop times: (loop_num-1){
      create simulation with: [seed::rnd(10000)];
    }
  }
}
```

2. The code which display a line graph which compares a single stochastic simulation to its deterministic counterpart with the number of stochastic individuals being displayed at each cycle.

```
experiment single_sim type: gui{
  output{
    monitor "Number of S" value: nb_S refresh: every(1
#cycle);
    monitor "Number of I" value: nb_I refresh: every(1
#cycle);
    monitor "Number of Z" value: nb_Z refresh: every(1
#cycle);
    display time_series refresh: every (1 #cycle){
      chart "Total Number of S and I" type: series{
        data "Number of S" value: nb_S color: #red;
        data "Number of I" value: nb_I color: #green;
        data "Number of Z" value: nb_Z color:#blue;
        data 'S_ODE' value: first(SIZ_ODE).S_ODE color:
#red;
        data 'I_ODE' value: first(SIZ_ODE).I_ODE color:
#green;
        data 'Z_ODE' value: first(SIZ_ODE).Z_ODE color:
#blue;
      }
    }
  }
}
```

3. The code which display a line graph which compares the mean of 100 stochastic simulations to its deterministic counterpart.

```
experiment multi_sim2 type: gui{
  init{
    loop times: (loop_num-1){
      create simulation with: [seed::rnd(10000),
theta::0.0];
    }
  }

  permanent{
    display mean_series{
      chart "Total Number of N" type: series{
        data "S Mean of " + loop_num +" simulations"
value: mean (simulations collect each.nb_S) color: #red;
        data "I Mean of " + loop_num +" simulations"
value: mean (simulations collect each.nb_I) color:
#green;
        data "Z Mean of " + loop_num +" simulations"
value: mean (simulations collect each.nb_Z) color: #blue;
        data 'S' value: first(SIZ_ODE).S_ODE color: #pink;
        data 'I' value: first(SIZ_ODE).I_ODE color: #gold;
        data 'Z' value: first(SIZ_ODE).Z_ODE color:
#black;
      }
    }
  }
}
```

Running the Simulations

After the "experiment" statements have been specified, a green play button with the respective "experiment" name will be shown in the row below the Modified_PA.gaml tab as shown in Fig. 6. By clicking on the desired "experiment" button, GAMA 1.8 will execute the respective simulation. For example, by clicking on the "single_sim" button, the user will be directed to the simulation page similar to the screen shown in Fig. 7. By clicking on the "green play" button on the top row, the simulation will then execute and automatically stopped after some time/cycles specified in the "global" section. The user should get a result similar to Fig. 8. (Note: the stochastic results will be different at each run.) If the user would like to save the graphical output, then the user can click on the "snapshot" button as shown in the red box in Fig. 8 and he or she can find the saved snapshot under the "snapshots" subfolder in the "models" folder as shown in Fig. 9. Another thing to note is that the simulations with graphical output will take a longer time to run compared to those without graphical output. Hence, the user should run the simulation without the graphical output to obtain the results in shorter time if his or her laptop/desktop is not powerful enough. The user can then graph the results from the save data files which can be found at the saved location specified in the "global" statement. For example, Fig. 10 was plotted to compare the averaged 100 stochastic simulations with the deterministic simulation. The averaged 100 stochastic simulations lines are less jagged and are more resemblance to the deterministic results compared to the plots in Fig. 8. The gap between the averaged stochastic and deterministic results is due to the inherent characteristic of the extinction probability in stochastic model [27].

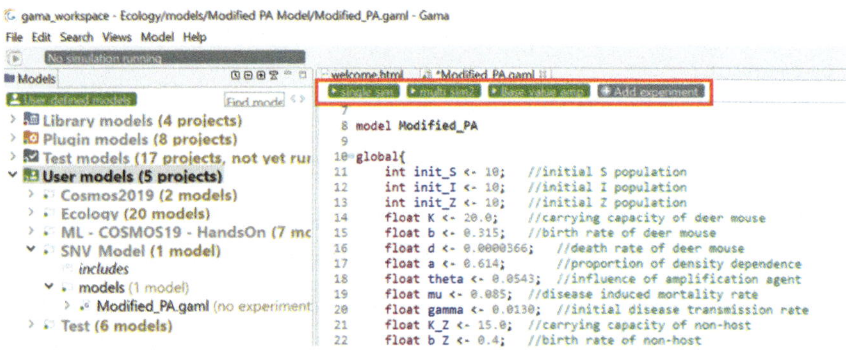

Fig. 6 The green boxes within the red box are the play button for the respective defined "experiment"

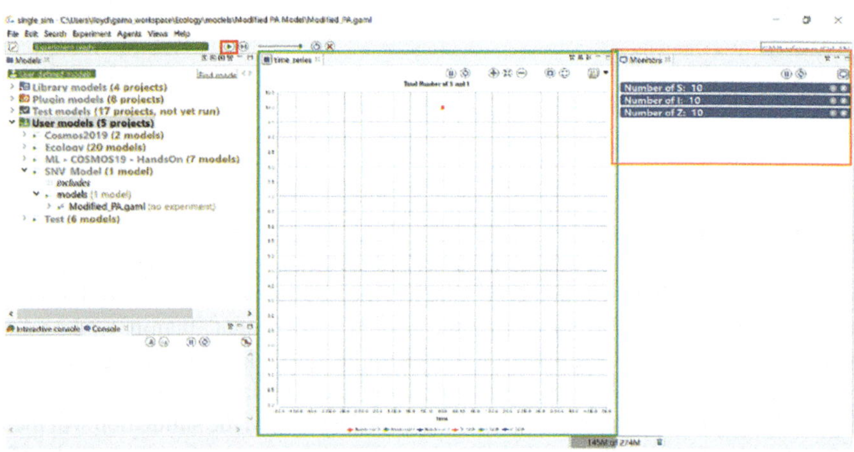

Fig. 7 The simulation page before execution. By clicking on the play button in the red box would start the simulation. The green box depicts the simulation graph while the orange box depicts the tracking monitor

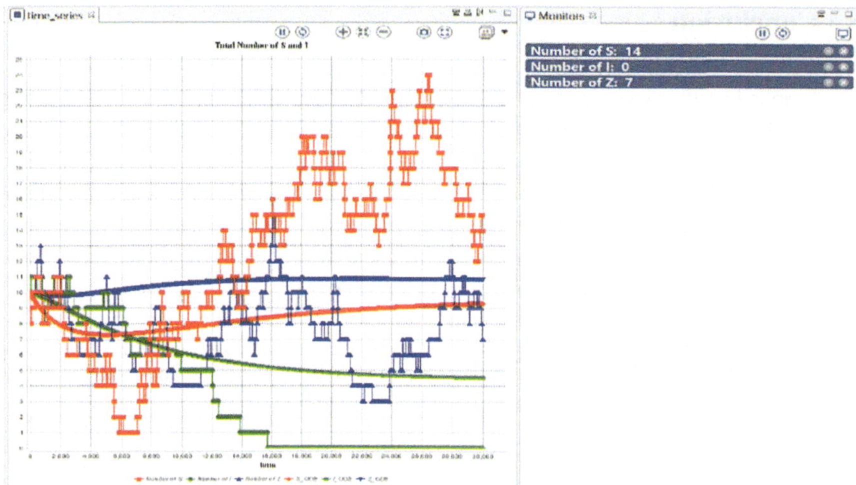

Fig. 8 Simulation results of the "single_sim" experiment. The smooth red, blue and green lines are the deterministic results while the jagged red, blue and green lines are the stochastic results in a single run

Fig. 9 The saved snapshot of the simulated graph can be found under the folder displayed in the red box

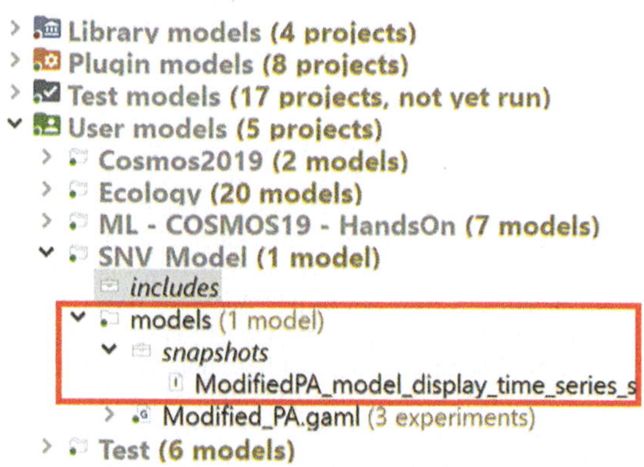

Fig. 10 A plot of an averaged 100 stochastic simulations with the deterministic simulation. The solid red, green and blue lines are the deterministic result while the dotted red, green and blue lines are the mean results of 100 stochastic simulations

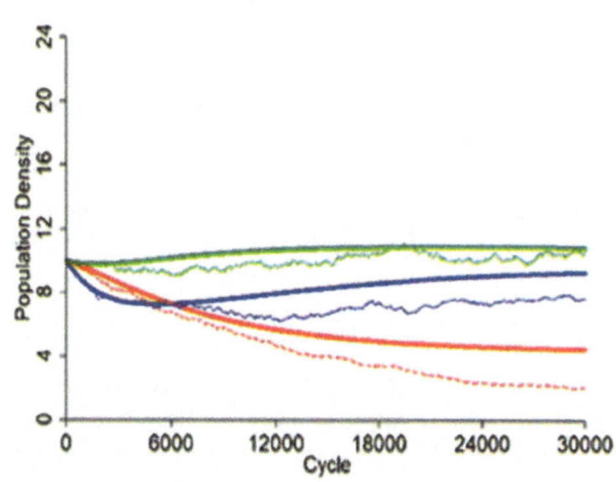

5 Conclusion

This paper has discussed on the techniques and codes to simulate an ABM using GAMA 1.8. The codes are provided with comments to aid the understanding of the contents. The example provided serves as a good reference for new users to simulate ABM in GAMA 1.8. For more tutorials and examples, users are asked to refer the GAMA website or explore more examples under the "Library Models". We hope that this paper will be beneficial to users in research especially to those early-stage postgraduate students in simulating their desired ABM.

Acknowledgements The authors would like to thank the School of Mathematical Sciences and the Universiti Sains Malaysia (USM) for the support. Mohd Hafiz Mohd is supported by the USM Fundamental Research Grant Scheme (FRGS) No. 203/PMATHS/6711645.

References

1. Taillandier, P., Gaudou, B., Grignard, A., Huynh, Q.-N., Marilleau, N., Caillou, P., Philippon, D., Drogoul, A.: Building, composing and experimenting complex spatial models with the GAMA platform. Geoinformatica, **23**(2), 299–322 (2019). https://doi.org/10.1007/s10707-018-00339-6
2. Touhbi, S., Babram, M.A., Nguyen-Huu, T., Marilleau, N., Hbid, M.L., Cambier, C., Stinckwich, S.: Time Headway analysis on urban roads of the city of Marrakesh. Procedia Comput. Sci. **130**, 111–118 (2018). https://doi.org/10.1016/j.procs.2018.04.019
3. Tsagkis, P., Photis, Y.N.: Using Gama platform and Urban Atlas Data to predict urban growth. The case of Athens. 13th International Conference of the Hellenic Geographical Society (2018). http://www.hellenicgeosociety.org/en/system/files/tsagkis_urban_growth_gama_athens.pdf. Accessed 1 Sept 2020
4. Galimberti, A., Alyokhin, A., Qu, H., Jason, R.O.S.E.: Simulation modelling of potato virus Y spread in relation to initial inoculum and vector activity. J. Integr. Agric. **19**(2), 376–388 (2020). https://doi.org/10.1016/S2095-3119(19)62656-0
5. Rodrique, K., Tuong, H., Manh, N., An agent-based simulation for studying air pollution from traffic in urban areas: the case of Hanoi City. Int. J. Adv. Comput. Sci. Appl. **10**(3) (2019). https://doi.org/10.14569/ijacsa.2019.0100376
6. Collier, R.: Debugging agents in agent factory. In: Bordini, R.H., Dastani, M., Dix, J., Seghrouchni, A.E.F. (eds.) Programming Multi-Agent Systems. ProMAS 2006. Lecture Notes in Computer Science, vol. 4411. Springer, Berlin, Heidelberg (2007). https://doi.org/10.1007/978-3-540-71956-4_14
7. Brazier, F., Mobach, D., Overeinder, B., Splunter, S.V., Steen, M.R., Wijngaards, N.: AgentScape: middleware, resource management, and services (2002)
8. Su, C.J., Wu, C.Y.: JADE implemented mobile multi-agent based, distributed information platform for pervasive health care monitoring. Appl. Soft Comput. **11**(1), 315–325 (2011). https://doi.org/10.1016/j.asoc.2009.11.022
9. Bordini, R.H., Hübner, J.F., Wooldridge, M.: Programming Multi-Agent Systems in AgentSpeak using Jason (Wiley Series in Agent Technology). John Wiley & Sons, Inc., Hoboken, NJ, USA (2007). https://doi.org/10.1002/9780470061848
10. Wilensky, U.: NetLogo. http://ccl.northwestern.edu/netlogo/. Center for Connected Learning and Computer-Based Modeling, Northwestern University, Evanston, IL (1999)

11. Mohd, M.H., Murray, R., Plank, M.J., Godsoe, W.: Effects of dispersal and stochasticity on the presence–absence of multiple species. Ecol. Model. **342**, 49–59 (2016). https://doi.org/10.1016/j.ecolmodel.2016.09.026
12. Mohd, M.H.B.: Modelling the presence-absence of multiple species. PhD Thesis. University of Canterbury (2016)
13. Mohd, M.: How can modelling tools inform environmental and conservation policies?. Int. J. Eng. Technol. **7**(4.28), 333–337 (2018). https://doi.org/10.14419/ijet.v7i4.28.22610
14. Feraud, M., Galland, S.: First comparison of SARL to other agent-programming languages and frameworks. Conference: international workshop on agent-based modeling and applications with SARL (SARL 2017). Procedia Comput. Sci. **109**, 1080–1085 (2017). https://doi.org/10.1016/j.procs.2017.05.389
15. Kravari, K., Bassiliades, N.: A survey of agent platforms. J. Artif. Soc. Soc. Simul. **18**(1), 11 (2015). https://doi.org/10.18564/jasss.2661
16. Lee, L.W.F., Mohd, H.M.: Modelling the Dilution and Amplification Effects on Sin Nombre Virus (SNV) in Deer Mouse in GAMA 1.8 (2020). Manuscript submitted for publication
17. Jiang, W., Wang, P.Z., Yu, H.T., Zhang, Y., Zhao, K., Du, H., Bai, X.F.: Development of a SYBR Green I based one-step real-time PCR assay for the detection of Hantaan virus. J. Virol. Methods **196**, 145–151 (2014). https://doi.org/10.1016/j.jviromet.2013.11.004
18. Jonsson, C.B., Figueiredo, L.T.M., Vapalahti, O.: A global perspective on hantavirus ecology, epidemiology, and disease. Clin. Microbiol. Rev. **23**(2), 412–441 (2010). https://doi.org/10.1128/CMR.00062-09
19. Dizney, L.J., Ruedas, L.A.: Increased host species diversity and decreased prevalence of Sin Nombre Virus. Emerg. Infect. Dis. **15**(7), 1012–1018 (2009). https://doi.org/10.3201/eid1507.081083
20. Suzan, G., Marce, E., Giermakowski, J.T., Mills, J.N., Ceballos, G., Ostfeld, R.S., Armien, B., Pascale, J.M., Yates, T.L.: Experimental evidence for reduced rodent diversity causing increased hantavirus prevalence. PLoS ONE **4**(5), (2009). https://doi.org/10.1371/journal.pone.0005461
21. Dearing, M.D., Clay, C., Lehmer, E., Dizney, L.: The roles of community diversity and contact rates on pathogen prevalence. J. Mammal. **96**(1), 29–36 (2015). https://doi.org/10.1093/jmammal/gyu025
22. Luis, A.D., Kuenzi, A.J., Mills, J.N.: Species diversity concurrently dilutes and amplifies transmission in a zoonotic host-pathogen system through competing mechanisms. Proc. Natl. Acad. Sci. **115**(31), 7979–7984 (2018). https://doi.org/10.1073/pnas.1807106115
23. Keesing, F., Holt, R.D., Ostfeld, R.S.: Effects of species diversity on disease risk. Ecol. Lett. **9**, 485–498 (2006). https://doi.org/10.1111/j.1461-0248.2006.00885.x
24. Lee, L.W.F., Mohd, M.H.: The effects of amplification and dilution agent in transmitting Sin Nombre virus (SNV) in deer mouse population. Malays. J. Ind. Applied Math. (MATEMATIKA) **36**(2), 85–98 (2020). https://doi.org/10.11113/matematika.v36.n2.1244
25. Botten, J., Mirowsky, K., Kusewitt, D., Ye, C., Gottlieb, K., Prescott, J., Hjelle, B.: Persistent Sin Nombre virus infection in the deer mouse (Peromyscus maniculatus) model:sites of replication and strand-specific expression. J. Virol. **77**(2), 1540–1550 (2003). https://doi.org/10.1128/jvi.77.2.1540-1550.2002
26. Cheever, E.: Fourth Order Runge-Kutta. https://lpsa.swarthmore.edu/NumInt/NumIntFourth.html. Accessed 1 Sept 2020
27. Allen, L.J., McCormack, R.K., Jonsson, C.B.: Mathematical models for hantavirus infection in rodents. Bull. Math. Biol. **68**(3), 511–524 (2006). https://doi.org/10.1007/s11538-005-9034-4

Deterministic Modelling and Simulation of Complex Systems

Mathematical Analysis of Fasciola Epidemic Model with Treatment and Quarantine

Mouhamadou Diaby, Oumar Diop, Estelle Nassouri, Abdou Sène, and Mariama Sène

Abstract Fasciola (hepatica and gigantica) is a common liver fluke that infects cattle, causing disease and considerable production losses and might be transmitted to humans. Current control methods rely on drugs designed to kill parasites. Epidemiological modelling can be a helpful instrument for evaluating parasite control strategies. In the present paper, a mathematical model of Fasciola is established as a multi-scale 9 dimensional system of ODEs that includes the intermediate host and the larval development according to the life cycle of the disease. The model includes quarantine as a control strategy. The main purpose of this work is to establish a more realistic new model of the disease and observe the role of quarantine strategy in such systems. The model is shown to be well-posed. The disease-free equilibrium is locally asymptotically stable whenever the basic reproduction number is less than unity and unstable otherwise. Numerical simulations have been performed to show the impact of certain parameters on the spread of the disease and to confirm the analytical results of the model.

Keywords Quarantine · Fasciola · Epidemic model · Reproduction number · Stability

M. Diaby (✉)
LANI, Université Virtuelle du Sénégal, Dakar, Senegal
e-mail: mouhamadou.diaby@uvs.edu.sn

O. Diop · A. Sène
PIED (Pôle d'Innovation et d'Expertise pour le Développement), Université Virtuelle du Sénégal, Dakar, Senegal
e-mail: oumar.diop@uvs.edu.sn

A. Sène
e-mail: abdou.sene@uvs.edu.sn

E. Nassouri
LAMI, Université Joseph Ki-Zerbo, Ouagadougou, Burkina Faso

M. Sène
UFR AGRO, Université Gaston Berger de Saint-Louis, Saint Louis, Senegal
e-mail: mariama-sene.wade@ugb.edu.sn

M. H. Mohd et al. (eds.), *Modelling, Simulation and Applications of Complex Systems*, Springer Proceedings in Mathematics & Statistics 359,
https://doi.org/10.1007/978-981-16-2629-6_7

1 Introduction

Fascioliasis is a parasitic disease caused by two species of the flatworm Fasciola (F. hepatica and F. gigantica). It is a global disease that occurs mainly in domestic animals and occasionally in humans when consuming metacercariae-contaminated uncooked vegetables or drinking infected water. Human infections are reported in very rainy areas, in places with inadequate drainage and in irrigated pastures. According to an estimation by WHO, in 2017 there were around 2.4 million infected people in the world and nearly 100 millions are at risk of infection [1–3]. Fasciola disease is found throughout the world, it is most common in some Asian and African countries particularly in West Africa (Benin, Ghana, Mali, Niger, Nigeria, Senegal, etc.) [4–6]. In Senegal, cases of human fascioliasis are reported by Ka et al. [7] and Seydi et al. [8]. The patient diagnosed by Seydi et al. using the ELISA method and coprology declares she was consuming wild plants. Fascioliasis disease has a serious economic impact: loss of quality and quantity of milk production, anaemia, lethargy, liver condemnation, etc. [9–11]. Losses in animal productivity due to fascioliasis is estimated at over US$ 3.2 billion per annum [9]. Both animals and humans are orally infected with encysted larvae (metacercariae) attached on vegetables or floated on water. The characteristic symptoms of fascioliasis disease are epigastric pain, upper abdominal pain, eosinophilia (high number of white blood cells), diarrhea, vomiting, nausea, fever and arthralgia [12].

The causal agents are the trematodes Fasciola hepatica (also known as the common liver fluke or the sheep liver fluke) and Fasciola gigantica. They are large liver flukes (F. hepatica: up to 30 mm by 15 mm; F. gigantica: up to 75 mm by 15 mm), primarily found in domestic and wild ruminants (their main definitive hosts), but are also causal agents of fascioliasis in humans. Although F. hepatica and F. gigantica are distinct species, "intermediate forms" that are thought to represent hybrids of the two species have been found in parts of Asia and Africa where both species are endemic. These forms usually have intermediate morphologic characteristics (e.g. overall size, proportions), possess genetic elements from both species, exhibit unusual ploidy levels (often triploid), and do not produce sperm. Further research into the nature and origin of these forms is ongoing.

The life cycles of the liver fluke and its intermediate host are illustrated in Fig. 1. Fasciola hepatica's life begins with eggs. If immersed in fresh water, the eggs develop and hatch. When it hatches, a small larva named miracidium appears. With her cilia, the miracidium swims looking for the snail (the intermediate host of the parasite). The miracidium penetrates in snails and immediately develops to its next form called sporocyst. The latter transforms into redia which, a few months later, allows the production of cercariae. These cercariae leave the snail to attach to the plant and form a cyst called metacercaria, which await being eaten by a mammalian host (a cow or a sheep). The fluke transforms into its adult form further when eaten. It spreads to the host's liver, feeding and producing eggs throughout its lifetime (which may take months or years). Eggs are excreted by the host, and if they fall into fresh water, they restart the cycle. Several studies are carried out [10, 13–15] in order to have

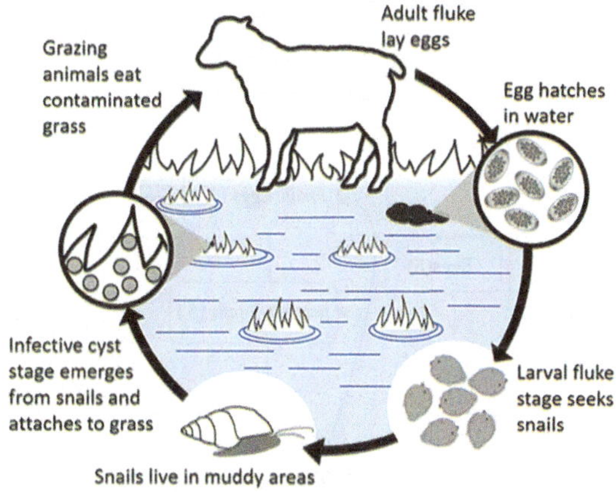

Fig. 1 Liver fluke (Gigantica/Hepatica) life cycle

a better understanding of the dynamics of transmission, to predict changes in the human and animal disease over time and to design optimal strategies for the control of the disease.

The main objective of this work is to develop a Fascioliasis transmission model incorporating both the final host and the parasite to describe the transmission dynamics of the disease and propose optimal public health strategy to fight against the disease. To do so, we use the SIR submodel for the dynamics of the hosts, with one of the variables being the quarantined hosts, in order to protect the cattle against contamination. The intermediate host (snails) of the parasite dynamics is a SI submodel, which is age-structured. This submodel's variables are eggs produced by the mature (uninfected snails), juvenile snails and mature snails. The model takes also into account the variation of parasites through the dynamics of Fasciola hepatic eggs and metacercariae. We finally obtain a mathematical model which consists of a 9 dimensional system of non linear differential equations. Then, we prove that the disease free equilibrium is stable if $\mathcal{R}_0 < 1$ and unstable if $\mathcal{R}_0 \geq 1$; and perform mathematical analysis and numerical simulations.

The paper is organised as follows. In Sect. 2, we present a mathematical formulation of the model among hosts and parasites. Some qualitative analysis are addressed in Sect. 3 including the positivity and boundedness of variables, the basic reproduction number and the stability analysis of the disease free equilibrium. Finally, Sect. 4 gives some numerical examples to confirm the theoretical results.

2 Model Formulation

Figure 2 shows the life cycles of the hepatic fluke and its intermediate host. The development of the worm is completed in 5 different stages: adult flukes in the biliary ducts of the mammalian host, eggs in the environment, miracidia in water, intra-molluscan

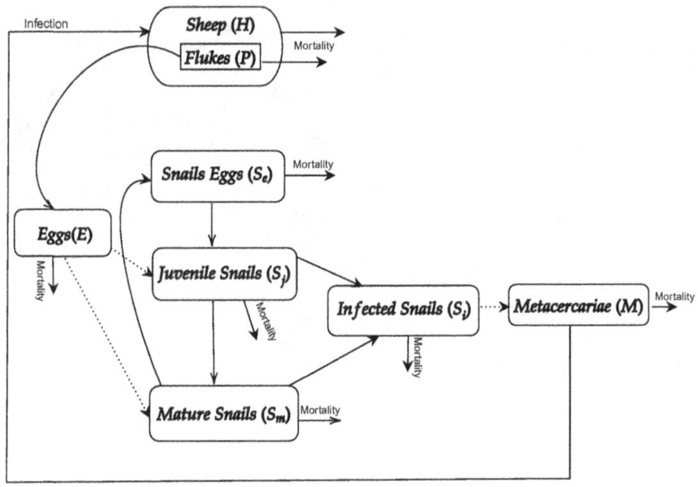

Fig. 2 Flow chart represents transmission routes and other processes modelled by our system

phases, and infectious cysts on aquatic vegetation. Between one subpopulation and another, the parasite flow through the system is led by a series of parameters such as death, birth and transmission rates. Transmission is like a movement among subgroups at the same time. We center around the improvement of a mathematical model for the free-living stages of Fasciola, utilising differential equations to depict different phase transitions. The accessibility of information on key parameters that impact on each step of the lifecycle is of great importance, because only the accessible reliable data are useful to be integrated into the model. The specificity of our modelling approach is that it takes into account the particular interactions among different development phases of the parasite and intermediate hosts, and incorporates quarantine strategy. Egg and metacercariae stages are explicitly taken into account in the modelling.

State Variables. We consider the parasite life-cycle (P) in many stages and locations: Susceptible cattle (H_s), Infected cattle (H_i), Quarantined cattle (Q_h), Fluke eggs on pasture (E), Snail eggs (S_e), Juvenile snails (S_j), Mature snails (S_m), Infected snails (S_i), Metacercariae on pasture (M).

Equations for Dynamics of Hosts. Changes in the total number of hosts (H) are given by the following equations:

$$\frac{\mathrm{d}H_s(t)}{\mathrm{d}t} = b_s\,(H_s(t) + \rho\,H_i(t)) - \mu_h\,H_s(t) - \beta\,H_s(t)\,M(t) + \eta_1\,H_i(t) + \eta_2\,Q_h(t),$$
$$\frac{\mathrm{d}H_i(t)}{\mathrm{d}t} = \beta\,H_s(t)\,M(t) - (\mu_h + \alpha_1)\,H_i(t) - (\eta_1 + q_h)\,H_i(t),$$
$$\frac{\mathrm{d}Q_h(t)}{\mathrm{d}t} = q_h\,H_i(t) - (\mu_h + \alpha_2)\,Q_h(t) - \eta_2\,Q_h.$$

Here, the number of susceptible host cattle increases through density-dependent births with maximum rate b_s. It is assumed that individuals are born uninfected and that the infection reduces fertility rate $0 \le \rho < 1$. Susceptible hosts are infected at the per capita infection rate β, by contact with metacercariae on pasture. Infected hosts die at an augmented death rate, $(\mu_h + \alpha_1)$, with μ_h the natural death rate and α_1 the death rate due to parasite virulence. They recover at the per capita rate η_1. Infected hosts are quarantined/isolated at a rate of q_h, and the quarantine compartment diminishes by recovery at a rate of η_2 and by death due to the disease at a rate of α_2.

Equations for Dynamics of Snails and Cysts. Our model formulation explicitly accounts the eggs and metacercariae. Since miracidia and the cercariae have a short lifespan ($\approx$ one day) [16], they are not expressly modeled. As for snails, we present their developmental phases in four stages: eggs, juvenile snails, mature snails and infected snails. And we use the fact that only mature snails lay eggs and that infected snails are castrated [17].

Changes in the number of parasite Eggs (E), Snail eggs (S_e), Juvenile Snail (S_j), Mature Snail (S_m), Infected Snail (S_i), and Metacercariae (M) are given by the following system:

$$\frac{\mathrm{d}E(t)}{\mathrm{d}t} = \sigma\,H_i(t) - (m_E + \mu_E)E(t), \tag{1a}$$
$$\frac{\mathrm{d}S_e(t)}{\mathrm{d}t} = \lambda\,S_m(t) - (r_{ej} + \mu_{se})S_e(t), \tag{1b}$$
$$\frac{\mathrm{d}S_j(t)}{\mathrm{d}t} = r_{ej}\,S_e(t) - (r_{jm} + \mu_{sj})S_j(t) - r_{ji}m_E E(t)S_j(t), \tag{1c}$$
$$\frac{\mathrm{d}S_m(t)}{\mathrm{d}t} = r_{jm}\,S_j(t) - \mu_{sm}S_m(t) - r_{mi}m_E E(t)S_m(t), \tag{1d}$$
$$\frac{\mathrm{d}S_i(t)}{\mathrm{d}t} = r_{ji}\,m_E E(t)S_j(t) + r_{mi}m_E E(t)S_m(t) - \mu_{si}\,S_i(t), \tag{1e}$$
$$\frac{\mathrm{d}M(t)}{\mathrm{d}t} = r_{im}\,S_i(t) - (\mu_m + \beta H(t))\,M(t). \tag{1f}$$

The first Eq. (1a) models the rate of change of the Fasciola eggs population. The release rate of eggs by infected mammals is σ. A fraction of eggs m_E infect (by absorption) juvenile and adult snails while eggs die at the rate of μ_E.

The average snail eggs production rate of a mature snail is given by λ. Snail eggs S_e can hatch, at a rate r_{ej} and develop into juvenile snails S_j, and their death rate is μ_{se}. Maturing of juvenile snails occurs at a rate r_{jm} and mature snails S_m die, at a rate of μ_{sj}. The infection rates of juvenile and mature snails are denoted by r_{ji} and r_{mi} respectively. The infected snails die at a rate μ_{si}. Metacercariae M are produced by infected snails at a rate r_{im}. They die at a rate of μ_m or end up in mammal hosts by ingestion at a rate β.

Then, the global model satisfies the following system of nonlinear ordinary differential equations:

$$
\begin{aligned}
\frac{\mathrm{d}H_s(t)}{\mathrm{d}t} &= \Lambda_h - \mu_h\, H_s(t) - \beta\, H_s(t)\, M(t) + \eta_1\, H_i(t) + \eta_2\, Q_h(t),\\
\frac{\mathrm{d}H_i(t)}{\mathrm{d}t} &= \beta\, H_s(t)\, M(t) - (\mu_h + \alpha_1)\, H_i(t) - (\eta_1 + q_h)\, H_i(t),\\
\frac{\mathrm{d}Q_h(t)}{\mathrm{d}t} &= q_h\, H_i(t) - (\mu_h + \alpha_2)\, Q_h(t) - \eta_2\, Q_h,\\
\frac{\mathrm{d}E(t)}{\mathrm{d}t} &= \sigma\, H_i(t) - (m_E + \mu_E)E(t),\\
\frac{\mathrm{d}S_e(t)}{\mathrm{d}t} &= \Lambda_e - (r_{ej} + \mu_{se})S_e(t),\\
\frac{\mathrm{d}S_j(t)}{\mathrm{d}t} &= r_{ej}\, S_e(t) - (r_{jm} + \mu_{sj})S_j(t) - r_{ji} m_E\, E(t) S_j(t),\\
\frac{\mathrm{d}S_m(t)}{\mathrm{d}t} &= r_{jm} S_j(t) - \mu_{sm} S_m(t) - r_{mi} m_E\, E(t) S_m(t),\\
\frac{\mathrm{d}S_i(t)}{\mathrm{d}t} &= r_{ji}\, m_E\, E(t) S_j(t) + r_{mi} m_E\, E(t)\, S_m(t) - \mu_{si}\, S_i(t),\\
\frac{\mathrm{d}M(t)}{\mathrm{d}t} &= r_{im} S_i(t) - (\mu_m + \beta(H_s(t) + H_i(t)))\, M(t).
\end{aligned}
\tag{2a}
$$

Here, Λ_h and Λ_e are the recruitment rate of the hosts and snails eggs, respectively.

Model Assumptions. The life span of miracidia is not expressly displayed and is considered to be one day. Consequently, the production of miracidia is represented by the quantity of eggs which are laid in every day.

It is fine to point out the supposition implicitly made about the life cycle of worms and the environment. In fact, a homogeneous distribution of eggs in the pasture is assumed. In addition, the model does not include a spatial structure that can measure snail density by location.

3 Basic Properties of the Model

It is obvious that system (2a) is a Lipschitzian system, and that the basic properties of local existence, uniqueness and continuity of solutions are satisfied. Then, we give the properties verified by its solutions.

3.1 Positivity of Solutions

Since system (2a) models variation in the population of cattle host, metacercariae and snails, it is essential that the solution of system (2a) be biologically valid at all times when the initial data are positive. We therefore show in this section that the biologically feasible region for system (2a) is given by:

$$\Gamma = \{(H_s, H_i, Q_h, E, S_e, S_j, S_m, S_i, M) \in \mathbb{R}^9_+ \text{ such that } H_s + H_i + Q_h \leq \frac{\Lambda_h}{\mu_h}, \tag{3a}$$

$$S_e + S_j + S_m + S_i \leq \frac{\Lambda_e}{\gamma},\ E \leq \frac{\sigma \Lambda_h}{\mu_h \mu_E},\ M \leq \frac{r_{im} \Lambda_s}{\mu_m \gamma}\},$$

with $\gamma = \min\{\mu_{se}, \mu_{sj}, \mu_{sm}, \mu_{si}\}$.

Let us give the following lemma to show the positivity of the solutions of system (2a) when $(H_s(0), H_i(0), Q_h(0), E(0), S_e(0), S_j(0), S_m(0), S_i(0), M(0))^T \in \mathbb{R}^9_+$.

Lemma 1 (see [18]).
Let n be a positive integer and $f_i(t, x)$, $(i = 1, ..., n)$, n, smooth functions. If $f_i(t, x) \mid_{x_i=0, x \in \mathbb{R}^n_+} \geq 0$, then, $\mathbb{R}^n_+$ is an invariant domain of the following equations:

$$\frac{dx_i}{dt} = f_i(t, x),\ i = 1, ..., n.$$

We have the following result.

Theorem 1. *Each solution $(H_s, H_i, Q_h, E, S_e, S_j, S_m, S_i, M)$ of system (2a) with non-negative initial values is non-negative for all $t \geq 0$.*

Proof. Note that

$$\begin{aligned}
&\frac{dH_s(t)}{dt} \mid_{H_s=0} = \Lambda_h + \eta_1 H_i(t) + \eta_2 Q_h(t) \geq 0,\\
&\frac{dH_i(t)}{dt} \mid_{H_i=0} = \beta H_s M(t) \geq 0,\\
&\frac{dQ_h(t)}{dt} \mid_{Q_h=0} = q_h H_i(t) \geq 0,\\
&\frac{dE(t)}{dt} \mid_{E=0} = \sigma H_i(t) \geq 0,\\
&\frac{dS_e(t)}{dt} \mid_{S_e=0} = \Lambda_e \geq 0,\\
&\frac{dS_j(t)}{dt} \mid_{S_j=0} = r_{ej} S_e(t) \geq 0,\\
&\frac{dS_m(t)}{dt} \mid_{S_m=0} = r_{jm} S_j(t) \geq 0,\\
&\frac{dS_i(t)}{dt} \mid_{S_i=0} = r_{ji}\, m_E\, E(t)\, S_j(t) + r_{mi}\, m_E\, E(t)\, S_m(t) \geq 0,
\end{aligned}$$

$$\frac{dM(t)}{dt}\ |_{M=0}= r_{im}\ S_i(t) \geq 0.$$

Then, it follows from Lemma 1 that each solution $(H_s, H_i, Q_h, E, S_e, S_j, S_m, S_i, M)$ of system (2a) is non-negative.

Let us give now the following theorem which proves that Γ is an invariant domain for system (2a).

Theorem 2. *The set Γ defined in (3a) is positively invariant and attracts all solutions of system (2a).*

Proof. Considering the total cattle population $H(t) = H_s(t) + H_i(t) + Q_h(t)$ and the total snail population $S = S_e + S_j + S_m + S_i$, direct calculation gives:

$$\frac{dH(t)}{dt} = \Lambda_h - \mu_h H - \alpha_1 H_i - \alpha_2 Q_h \leq \Lambda_h - \mu_h H,$$

$$\frac{dS(t)}{dt} = \Lambda_s - \mu_{se} S_e - \mu_{sj} S_j - \mu_{sm} S_m - \mu_{si} S_i \leq \Lambda_e - \gamma\ S.$$

From eggs and metacercariae equations, we obtain:

$$\frac{dE(t)}{dt} = \sigma H_i(t) - (m_E + \mu_E)E(t) \leq \sigma H_i(t) - \mu_E E(t),$$

and

$$\frac{dM(t)}{dt} = r_{im} S_i(t) - (\mu_m + \beta(H_s(t) + H_i(t)))M(t) \leq r_{im} S_i(t) - \mu_m\ M(t).$$

It follows that the set Γ is positively invariant and attracts all solutions of system (2a).

3.2 Reproduction Number and Disease-Free Equilibrium (DFE) Analysis

In what follows, we start with an analysis of the possible equilibrium point of system (2a).

The system has a disease-free-equilibrium $E_{DFE} = (H_s^*, 0, 0, 0, S_e^*, S_j^*, S_m^*, 0, 0)$, where:
$H_s^* = \frac{\Lambda_h}{\mu_h}; S_e^* = \frac{\Lambda_e}{(r_{ej} + \mu_{se})}; S_j^* = \frac{r_{ej}\ \Lambda_e}{(r_{jm} + \mu_{sj})(r_{ej} + \mu_{se})}; S_m^* = \frac{r_{ej}\, r_{jm}\ \Lambda_e}{(r_{jm} + \mu_{sj})(r_{ej} + \mu_{se})\mu_{sm}}.$

We use the next generation matrix approach [19] to compute the basic reproduction number $\mathcal{R}_0$ of the system. We consider only the equations where the infection progresses, namely

$$\frac{dX}{dt} = \mathcal{F}(X) - \mathcal{V}(X),$$

where, in this case $X = (H_i, Q_h, E, S_i, M)^t$,

$$\mathcal{F}(X) = \begin{pmatrix} \beta\, H_s(t)\, M(t) \\ q_h\, H_i(t) \\ \sigma\, H_i(t) \\ r_{ji} m_E\, E(t) S_j(t) + r_{mi} m_E\, E(t) S_m(t) \\ r_{im}\, S_i(t) \end{pmatrix} \text{ and}$$

$$\mathcal{V}(X) = \begin{pmatrix} (\mu_h + \alpha_1)\, H_i(t) + (\eta_1 + q_h)\, H_i(t) \\ (\mu_h + \alpha_2)\, Q_h(t) + \eta_2\, Q_h \\ (m_E + \mu_E) E(t) \\ \mu_{si}\, S_i(t) \\ (\mu_m + \beta(H_s(t) + H_i(t))) M(t) \end{pmatrix}.$$

The matrices $F = \left[\frac{\partial \mathcal{F}(x_i)}{\partial x_j}\right]$ and $V = \left[\frac{\partial \mathcal{V}(x_i)}{\partial x_j}\right]$, for $1 \le i, j \le 5$, are calculated as follows:

$$F = \begin{pmatrix} 0 & 0 & 0 & 0 & H_s \beta \\ q_h & 0 & 0 & 0 & 0 \\ \sigma & 0 & 0 & 0 & 0 \\ 0 & 0 & S_j\, m_E\, r_{ji} + S_m\, m_E\, r_{mi} & 0 & 0 \\ 0 & 0 & 0 & r_{im} & 0 \end{pmatrix} \text{ and}$$

$$V = \begin{pmatrix} q_h + \alpha_1 + \eta_1 + \mu_h & 0 & 0 & 0 & 0 \\ 0 & \alpha_2 + \eta_2 + \mu_h & 0 & 0 & 0 \\ 0 & 0 & m_E + \mu_e & 0 & 0 \\ 0 & 0 & 0 & \mu_{si} & 0 \\ M\,\beta & 0 & 0 & 0 & (H_i + H_s)\beta + \mu_m \end{pmatrix}.$$

The basic reproduction number $\mathcal{R}_0$ is obtained as follows: $\mathcal{R}_0 = \rho(F V^{-1})$, then

$$\mathcal{R}_0^4 = \frac{\beta\, \sigma\, \Lambda_e\, m_e\, r_{ej}\, \Lambda_h\, r_{im} \left(r_{ji}\, \mu_{sm} + r_{jm}\, r_{mi}\right)}{\mu_{si}\, \mu_{sm}\, (\mu_e + m_e) \left(r_{ej} + \mu_{se}\right) \left(r_{jm} + \mu_{sj}\right) (\alpha_1 + \eta_1 + q_h + \mu_h)(\beta\, \Lambda_h + \mu_h\, \mu_m)}.$$

Now, we introduce the following result which shows that $\mathcal{R}_0$ is a significant factor of the development or the extinction of the disease.

Theorem 3. *The disease-free equilibrium E_{DFE} of system (2a) is locally asymptotically stable if and only if $\mathcal{R}_0 < 1$ and unstable if $\mathcal{R}_0 \ge 1$.*

Proof. The Jacobian matrix (J) for system (2a) is given by:

$$\begin{pmatrix}
-\mu_h & \eta_1 & \eta_2 & 0 & 0 & 0 & 0 & 0 & -\frac{\beta\Lambda_h}{\mu_h} \\
0 & -k_1 & 0 & 0 & 0 & 0 & 0 & 0 & \frac{\beta\Lambda_h}{\mu_h} \\
0 & q_h & -k_2 & 0 & 0 & 0 & 0 & 0 & 0 \\
0 & \sigma & 0 & -(m_E+\mu_E) & 0 & 0 & 0 & 0 & 0 \\
0 & 0 & 0 & 0 & -(r_{ej}+\mu_{se}) & 0 & 0 & 0 & 0 \\
0 & 0 & 0 & -r_{ji}m_E S_j^* & r_{ej} & -(r_{jm}+\mu_{sj}) & 0 & 0 & 0 \\
0 & 0 & 0 & -r_{mi}m_E S_m^* & 0 & r_{jm} & -\mu_{sm} & 0 & 0 \\
0 & 0 & 0 & r_{ji}m_E S_j^* + r_{mi}m_E S_m^* & 0 & 0 & 0 & -\mu_{si} & 0 \\
0 & 0 & 0 & 0 & 0 & 0 & 0 & r_{im} & -(\mu_m+\frac{\beta\Lambda_h}{\mu_h})
\end{pmatrix}$$

For the sake of simplicity, we denote $k_1 = (\mu_h + \alpha_1 + \eta_1 + q_h)$ and $k_2 = (\mu_h + \alpha_2 + \eta_2)$. Its characteristic equation is $P(x) = Q(x)R(x)$ with

$$Q(x) = -(x+\mu_h)(x+k_2)(x+\mu_{sm})(x+r_{ej}+\mu_{se})(x+r_{jm}+\mu_{sj}),$$

$$\begin{aligned}
R(x) &= (x+k_1)\left(\mu_m + \frac{\beta\Lambda_h}{\mu_h} + x\right)(x+\mu_{si})(\mu_E+m_E+x) - \frac{\beta\Lambda_h r_{im}\sigma m_E\left(r_{mi}S_m^* + r_{ji}S_j^*\right)}{\mu_h} \\
&= \gamma_0 x^4 + \gamma_1 x^3 + \gamma_2 x^2 + \gamma_3 x + \gamma_4,
\end{aligned}$$

where $\gamma_0 = 1$,

$$\gamma_1 = (\mu_{si} + \mu_E + m_E) + (\mu_h + \alpha_1 + \eta_1 + q_h) + (\mu_m + \frac{\beta\Lambda_h}{\mu_h}),$$

$$\begin{aligned}
\gamma_2 &= \mu_{si}(\mu_E + m_E) + (\mu_h + \alpha_1 + \eta_1 + q_h)(\mu_m + \frac{\beta\Lambda_h}{\mu_h}) \\
&\quad + (\mu_m + \frac{\beta\Lambda_h}{\mu_h} + (\mu_h + \alpha_1 + \eta_1 + q_h))(\mu_{si} + \mu_E + m_E),
\end{aligned}$$

$$\begin{aligned}
\gamma_3 &= \mu_{si}(\mu_E + m_E)(\mu_m + \frac{\beta\Lambda_h}{\mu_h} + (\mu_h + \alpha_1 + \eta_1 + q_h)) \\
&\quad + (\mu_h + \alpha_1 + \eta_1 + q_h)(\mu_m + \frac{\beta\Lambda_h}{\mu_h})(\mu_{si} + \mu_E + m_E),
\end{aligned}$$

$$\begin{aligned}
\gamma_4 &= (\mu_h + \alpha_1 + \eta_1 + q_h)\mu_{si}(\mu_E + m_E)(\mu_m + \frac{\beta\Lambda_h}{\mu_h}) - \frac{\beta\Lambda_h r_{im}\sigma\left(r_{mi}S_m^* m_E + r_{ji}S_j^* m_E\right)}{\mu_h} \\
&= (\mu_h + \alpha_1 + \eta_1 + q_h)\mu_{si}(\mu_E + m_E)(\mu_m + \frac{\beta\Lambda_h}{\mu_h}) - \frac{\beta\Lambda_h r_{im}\sigma m_E r_{ej}\Lambda_e\left(r_{ji}\mu_{sm} + r_{mi}r_{jm}\right)}{\mu_h\mu_{sm}(r_{ej}+\mu_{se})(r_{jm}+\mu_{sj})} \\
&= 1 - \frac{\beta\Lambda_h r_{im}\sigma m_E r_{ej}\Lambda_e\left(r_{ji}\mu_{sm} + r_{mi}r_{jm}\right)}{\mu_{sm}(r_{ej}+\mu_{se})(r_{jm}+\mu_{sj})(\mu_h + \alpha_1 + \eta_1 + q_h)\mu_{si}(\mu_E + m_E)(\mu_m\mu_h + \beta\Lambda_h)} \\
&= 1 - \mathcal{R}_0^4.
\end{aligned}$$

If $\mathcal{R}_0 < 1$, we see that $\gamma_1, \gamma_2, \gamma_3$ and γ_4 are positive and $(\gamma_1\gamma_2 - \gamma_0\gamma_3)\gamma_3 - \gamma_1^2\gamma_4 > 0$. So, thanks to Routh-Hurwitz criterion, all solutions of the equation $R(x) = 0$ have non-positive real part. Also we see that all solutions of the equation $Q(x) = 0$ are non-positive. Then, if $\mathcal{R}_0 < 1$, E_{DFE} is locally asymptotically stable. And one can easily prove that the solution is unstable if $\mathcal{R}_0 \geq 1$.

4 Numerical Simulations

4.1 *Quarantine, DFE and Endemic Scenarios Simulation*

Numerical simulations of system (2a) are carried out to illustrate some of the analytical results. Table 1 provides the parameter values used for model simulation. Some of those values are obtained from the literature, while others are assumed (within realistic range) for the purpose of simulations. The assumed values of the parameters are the recruitment rates Λ_h and Λ_e as they depend on the number and geographical distribution of the livestock concerned (Figs. 3 and 4).

4.2 *Sensitivity Analysis of $\mathcal{R}_0$*

We analyse how a change in the value of the parameters of the system can affect the reproduction number. Chitnis et al. [26] described a sensitivity analysis called normalised forward sensitivity index that allows us to point out parameters that have a high impact on $\mathcal{R}_0$. The normalised forward sensitivity index of a variable Φ, that depends on a parameter p, is defined as:

$$\Upsilon_p^{\Phi} = \frac{\partial \Phi}{\partial p} \times \frac{p}{\Phi}.$$

Thus, we use it to identify the parameters having a significant influence on $\mathcal{R}_0$ and which can therefore be targeted in control strategies. The sensitivity indices obtained from the parameter values in Table 1 are summarised in Table 2. The most sensitive parameters to $\mathcal{R}_0$ are the eggs production rate σ, the shedding rate of metacercariae r_{im}, the host recruitment rate Λ_h, the recruitment rate of snail eggs Λ_e and the death rate of infected snails μ_{si}. They are closely followed by the per capita rate of infection β, the death rate of metacercaria μ_m, the death rate of juvenile snail μ_{sj}, the migration rate of parasite eggs m_E, the death rate of parasite eggs μ_e, the recovery rate of infected host η_1 and the proportion of quarantined infected host q_h. Note that the effectiveness of disease management through quarantine of infected hosts may be similar to that achieved by medical treatment for the hosts population.

It can be seen from the analysis of $\mathcal{R}_0$ and simulations that if quarantine and hosts treatment are appropriately combined, the Fasciola disease tends to disappear (Figs. 5, 6, 7 and 8).

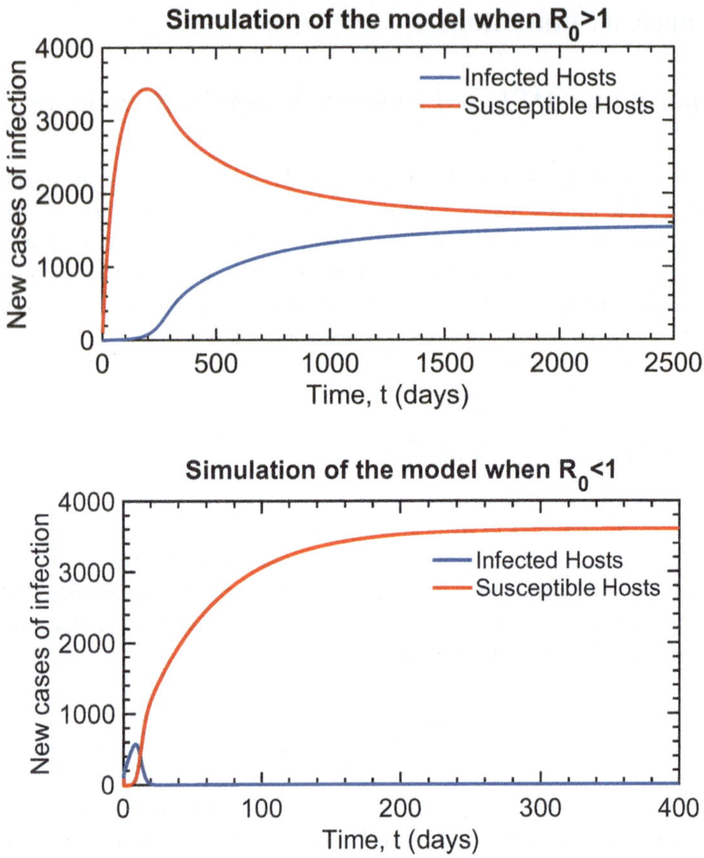

Fig. 3 Simulation model showing the relationship between the number of hosts and time when $\mathcal{R}_0 > 1$ or $\mathcal{R}_0 < 1$. Values of parameters for $\mathcal{R}_0 > 1$: see Table 1. For $\mathcal{R}_0 < 1$ scenario, we take: $\Lambda_h = 67$; $\mu_h = 100 * 0.000185$; $\beta = 0.0000005$; $\eta_1 = 0.9567$; $\eta_2 = 0.9507$; $\alpha_2 = 0.00001/10$; $\alpha_1 = 0.00001$; $\sigma = 150$; $m_e = 0.0025$; $\mu_e = 10 * 2/365$; $\Lambda_e = 200$; $r_{ej} = 0.01$; $\mu_{se} = 0.0025$; $r_{jm} = 0.03/365$; $\mu_{sj} = 20 * 0.3/365$; $r_{ji} = 0.0005/365$; $\mu_{sm} = 0.3/365$; $r_{mi} = 0.0005/365$; $\mu_{si} = 0.6/365$; $r_{im} = 8.12$; $\mu_m = 20 * 1/31$

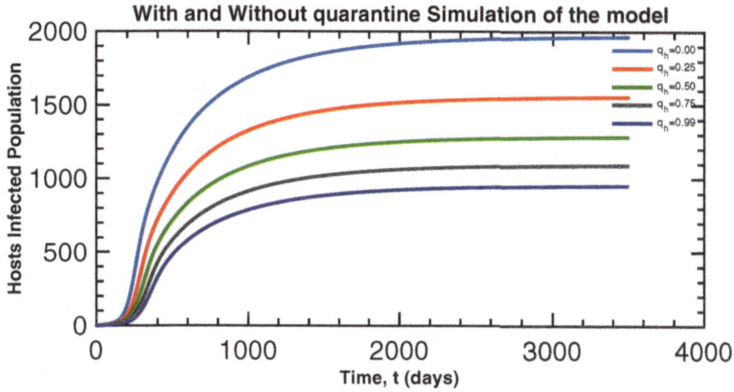

Fig. 4 The effects of variation in the proportion of infected quarantined

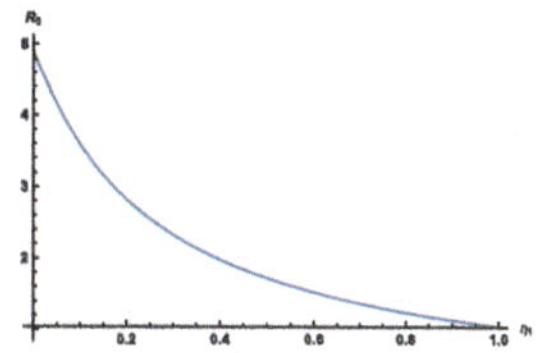

Fig. 5 $\mathcal{R}_0$ as a function of the recovery rate η_1

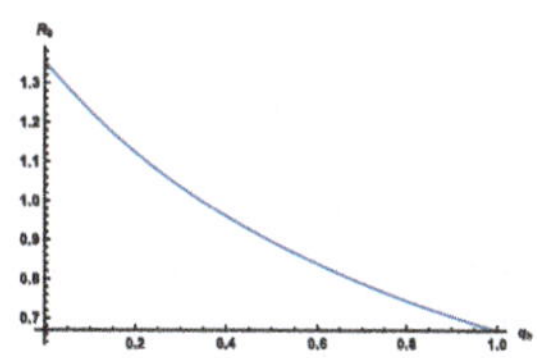

Fig. 6 $\mathcal{R}_0$ as a function of the quarantine rate q_h

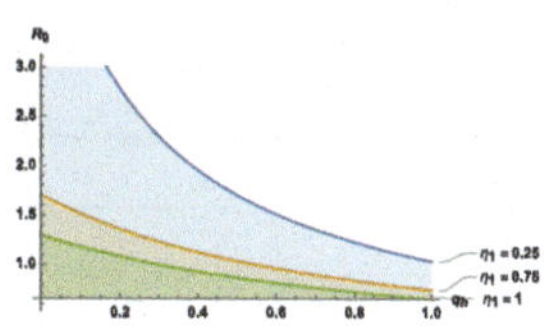

Fig. 7 $\mathcal{R}_0$ as a function of the quarantine rate q_h for different values of recovery rate η_1

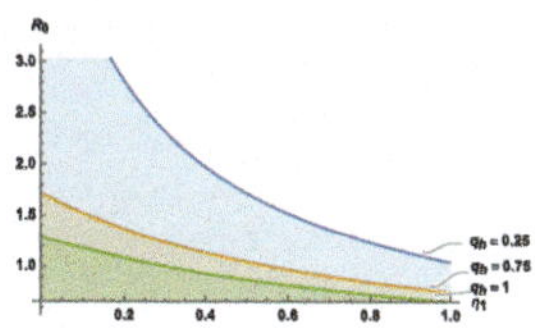

Fig. 8 $\mathcal{R}_0$ as a function of the recovery rate η_1 for different values of quarantine rate q_h

Table 1 Summary of parameters of the model

Name		
Parameters	Description	Default values assigned (sources)
Λ_h	Recruitment rate	67 (assumed)
μ_h	Death rate of hosts	0.000185/cattle/day [20]
α_1	Disease-induced death rate of infected hosts	0.00001/host/fluke/day [20]
β	Per capita rate of infection	0.0000005/cattles/cyst/day [20]
η_1	Recover rate of infected hosts	0.9567 (assumed)
η_2	Recover rate of quarantined hosts	0.9507 (assumed)
q_h	Proportion of infected quarantined hosts	0.25 (assumed)
α_2	Disease-induced death rate of quarantined hosts	$\alpha_1/10$ (assumed)
σ	Eggs production rate	24590 [20]
Λ_e	Recruitment rate of snail eggs	2000 (assumed)
m_E	Absorption rate of parasites by snails	0.25/day [25]
μ_e	Death rate of parasite eggs	(2–6)/365 larvae/day [23]
r_{ej}	Development rate of eggs to juvenile snails	0.01 larvae/day [24]
r_{jm}	Development rate of juvenile to mature snails	0.03/365/day [22]
μ_{sj}	Death rate of juvenile snails	0.3/365/days [22]
μ_{sm}	Death rate of mature snails	0.3/365/days [22]
r_{ji}	Infection rate of juvenile snails	0.0005/365/days [22]
r_{mi}	Infection rate of mature snails	0.0005/365/days [22]
μ_{si}	Death rate of infected snails	0.6/365/days [22]
r_{im}	Shedding rate of metacercariae	8.12/snail/day [20]
μ_m	Death rate of metacercariae	1/31/day [21]

Table 2 Normalized forward sensitivity indices of $\mathcal{R}_0$ calculated at parameters near $\mathcal{R}_0 = 1$

$\mathcal{R}_0$		
Parameter	Υ_p^{Φ}	Sensitivity index
β	$\frac{\mu_h\mu_m}{\beta\Lambda_h+\mu_h\mu_m}$	0.997201
σ	1	1
m_E	$\frac{\mu_e}{\mu_e+m_e}$	0.956366
r_{ej}	$\frac{\mu_{se}}{r_{ej}+\mu_{se}}$	0.2
r_{im}	1	1
Λ_e	1	1
Λ_h	1	1
r_{jm}	$\frac{r_{jm}r_{mi}}{r_{ji}\mu_{sm}+r_{jm}r_{mi}}-\frac{r_{jm}}{r_{jm}+\mu_{sj}}$	0.085934
r_{mi}	$\frac{r_{mi}r_{jm}}{r_{ji}\mu_{sm}+r_{mi}r_{jm}}$	0.0909091
r_{ji}	$\frac{r_{ji}\mu_{sm}}{r_{jm}r_{mi}+r_{ji}\mu_{sm}}$	0.0909091
μ_{sm}	$\frac{\mu_{sm}r_{ji}}{\mu_{sm}r_{ji}+r_{jm}r_{mi}}-1$	0.0909091
μ_e	$-\frac{\mu_e}{m_e+\mu_e}$	−0.956366
μ_{si}	−1	−1
μ_{sj}	$-\frac{\mu_{sj}}{r_{jm}+\mu_{sj}}$	−0.995025
q_h	$-\frac{q_h}{\alpha_1+\eta_1+\mu_h+q_h}$	−0.204047
α_1	$-\frac{\alpha_1}{\alpha_1+\eta_1+q_h+\mu_h}$	$-8.16187 * 10^{-6}$
η_1	$-\frac{\eta_1}{\eta_1+\alpha_1+q_h+\mu_h}$	−0.780846
μ_h	$\frac{\beta\Lambda_h}{\beta\Lambda_h+\mu_h\mu_m}-\frac{\mu_h}{\alpha_1+\eta_1+q_h+\mu_h}-1$	−1.0123
μ_m	$-\frac{\mu_m\mu_h}{\beta\Lambda_h+\mu_m\mu_h}$	−0.997201
μ_{se}	$-\frac{\mu_{se}}{r_{ej}+\mu_{se}}$	−0.2

5 Conclusions

In this work, a mathematical model is developed and analysed to study the transmission dynamics of fascioliasis desease. The model is given by a 9 dimensional system of non-linear ordinary differential equations involving the dynamics of cattle, snails and metacercariae. Dynamics' of cattle and snails are given by SIR and age-structure population submodels respectively. The model considers two intervention strategies, namely quarantine and medical treatment. We have first shown that there exists a domain where the model is well-posed mathematically and epidemiologically. As for stability, mathematical analysis shows that the DFE is locally stable if $\mathcal{R}_0 < 1$ and unstable if $\mathcal{R}_0 \geq 1$. Numerical simulations, using the parameter values in Table 2, are done to underpin the theoretical results. Using the sensitivity indices of the reproduction number, we have identified the most sensitive parameters involved in the expression of $\mathcal{R}_0$. We have found that the treatment rate η_1 and the proportion of quarantined hosts q_h are not the most sensitive. However, it has been shown that combining quarantine and medical treatment may lead to disease control.

The present work provides three majors perspectives.

1. Data should be collected in Senegal and used for a more specific simulation.
2. A more detailed study can also be considered by performing further control strategies
3. From a mathematical analysis point of view, it will be very interesting to carry out the global stability analysis.

Acknowledgments We would like to acknowledge the support of the CEA-MITIC (Centre d'Excellence Africain en Mathématiques, Informatique et Technologie de l'Information et de la Communication) and PIED-UVS (Pôle d'Innovation et d'Expertise pour le Développement).

References

1. Cwiklinski, K., Donnelly, S., Drysdale, O., Jewhurst, H., Smith, D., Verissimo, C.D.M., Pritsch, I.C., O'Neill, S., Dalton, J.P., Robinson, M.W.: The cathepsin-like cysteine peptidases of trematodes of the genus Fasciola. Adv. Parasitol. **104**, 113–164 (2019)
2. Krauss, H., Weber, A., Appel, M., Enders, B., Isenberg, H.D., Schiefer, H.G., Slenczka, W., von Graevenitz, A., Zahner, H.: Zoonoses: Infectious Diseases Transmissible from Animals to Humans. ASM Press, Washington, DC (2003)
3. WHO: Trématodoses d'origine alimentaires. http://www.who.int/mediacentre/factsheets/fs368/fr/Aidem'emoire
4. Farrar, J., Hotez, P.J., Junghanss, T., Kang, G., Lalloo, D., White, N.J.: Manson's Tropical Diseases E-Book. Elsevier Health Sciences (2013)
5. Schillhorn, V.V.T.: Fasciolosis (f. gigantica) in West Africa. Rev. Vet. Bull **5**, 229–233 (1980)
6. Assogba, M.N., Youssao, A.K.I.: Epidémiologie de la fasciolose à fasciola gigantica (cobbold, 1885), de la dicrocoeliose et de la paramphistomose bovines au bénin. Ann. Méd. Vét **145**, 260–268 (2001)
7. Ka, M., Mbengue, M., Diop, B., Pouye, A., Da, J.V., Dia, D., Welle, A., Ndir, O., Diop, T.: Two unexpected cases of hepatobiliary fascioliasis in Dakar (Senegal). Dakar Med. **47**(2), 202–205 (2002)

8. Seydi, M., Soumare, M., Sow, M., Ka, E., Diop, S., Diop, B., Sow, P.: La distomatose hepatobiliaire a fasciola hepatica: a propos du premier cas observe a la clinique des maladies infectieuses ibrahima diop mar du chu de fann a Dakar (Senegal). Med. Trop. **68**(4), 431 (2008)
9. Spithill, T.W.: Fasciola gigantica: epidemiology, control, immunology and molecular biology. Fasciolosis, pp. 465–525 (1999)
10. Kaplan, R.M.: Fasciola hepatica: a review of the economic impact in cattle and considerations for control. Vet Ther **2**(1), 40–50 (2001)
11. Malone, J., Loyacano, A., Armstrong, D., Archbald, L.: Bovine fascioliasis: economic impact and control in gulf coast cattle based on seasonal transmission [fasciola hepatica, louisiana]. J. Am. Assoc. Bovines Pract. (1982)
12. Saba, R., Korkmaz, M., Inan, D., Mamikoğlu, L., Turhan, Ö., Günseren, F., Cevikol, C., Kabaalioğlu, A.: Human fascioliasis (2004)
13. Hope-Cawdery, M.: Changing temperatures and prediction models for the liver fluke (Fasciola hepatica). J. Therm. Biol **6**(4), 403–408 (1981)
14. Ashrafi, K., Valero, M., Panova, M., Periago, M., Massoud, J., Mas-Coma, S.: Phenotypic analysis of adults of Fasciola hepatica, Fasciola gigantica and intermediate forms from the endemic region of Gilan, Iran. Parasitol. Int. **55**(4), 249–260 (2006)
15. Gettinby, G., Hope-Cawdery, M., Grainger, J.: Forecasting the incidence of fascioliasis from climatic data. Int. J. Biometeorol. **18**(4), 319–323 (1974)
16. Ciddio, M., Mari, L., Sokolow, S.H., De Leo, G.A., Casagrandi, R., Gatto, M.: The spatial spread of schistosomiasis: a multidimensional network model applied to Saint-Louis region, Senegal. Adv. Water Resour. **108**, 406–415 (2017)
17. Wilson, R., Denison, J.: The parasitic castration and gigantism of Lymnaea truncatula infected with the larval stages of Fasciola hepatica. Z. Parasitenkd. **61**(2), 109–119 (1980)
18. Yang, X., Chen, L., Chen, J.: Permanence and positive periodic solution for the single-species nonautonomous delay diffusive models. Comput. Math. Appl. **32**(4), 109–116 (1996)
19. Van den Driessche, P., Watmough, J.: Reproduction numbers and sub-threshold endemic equilibria for compartmental models of disease transmission. Math. Biosci. **180**(1–2), 29–48 (2002)
20. Smith, G.: Density-dependent mechanisms in the regulation of Fasciola hepatica populations in sheep. Parasitology **88**(3), 449–461 (1984)
21. Meek, A., Morrist, R.: The longevity of Fasciola hepatica metacercariae encysted on herbage. Aust. Vet. J. **55**(2), 58–60 (1979)
22. Feng, Z., Li, C.-C., Milner, F.A.: Schistosomiasis models with density dependence and age of infection in snail dynamics. Math. Biosci. **177**, 271–286 (2002)
23. Jain, S.K., Botsford, L.W.: Applied Population Biology, vol. 67. Springer, Heidelberg (2007)
24. Grenfell, B.T., Wilson, K., Isham, V., Boyd, H., Dietz, K.: Modelling patterns of parasite aggregation in natural populations: trichostrongylid nematode-ruminant interactions as a case study. Parasitology **111**(S1), S135–S151 (1995)
25. Bolajoko, M.-B., Rose, H., Musella, V., Bosco, A., Rinaldi, L., van Dijk, J., Cringoli, G., Morgan, E.R.: The basic reproduction quotient (Q0) as a potential spatial predictor of the seasonality of ovine haemonchosis. Geospatial Health 333–350 (2015)
26. Chitnis, N., Hyman, J.M., Cushing, J.M.: Determining important parameters in the spread of malaria through the sensitivity analysis of a mathematical model. Bull. Math. Biol. **70**(5), 1272 (2008)

Control, Sensitivity and Identification of a Cardiovascular-Respiratory System Model

Pio Gabrielle B. Calderon, Lean V. Palma, Franz Kappel, and Aurelio A. de los Reyes V

Abstract This work examines a cardiovascular-respiratory system model capable of describing its dynamic response to a constant workload. The heart rate and alveolar ventilations are considered fundamental controls of the system which represent the baroreceptor and chemoreceptor loops, respectively. Multi-method sensitivity analysis is performed to quantify how variations in the parameters influence the model output. Sensitivity approaches include the traditional sensitivity analysis, partial rank correlation coefficient and extended Fourier amplitude sensitivity test. A set of parameters which can be reliably estimated from given measurements is determined from subset selection. Hemodynamic and respiratory data acquired from ergometric exercise are used for model identification and validation obtaining subject-specific parameter estimates.

Keywords Cardiovascular-respiratory system model · Multi-method sensitivity analysis · Parameter estimation · Optimal control · Hemodynamic and respiratory data

1 Introduction

Mathematical models are simplifications or approximations of complex and complicated reality and hence characterized by certain degree of uncertainties. These uncertainties often arise from modeling assumptions due to incomplete or even lack

---

P. G. B. Calderon · L. V. Palma · A. A. de los Reyes V (✉)
Institute of Mathematics, University of the Philippines Diliman, C.P. Garcia St., U.P. Campus Diliman, 1101 Quezon City, Philippines
e-mail: adlreyes@math.upd.edu.ph

P. G. B. Calderon
e-mail: piocalderon@gmail.com

L. V. Palma
e-mail: lean.palma@obf.ateneo.edu

F. Kappel
Institute of Mathematics and Scientific Computing, University of Graz, Heinrichstrasse 36, A, 8010 Graz, Austria
e-mail: franz.kappel@uni-graz.at

M. H. Mohd et al. (eds.), *Modelling, Simulation and Applications of Complex Systems*, Springer Proceedings in Mathematics & Statistics 359,
https://doi.org/10.1007/978-981-16-2629-6_8

of information on the underlying mechanisms, difficulty or sometimes impossibility of obtaining experimental measurements, ambiguous and contradictory observations, and structural and numerical errors of the model. An integral part of the modeling process includes assessing uncertainties associated with the model outcome. It is essential in model building to have better understanding of the influence of different uncertainties of model factors on the outcome [1]. Uncertainty and sensitivity analysis methods provide insights on how to control these uncertainties. Several approaches are developed to quantify uncertainty and sensitivity issues including differential analysis, response surface methodology, and variance decomposition methods [2–6]. The main task of sensitivity analysis (SA) is to assess how variations in the model outputs can be qualitatively or quantitatively apportioned and attributed to different input sources [7, 8]. The type of approach, complexity level, and goals of SA vary significantly depending on modeling domain and objectives. Other SA approaches will perform better for certain types of models, and could lead to different ranking of importance of model factors [9–11]. As SA methods differ on the underlying assumptions, it is generally recommended to use several methods, preferably with dissimilar foundations. Multi-method SA can be used as exploratory tool to provide improved understanding and deeper insights into the model structure. It provides a practical way to validate, reject, or reinforce sensitivity conclusions increasing robustness measures in identifying key inputs [9, 11–14].

In this work, a cardiovascular-respiratory system (CVRS) model that describes its reaction to a constant workload is validated. First, three different SA techniques are performed to identify which parameters are more/less influential to the model output, in particular, arterial systemic pressure P_{as} (and also alveolar ventilation $\dot{V}_A$). Then, a (partial) subset selection is performed to assess the best possible parameter set that could be estimated given specific measurement. Lastly, a set of model parameters is identified using the heart rate H, P_{as} and $\dot{V}_A$ profiles obtained during exercise on an ergometer. This is the first CVRS model validated utilizing $\dot{V}_A$ as an additional information to describe the system's dynamics.

The CVRS model under consideration includes the baroreceptor and chemoreceptor loop as essential controls of the system [15]. The interactions of the control mechanisms are modelled using an optimal control approach. This method has been applied to predict physiological markers within the system, such as, investigating the transition dynamics from rest to exercise under a constant workload [16–18], describing CVRS response under orthostatic stress conditions [19, 20], and modeling congestive heart failure in humans [21]. In Calderon, et al. [15], the dynamic response of a CVRS to a nonconstant workload is investigated by formulating an optimal control problem utilizing the Euler-Lagrange approach. In the current study, the modeling framework presented in [15] is validated to characterize the reaction of the CVRS under a constant workload.

The SA methods applied in this study are the traditional sensitivity analysis, partial rank correlation coefficient (PRCC), and extended Fourier Amplitude Sensitivity Test (eFAST). These techniques are chosen due to their varying assumptions and foundations. The first SA technique is a modification of the basic differential equations approach described in [22, 23]. This has been applied to a cardiovascular model in [24] utilizing computational approach, for instance, automatic differentiation, to derive the sensitivities of their model parameters describing its dynamic response from sitting to standing transition. While PRCC is an efficient sampling-based index method, eFAST is a reliable variance-based approach [25, 26]. Importantly, PRCCs provide a measure of monotonicity when the linear effects of other variables are removed, and eFAST measures fractional variance accounted for by individual and groups of variables [10]. The mentioned SA approaches assess and quantify different model properties.

Parameter values should be identified using observed measurements from specific subject since physiological quantities vary significantly between individuals. Measurements for H, P_{as}, and $\dot{V}_{\mathrm{A}}$ are obtained during exercise on an ergometer with moderate workload, for different subjects. In order to capture between-subject variabilities, model parameters are identified to characterize the given time-series dataset. Numerical results are presented for the parameter values obtained for three subjects showing different profiles.

Section 2 reviews the framework and optimal control formulation of the CVRS model under study. The algorithms for the three different SA techniques are briefly summarized in Sect. 3. Results of the SA approaches for CVRS model are presented in Sect. 4. In this section, subset selection and parameter identification procedures are discussed. Moreover, model validation is presented using hemodynamic and respiratory data obtained from ergometric exercise for different subjects. The conclusion section considers the key points in this work and outlook for future research directions.

2 Mathematical Model and Optimal Control Formulation

2.1 Cardiovascular-Respiratory System Model

This section presents a cardiovascular-respiratory system (CVRS) model describing its overall response under constant (also nonconstant) workload adapted from Calderon, et al. [15]. A global four-compartment cardiovascular system (CVS) component of the model is depicted in Fig. 1. For instance, the arterial systemic compartment is represented by lumped arteries of the systemic circuit. The arterial systemic, together with the venous systemic and the two pulmonary compartments are compliant vessels connected to each other via resistance vessels composed of the systemic and pulmonary peripheral region. The CVS dynamics is governed by nine state variables describing the pressures in the compartments, the contractilities of the ventricles and associated synthetic variables, and the heart rate. Refer to equations (1)–(9). On the other hand, the respiratory system (RS) component of the model is divided into the lung and tissue compartments. It is characterized by five state variables consisting of arterial and venous gas concentrations and alveolar ventilation given by Eqs. (10)–(14).

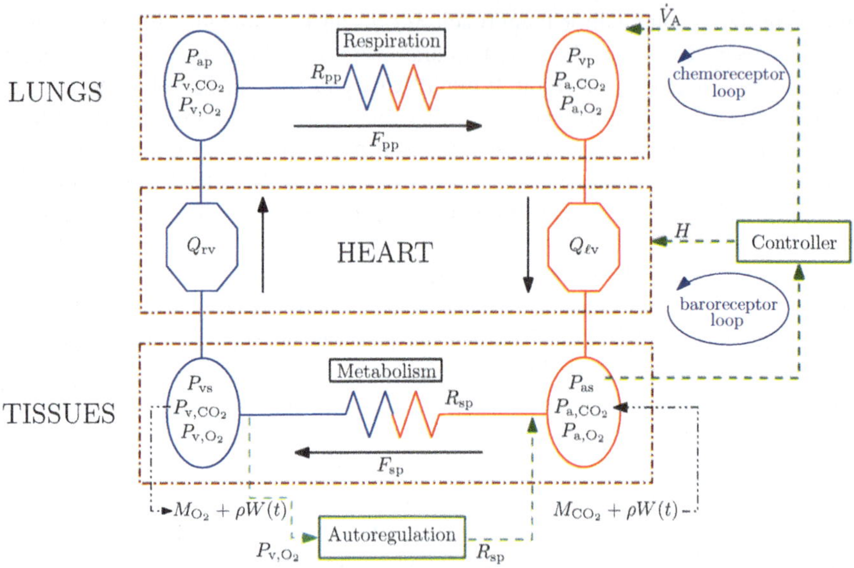

Fig. 1 Block diagram of a CVRS model depicting blood flow and control loops [15].

The control variable $u_1(t)$ represents the rate at which the heart frequency varies describing the baroreceptor loop which includes actions of the baroreceptors measuring the P_{as} and adjusting the heart rate. The control variable $u_2(t)$ denotes the rate at which ventilation rate changes representing the chemoreceptor loop regulating the partial pressure of oxygen P_{a,O_2} and carbon dioxide P_{a,CO_2} in arterial blood. The set of differential equations describing the CVRS model is given by

$$\frac{dP_{as}(t)}{dt} = \frac{1}{c_{as}}\big(Q_{\ell v}(t) - F_{sp}(t)\big), \tag{1}$$

$$\frac{dP_{vs}(t)}{dt} = \frac{1}{c_{vs}}\big(F_{sp}(t) - Q_{rv}(t)\big), \tag{2}$$

$$\frac{dP_{ap}(t)}{dt} = \frac{1}{c_{ap}}\big(Q_{rv}(t) - F_{pp}(t)\big), \tag{3}$$

$$\frac{dP_{vp}(t)}{dt} = \frac{1}{c_{vp}}\big(F_{pp}(t) - Q_{\ell v}(t)\big), \tag{4}$$

$$\frac{dS_{\ell v}(t)}{dt} = \sigma_{\ell v}(t), \tag{5}$$

$$\frac{d\sigma_{\ell v}(t)}{dt} = -\gamma_{\ell v}\sigma_{\ell v}(t) - \alpha_{\ell v}S_{\ell v}(t) + \beta_{\ell v}H(t), \tag{6}$$

$$\frac{dS_{rv}(t)}{dt} = \sigma_{rv}(t), \tag{7}$$

$$\frac{d\sigma_{\mathrm{rv}}(t)}{dt} = -\gamma_{\mathrm{rv}}\sigma_{\mathrm{rv}}(t) - \alpha_{\mathrm{rv}} S_{\mathrm{rv}}(t) + \beta_{\mathrm{rv}} H(t), \tag{8}$$

$$\frac{dH}{dt} = u_1(t), \tag{9}$$

$$V_{\mathrm{A,CO_2}} \frac{dP_{\mathrm{a,CO_2}}(t)}{dt} = 863 F_{\mathrm{pp}}(t)\big(C_{\mathrm{v,CO_2}} - K_{\mathrm{CO_2}} P_{\mathrm{a,CO_2}}(t) - K_{\mathrm{CO_2}}\big) + \dot{V}_{\mathrm{A}}(t)\big(P_{\mathrm{I,CO_2}} - P_{\mathrm{a,CO_2}}(t)\big), \tag{10}$$

$$V_{\mathrm{A,O_2}} \frac{dP_{\mathrm{a,O_2}}(t)}{dt} = 863 F_{\mathrm{pp}}(t)\left(C_{\mathrm{v,O_2}}(t) - K_{\mathrm{a,1}}\left(1 - e^{-K_{\mathrm{a,2}} P_{\mathrm{a,O_2}}(t)}\right)^2\right) + \dot{V}_{\mathrm{A}}(t)\big(P_{\mathrm{I,O_2}} - P_{\mathrm{a,O_2}}(t)\big), \tag{11}$$

$$V_{\mathrm{T,CO_2}} \frac{dC_{\mathrm{v,CO_2}}(t)}{dt} = M_{\mathrm{CO_2}} + \rho_{\mathrm{CO_2}} W(t) + F_{\mathrm{sp}}(t)\big(K_{\mathrm{CO_2}} P_{\mathrm{a,CO_2}}(t) + K_{\mathrm{CO_2}} - C_{\mathrm{v,CO_2}}(t)\big), \tag{12}$$

$$V_{\mathrm{T,O_2}} \frac{dC_{\mathrm{v,O_2}}(t)}{dt} = -M_{\mathrm{O_2}} - \rho_{\mathrm{O_2}} W(t) + F_{\mathrm{sp}}(t)\left(K_{\mathrm{a,1}}\left(1 - e^{-K_{\mathrm{a,2}} P_{\mathrm{a,O_2}}(t)}\right)^2 - C_{\mathrm{v,O_2}}(t)\right), \tag{13}$$

$$\frac{d\dot{V}_{\mathrm{A}}(t)}{dt} = u_2(t). \tag{14}$$

The left and right ventricular cardiac output are represented respectively, as

$$Q_{\ell\mathrm{v}} = H \frac{c_{\ell\mathrm{v}} a_{\ell\mathrm{v}}(H) P_{\mathrm{vp}} S_{\ell\mathrm{v}}}{a_{\ell\mathrm{v}}(H) P_{\mathrm{as}} + k_{\ell\mathrm{v}}(H) S_{\ell\mathrm{v}}}, \qquad Q_{\mathrm{rv}} = H \frac{c_{\mathrm{rv}} a_{\mathrm{rv}}(H) P_{\mathrm{vs}} S_{\mathrm{rv}}}{a_{\mathrm{rv}}(H) P_{\mathrm{ap}} + k_{\mathrm{rv}}(H) S_{\mathrm{rv}}},$$

where

$$k_{\ell\mathrm{v}}(H) = e^{-(c_{\ell\mathrm{v}} R_{\ell\mathrm{v}})^{-1} t_{\mathrm{d}}(H)}, \qquad a_{\ell\mathrm{v}}(H) = 1 - k_{\ell\mathrm{v}}(H),$$
$$k_{\mathrm{rv}}(H) = e^{-(c_{\mathrm{rv}} R_{\mathrm{rv}})^{-1} t_{\mathrm{d}}(H)}, \qquad a_{\mathrm{rv}}(H) = 1 - k_{\mathrm{rv}}(H),$$

and the diastolic duration t_{d} is expressed as

$$t_{\mathrm{d}}(H) = \frac{1}{H^{1/2}}\left(\frac{1}{H^{1/2}} - \kappa\right).$$

The blood flow across the systemic and pulmonary regions are given by

$$F_{\mathrm{sp}} = \frac{1}{R_{\mathrm{sp}}}(P_{\mathrm{as}} - P_{\mathrm{vs}}) \quad \text{and} \quad F_{\mathrm{pp}} = \frac{1}{R_{\mathrm{pp}}}(P_{\mathrm{ap}} - P_{\mathrm{vp}}),$$

respectively, where the arterial systemic resistance R_{sp} is dependent on the venous oxygen concentration $C_{\mathrm{v,O_2}}$, that is,

$$R_{\mathrm{sp}} = A_{\mathrm{pesk}} C_{\mathrm{v,O_2}}.$$

Further details on model development can be found in [15, 27]. For convenience of the reader, a table listing all the model parameters, its description, nominal values and units is provided in Appendix A.1.

2.2 Control Formulation

The optimal control problem seeks to obtain time-dependent controls $u_1(t)$ and $u_2(t)$ that minimizes the cost functional

$$J(u) = \frac{1}{2}\int_0^T \Big(\big(P_{\text{as}}(t) - P_{\text{as}}^{\text{exer}}\big)^2 + \big(P_{\text{a,CO}_2}(t) - P_{\text{a,CO}_2}^{\text{eq}}\big)^2 + \omega_1 u_1(t)^2 + \omega_2 u_2(t)^2 \Big)\, dt\,, \tag{15}$$

subject to the differential equations (1)–(14) including the auxilliary equations which can be compactly written as

$$\frac{d\,x(t)}{dt} = \mathcal{F}\big(x(t), p, W(t)\big) + Bu(t), \quad x(0) = x^{\text{rest}}\,, \tag{16}$$

where the state and parameter vector are given, respectively, by

$$x = \Big(P_{\text{as}}, P_{\text{vs}}, P_{\text{ap}}, P_{\text{vp}}, S_{\ell\text{v}}, \sigma_{\ell\text{v}}, S_{\text{rv}}, \sigma_{\text{rv}}, H, P_{\text{a,CO}_2}, P_{\text{a,O}_2}, C_{\text{v,CO}_2}, C_{\text{v,O}_2}, \dot{V}_{\text{A}} \Big)^{\mathsf{T}} \in \mathbb{R}^{14}, \tag{17}$$

and

$$p = \Big(c_{\text{as}}, c_{\text{vs}}, c_{\text{ap}}, c_{\text{vp}}, c_{\ell\text{v}}, c_{\text{rv}}, R_{\text{pp,rest}}, R_{\text{pp,exer}}, R_{\ell\text{v}}, R_{\text{rv}}, \alpha_{\ell\text{v}}, \alpha_{\text{rv}}, \beta_{\ell\text{v}}, \beta_{\text{rv}}, \gamma_{\ell\text{v}}, \gamma_{\text{rv}}, \kappa, A_{\text{pesk,rest}}, A_{\text{pesk,exer}}, M_{\text{O}_2}, M_{\text{CO}_2}, \rho_{\text{O}_2}, \rho_{\text{CO}_2}, P_{\text{I,O}_2}, P_{\text{I,CO}_2}, V_{\text{A,O}_2}, V_{\text{A,CO}_2}, V_{\text{T,O}_2}, V_{\text{T,CO}_2}, K_{\text{CO}_2}, K_{\text{CO}_2}, K_{\text{a,1}}, K_{\text{a,2}} \Big)^{\mathsf{T}} \in \mathbb{R}^{33}. \tag{18}$$

Here, $W(t)$ denotes the imposed workload, $u(t) = \big(u_1(t), u_2(t)\big)^{\mathsf{T}}$ is the control vector and the matrix $B = (b_{i,j}) \in \mathbb{R}^{14\times 2}$ such that $b_{9,1} = b_{14,2} = 1$ and 0 otherwise. This optimal control problem is solved using Euler-Lagrange approach. Details on the formulation and numerical framework are presented in [15].

3 Sensitivity Analysis

Sensitivity analysis (SA) quantifies the contribution of different input sources to the variations in the model outputs. It investigates significant inputs (such as model parameters and initial conditions) and assesses how the degree of uncertainty influences the model outcome(s). It sheds light on the relevance of the input factors inducing largest/least variations in the model output, as well as possible interactions that could amplify or diminish the variability caused by individual factors [7, 11]. Different SA approaches will naturally perform differently for specific model structures. The objective of performing different SA methods is to verify the consistency of the model behaviour and/or to assess the robustness of the simulation results to uncertain inputs or model assumptions (see for instance [28]). This section briefly presents three different methods – traditional sensitivity analysis, partial rank correlation coefficient (PRCC), and extended Fourier amplitude sensitivity test (eFAST).

3.1 *Traditional Sensitivity Analysis*

The traditional sensitivity analysis is implemented following the procedure presented in [24]. Let $X = (X_1, X_2, ..., X_{n_S})$ and $\mu = (\mu_1, ..., \mu_{n_P})$ be the state and parameter vectors, respectively, where n_S and n_P are the number of states and parameters, respectively. Let $\dot{X} = F(X; \mu)$ be the system of ordinary differential equations which models the dynamics of X. Suppose we are interested in the dynamics of X during the time interval $[0, T]$. The algorithm to compute traditional sensitivities is as follows:

1. Discretize the time interval $[0, T]$ into $N + 1$ equally spaced points separated by the length $h = T/N$. Given the initial condition X_0 for the states and the set of nominal parameters μ_0, solve the system $\dot{X} = F(X; \mu)$ numerically from $t = 0$ to $t = T$.
2. Form the system $\frac{d}{dt}\frac{\partial X_i}{\mu_j}$. Assuming that the partial derivatives commute, this is equivalent to $\frac{\partial}{\partial \mu_j} F_i(X; \mu)$. By the chain rule, we have
$$\frac{d}{dt}\frac{\partial X_i}{\partial \mu_j} = \frac{\partial}{\partial \mu_j} F_i(X; \mu) = \sum_{k=1}^{n_S} \left(\frac{\partial F_i}{\partial X_k}\frac{\partial X_k}{\mu_j} \right) + \frac{\partial F_i}{\partial \mu_j}, \tag{19}$$
where $\frac{\partial F}{\partial X}$ and $\frac{\partial F}{\partial \mu}$ are calculated using automatic differentiation [29].
3. Solve the system in Step 2 using backward Euler scheme with the information from Step 1 .
4. Compute the (non-dimensional) sensitivity of state X_i to parameter μ_j as:

$$S_{ij}(t,\mu)|_{\mu=\mu_0} = \left.\frac{\partial X_i(t,\mu)}{\partial \mu_j}\frac{\mu_j}{X_i(t,\mu)}\right|_{\mu=\mu_0}, \quad \mu_j, X_i(t,\mu) \neq 0. \tag{20}$$

5. Sensitivities of X_i with respect to the model parameter μ_j can then be computed as

$$\max_{k\in 1,\ldots,N+1} S_{ij}(t_k,\mu)|_{\mu=\mu_0}. \tag{21}$$

The preceding algorithm will output the sensitivity of X_i with respect to each model parameter. With this information, parameters can then be ranked according to their X_i sensitivities.

3.2 *Partial Rank Correlation Coefficient (PRCC)*

A metric which quantifies the linear relationship between the output y and an input μ_j when the linear effects on y of the other inputs μ_k, $k \neq j$ are removed is referred to as Partial correlation coefficient (PCC) . When PCC is performed on the rank-transformed input and output, it yields the Partial Rank Correlation Coefficient (PRCC).

In order to obtain PRCC, *Latin Hypercube Sampling* (LHS), which is a stratified sampling without replacement technique, is selected as sampling method for the input parameters. Here, a nominal distribution to each parameter of the model is assigned, and sampling for each parameter independently is carried out. We then perform M simulations of the model, wherein at each simulation we draw a parameter value from the distribution. From these simulations, we consider time 'slices' which are points of interest where changes in parameter affect the model output. We then obtain the PRCC of the model output at this slice with respect to each parameter.

The PRCC lies between –1 and 1. The higher the absolute value of the PRCC, the larger the correlation of the parameter on the output. Thus, when we rank the parameters, we will consider the absolute value of the PRCC. As it is a sampling based method, the results are highly dependent on the distribution we select for each parameter. In addition, significance of results will be higher when a larger M is selected.

3.3 *Extended Fourier Amplitude Sensitivity Test (eFAST)*

Extended Fourier Amplitude Sensitivity Test (eFAST) is a variance decomposition method. The output variation is quantified in terms of statistical variance. The eFAST algorithm is able to partition the output variance in a way that we can determine what fraction of the output variance can be associated to the variation of each input parameter.

In eFAST, a sinusoidal function (search curve) of a particular frequency is assigned to each parameter. This function is dependent on the distribution we set for the parameter. N_S samples of each parameter are taken from this function. Since the sinusoidal function has certain symmetry properties, repeated samples will eventually be obtained. To circumvent this issue, we use a resampling scheme: we run the eFAST algorithm N_R times, each time adding a random phase shift to the search curve. Thus, the total number of model simulations is given by $N_S \times N_R \times k$, where k is the number of parameters. An analysis of the effect of input parameter variation on the model output is done using Fourier analysis.

For each parameter i, we obtain two types of sensitivities from eFAST analysis: first-order sensitivity S_i and total-order sensitivity S_{ti}. The first-order sensitivity is the fraction of model output variance attributed to the input variation of the given parameter. On the other hand, the total-order sensitivity of parameter i is the variance remaining after removing the contribution of the complementary set S_{ci}, where S_{ci} represents the summed sensitivity index of the entire complementary set of parameters. The total-order sensitivity includes higher order nonlinear interactions between the parameter of interest and the complementary set of parameters.

Significance of these sensitivities can be obtained by comparison with the sensitivities of a dummy parameter, which does not appear in the model equations. A t-test is performed between the sensitivity of a parameter and the dummy to test for significant difference.

4 Results and Discussion

4.1 CVRS Model Sensitivity

In order to identify the best subset of model parameters to estimate given experimental data, sensitivity analysis is performed *a priori*. In the CVRS model under study, we consider P_{as} as our target variable or the model output of interest. We apply the three SA methods discussed in the previous section to our CVRS model. The first one, traditional sensitivity analysis, looks at the variation of the output, through its derivative, when we apply small changes to the parameters set at their nominal values. This method requires the use of an efficient numerical scheme, automatic differentiation, to compute the derivatives of the model equations. The second and third scheme, PRCC and eFAST, are methods that rely on sampling. We assume that each parameter is distributed according to some known distribution. We then run the model for various samples from these distributions, and look at their effect on the model output. In PRCC, we compute correlation between the parameter and the output, while in eFAST, we perform Fourier analysis on the output to detect the input parameters' signatures.

In contrast to traditional sensitivity analysis, for PRCC and eFAST we have to select time points within the model run wherein we want to compute the sensitivities.

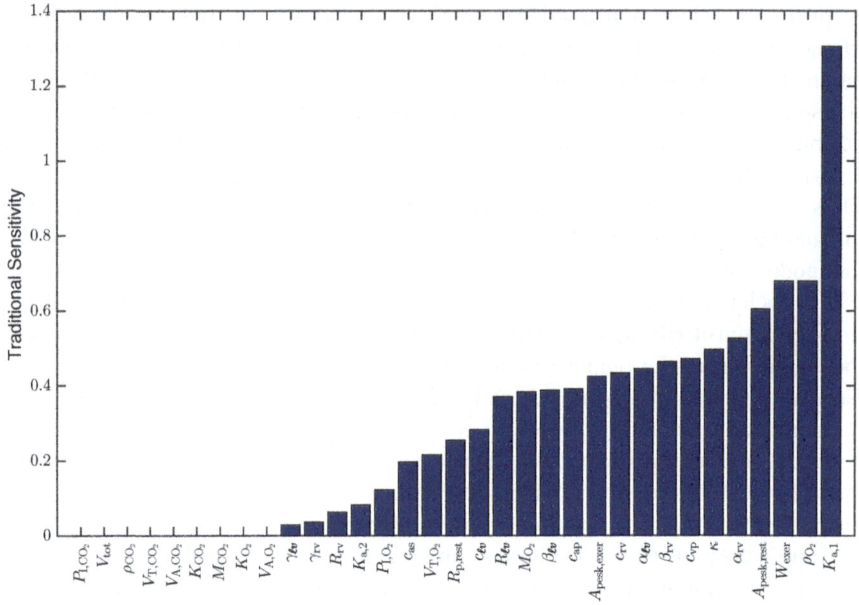

Fig. 2 Bar graph of traditional sensitivity of P_{as} with respect to model parameters showing $K_{a,1}$ as the most sensitive parameter.

For our analysis, we chose three time points corresponding to different conditions of the test person: $t = 4$ min corresponding to the rest phase, $t = 6$ min corresponding to the point just after the workload is introduced, and $t = 12$ min corresponding to the exercise phase. For each parameter, we assign a uniform distribution as the sampling distribution. The minimum and maximum values of the uniform distribution are set to -10% and 10% of the nominal values. We tested the robustness of the results for a larger range, namely -20% and 20%, and we found similar qualitative results.

Figure 2 shows the traditional sensitivity of P_{as} to each model parameter. For our PRCC analysis, we set the number of samples to $M = 1000$. For significance, we set the confidence level to be $\alpha = 0.05$. Figure 3 shows PRCC for each time point and each parameter and Fig. 4 depicts the total sensitivity obtained from eFAST.

The three methods agree in the general rankings of the parameters where $K_{a,1}$ appears at the top of the sensitivity rankings, followed by ρ_{O_2} and M_{O_2}. Note that $K_{a,1}$ appears in the model equations as the coefficient that governs the dissociative law for O_2, while ρ_{O_2} and M_{O_2} appear as the multiplier to the workload and the at-rest metabolic rate for O_2. Since the scenario we are modelling is the transition from rest to exercise, we expect that these three parameters have high sensitivity. These three influence how the body expends O_2, which increases significantly during exercise. Notice however that ρ_{O_2} appears low in the ranking for the rest time point ($t = 4$ min) for the PRCC and eFAST analyses. This is to be expected since ρ_{O_2} does not influence rest dynamics since the workload is null at this point. High in the rankings as well

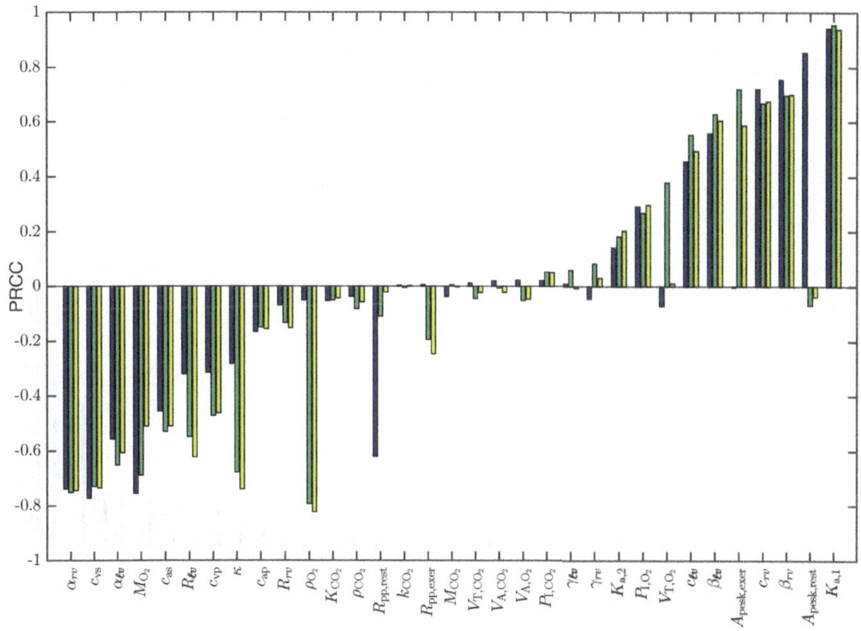

Fig. 3 Bar graphs of PRCCs of model parameters across the three time points: $t = 4, 6, 12$ min corresponding to rest, transition from rest to exercise, and exercise phase, respectively. Here, P_{as} is considered as the target output. Positive (negative) PRCC values signify positive (negative) correlations of the parameter with the model output. For instance, an increase in $K_{a,1}$ will result in an increase in P_{as}, while an increase in ρ_{O_2} will reduce P_{as}. The difference of the PRCC values across time points indicates variable sensitivities for different phases.

are $A_{pesk,rest}$ and $A_{pesk,exer}$, which are connected to local metabolic control, i.e., when energy demand in a tissue region is high, blood flow is increased to supply more O_2. Again, we see in PRCC and eFAST that the former is high during rest, while the latter is high during exercise. High in the rankings as well are the synthetic variables $\alpha_{\ell v}, \beta_{\ell v}, \alpha_{rv}, \beta_{rv}$. These parameters are part of the feedback loop which regulates heart rate depending on the arterial pressure P_{as}. Since the synthetic parameters are not observable, we did not include them in the parameter set of interest.

4.2 *Subset Selection*

We performed subset selection in order to see how our selected set of five parameters fared against other sets of parameters. Suppose Y_i is the model output and $z(t_i; \theta)$, $\theta \in \mathbb{R}^u$ represents the model dynamics under the parameter set θ. Here, t_i represents the ith time point and u is the number of parameters in the model. We follow the algorithm presented in [30]:

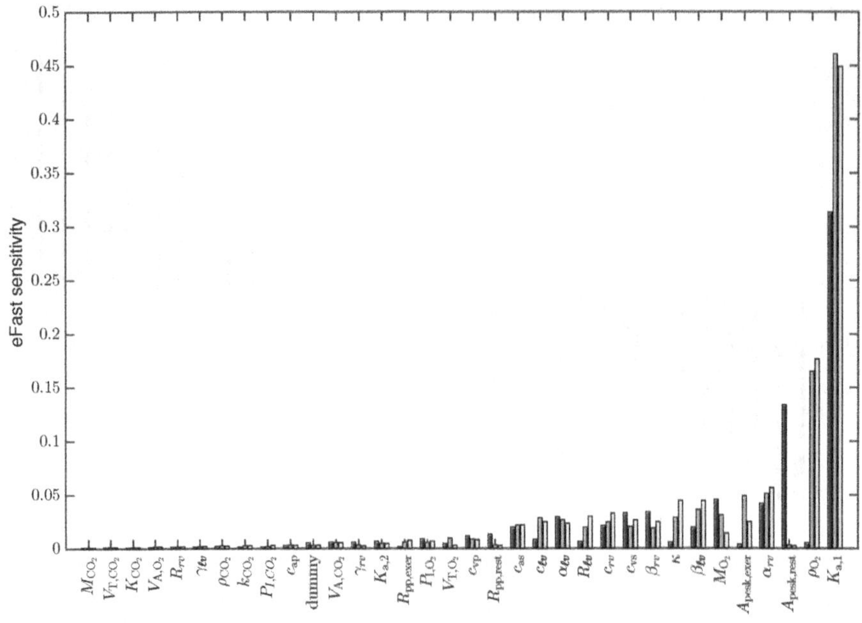

Fig. 4 Bar graphs of the total sensitivity-order S_{ti} obtained from eFAST across the three time points: $t = 4, 6, 12$ min corresponding to rest, transition from rest to exercise, and exercise phase, respectively. The total sensitivity-order indicates the amount of variation in the model output (P_{as}) in conjunction with all other parameters. As depicted, $K_{\text{a},1}$ and ρ_{O_2} are top two dominant parameters influencing variations in P_{as}.

1. For a fixed value of $p < u$, set $u - p$ parameters to their nominal values. Form the set S_p consisting of p-length subsets of the T parameters which we want to vary.
2. Calculate $\Theta_p = \{\theta | \theta \in S_p, rank(\chi(\theta)) = p\}$ where $\chi_{i,j}(\theta) = \dfrac{\partial z(t_i;\theta)}{\partial \theta_j}$ is the sensitivity matrix.
3. For each $\theta \in \Theta_p$, calculate $\sqrt{(\Sigma(\theta)_{ii})}$, where $\Sigma(\theta) = \sigma_0^2[\chi(\theta)^T\chi(\theta)]^{-1}$ and σ_0^2 is estimated as $\dfrac{1}{n-p}\sum_{i=1}^{n}[y_i - z(t_i;\theta)]^2$. The quantity $\sqrt{(\Sigma(\theta)_{ii})}$ is the asymptotic standard error associated with estimating parameter i.
4. Calculate the condition number of the Fisher information matrix $\chi(\theta)^T\chi(\theta)$, denoted by $\text{cond}\big(F(\theta)\big)$, and the selection score $\alpha(\theta) = |\nu(\theta)|$, where $\nu_i(\theta) = \dfrac{\sqrt{(\Sigma(\theta)_{ii})}}{\theta_i}$.
5. Rank the members of Θ_p from lowest selection score to highest.

Table 1 Top 10 subsets of S_p indicating the condition number $\text{cond}\big(F(\theta)\big)$ and ranked according to selection score $\alpha(\theta)$.

Rank	Parameter vector θ	$\text{cond}\big(F(\theta)\big)$	$\alpha(\theta)$
1	κ, ρ_{O_2}, $A_{\text{pesk,rest}}$, $K_{a,1}$, $A_{\text{pesk,exer}}$	123787673755.89	0.02094727
2	κ, ρ_{O_2}, $R_{\text{pp,rest}}$, $K_{a,1}$, $A_{\text{pesk,exer}}$	91304443302.66	0.02929534
3	M_{O_2}, ρ_{O_2}, $A_{\text{pesk,rest}}$, $K_{a,1}$, $R_{\text{pp,rest}}$	11215255210.14	0.04414743
4	κ, $R_{\text{pp,rest}}$, $A_{\text{pesk,rest}}$, $K_{a,1}$, $A_{\text{pesk,exer}}$	9104217027.98	0.04584029
5	M_{O_2}, ρ_{O_2}, $R_{\text{pp,rest}}$, $K_{a,1}$, $R_{\text{pp,exer}}$	15527731.93	0.04704955
6	M_{O_2}, ρ_{O_2}, $A_{\text{pesk,rest}}$, $K_{a,1}$, $A_{\text{pesk,exer}}$	239316853176.63	0.05771713
7	M_{O_2}, ρ_{O_2}, $R_{\text{pp,rest}}$, $K_{a,1}$, $A_{\text{pesk,exer}}$	235373492836.74	0.06046162
8	M_{O_2}, $R_{\text{pp,rest}}$, $A_{\text{pesk,rest}}$, $K_{a,1}$, $R_{\text{pp,exer}}$	2949490093.64	0.07996181
9	κ, ρ_{O_2}, $A_{\text{pesk,rest}}$, $K_{a,1}$, $R_{\text{pp,exer}}$	16115906426.15	0.08112010
10	M_{O_2}, $R_{\text{pp,rest}}$, $A_{\text{pesk,rest}}$, $K_{a,1}$, $A_{\text{pesk,exer}}$	4696022262.42	0.09759110

We consider S_p to be the collection of subsets which include:

1. $S_p^{\text{int}} = \{A_{\text{pesk,rest}}, A_{\text{pesk,exer}}, M_{O_2}, \rho_{O_2}, K_{a,1}\}$;
2. subsets wherein one parameter in S_p^{int} is replaced by one member of $\{R_{\text{pp,rest}}, R_{\text{pp,exer}}, \kappa\}$; and
3. subsets wherein two parameters in S_p^{int} are replaced by two members of $\{R_{\text{pp,rest}}, R_{\text{pp,exer}}, \kappa\}$.

In total, there are 46 subsets included in S_p.

Note that the $\text{cond}\big(F(\theta)\big)$ indicates the ratio between the largest and the smallest eigenvalue of the Fisher information matrix. The larger $\text{cond}\big(F(\theta)\big)$ is, the harder it is to identify such parameter subset. Hence, it is recommended to choose a parameter subset with good selection score and reasonable condition number in identifying such subset.

Table 1 shows the top 10 subsets in S_p according to selection score. Note that our parameter set of interest S_p^{int} is ranked 6 out of 46 which indicates local identifiability.

In this study, PRCC is also performed using $\dot{V}_A$ as the target output. In order to perform sensitivity analysis, we set H as a parameter. The control is solved once with the nominal parameter values and used it throughout each run. As shown in Appendix Fig. 7.2, M_{CO_2} is the most significant parameter influencing the behaviour of $\dot{V}_A$. Hence, we included M_{CO_2} in our parameter set to be estimated which is presented in the succeeding section.

4.3 Model Identification

The identifiability of a model is assessed to determine parameters from measured data. Model identifiability refers to the goal of ascertaining unambiguous and accurate parameter estimation [31]. Parameter estimation is a process that identifies values in a parameter set such that the model output is as close as possible to the corresponding set of measurements. This involves minimizing an error between model output and observations. It should be noted that the quality of the parameter estimates depends, among others, on the model structure, reliability of the available data, and error criterion involved in the numerical computation [32].

The data used in this study is obtained from [33] where 8 test subjects (5 males and 3 females) with ages ranging from 25 to 30 years were recruited as volunteers to participate into three stress ergometer challenges with different workload sequences. The study focused on the dynamics of volatile organic compounds (VOCs), in particular, isoprene and acetone during ergometric exercise. Moreover, a real-time recording setup had been developed to monitor hemodynamic, respiratory, and VOC-related profiles during workload scenarios. Details of data acquisition and experimental setup are documented in [34]. Since the current work investigates the short-term regulation of CVRS under constant workload, only the first 15-min hemodynamic (i.e., heart rate H and P_{as}) and respiratory (i.e., alveolar ventilation) data are considered where the first 5-min measurement record the rest phase and the next 10-min monitor the exercise phase with workload of $W = 75$ W.

In order to estimate the parameter set of interest, the following procedures are taken. Let $t_{\text{init}}, t_{\text{trans}}, t_{\text{end}}$ be the initial time, transition time when the subject begins exercising, and the terminal time, respectively. If $t \in [t_{\text{init}}, t_{\text{trans}})$, the subject is at rest. At time t_{trans}, exercise begins, and the parameters instantaneously change from rest to exercise values. In the following, we set $t_{\text{init}} = 0$ and $t_{\text{end}} = 15$ and t_{trans} was adjusted based on the data ($t_{\text{trans}} \approx 5$min).

Given raw data for the heart rate H_{data}, systolic arterial systemic pressure $P^{\text{data}}_{\text{as,dias}}$, diastolic arterial systemic pressure $P^{\text{data}}_{\text{as,dias}}$, and alveolar ventilation $\dot{V}^{\text{data}}_{\text{A}}$ over the whole 15 minutes, we perform the following steps:

1. Smoothen the given raw data by calculating their moving averages, and select the time points $\{t_0, t_1, ..., t_N\}$ which matches the discretization N of the time interval used to numerically solve the model.
2. Using H_{data}, calculate the average H over the rest period, i.e., the first 5 minutes, and the average H over the exercise period, i.e., the last 10 minutes. Set these values as the equilibrium H^{rest} and H^{exer}, respectively. We use these to calculate the equilibrium values of all the other state variables.
3. Calculate the mean arterial pressure from the data as

$$P^{\text{data}}_{\text{as}} = P^{\text{data}}_{\text{as,dias}} + \frac{1}{3}(P^{\text{data}}_{\text{as,dias}} - P^{\text{data}}_{\text{as,dias}}). \tag{22}$$

4. Initialize the parameters of the model to the nominal values μ_0. Let Q be the set of parameters to be estimated. We find the values of Q which minimize the deviation of the model output P_{as} from the data. That is, we minimize the following cost functional,

$$
\begin{aligned}
G(Q) = a_1^{\text{rest}} & \sum_{t_i \in rest} \left(P_{\text{as}}(t_i; Q) - P_{\text{as,i}}^{\text{data}}\right)^2 \\
+ a_1^{\text{exer}} & \sum_{t_i \in exer} \left(P_{\text{as}}(t_i; Q) - P_{\text{as,i}}^{\text{data}}\right)^2 \\
+ a_3^{rest} & \sum_{t_i \in rest} \left(\dot{V}_{\text{A}}(t_i; Q) - \dot{V}_{\text{A}}^{\text{data}}\right)^2 \\
+ a_3^{exer} & \sum_{t_i \in exer} \left(\dot{V}_{\text{A}}(t_i; Q) - \dot{V}_{\text{A}}^{\text{data}}\right)^2 .
\end{aligned}
\tag{23}
$$

In each iteration of the optimization, we solve the model given the parameters in Q and compare the simulated P_{as} values with the data. Note that we do not include H in the cost functional since we used H_{data} to define our equilibrium H^{rest} and H^{exer}. We use $(a_1^{\text{rest}}, a_1^{\text{exer}}) = (33.11, 3.30)$ as in [27]. Further, we set $(a_3^{\text{rest}}, a_3^{\text{exer}}) = 5(a_1^{\text{rest}}, a_1^{\text{exer}})$.

5. Solve the model using the optimal values of Q and plot the model output versus the data.

We performed parameter estimation on eight data sets, but will show the results on three sets namely: Subject 1, Subject 2, and Subject 3. We chose set Q based on the results from our sensitivity analysis and subset selection. Recall that we narrowed down the parameter space down to 6, and so we set

$$
Q = \left\{K_{\text{a,1}}, \rho_{\text{O}_2}, M_{\text{O}_2}, A_{\text{pesk,rest}}, A_{\text{pesk,exer}}, M_{\text{CO}_2}\right\}.
$$

The numerical results consist of two-level simulations: solving the control and adjusting the parameters given measurement points. Figures 5–7 depict the data for H, P_{as}, $\dot{V}_{\text{A}}$ (blue), corresponding moving averages (yellow) and simulation results with identified parameters (red). The solution captures the rest to exercise transition dynamics of the heart rate for the three subjects (see Figs. 5(a), 6(a) and 7(a)). An increase in P_{as} is evident during exercise when ergometric workload is $W = 75$ watts for subjects 1 and 3 (refer to Figs. 5(b) and 7(b)), while the collected data for subject 2 (Fig. 6(b)) do not clearly show this trend. Further, note the different physiological response of individuals to the same workload. However, it can be observed that simulation results reasonably agree with the measurement. On the other hand, it is depicted in the figures that $\dot{V}_{\text{A}}$ measurement also increase during exercise. Although this transition dynamics are illustrated by the simulations, only the results for subject 3 fits the data well as illustrated in Fig. 7(c). The unsatisfactory outcome could include, among others, model structure needing revision, accuracy of measurement, and data integrity.

Table 2 shows the optimal parameter values obtained using the algorithm presented in the previous section. We specified here, as well, the asymptotic standard errors of the estimate of each parameter i in Q, given by $\sqrt{(\Sigma(\theta)_{ii})}$, as discussed in Sect. 4.2. Unfortunately, standard errors are not obtained for M_{CO_2} since the reference output, $\dot{V}_{\text{A}}$ is attained via solving the control problem. This entails more involved numerical

Table 2 Optimal parameter values in set Q and asymptotic standard errors

Parameter	Nominal	Subject 1	Subject 2	Subject 3
$A_{pesk,rest}$	177.68	178.44 ± 4.7749	176.51 ± 5.1744	184.67 ± 5.3360
$A_{pesk,exer}$	270.00	259.07 ± 22.8044	257.87 ± 24.712	280.36 ± 25.4841
ρ_{O_2}	0.0110	0.0193 ± 0.0017	0.0144 ± 0.0018	0.0189 ± 0.0019
$K_{a,1}$	0.2000	0.2439 ± 0.0277	0.2031 ± 0.0300	0.2067 ± 0.0310
M_{O_2}	0.3500	0.3571 ± 0.1122	0.3510 ± 0.1216	0.2806 ± 0.1254
M_{CO_2}	0.2800	0.3143	0.5605	0.2806

considerations. Note the variation in the achieved values describing CVRS reaction to a constant workload for different individuals based on the collected data.

5 Conclusion

The cardiovascular-respiratory system (CVRS) model considered in this study works on specific domain of validity. Several simplifying assumptions are taken into account to describe an aspect of CVRS response to constant (moderate) workload. Optimal control theory is utilized to characterize the short-term regulation and interactions of baroreceptor and chemoreceptor loops. The modeling approach employs the Euler-Lagrange formulation of the optimal control problem. Due to simplifications, the CVRS model like any other models, is subject to certain degree of uncertainties.

Assessing model uncertainties is an integral part of modeling process. Several SA methods are proposed in various literature to quantify these uncertainties. Since different SA approaches vary significantly depending on its framework, modeling domain, and objectives, a multi-method SA can be performed. It provides better and deeper understanding of the model structure. It further supports the identified key model inputs to be accepted or rejected, and validates its significant influence in the outcome(s).

In this work, three different SA techniques are performed to determine which model parameters essentially contribute to the variations in the model output. Considering various schemes, assumptions, and foundations, traditional sensitivity analysis, partial rank correlation coefficient, and extended Fourier Amplitude Sensitivity Test are carried out. The parameter $K_{a,1}$ which is the coefficient governing the dissociative law for O_2 consistently topped the sensitivity rankings across the various methods. This indicates that $K_{a,1}$ has crucial effect on the reference output, arterial pressure P_{as}. Though different in rankings among the SA approaches, several parameters showed consistent influence in the output. The next concern is to determine a set of sensitive parameters influential to the output, which can be identified from a given dataset. In order to address this issue, a (partial) subset selection is performed. Here, synthetic sensitive parameters are disregarded and those with physiological impor-

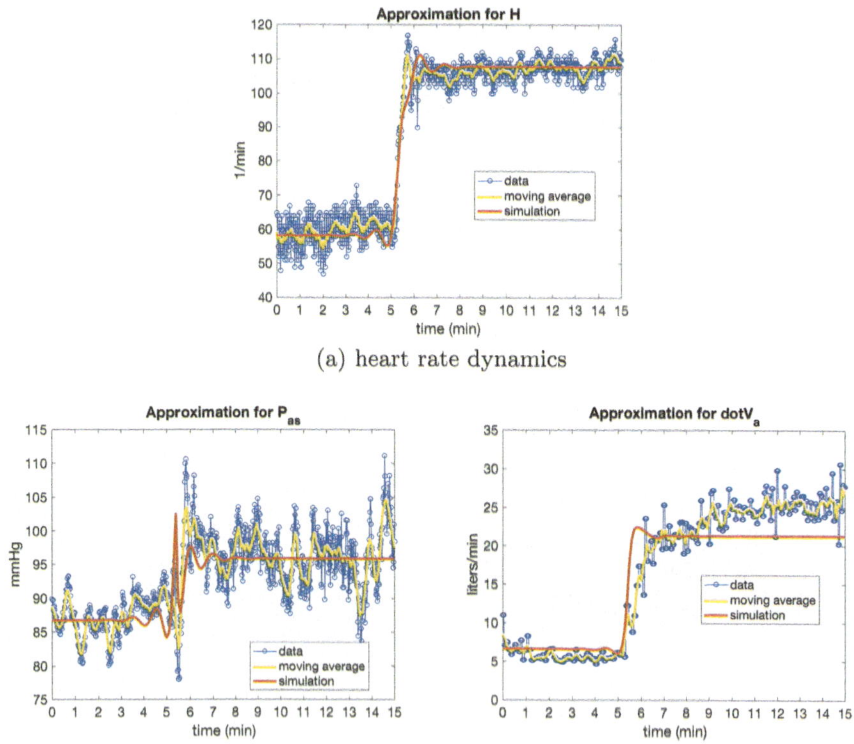

(a) heart rate dynamics

(b) arterial pressure dynamics

(c) alveolar ventilation dynamics

Fig. 5 The (a) heart rate H, (b) systemic arterial pressure P_{as}, and (c) alveolar ventilation $\dot{V}_{\mathrm{A}}$ dynamics obtained from the data (connected blue circles), moving average (yellow curve), and model simulation with the identified parameters (red curve) of Subject 1.

tance are included. This method streamlined the set of parameters to be identified from available measurements. Real-time hemodynamic and respiratory measurements obtained from ergometric exercise is used to validate the CVRS model. The parameter set $\{K_{\mathrm{a},1}, \rho_{\mathrm{O}_2}, M_{\mathrm{O}_2}, A_{\mathrm{pesk,rest}}, A_{\mathrm{pesk,exer}}, M_{\mathrm{CO}_2}\}$ is identified for different subjects showing between-subject variabilities.

The results presented here signify satisfactory work of the optimal control formulation for the CVRS model. Methods on parameter sensitivities indicate the robustness of model structure and to some degree, validates the chosen parameter set to be estimated. To the author's best knowledge, this is the first time alveolar ventilation $\dot{V}_{\mathrm{A}}$ is considered to identify parameters for CVRS model. However, several points need careful attention for further studies. For instance, a more exhaustive and reliable $\dot{V}_{\mathrm{A}}$ measurement could give dependable estimates. Within-subject variabilities can also be explored if repeated data acquisition for the same subjects can be done. This would further limit the uncertainties in the estimates. Different SA approaches can be investigated for a more substantial conclusions. Hemodynamic and respiratory

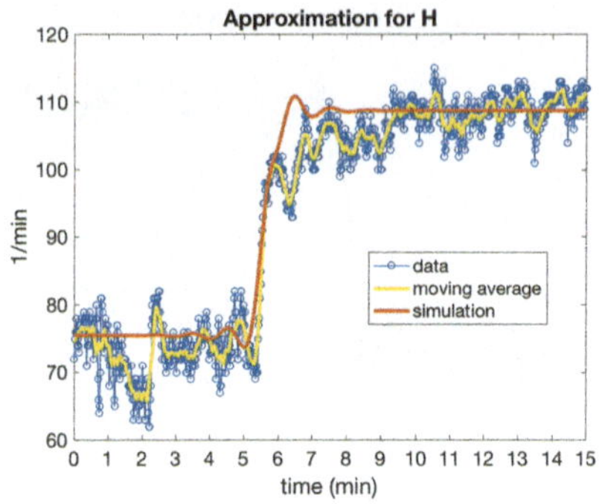

(a) heart rate dynamics

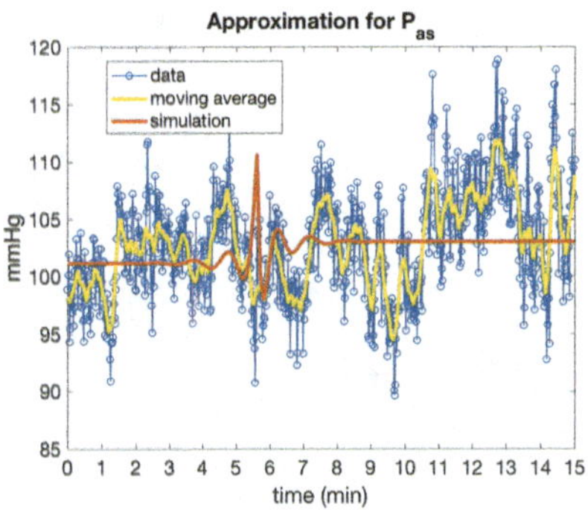

(b) arterial pressure dynamics

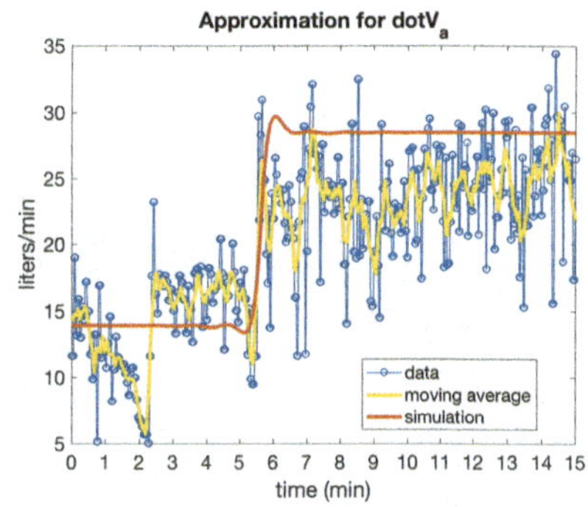

(c) alveolar ventilation dynamics

Fig. 6 The (a) heart rate H, (b) systemic arterial pressure P_{as}, and (c) alveolar ventilation $\dot{V}_{\mathrm{A}}$ dynamics obtained from the data (connected blue circles), moving average (yellow curve), and model simulation with the identified parameters (red curve) of Subject 2.

measurements for time-varying workload protocol can be designed to acquire data to verify and validate the reaction of CVRS under nonconstant workload.

Acknowledgments We acknowledge the support given by the Philippine Council for Health Research and Development (PCHRD) of the Department of Science and Technology (DOST), and the Office of the Vice President for Academic Affairs (OVPAA) of the University of the Philippines. We also thank Karl Unterkofler of the Institute for Breath Research, University of Innsbruck, Austria for sharing their ergometric hemodynamic and respiratory data, and insightful discussion with Julian King.

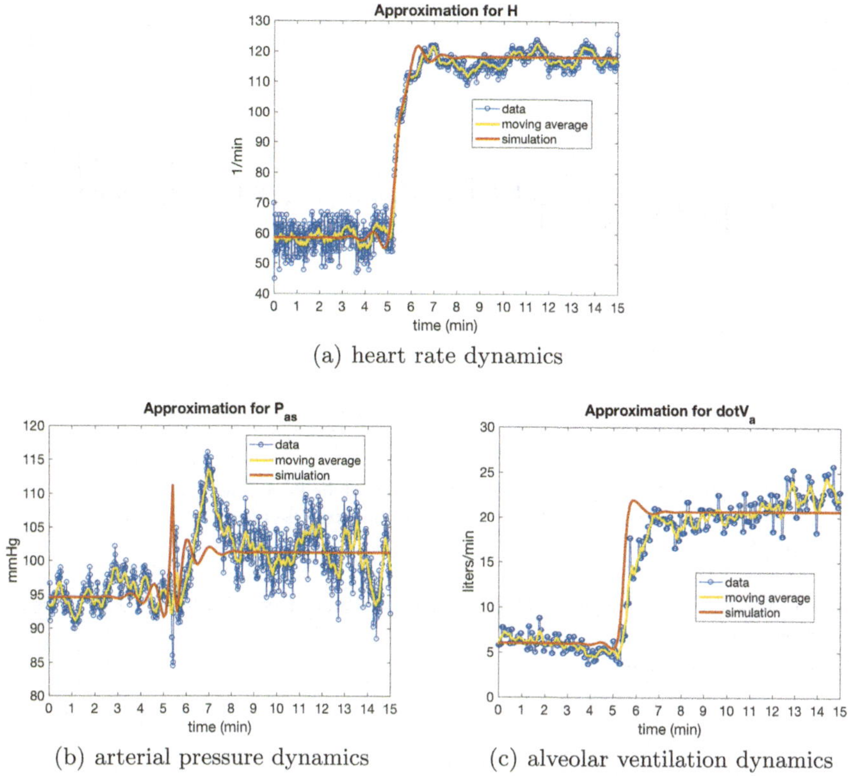

Fig. 7 The (a) heart rate H, (b) systemic arterial pressure P_{as}, and (c) alveolar ventilation $\dot{V}_{\mathrm{A}}$ dynamics obtained from the data (connected blue circles), moving average (yellow curve), and model simulation with the identified parameters (red curve) of Subject 3.

A Appendix

A.1 Parameters in the CVRS Model

Table 3 lists the model parameters, description, values and corresponding units used in the simulation. Values of the parameters are taken from [35].

A.2 Sensitivity of Alveolar Ventilation

Figure 8 depicts the sensitivity of alveloar ventilation $\dot{V}_{\mathrm{A}}$ with respect to the model parameters. As can be seen, $M_{\mathrm{CO_2}}$ have the highest PRCC value signifying its major influence in $\dot{V}_{\mathrm{A}}$.

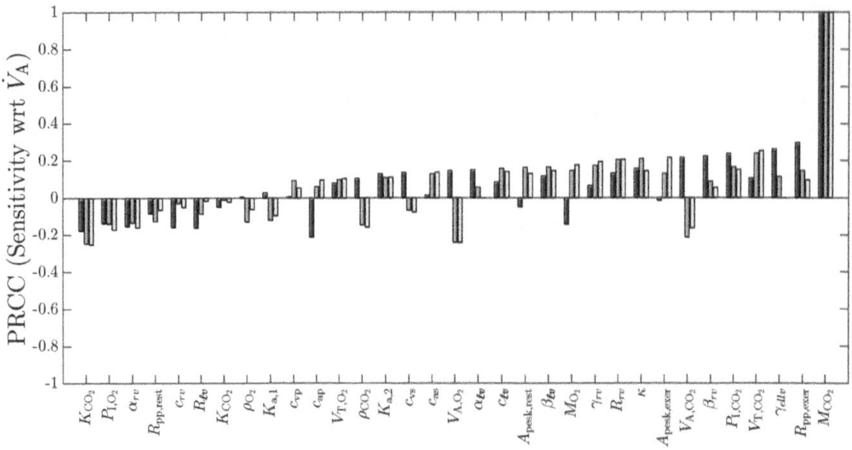

Fig. 8 Bar graphs of PRCCs of model parameters across the three time points: $t = 4, 6, 12$ min corresponding to rest, transition from rest to exercise, and exercise phase, respectively, with $\dot{V}_{\text{A}}$ as target output.

Table 3 Nominal parameter values of the CVRS model. Parameters with asterisk (*) are estimated in this work

Parameter	Description	Value	Unit
	Compliances		
c_{as}	Arterial systemic compartment	0.01016	L/mmHg
c_{vs}	Venous systemic compartment	0.6500	L/mmHg
c_{ap}	Arterial pulmonary compartment	0.03608	L/mmHg
c_{vp}	VENOUS pulmonary compartment	0.1408	L/mmHg
$c_{\ell\text{v}}$	Left ventricle	0.02305	L/mmHg
c_{rv}	Right ventricle	0.04413	L/mmHg
	Resistances		
$R_{\text{pp,rest}}$	Peripheral region of the pulmonary circuit corresponding to rest phase	1.5446	mmHg min/L
$R_{\text{pp,exer}}$	Peripheral region of the pulmonary circuit corresponding to exercise phase	0.3	mmHg min/L
$R_{\ell\text{v}}$	Inflow valve of the left ventricle	0.2671	mmHg min/L
R_{rv}	Inflow valve of the right ventricle	0.04150	mmHg min/L
κ	Coefficient in Bazett's formula	0.05164	$\text{min}^{1/2}$
	Peskin's constant		
$^*A_{\text{pesk,rest}}$	Corresponding to rest phase	177.682	mmHg min/L
$^*A_{\text{pesk,exer}}$	Corresponding to exercise phase	270	mmHg min/L

(*continued*)

Table 3 (*continued*)

Parameter	Description	Value	Unit
	Metabolic rates		
$^*M_{O_2}$	O_2 in the systemic tissue region corresponding to zero workload	0.35	L/min
$^*M_{CO_2}$	CO_2 in the systemic tissue region corresponding to zero workload	0.28	L/min
	Coefficients of the workload $W(t)$		
$^*\rho_{O_2}$	In Eq. (13)	0.011	L/(min W)
ρ_{CO_2}	In Eq. (12)	0.009	L/(min W)
	Parameters in the derivative of $\sigma_{\ell v}$ and σ_{rv}		
$\alpha_{\ell v}$	Quad coefficient of $S_{\ell v}$	30.5587	min^{-2}
α_{rv}	Coefficient of S_{rv}	28.6785	min^{-2}
$\beta_{\ell v}$	Coefficient of H	25.0652	mmHg/min
β_{rv}	Coefficient of H	1.4132	mmHg/min
$\gamma_{\ell v}$	Coefficient of $\sigma_{\ell v}$	−1.6744	min^{-1}
γ_{rv}	Coefficient of σ_{rv}	−1.8607	min^{-1}
	Partial pressures		
P_{I,O_2}	O_2 in inhaled air	150	mmHg
P_{I,CO_2}	CO_2 in inhaled air	0	mmHg
	Effective volumes		
V_{A,O_2}	O_2 in the lung compartment	2.5	L
V_{A,CO_2}	CO_2 in the lung compartment	3.2	L
V_{T,O_2}	O_2 in the tissue compartment	6.0	L
V_{T,CO_2}	CO_2 in the tissue compartment	15.0	L
	COEFFICIENTS in the dissociative laws		
K_{CO_2}	for CO_2	0.0057	$mmHg^{-1}$
K_{CO_2}	for CO_2	0.224	Dimensionless
$^*K_{a,1}$	for O_2	0.2	Dimensionless
$K_{a,2}$	for O_2	0.05	$mmHg^{-1}$

References

1. Ratto, M., Tarantola, S., Saltelli, A.: Sensitivity analysis in model calibration: GSA-GLUE approach. Comput. Phys. Commun. **136**(3), 212–224 (2001)
2. Draper, D.: Assessment and propagation of model uncertainty. J. Roy. Stat. Soc. Ser. B (Methodological) **57**(1), 45–97 (1995)
3. Helton, J.C.: Uncertainty and sensitivity analysis techniques for use in performance assessment for radioactive waste disposal. Reliab. Eng. Syst. Saf. **42**(2), 327–367 (1993)
4. Iman, R.L., Helton, J.C.: An investigation of uncertainty and sensitivity analysis techniques for computer models. Risk Anal. **8**(1), 71–90 (1988)
5. Ionescu-Bujor, M., Cacuci, D.G.: A comparative review of sensitivity and uncertainty analysis of large-scale systems—i: deterministic methods. Nucl. Sci. Eng. **147**(3), 189–203 (2004)

6. Saltelli, A., Ratto, M., Tarantola, S., Campolongo, F.: Sensitivity analysis for chemical models. Chem. Rev. **105**(7), 2811–2828 (2005)
7. Saltelli, A., Campolongo, F., Cariboni, J., Gatelli, D., Pennoni, F., Ratto, M., Saisana, M., Tarantola, S.: Global sensitivity analysis. The Primer. Wiley (2008)
8. Saltelli, A., Chan, K., Scott, E.M.: Sensitivity Analysis. Wiley Series in Probability and Statistics. Wiley, Chichester (2000)
9. Christopher Frey, H., Patil, S.R.: Identification and review of sensitivity analysis methods. Risk Anal. **22**(3), 553–578 (2002)
10. Marino, S., Hogue, I.B., Ray, C.J., Kirschner, D.E.: A methodology for performing global uncertainty and sensitivity analysis in systems biology. J. Theoret. Biol. **254**(1), 178–196 (2008)
11. Pianosi, F., Beven, K., Freer, J., Hall, J.W., Rougier, J., Stephenson, D.B., Wagener, T.: Sensitivity analysis of environmental models: a systematic review with practical workflow. Environ. Modell. Softw. **79**, 214–232 (2016)
12. Pappenberger, F., Beven, K.J., Ratto, M., Matgen, P.: Multi-method global sensitivity analysis of flood inundation models. Adv. Water Resourc. **31**(1), 1–14 (2008)
13. Pianosi, F., Sarrazin, F., Wagener, T.: A matlab toolbox for global sensitivity analysis. Environ. Modell. Softw. **70**, 80–85 (2015)
14. Saltelli, A., Tarantola, S., Campolongo, F.: Sensitivity analysis as an ingredient of modeling. Stat. Sci. **15**(4), 377–395 (2000)
15. Calderon, P.G.B., Habib, M., Kappel, F., de los Reyes V, A.A.: Control aspects of the human cardiovascular-respiratory system under a nonconstant workload. Math. Biosci. **289**, 142–152 (2017)
16. Kappel, F., Peer, R.: A mathematical model for fundamental regulation processes in the cardiovascular system. J. Math. Biol. **31**, 611–631 (1993)
17. de los Reyes V, A.A., Jung, E., Kappel, F.: Stabilizing control for a pulsatile cardiovascular mathematical model. Bull. Math. Biol. **76**(6), 1306–1332 (2014)
18. Timischl, S.: A global model of the cardiovascular and respiratory system. Ph.D. thesis, University of Graz, Institute for Mathematics and Scientific Computing (1998)
19. Fink, M., Batzel, J., Kappel, F.: An optimal control approach to modeling the cardiovascular-respiratory system: an application to orthostatic stress. Cardiovasc. Eng. **4**(1), 27–38 (2004)
20. Kappel, F., Fink, M., Batzel, J.: Aspects of control of the cardiovascular-respiratory system during orthostatic stress induced by lower body negative pressure. Math. Biosci. **206**(2), 273–308 (2007)
21. Batzel, J., Kappel, F., Timischl-Teschl, S.: A cardiovascular-respiratory control system model including state delay with application to congestive heart failure in humans. Journal of Mathematical Biology **50**, 293–335 (2005)
22. Eslami, M.: Theory of Sensitivity in Dynamic Systems: An Introduction, 1 edn. Springer, Heidelberg (1994)
23. Frank, P.M.: Introduction to System Sensitivity Theory. Academic Press (1978)
24. Ellwein, L., Tran, H., Zapata, C., Novak, V., Olufsen, M.: Sensitivity analysis and model assessment: mathematical models for arterial blood flow and blood pressure. Cardiovasc. Eng. **8**, 94–108 (2008)
25. Saltelli, A., Marivoet, J.: Non-parametric statistics in sensitivity analysis for model output: a comparison of selected techniques. Reliab. Eng. Syst. Saf. **28**(2), 229–253 (1990)
26. Saltelli, A., Tarantola, S., Campolongo, F., Ratto, M.: Sensitivity Analysis in Practice: A Guide to Assessing Scientific Models. Wiley (2004)
27. Batzel, J., Kappel, F., Schneditz, D., Tran, H.: Cardiovascular and Respiratory Systems: Modeling, Analysis and Control. Frontiers in Applied Mathematics, vol. 34. SIAM, Philadelphia, PA (2007)
28. Paton, F.L., Maier, H.R., Dandy, G.C.: Relative magnitudes of sources of uncertainty in assessing climate change impacts on water supply security for the southern Adelaide water supply system. Water Resourc. Res. **49**(3), 1643–1667 (2013)
29. Fink, M.: Automatic differentiation for Matlab. Software (2006)

30. Banks, H.T., Cintrón-Arias, A., Kappel, F.: Parameter selection methods in inverse problem formulation. In: Batzel, J.J., Bachar, M., Kappel, F. (eds.) Mathematical Modeling and Validation in Physiology. Lecture Notes in Mathematics, pp. 43–73. Springer, Heidelberg (2013)
31. Batzel, J.J., Bachar, M., Karemaker, J.M., Kappel, F.: Merging mathematical and physiological knowledge: dimensions and challenges. In: Batzel, J.J., Bachar, M., Kappel, F. (eds.) Mathematical Modeling and Validation in Physiology. Lecture Notes in Mathematics, pp. 3–19. Springer, Heidelberg (2013)
32. Heldt, T., Verghese, G.C., Mark, R.G.: Mathematical modeling of physiological systems. In: Batzel, J.J., Bachar, M., Kappel, F. (eds.) Mathematical Modeling and Validation in Physiology: Applications to the Cardiovascular and Respiratory System, Chap. 2, pp. 21–41. Springer, Heidelberg (2013)
33. King, J., Kupferthaler, A.: VOCErgo: isoprene and acetone concentration profiles during exercise on an ergometer. Technical report, Breath Research Unit, Austrian Academy of Sciences, Innsbruck, Austria. http://realtime.voc-research.at
34. King, J., Kupferthaler, A., Unterkofler, K., Koc, H., Teschl, S., Teschl, G., Miekisch, W., Schubert, J., Hinterhuber, H., Amann, A.: Isoprene and acetone concentration profiles during exercise on an ergometer. J. Breath Res. **3**(2), 027006 (2009)
35. Habib, M.: Control of the human cardiovascular-respiratory system under a time-varying ergometric workload. Ph.D. thesis, Karl-Franzens-Universität Graz (2011)

Cytotoxic Activity of *Raphanus sativus* Linn. on Selected Cancer Cell Lines and Mechanistic Pathways Predicted Through Mathematical Modeling

Angelyn Lao, Jan Marie Claire Edra, Kathleen Dane Talag, Daisylyn Senna Tan, Glenn Oyong, Marissa Noel, Ma. Luisa Enriquez, and Maria Carmen Tan

Abstract Studies have shown different effects of *Raphanus sativus* L. (RS) extract on different cancer cell types. However, the dynamics of gene regulation between RS and cancer is still unknown. In this study, a mathematical model was incorporated to link cell cycle regulation pathways associated with the pharmacognosy of radish extracts towards breast cancer (BC), chronic myelogenous leukemia (CML), and colorectal cancer (CC). The cell cycle regulation pathways considered were MAPK/ERK and PI3K/Akt signaling pathways. A model created using ordinary differential equations was used to simulate the steady state concentrations of the proteins in both pathways, before and after exposure to radish extracts. Among the proteins ubiquitous in these pathways, Cyclin D1-CDK complex and p53 were found to be predominantly dysregulated in most cancers. The expression of these proteins was used as the benchmark of the pharmacognosy of radish extracts. Our simulation showed a decreasing trend in the binding specificities (K_M) of RS with Cyclin D1-CDK complex and p53 respectively from CML [5 μM, 100 μM], CC [2.5×10^{-2} μM, 10 μM] up to BC [10^{-6} μM, 10^{-3} μM] which led to a decrease in Cyclin D1-CDK complex (90.74% for BC, 73.42% for CC, 1.10% for CML) and an

A. Lao (✉) · J. M. C. Edra · K. D. Talag
Mathematics and Statistics Department, De La Salle University, 2401 Taft Avenue, 0922 Manila, Philippines
e-mail: angelyn.lao@dlsu.edu.ph

D. S. Tan · M. Noel · M. C. Tan
Chemistry Department, De La Salle University, 2401 Taft Avenue, 0922 Manila, Philippines

G. Oyong
Molecular Science Unit Laboratory, Center for Natural Sciences and Environmental Research (CENSER), De La Salle University, 2401 Taft Avenue, 0922 Manila, Philippines

Ma. L. Enriquez
Biology Department, De La Salle University, 2401 Taft Avenue, 0922 Manila, Philippines

Research and Biotechnology Division, St. Luke's Medical Center, 279 E. Rodriguez Sr. Avenue, Quezon City, Metro Manila, Philippines

M. H. Mohd et al. (eds.), *Modelling, Simulation and Applications of Complex Systems*, Springer Proceedings in Mathematics & Statistics 359,
https://doi.org/10.1007/978-981-16-2629-6_9

increase in p53 (83.47% for BC, 62.75% for CC, 5.69% for CML). This suggests that the anticancer activity of RS was through the moderation of cell proliferation and induction of apoptosis in malignant neoplasms. Hence, the model proposed that RS had high chemotherapeutic activity on BC while having moderate efficacy on CC and insignificant efficacy on CML as substantiated by cytotoxicity assays on MCF-7 and HT-29 which both exhibited $IC_{50} < 20\ \mu g/mL$ and K562 trials which gave $IC_{50} > 100\ \mu g/mL$.

Keywords Chronic myelogenous leukemia · Breast cancer · Colorectal cancer · Mathematical modeling · Systems biology · *Raphanus sativus* L.

1 Introduction

Cancer is one of the leading causes of death worldwide, and chronic myelogenous leukemia (CML), breast cancer (BC), and colorectal cancer (CC) are three of the most prevalent types [46]. Annually, there are about 100 thousand, 2.09 million, and 1.1 million cases of CML, BC, and CC, respectively [5, 31]. In most of these cases, cancer is only detected in its advanced stage due to the lack of early detection and limited options of targeted therapies [47, 66]. Cancer cells has disturbed homeostasis which leads to uncontrolled proliferation and reduced apoptosis. Cancer cells have an acquired ability to elude apoptosis through a variety of ways. Hence, induction of apoptosis provides an important valuable strategy for the management of cancer [3].

Raphanus sativus L. (RS) is a root vegetable of the Cruciferae family, commonly known as radish. Some studies described cancer preventive effects of this natural product. For example, Kim et al. [34] studied on the effects of ethanol extract of aerial parts of RS leaf (ERL) on BC cell proliferation and gene expression and was able to show that it suppresses the EGFR-Akt pathway which includes Ras by culturing human BC cells in the absence and presence of ERL.

On the other hand, analysis of the inhibitory effects of RS root extract in human cancer was the focus of the study by Beevi et al. [3]. They discovered that the hexane extract of RS root had interactions with p53 and with Bcl-2 family of genes, which includes Bad. In an article by Cragg and Newman [8], they stated that olomoucine, isolated from the cotyledons of radish, was shown to inhibit cyclin dependent kinases (CDKs) in human cancer cell lines.

In a study by Tan et al. [61], the genotoxic effects of RS extract (juiced and semi—purified preparations) were tested on four cell lines, namely, MCF-7, HT-29, K562, and a normal cell line (PAE) using the Comet Assay. Constituents of RS were observed to have caused significant DNA migrations in all three parameters: tail length (TL), percent DNA in the tail (D), and tail moment (TM). DNA fragmentation for MCF-7 cells treated with juice and semi-purified preparations of the white RS tuber were found to be similar with 95% CI for mean differences of $[-112.81, 24.21]$

for TL, $[-18.03, 5.93]$ for D and $[-56.72, 52.15]$ for TM. It was concluded that preparations of RS can be a substantial source of biologically active constituents that can cleave DNA in mutant cell lines, while being non-genotoxic to wild type immortalized cells.

Using naturally occurring or synthetic chemicals as a strategy for chemotherapeutic protocols have been found to inhibit, delay or reverse human carcinogenesis [44]. The best way to ascertain the chemopreventive potential of dietary substances can be correlated to their additive or synergistic interactions that may act on the conduits of carcinogenesis [3]. Comprehension of the complexity of disease dynamics, observed patterns visualized by mechanistic hypothesis, and testing feasible control measures can be facilitated by mathematical modeling. It plays an important role in interpreting complex systems in biology [62].

Several studies have already applied mathematical modeling to simulate cell signaling pathways. Wee and Aguda [65] used ordinary differential equations in examining the dynamics of the p53-Mdm2 regulatory network which contains the tumor suppressor protein p53 and oncoprotein Akt. Their model predicted the existence of the robust phenomenon of bistability in the p53-Akt network. On the other hand, Gong et al. [20] analyzed the components of the HMGB1 pathway, including Ras, Raf, Mek, ERK, Akt, and p53, through formulation of ordinary differential equations and discrete stochastic simulation to investigate on tumorigenesis. They have concluded that Cyclin D was overexpressed and p53 was inhibited if HMGB1 was upregulated. Furthermore, HMGB1 signaling influenced the cancer cell proliferation through the meditations of Ras, ARF, and p21. A similar study was done by Kang et al. [30] wherein they created a mathematical model comprising differential equations pertaining to certain genes in lung cancer signaling which includes moderation of Ras, ERK, Myc etc. In simulating the equations, they were able to show that the genes for Ras, ERK, and Myc were upregulated in lung cancer.

In this study, we further the knowledge on *R. sativus* efficacy as a chemopreventive agent. The experimental design and analyses of the cytotoxic activity of dichloromethane extracts of *R. sativus* on MCF-7, HT-29 and K562 were discussed and performed (Sects. 2 and 4, respectively) as a corollary with the mathematical modeling of the proteins (shown in Sect. 3) which could be significant biomarkers in the cancers associated with these cell lines (as analyzed and discussed in Sects. 5 and 6). For this reason, scientists may use this study to possibly hone in on which proteins to focus on for drug discovery as well as a method of comparison on possible mechanistic pathways. To the best of our knowledge, this is the first reported study using this mathematical model methodology of analyses on RS against the aforementioned cancers.

2 Materials and Methods

2.1 Preparation of Compounds for Bioassay

The dried dichloromethane (DCM) extracts of RS were dissolved in dimethyl sulfoxide (DMSO) to make a 4 mg/mL stock solution. Working solutions were prepared in complete cell culture medium to a final non-toxic DMSO concentration of 0.2%.

2.2 Culture of Cell Lines

The bioactivity of the DCM extract from *R. sativus* was tested on the following human cell lines (ATCC, Manassas, Virginia, USA): BC (MCF-7), CC (HT- 29), CML (K562); and a primary culture of normal human dermal fibroblast, neonatal (HDFn) (ThermoFisher Scientific, Gibco®, USA). All cells were maintained and cultured at the Cell and Tissue Culture Laboratory, Molecular Science Unit, Center for Natural Sciences and Environmental Research, De La Salle University, using complete growth medium consisting of Dulbecco's Modified Eagle's Medium (DMEM, ThermoFisher Scientific Gibco®, USA), 10% fetal bovine serum (FBS, ThermoFisher Scientific, Gibco®, USA) and 1x antibiotic-antimycotic (ThermoFisher Scientific, Gibco®, USA). Culture conditions were achieved at 37 °C, 5% CO_2 and 95% relative humidity (Rh). The cells were harvested after reaching 90% confluence by two rounds of washing with phosphate-buffered saline (PBS, pH 7.4, ThermoFisher Scientific, Gibco®, USA) and trypsinization (0.05% Trypsin-EDTA, ThermoFisher Scientific, Gibco®, USA) followed by deactivation of trypsin by resuspension with fresh complete media. Viable cells were counted via 0.4% trypan blue (ThermoFisher Scientific, Gibco®, USA) and seeded in 96-well plates (FalconTM, USA) in 100 μL aliquots of 1×10^4 cells/well. The plates were further incubated overnight at 37 °C with 5% CO_2 in a 95% humidified incubator to allow attachment and monolayer formation of the adherent cells (BC, CC and HDFn) prior to succeeding cell viability assays.

2.3 Cell Viability Assay

A rezasurin-based assay, PrestoBlue® (ThermoFisher Scientific, Molecular Probes®, Invitrogen, USA), was applied to assess the effects of *R. sativus* DCM extract on the viability of BC, CC, CML and HDFn cells, respectively. The assay works behind mitochondrial reductases in viable cells which convert rezasurin to resorufin the amount of which can be spectrophotometrically measured at 570 nm as a function proportional to the relative number of metabolically active viable cells.

A total of 100 μL filter-sterilized RS extract (100 μg/mL) was respectfully added into the 96-well plates following two-fold serial dilution resulting to final screening concentrations of 50, 25, 12.5, 6.25, 3.12, 1.56, 0.78 and 0.39 μg/mL, respectively. Wells without RS served as untreated negative controls while wells with bleomycin served as positive cytotoxic controls. After the plates underwent further incubation for 72 h at 37 °C, 5% CO_2 and 95% Rh, 20 μL of PrestoBlue® were added to each well followed by further incubation for 1 h at similar conditions. Optical density measurements were carried out at 570 nm normalized at 600 nm reference wavelength (BioTek ELx800, BioTek® Instruments, USA) from which cell viability index values were derived.

The half-maximal inhibitory concentration (IC_{50}) values of RS across all cell lines were extrapolated via nonlinear regression and statistical analyses computed using GraphPad Prism 7.01 (GraphPad Software, California, USA). All tests were accomplished in three replicates and expressed as mean ± SEM. The extra sum-of-squares F-test (Brown-Forsythe) was used to assess the best-fit parameter differences (IC_{50}) among treatments and to evaluate the differences among the dose-response curve fits based on the software's suggested approach. One-way ANOVA ($p < 0.05$) and unpaired two-tailed T-test at 95% CI were also performed to calculate significant differences among group variables, followed by Tukey's post hoc multiple comparison ($p < 0.05$) among different pairs of data sets. Data were regarded as significant at $p < 0.05$.

3 Mathematical Model

In discerning the complex relationships of these cancers, we first identified proteins that were common and different among the pathways of the three types of cancer, as shown in Fig. 1. For CML (Fig. 1A) and CC (Fig. 1C), possible KEGG pathways were also collected [25–29]. For BC (Fig. 1B), the study by Crown et al. [9] was the prototype for the possible mechanisms involved [9]. After establishing the signaling pathways of the three diseases, RS was integrated into the model. The resulting signaling pathways for CML, BC, and CC are shown in Figs. 1A–C.

RS inhibited Ras and Cyclin D1-CDK complex but activated p53 for all the three diseases. RS inhibited BCR-ABL for both CML and BC. RS activated Bad in CML as well as Bad and DCC in CC.

Ras-Raf-MEK-ERK are series of proteins that are part of the MAPK pathway. The MAPK pathway proceeds to influence cell cycle progression, cell survival and cell proliferation [41]. These proteins play an important role in human oncogenesis [53, 57]. Myc, is then activated by ERK 1/2, which is a proto-oncogene found in many signaling pathways that are involved in the cellular microenvironment and is deregulated in cancer [10, 48]. The activity of DCC is unique in CC as shown in Fig. 1C. DCC, which is activated by Ras, belongs to a family of receptors which is in control of cell survival or apoptosis. Loss of this is implicated in the progressing phase of CC [36].

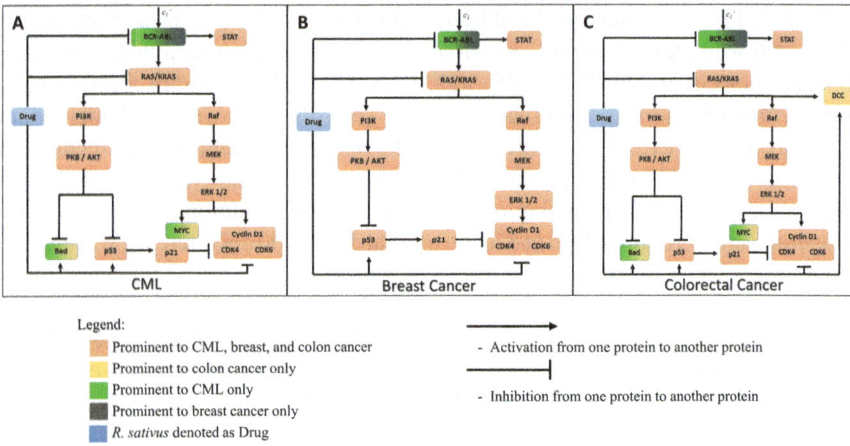

Fig. 1 Signalling pathways of CML (A), BC (B), and CC (C) in the presence of the RS. The pathways of CML, BC and CC in the absence and presence of drug is differentiated by letting the concentration value of *Raphanus sativus* L variable to be zero or nonzero, respectively. A pathway starting from an activation rate c1 and ending with Bad, Cyclin D1-CDK complex, and Myc complex is given in (A). Furthermore, a pathway shown in (B) is beginning from an activation rate c1′ and ending with p53 and Cyclin D1-CDK complex. On the other hand, a pathway given in (C) starts from an activation rate c2 and ends with Bad, Cyclin D1-CDK complex, Myc, and DCC.

In BC and CC, Ras was activated by BCR-ABL (Figs. 1A and B) [15]. BCR-ABL comes from an oncogene that contributes to a chromosome called Philadelphia-chromosome which causes leukemia especially CML and a subset of acute leukaemia [12]. Signal Transducer and Activator of Transcription (STAT) proteins are a group of cytoplasmic interpretation factors [24]. It was found that STAT was activated by BCR-ABL in BC and CML while directly activated by Ras in CC [25–29, 55, 64].

PI3K—Akt/PKB pathway was involved in cancer by mediating transformations in cellular growth, metabolism, and apoptosis of cells. PI3K/Akt survival pathway contributes to understanding BC development and formation of tumors because PI3K was found to be mutated in 26% of invasive BCs [40]. The tumor suppressor and pro-apoptotic proteins, Bad and p53 was inhibited by PKB/Akt [65].

p53 is one of the most studied proteins with mutations frequently found in various tumor types [45]. It is degraded and inactivated in most cancers. Because of its near universal alteration in cancer, p53 is an attractive target for the development of new targeted therapies for this disease [13]. It is known to induce apoptosis and was found to be upregulated by RS [3, 61].

Cyclin D1 binds and activates with cyclin dependent kinases, CDK4/6. These complexes are pivotal in the advancement of cell cycle progression and proliferation [22]. However, Cyclin D1 upregulation have been prevalent in most CC and 15% of BC cases [43]. p21, which causes tumor suppression through p53 activation, binds to and inhibits the activity of cyclin-CDK complexes [1].

A mathematical model was used to represent the steady state concentrations of the components found in the signaling pathways. Each component of the signaling pathways was translated to ordinary differential equations by applying concepts of enzyme kinetics. We calibrated the parameter values in order to satisfy the effect of the drug observed in the cytotoxicity assays and the study by Tan et al. [61]. MATLAB's ode solver, ode15s, was used to generate the solutions for the system of ordinary differential equations and to produce simulations which will explain the behavior of the systems.

In this study, we established a mathematical model that represents the interaction of the proteins in the three signaling pathways. We do this by generating three systems of ordinary differential equations, one for each of the three types of cancer (See Table 1). For CML, these are the equations shown in Table 1A–D, and G. For BC, equations included in Table 1A, B, D and G make up the system. In CC, we have the equations presented in Table 1A, C, F, and G. A summary table of the notations and parameter values may be found in Table 3.

Without RS, the concepts of first order irreversible reaction and linear mass kinetics were used in forming some equations. The rate constants are represented by the activation rate μ and deactivation rate δ of the proteins involved in the signaling pathway. Describing these equations using the linear mass action kinetics, we create the equation for PI3K as

$$\frac{dP}{dt} = \mu - \delta.$$

Using the fact that Ras activates PI3K in the three cancers, we replace μ by $\mu_P \cdot R$ and δ by $\delta_P \cdot P$ and we get Eq. (1). Similarly, Eqs. (4)–(6), (9)–(11), (15) and (17) were formed in the same manner.

The activation rate of Akt is described using the Michaelis-Menten kinetics as shown in Eq. (2). Formation of Eq. (8) was also established using a similar concept. Michaelis-Menten kinetics was also used to represent the amount of p53 that went into activating p21. Since Akt inhibits Bad and p53, we used the concept of competitive inhibition in forming the equations for these two proteins. The deactivation rate is expressed by an inhibition factor, as seen in Eqs. (3) and (12).

Equations for the proteins inhibited by the drug were also generated. The equation for the drug contains a negative deactivation rate denoted by $-\delta_Z \cdot Z$ since its concentration is being consumed and thus is decreased when it is distributed to some of the genes.

The inhibition of Ras by the drug is described in Eq. (10) as:

$$\mu_{RZ} - \delta_{RZ} \cdot R \cdot \frac{Z}{Z + \kappa_{RZ}}.$$

In the absence of RS, the above expression is 0. The added expression in Eqs. (7), following an inhibition by the drug (i.e. RS in this study), were similarly formed as in Eq. (10).

Table 1 Model equations associated to all three cancers (A), to CML and BC (B), to CML and CC only (C), to CML only (D), to BC only (E), to CC only (F) and to RS (G)

A	$\frac{dP}{dt} = \mu_P \cdot R - \delta_P \cdot P$	(1)
	$\frac{dA}{dt} = \mu_A \cdot \frac{P}{P + \kappa_A} - \delta_A \cdot A$	(2)
	$\frac{dF}{dt} = \mu_F - \delta_F \cdot F \cdot \frac{A}{A + \kappa_F} - \mu_W \cdot \frac{F}{F + \kappa_W} + \delta_{FZ} \cdot F \cdot \frac{Z}{Z + \kappa_{FZ}}$	(3)
	$\frac{dRf}{dt} = \mu_{Rf} \cdot R - \delta_{RF} \cdot Rf$	(4)
	$\frac{dM}{dt} = \mu_M \cdot Rf - \delta_M \cdot M$	(5)
	$\frac{dE}{dt} = \mu_E \cdot M - \delta_E \cdot E$	(6)
	$\frac{dD}{dt} = \mu_D \cdot E - \delta_D \cdot D \cdot \frac{W}{W + \kappa_D} + \mu_{DZ} - \delta_{DZ} \cdot D \cdot \frac{Z}{Z + \kappa_{DZ}}$	(7)
	$\frac{dW}{dt} = \mu_W \cdot \frac{F}{F + \kappa_W} - \delta_W \cdot W$	(8)
B	$\frac{dS}{dt} = \mu_S \cdot Br - \delta_S \cdot S$	(9)
	$\frac{dR}{dt} = \mu_R \cdot Br - \delta_R \cdot R + \mu_{RZ} - \delta_{RZ} \cdot R \cdot \frac{Z}{Z + \kappa_{RZ}}$	(10)
C	$\frac{dMC}{dt} = \mu_{MC} \cdot E - \delta_{MC} \cdot MC$	(11)
	$\frac{dB}{dt} = \mu_B - \delta_B \cdot B \cdot \frac{A}{A + \kappa_B} + \delta_{BZ} \cdot B \cdot \frac{Z}{Z + \kappa_{BZ}}$	(12)
D	$\frac{dBr}{dt} = c_1' - \delta_{Br} \cdot Br - \delta_{BrZ} \cdot Br \cdot \frac{Z}{Z + \kappa_{BrZ}}$	(13)
E	$\frac{dBr}{dt} = c_1 - \delta_{Br} \cdot Br - \delta_{BrZ} \cdot Br \cdot \frac{Z}{Z + \kappa_{BrZ}}$	(14)
F	$\frac{dS}{dt} = \mu_S \cdot R - \delta_S \cdot S$	(15)
	$\frac{dR}{dt} = c_2 - \delta_R \cdot R - \delta_{RZ} \cdot R \cdot \frac{Z}{Z + \kappa_{RZ}}$	(16)
	$\frac{dDC}{dt} = \mu_{DC} \cdot R - \delta_{DC} \cdot DC + \delta_{DCZ} \cdot DC \cdot \frac{Z}{Z + \kappa_{DCZ}}$	(17)
G	$\frac{dZ}{dt} = -\delta_Z \cdot Z$	(18)

Notations: Ras, P—PI3K, Rf—Raf, A—PKB/Akt, M—MEK, B—Bad, F—p53, E—ERK, MC—Myc, D—Cyclin D1—CDK complex, Br—BCR-ABL, S—STAT, DC—DCC, W—p21, Z—Drug. μ—catalytic production rate, δ—transition rate, κ—binding specificity

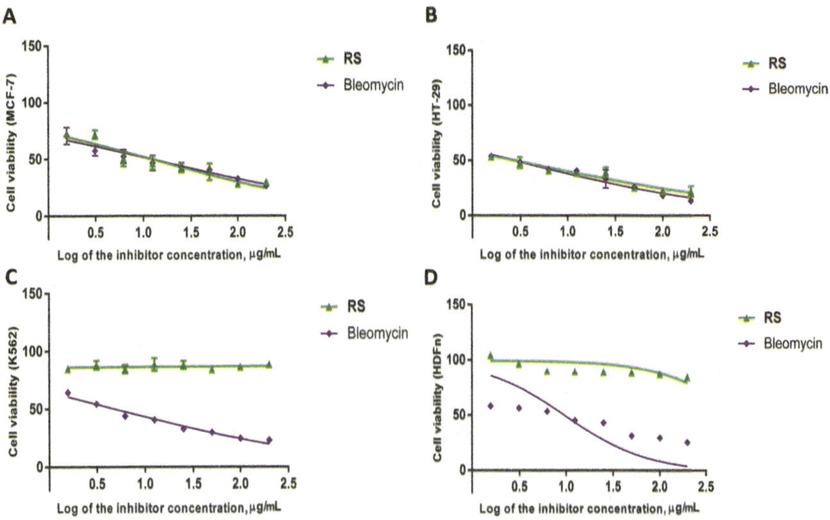

Fig. 2 Dose-response curves showing the cytotoxic activities of RS and Bleomycin on the cell viability of MCF-7 (A), HT-29 (B), K562 (C) and HDFn (D): Each plot displays the effect of RS and Bleomycin against each cell line. Data are shown as mean *pm* SEM. GraphPad Prism 7.01 was used to perform extra sum-of-squares F-test to (A) evaluate the significance of the best-fit-parameter (half-maximal inhibitory concentration) among different treatments, and to (B) determine the differences among the dose-response curve fits. MCF-7 (A) $F(Dfn, DFd) = F(1, 44) = 1.045$, $p = 0.3122$ and (B) $F(7, 7) = 1.460$, $p = 0.6301$; HT-29 (A) $F(1, 40) = 0.8249$, $p = 0.3692$ and (B) $F(7, 7) = 1.476$, $p = 0.6199$; K562 (A) $F(1, 44) = 327.2$, $p < 0.0001$ and (B) $F(7, 7) = 40.89$, $p < 0.0001$; HDFn (A) $F(1, 46) = 185.0$, $p < 0.0001$ and (B) $F(7, 7) = 2.462$, $p = 0.2575$.

Incorporating the activation of p53 by RS, the same principle as in forming Eq. (10) was used in Eq. (3). We denote this expression as

$$\delta_{FZ} \cdot F \cdot \frac{Z}{Z + \kappa_{FZ}}.$$

Using the same approach in forming Eq. (3) with the presence of RS, Eqs. (12), (13), (14), (16) and (17) were created.

Furthermore, in order for our results to agree with what has been observed in the cytotoxicity assays and the study by Tan et al. [61], we had to take into consideration the following conditions: The binding specificity (K_M) between the drug with Cyclin D1-CDK complex and p53, noted as κ_{DZ} and κ_{FZ}, have different values for CML, BC, and CC. This is due to the fact that K_M value for an enzyme depends on the particular substrate and on environmental conditions [4].

The activation rates c_1' and c_1 activating BCR-ABL in CML and BC, respectively, have different values. A basis for this is a statement made by Srinavasan and Plattner [58] saying that Abl kinases, which include BCR-ABL, in leukemia are activated by translocation or gene amplification. In BC, Abl kinases are activated by deregulated EGFR, HER-2, and Src kinases.

4 Cytotoxic Activity of RS on the Immortalized Cell Lines

The cytotoxicity of the crude extract of RS was investigated on immortalized MCF-7, HT-29 and K562, including HDFn primary cells. Analyses of the cytotoxicity of RS obtained from wild type HDFn resulted in IC_{50} values exceeding 100 μg/mL. Bleomycin, a cytotoxic antibiotic which instigates DNA fragmentation, was used as the positive control in all the trials. An illustration of the percent cell viability as a function of the logarithmic function of the concentrations used is presented in Figs. 2 and 3. The characteristic inhibitory dose-response, which is a sigmoidal curve or function, was found in generally most of the charts. Plots comparing the cytotoxic properties of RS and bleomycin on the viability of each specific cell line are found in Fig. 2. The effects of RS and bleomycin on all the cell lines are shown in Fig. 3. The IC_{50} values of RS and the positive control are synopsized in Table 2.

Table 2 Cytotoxic activities (IC_{50}) of RS and Bleomycin against MCF-7, HT-29, K562 and HDFn.

Sample	IC^*_{50}(μg/mL)			
	MCF-7	HT-29	K652	HDFn
RS	13.79	2.63	>100	>100
Bleomycin	11.35	3.2	5.081	9.82

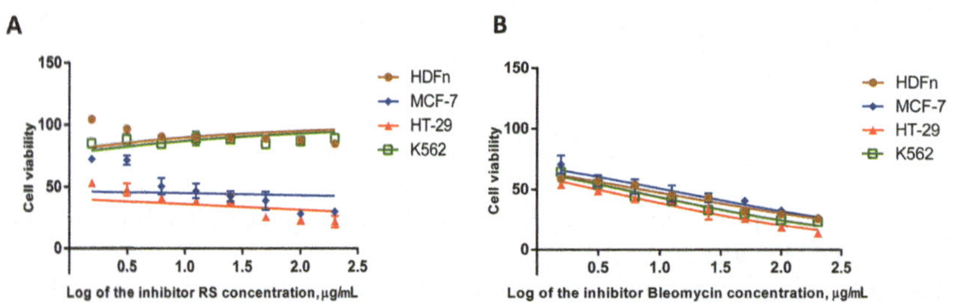

Fig. 3 Dose-response curves showing the cytotoxic activities of RS on the cell viability of MCF-7, HT-29, K562 and HDFn. Each plot displays the effect of RS (A) and Bleomycin (B) against each cell line: Data are shown as mean ± SEM. GraphPad Prism 7.01 was used to perform extra sum-of-squares F-test to (A) evaluate the significance of the best-fit-parameter (half-maximal inhibitory concentration) among different treatments, and to (B) determine the differences among the dose-response curve fits. The results are: RS, (A) $F(3, 80) = 103.5$, $p < 0.0001$ (B) $F(3, 28) = 4.211$, $p = 0.0141$; Bleomycin (A) $F(3, 84) = 23.86$, $p < 0.0001$ (B) $F(3, 28) = 0.0.06684$, $p = 0.9771$.

Crude extracts of RS gave a half-maximal inhibitory concentration of 2.63 μg/mL for HT-29, and 13.79 μg/mL for MCF-7. Wild type HDFn and mutant cells K562 exhibited IC_{50} values of greater than 100 μg/mL and were considered noncytotoxic. Substantial distinctions, ascertained through Tukey's post hoc multiple comparison, was displayed between MCF-7 vs. HDFn, HT-29 vs. HDFn, K562 vs. MCF-7, and K562 vs. HT-29 at 95% CI of difference ($p < 0.05$); while K562 vs. HDFn and HT-29 vs. MCF-7 paired treatments were found to be not significantly different ($p > 0.05$). One-way ANOVA revealed that the differences among the means of the data subsets of all the cancer cell line treatments with RS (Fig. 2) were statistically significant ($p < 0.0001$) and Bleomycin demonstrated that the paired treatments were statistically similar ($p > 0.05$).

5 Simulation

Our model simulations are shown in Figs. 5, 6 and 7. The K_M values of RS with Cyclin D1-CDK complex (κ_{DZ}) are—5 μM, 2.5×10^{-2} μM, and 1×10^{-6} μM, whereas the K_M values for RS with p53 (κ_{FZ}) are—100 μM, 10 μM, and 1×10^{-3} μM for CML, CC, and BC, respectively (Table 3). These values were estimated to fit the percent change (after drug insertion) of Cyclin D1-CDK complex and p53 that is consistent to what is observed in the cytotoxicity assays and the study by Tan et al. [7].

With the presence of RS, the steady-state concentration of the proteins is expected to control regulation and is dependent on their behavior in the absence of the compound. Upon insertion of RS, proteins that influence cell proliferation in cancer such as Cyclin D1-CDK Complex, Ras, Raf, ERK, MEK, Akt, BCR-ABL, STAT, PI3K, and Myc decreased in concentration in all three cancers (Figs. 5A, D–K, M; 6A, D–K; 7A, D–H, J, K, M) On the other hand, insertion of RS had a positive effect on pro-apoptotic and tumor suppressor proteins such as p53, p21, Bad, DCC (Figs. 5B–C, L; 6B–C; 7B–C, 7I, 7L).

The simulation for Cyclin D1-CDK complex and p53 are highlighted in Fig. 4. The largest decrease for both BC and CC was in Cyclin D1-CDK complex, having decreased 90.74% and 73.42%, respectively. On the contrary, Cyclin D1-CDK complex barely decreased in CML (1.10%). The same trend is observed for the increase of p53. p53 had increased by 83.47% for BC and 62.75% for CC, while increasing minutely by 5.69% for CML.

The proteins in the MAPK/ERK pathway (Ras, Raf, ERK, and MEK) along with Akt and PI3K dropped in a uniform amount of around 75% in CML (Fig. 5D–H, K), and BC (Fig. 6D–5H, 5K) whereas around 50% in CC (Fig. 7D–H, K). STAT and BCR-ABL behaved similarly in all the cancer types with roughly a 50% drop as shown in Figs. 5J–7J and Figs. 5I–6I, respectively.

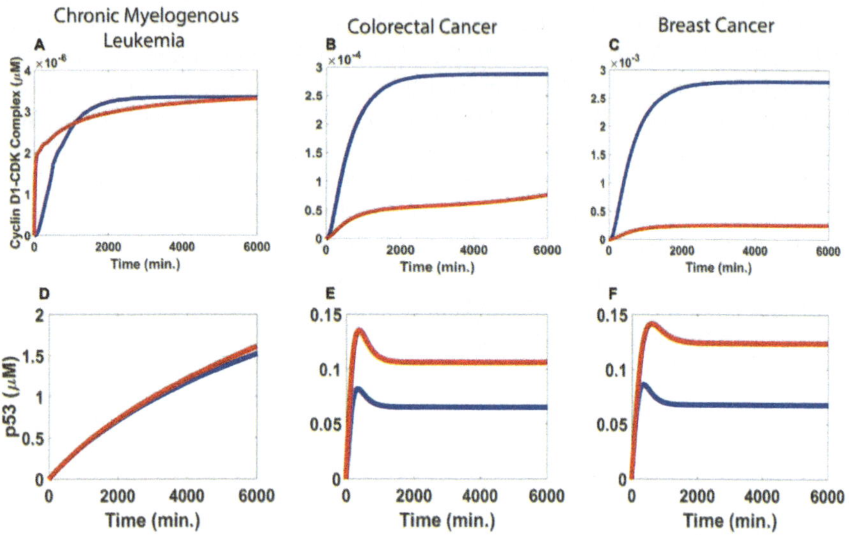

Fig. 4 Simulation results for the absence and presence of drug for Cyclin D1–CDK complex (A–C) and p53 (D– F) in CML, colon cancer, and BC: Each of the proteins' varying concentration level, together with RS, is shown with respect to time. In A-F, blue colored simulations represented the behavior of the proteins without the presence of drug while red colored simulations demonstrate the behavior of the proteins after RS was inserted in the model. Initial concentrations are set to 0

6 Discussion

The optimum IC_{50} for cancer cell viability for natural products is 20 μg/mL or less for crude extracts and 4 μg/mL or less for pure isolates [16] and compounds that have been purified with low concentrations of cytotoxicity can be proposed for further drug development [21]. RS was found to be cytotoxic to BC and CC cell lines, whereas having no significant cytotoxic activity on CML and normal HDFn. The findings in this work disclosed that crude dichloromethane extracts from RS can be candidates for chemotherapeutic drugs or used as a corollary for medical protocols in dealing with the management of human CC and human BC, while being noncytotoxic to normal cells.

Reported researches on the cytotoxic and chemotherapeutic activity of the crude dichloromethane extracts from RS have been conducted; however, to our knowledge no literature on the parameters and activities in these cell lines have been cited.

Glucoraphasatin, a GLS, has been reported in the tubers of several radish cultivars [52]. The hydrolysis product 4-(methylthio)-3-butenyl isothiocyanate (MTBITC),a constituent capable of anti-microbial, anti- mutagenic, and anti-carcinogenic activity, has been purified from radish tubers [39]. Isothiocyanates have been found to protect wild type cells from DNA injury and induce malignant cancer cells to undergo apoptosis.

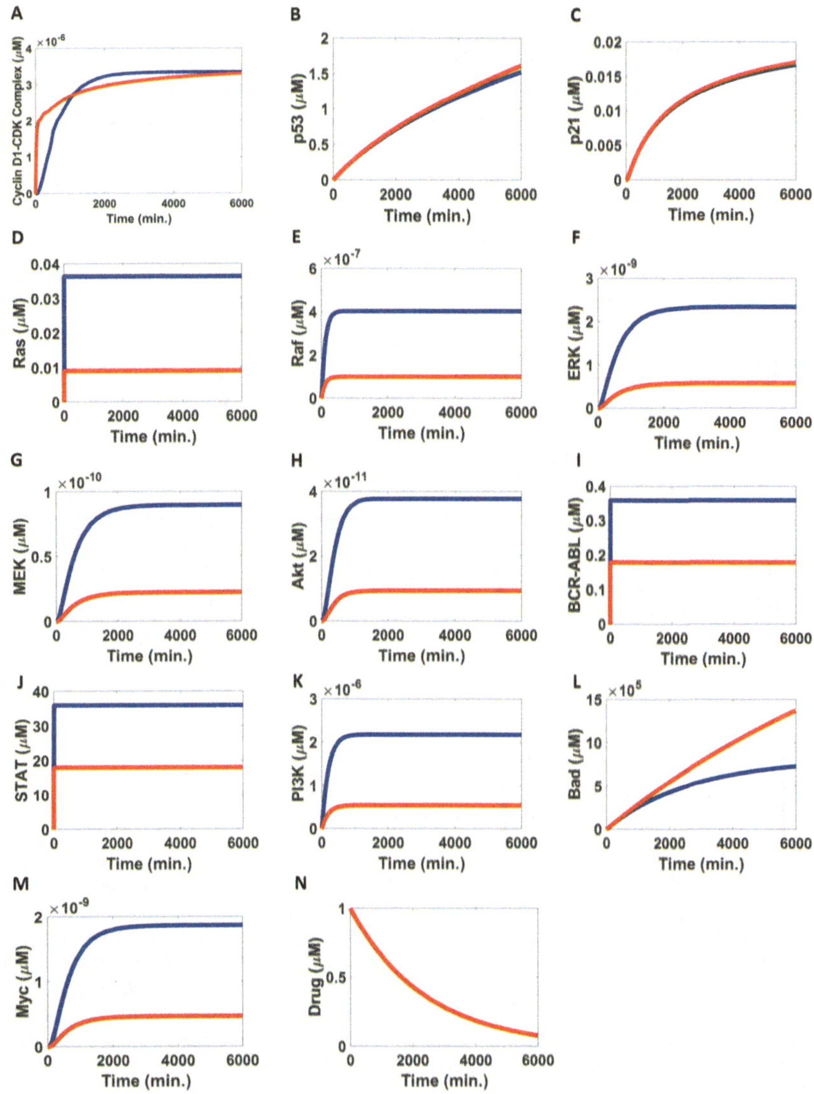

Fig. 5 Simulation results for the CML mathematical model for the absence and presence of drug: Each of the proteins' varying concentration level, together with RS, is shown with respect to time. In A-M, blue colored simulations represented the behavior of the proteins without the presence of drug while red colored simulations demonstrate the behavior of the protein after RS was inserted in the model. The behavior of the RS is shown in N. Initial concentrations in A-M are set to 0. For N, initial concentration is set to 1 to show the decreasing behavior of RS

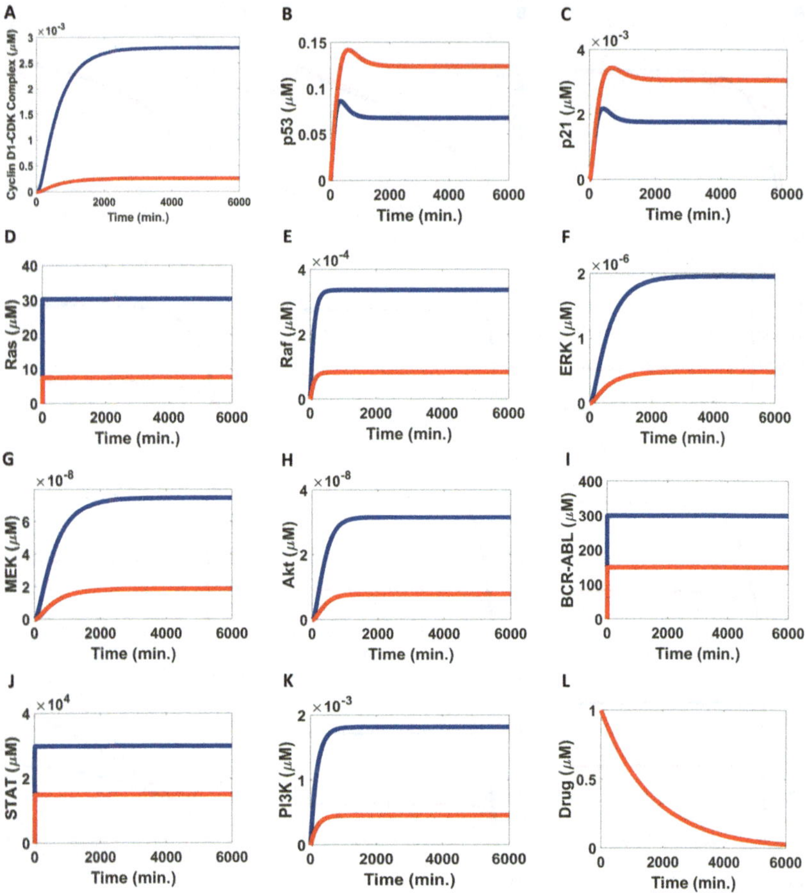

Fig. 6 Simulation results for the BC mathematical model for the absence and presence of drug: Each of the proteins' varying concentration level, together with RS, is shown with respect to time. In A-K, blue colored simulations represented the behavior of the proteins without the presence of RS while red colored simulations demonstrate the behavior of the proteins after RS was inserted in the model. The behavior of RS is shown in L Initial concentrations in A-J are set to 0. For L, initial concentration is set to 1 to show the decreasing behavior of the RS

With CML, BC and CC as some of the most predominant types of cancer, understanding the disease dynamics is important for effective prognosis and diagnosis. Through mathematical modeling, more comprehensive knowledge about this disease can be achieved. In this paper, we have proposed a mathematical model which would explain the behavior of the components of the signaling pathways of CML, BC and CC in the absence and presence of a chemotherapeutic agent, RS extract.

Based on the literature, we presented signaling pathways for each of the three diseases. To be able to mimic the effect of RS into the three cancers as observed in the study by Tan et al. [61] and supported by the cytotoxicity assays, we hypothesize

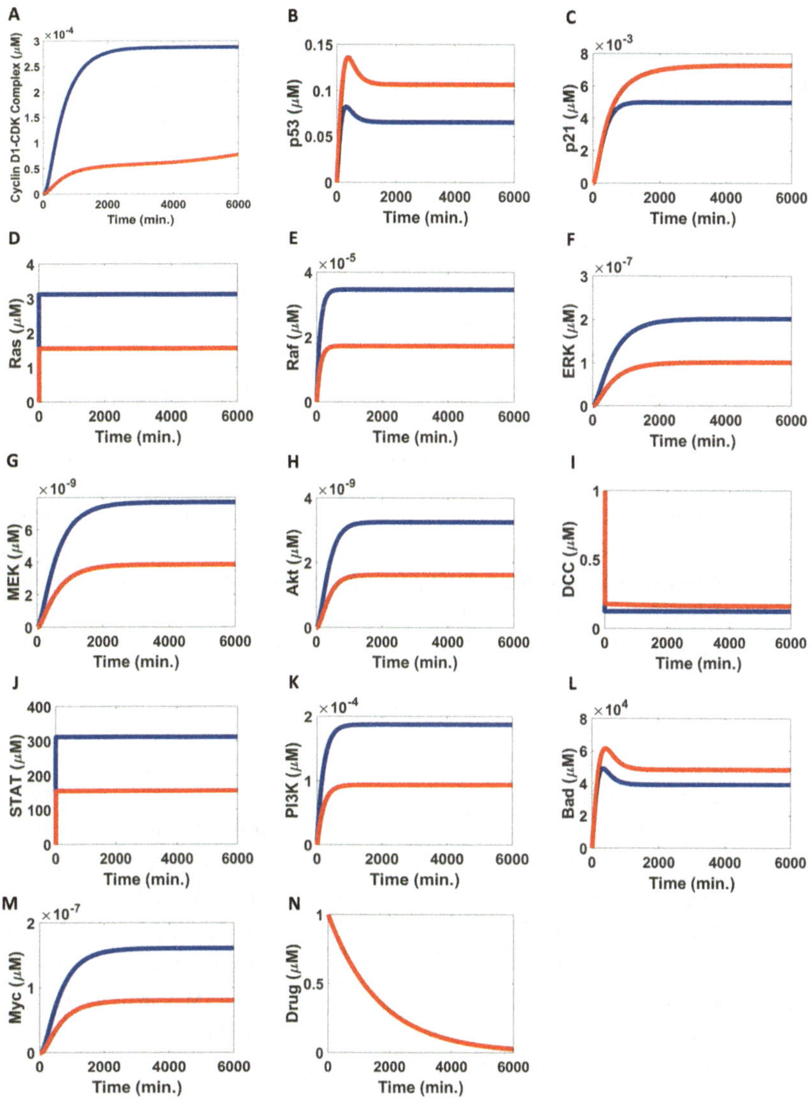

Fig. 7 Simulation results for the CC mathematical model for the absence and presence of drug: Each of the proteins' varying concentration level, together with RS, is shown with respect to time. In A-M, blue colored simulations represented the behavior of the proteins without the presence of RS while red colored simulations demonstrate the behavior of the proteins after RS was inserted in the model. The behavior of RS is shown in N. Initial concentrations in A-H and J-M are set to 0. For I, initial concentrations are set to 1 to show the decreasing behavior of the protein involved. Initial concentration for N is set to 1

Table 3 Notations for Species Concentrations and parameter values. The tables summarize the notations for the protein concentrations that are involved in the three mathematical models.

Notation	Description	Parameter Value	Reference
μ_R	Catalytic production rate of Ras	32.344 μM/min	[12]
μ_P	Catalytic production rate of PI3K	3×10^{-7} μM/min	[11]
μ_{Rf}	Catalytic production rate of Raf	1×10^{-7} μM/min	[11]
μ_A	Catalytic production rate of Akt	0.0566279 μM/min	[47]
μ_M	Catalytic production rate of MEK	4×10^{-7} μM/min	[11]
μ_B	Catalytic production rate of Bad	300 μM/min	Estimated
μ_E	Catalytic production rate of ERK 1/2	49.2683 μM/min	[12]
μ_{MC}	Catalytic production rate of Myc	0.0184 μM/min	[12]
μ_F	Catalytic production rate of p53	0.005 μM/min	[48]
μ_S	Catalytic production rate of STAT	50 μM/min	Estimated
μ_D	Catalytic production rate of Cyclin D1-CDK Complex	50 μM/min	Estimated
μ_{DC}	Catalytic production rate of DCC	0.2 μM/min	Estimated
μ_W	Catalytic production rate of p21	5×10^{-4} μM/min	Estimated
μ_{DZ}	Catalytic production rate of Cyclin D1-CDK Complex in the presence of drug	9×10^{-8} μM/min	Estimated
μ_{RZ}	Catalytic production rate of Ras in the presence of drug	1×10^{-7} μM/min	Estimated
μ_{DCZ}	Catalytic production rate of DCC in the presence of drug	0.5 μM/min	Estimated
δ_R	Transition rate of Ras	319.9672 μM/min	[12]
δ_P	Transition rate of PI3K	0.005 μM/min	[11]
δ_{Rf}	Transition rate of Raf	0.009 μM/min	[11]
δ_A	Transition rate of Akt	0.005 μM/min	[47]
δ_M	Transition rate of MEK	0.0018 μM/min	[11]
δ_B	Transition rate of Bad	0.01 μM/min	Estimated
δ_E	Transition rate of ERK 1/2	1.8848 μM/min	[12]
δ_{MC}	Transition rate of Myc	0.0231 μM/min	[12]
δ_F	Transition rate of p53	0.01 μM/min	[48]
δ_{Br}	Transition rate of BCR-ABL	1 μM/min	Estimated
δ_S	Transition rate of STAT	0.5 μM/min	Estimated
δ_D	Transition rate of Cyclin D1-CDK Complex	0.035 μM/min	Estimated
δ_{DC}	Transition rate of DCC	5 μM/min	Estimated
δ_W	Transition rate of p21	1.8×10^{-2} μM/min	[1]
δ_Z	Transition rate of drug	6×10^{-2} μM/min	Estimated
δ_{DZ}	Transition rate of Cyclin D1-CDK Complex in the presence of drug	0.06 μM/min	Estimated
δ_{BrZ}	Transition rate of BCR-ABL in the presence of drug	1 μM/min	Estimated
δ_{BZ}	Transition rate of Bad in the presence of drug	1×10^{-7} μM/min	Estimated

(*continued*)

Table 3 (*continued*)

Notation	Description	Parameter Value	Reference
δ_{DCZ}	Transition rate of DCC in the presence of drug	$5\,\mu M/min$	Estimated
δ_{RZ}	Transition rate of Ras in the presence of drug	$319.9672\,\mu M/min$	Estimated
δ_{FZ}	Transition rate of p53 in the presence of drug	$2 \times 10^{-5}\,\mu M/min$	Estimated
κ_A	Binding specificity of PI3K and PKB/Akt	$653951\,\mu M/min$	[63]
κ_W	Binding specificity of p53 and p21	$1\,\mu M$	Estimated
κ_B	Binding specificity of PKB/Akt and Bad	$1 \times 10^{-9}\,\mu M$	Estimated
κ_F	Binding specificity of PKB/Akt and p53	$1.4 \times 10^{-8}\,\mu M$	Estimated
κ_D	Binding specificity of p21 and Cyclin D1	$1 \times 10^{-9}\,\mu M$	Estimated
κ_B	Binding specificity of PKB/Akt and Bad	$1 \times 10^{-9}\,\mu M$	Estimated
κ_{RZ}	Binding specificity of drug and Ras	$1 \times 10^{-4}\,\mu M$	Estimated
κ_{DCZ}	Binding specificity of drug and DCC	$10\,\mu M$	Estimated
κ_{BrZ}	Binding specificity of drug and BCR-ABL	$1 \times 10^{-9}\,\mu M$	Estimated
κ_{DZ}	Binding specificity of drug and Cyclin D1 in CML	$5\,\mu M$	Estimated
	Binding specificity of drug and Cyclin D1 in CC	$2.5 \times 10^{-2}\,\mu M$	Estimated
	Binding specificity of drug and Cyclin D1 in BC	$1 \times 10^{-6}\,\mu M$	Estimated
κ_{FZ}	Binding specificity of drug and p53 in CML	$100\,\mu M$	Estimated
	Binding specificity of drug and p53 in CC	$10\,\mu M$	Estimated
	Binding specificity of drug and p53 in BC	$1 \times 10^{-3}\,\mu M$	Estimated
c_1	Rate constant activating BCR-ABL in BC	$300\,\mu M/min$	Estimated
c_1'	Rate constant activating BCR-ABL in CML	$0.36\,\mu M/min$	Estimated
c_2	Rate constant activating Ras in CC	$1000\,\mu M/min$	Estimated

that RS as a drug inserted in all three disease pathways could inhibit Ras, and Cyclin D1-CDK complex while activating p53. BCR-ABL was inhibited in the CML and BC pathways. Bad is then activated for CML and CC and DCC was activated only for CC.

Cyclin D1-CDK complex is one of the more dysregulated cell cycle mediator in aberrant cell growth [33]. Cyclin D1-CDK 4/6 becomes an oncoprotein when overexpressed in most neoplasms [38]. This is because Cyclin D1-CDK complexes plays an important role in the progression of the cell cycle. There is a strong selective pressure for p53 inactivation during tumor development [7]. The almost universal loss of p53 activity in tumors has pushed a tremendous effort to develop p53 based cancer therapies. Transfection of the wild-type p53 gene into a variety of human tumor cells was shown to have a curative response even in advanced tumors [37]. This makes these two protein ideal targets for cancer therapy.

In this study, we used Cyclin D1-CDK complex and p53 as a benchmark for the chemotherapeutic efficacy of RS. This is due to findings that the active compound in RS, MTBITC, was found to effectively inhibit cell proliferation and induce apoptosis through pathways that involved p53 and Cyclin D1-CDK [3, 61]. As shown in Fig. 4, our simulations led to a decrease in Cyclin D1-CDK complex (90.74% for BC, 73.42% for CC ,1.10% for CML) and an increase p53 (83.47% for BC ,62.75% for CC, 5.69% for CML). This shows that there were only significant changes in the concentration of Cyclin D1-CDK complex and p53 for CC and BC.

The simulations suggest that RS was found to be significantly effective for BC and CC. This was due to the varying binding specificity (K_M) of RS with these proteins in different environments [4]. There was a decreasing trend in the K_M values for both Cyclin D1-CDK complex and p53 from CML [$5\,\mu$M, $100\,\mu$M], CC [$2.5 \times 10^{-2}\,\mu$M, $10\,\mu$M] up to BC [$10^{-6}\,\mu$M, $10^{-3}\,\mu$M]. For enzyme catalyzed reactions, K_M is indicative of the biological function. K_M values are inversely proportional to the binding strength of an enzyme and substrate [4]. This suggest that RS had the greatest affinity towards Cyclin D1-CDK complex and p53 in BC followed by CC.

p21, Bad and DCC are also pro-apoptotic proteins that influence cell cycle arrest [3, 42]. p21 acts as a master effector of multiple tumor suppressor pathways for promoting anti-proliferative activities [36]. In a study by Goi et al. [17] and Peltom*ä*ki [49], a decreased expression of DCC has been implicated in CC. The increase of the steady state concentration of these proteins in the simulation of exposure to RS in all three cancers support that RS has chemotherapeutic effects.

The upregulation of proteins in the Raf/MEK/ERK or MAPK signaling pathway can be seen in approximately 30% of human cancers [18, 35]. It has been shown that upregulating Ras can promote BC and CC development [14]. A decrease to the concentration of MEK can slow down the increasing rate of BC cells and can provide additional life time on patients with metastatic diseases [56]. A 75% reduction of these proteins in BC, and CML has been observed after the application of RS as well as a 50% reduction in CC.

Akt is often overexpressed in cancer, which correlates with a poor prognosis for patients [60]. High PI3K activity has been observed in cell transformation and tumor progression in several human cancers including BC [51]. This high PI3K activity induces drug resistance in CML [50] and has been considered for colorectal neoplasms [23]. In our results, Akt and PI3K concentration level declined by 75% in CML and BC, and by 49.98% in colon cancer after the drug was induced.

Xiong et al. [67] reported that STAT3, a subfamily of STAT, has been overexpressed in BC and CC. Kaymaz et al. [32] also mentioned that the upregulation of STAT plays an important role in progression of CML. BCR-ABL is highly expressed in CML [11] and in many invasive BC cell lines [59]. Myc is overexpressed in CML patients who were untreated [19]. This behavior is also observed in human CC [54]. BCR-ABL decreased by 50% in CML and BC while STAT decreased by the same amount in all three cancer types. These observations suggest that RS may be a candidate for chemotherapeutic drug due to its anti-cancer effects on CML, BC and CC.

The role of RS as a chemotherapeutic drug was demonstrated in our results. As shown in Fig. 2, The cytotoxicity assays of RS towards CML, BC, and CC cell lines showed that RS was cytotoxic to CC cell line (HT-29) and BC cell line (MCF-7)—IC_{50} of 13.79 and $2.63\mu g/mL$, respectively, but not to normal cells HDFn. This contrasted to results using the cancer medication Bleomycin that was cytotoxic to all cancers as well as to normal cells. In addition, Tan et al., 2017 [61] found that RS caused the most DNA damage in the BC cell line, followed by CC with minimal genotoxicity to CML. These findings were also observed in the simulation of our model (Fig. 4) which displayed a decrease in Cyclin D1-CDK complex (90.74% for BC, 73.42% for CC, 1.10% for CML) and an increase p53 (83.47% for BC, 62.75% for CC, 5.69% for CML). In order to simulate this, different values binding specificities (K_M) of RS towards Cyclin D1-CDK complex and p53 estimated in the model. The estimates showed a decreasing trend of the K_M of RS with Cyclin D1-CDK complex and p53 respectively from CML [5 μM, 100 μM], CC [2.5×10^{-2} μM, 10 μM] up to BC [10^{-6} μM, 10^{-3} μM]. This trend affirms the findings of Tan et al. [61] and cytotoxicity analyses that RS as a drug was highly effective in BC, moderately effective in CC and minimal in CML.

7 Conclusion

All the behaviors of the proteins that were studied were consistent with what had been reported in previous studies. Inclusion of RS in the model exhibited a general decrease in the proteins that promoted the progression of CML, BC and CC. RS was observed to down regulate proteins that influence cell proliferation in cancer such as Cyclin D1-CDK Complex, Ras, Raf, ERK, MEK, Akt, BCR-ABL, STAT, PI3K, and Myc while upregulating pro-apoptotic proteins (Bad, DCC, p21 and p53). This suggests that the anticancer activity of RS was through the moderation of cell proliferation and induction of apoptosis in malignant neoplasms BC, CC and CML. We have shown that RS is found to be significantly effective for BC and CC.

References

1. Abbas, T., Dutta, A.: p21 in cancer: intricate networks and multiple activities. Nature Rev. Cancer **9**(6), 400–414 (2009). https://doi.org/10.1038/nrc2657
2. Au, W.Y., et al.: Chronic myeloid leukemia in Asia. Int. J. Hematol. **89**(1), 14–23 (2009). https://doi.org/10.1007/s12185-008-0230-0
3. Beevi, S.S., Mangamoori, L.N., Subathra, M., Edula, J.R.: Hexane extract of Raphanus sativus L. roots inhibits cell proliferation and induces apoptosis in human cancer cells by modulating genes related to apoptotic pathway. Plant Foods Hum. Nutr. **65**(3), 200–209 (2010). https://doi.org/10.1007/s11130-010-0178-0
4. Berg, J.M.J.M., Tymoczko, J.L., Stryer, L., Stryer, L.: Biochemistry. W.H. Freeman, New York (2002)

5. Bray, F., Ferlay, J., Soerjomataram, I., Siegel, R.L., Torre, L.A., Jemal, A.: Global cancer statistics 2018: Globocan estimates of incidence and mortality worldwide for 36 cancers in 185 countries. CA Cancer J. Clin. **68**(6), 394–424 (2018)
6. Buban, C.E.: Colorectal cancer curable if detected early | Inquirer Business (2013)
7. Bykov, V.J., Eriksson, S.E., Bianchi, J., Wiman, K.G.: Targeting mutant p53 for efficient cancer therapy. Nat. Rev. Cancer **18**(2), 89 (2018)
8. Cragg, G.M., Newman, D.J.: Natural Products as Sources of Antitumor Agents
9. Crown, J., O'Shaughnessy, J., Gullo, G.: Emerging targeted therapies in triple-negative breast cancer. Ann. Oncol. **23**(suppl 6), vi56–vi65 (2012). https://doi.org/10.1093/annonc/mds196
10. Dang, C.V.: MYC, metabolism, cell growth, and tumorigenesis. Cold Spring Harbor Perspect. Med. **3**(8), a014,217 (2013). https://doi.org/10.1101/cshperspect.a014217
11. Donato, N.J., et al.: BCR-ABL independence and LYN kinase overexpression in chronic myelogenous leukemia cells selected for resistance to STI571. Blood **101**(2), 690–698 (2003). https://doi.org/10.1182/blood.V101.2.690
12. Druker, B.J., et al.: Efficacy and safety of a specific inhibitor of the BCR-ABL tyrosine kinase in chronic myeloid leukemia. N. Engl. J. Med. **344**(14), 1031–1037 (2001). https://doi.org/10.1056/NEJM200104053441401
13. Duffy, M.J., Synnott, N.C., McGowan, P.M., Crown, J., O'Connor, D., Gallagher, W.M.: p53 as a target for the treatment of cancer. Cancer Treat. Rev. **40**(10), 1153–1160 (2014). https://doi.org/10.1016/j.ctrv.2014.10.004
14. Eckert, L.B., et al.: Involvement of ras activation in human breast cancer cell signaling, invasion, and Anoikis. Cancer Res. **64**(13), 4585–4592 (2004). https://doi.org/10.1158/0008-5472.CAN-04-0396
15. Furqan, M., Akinleye, A., Mukhi, N., Mittal, V., Chen, Y., Liu, D.: STAT inhibitors for cancer therapy. J. Hematol. Oncol. **6**(1), 90 (2013). https://doi.org/10.1186/1756-8722-6-90
16. Geran, R.I.: Protocols for screening chemical agents and natural products against animal tumors and other biological systems (3rd edn.) Cancer Chemother. Rep. **3**, 51–61 (1972)
17. Goi, T., Yamaguchi, A., Nakagawara1, G., Urano, T., Shiku, H., Furukawa, K.: Reduced expression of deleted colorectal carcinoma (DCC) protein in established colon cancers. Br. J. Cancer **77**(3), 466–471 (1998)
18. Gollob, J.A., Wilhelm, S., Carter, C., Kelley, S.L.: Role of Raf kinase in cancer: therapeutic potential of targeting the Raf/MEK/ERK signal transduction pathway. Semin. Oncol. **33**(4), 392–406 (2006). https://doi.org/10.1053/j.seminoncol.2006.04.002
19. Gómez-Casares, M.T., et al.: MYC antagonizes the differentiation induced by imatinib in chronic myeloid leukemia cells through downregulation of p27KIP1. Oncogene **32**(17), 2239–2246 (2013). https://doi.org/10.1038/onc.2012.246
20. Gong, H., Zuliani, P., Komuravelli, A., Faeder, J.R., Clarke, E.M.: Analysis and verification of the HMGB1 signaling pathway. BMC Bioinform. **11**(Suppl 7), S10 (2010). https://doi.org/10.1186/1471-2105-11-S7-S10
21. Jacinto, S.D., Chun, E.A.C., Montuno, A.S., Shen, C.C., Espineli, D.L., Ragasa, C.Y.: Cytotoxic cardenolide and sterols from Calotropis gigantea. Nat. Prod. Commun. **6**(6), 803–6 (2011)
22. John, P.C.L., Mews, M., Moore, R.: Cyclin/CDK complexes: their involvement in cell cycle progression and mitotic division. Protoplasma **216**(3–4), 119–142 (2001). https://doi.org/10.1007/BF02673865
23. Johnson, S.M., et al.: Novel expression patterns of PI3K/Akt/mTOR signaling pathway components in colorectal cancer. J. Am. Coll. Surg. **210**(5), 767–776 (2010). https://doi.org/10.1016/j.jamcollsurg.2009.12.008
24. Kamran, M.Z., Patil, P., Gude, R.P.: Role of STAT3 in cancer metastasis and translational advances. BioMed Res. Int. **2013** (2013). https://doi.org/10.1155/2013/421821
25. Kanehisa, M., Furumichi, M., Tanabe, M., Sato, Y., Morishima, K.: KEGG: new perspectives on genomes, pathways, diseases and drugs. Nucleic Acids Res. **45**(D1), D353–D361 (2017). https://doi.org/10.1093/nar/gkw1092
26. Kanehisa, M., Goto, S.: KEGG: Kyoto encyclopedia of genes and genomes. Nucleic Acids Res. **28**(1), 27–30 (2000). https://doi.org/10.1093/nar/28.1.27

27. Kanehisa, M., Sato, Y., Kawashima, M., Furumichi, M., Tanabe, M.: KEGG as a reference resource for gene and protein annotation. Nucleic Acids Res. **44**(D1), D457–D462 (2016). https://doi.org/10.1093/nar/gkv1070
28. Kanehisa Laboratories: KEGG PATHWAY: Colorectal cancer - Homo sapiens (human)
29. Kanehisa Laboratories: KEGG PATHWAY: Chronic myeloid leukemia - Homo sapiens (human) (2017)
30. Kang, H.W., et al.: A mathematical model for microrna in lung cancer. PLoS ONE **8**(1), e53,663 (2013). https://doi.org/10.1371/journal.pone.0053663
31. Kasper, D., Fauci, A., Hauser, S., Longo, D., Jameson, J., Loscalzo, J.: Harrison's principles of internal medicine, 19e, vol. 1. Mcgraw-Hill (2015)
32. Kaymaz, B.T., et al.: Repression of STAT3, STAT5A, and STAT5B expressions in chronic myelogenous leukemia cell line K-562 with unmodified or chemically modified siRNAs and induction of apoptosis. Ann. Hematol. **92**(2), 151–162 (2013). https://doi.org/10.1007/s00277-012-1575-2
33. Kim, J.K., Diehl, J.A.: Nuclear cyclin D1: an oncogenic driver in human cancer. J. Cell. Physiol. **220**(2), 292–296 (2009). https://doi.org/10.1002/jcp.21791
34. Kim, W.K., et al.: Radish (Raphanus sativus L. leaf) ethanol extract inhibits protein and mRNA expression of ErbB(2) and ErbB(3) in MDA-MB-231 human breast cancer cells. Nutr. Res. Pract. **5**(4), 288–93 (2011). https://doi.org/10.4162/nrp.2011.5.4.288
35. Koh, G., Teong, H.F.C., Clement, M.V., Hsu, D., Thiagarajan, P.: A decompositional approach to parameter estimation in pathway modeling: a case study of the Akt and MAPK pathways and their crosstalk. Bioinformatics **22**(14), e271–e280 (2006). https://doi.org/10.1093/bioinformatics/btl264
36. Krimpenfort, P., Song, J.Y., Proost, N., Zevenhoven, J., Jonkers, J., Berns, A.: Deleted in colorectal carcinoma suppresses metastasis in p53-deficient mammary tumours. Nature **482**(7386), 538–541 (2012). https://doi.org/10.1038/nature10790
37. Lane, D.P., Cheok, C.F., Lain, S.: p53-based cancer therapy. Cold Spring Harbor Perspect. Biol. **2**(9), a001,222 (2010). https://doi.org/10.1101/cshperspect.a001222
38. Li, C., et al.: PD-0332991 induces G1 arrest of colorectal carcinoma cells through inhibition of the cyclin-dependent kinase-6 and retinoblastoma protein axis. Oncol. Lett. **7**(5), 1673–1678 (2014). https://doi.org/10.3892/ol.2014.1957
39. Li, Y., Chinni, S.R., Sarkar, F.H.: Selective growth regulatory and pro-apoptotic effects of DIM is mediated by AKT and NF-kappaB pathways in prostate cancer cells. Front Biosci **10**, 236–43 (2005)
40. Mayer, I.A., Arteaga, C.L.: PIK3CA activating mutations: a discordant role in early versus advanced hormone-dependent estrogen receptor-positive breast cancer? J. Clin. Oncol. **32**(27), 2932–2934 (2014). https://doi.org/10.1200/JCO.2014.55.9591
41. McCain, J.: The MAPK (ERK) pathway: investigational combinations for the treatment of BRAF-mutated metastatic melanoma. P T Peer-Review. J. Formulary Manage. **38**(2), 96–108 (2013)
42. Mehlen, P., Rabizadeh, S., Snipas, S.J., Assa-Munt, N., Salvesen, G.S., Bredesen, D.E.: The DCC gene product induces apoptosis by a mechanism requiring receptor proteolysis. Nature **395**(6704), 801–804 (1998). https://doi.org/10.1038/27441
43. Moreno-Bueno, G., et al.: Cyclin D1 gene (CCND1) mutations in endometrial cancer. Oncogene **22**(38), 6115–6118 (2003). https://doi.org/10.1038/sj.onc.1206868
44. Morse, M.A., Stoner, G.D.: Cancer chemoprevention: principles and prospects. Carcinogenesis **14**(9), 1737–1746 (1993)
45. Muller, P.A.J., Vousden, K.H.: Mutant p53 in cancer: new functions and therapeutic opportunities. Cancer Cell **25**(3), 304–17 (2014). https://doi.org/10.1016/j.ccr.2014.01.021
46. Ngelangel, C.A., Wang, E.H.M.: Cancer and the Philippine cancer control program. Jpn. J. Clin. Oncol. **32**, 52–61 (2002)
47. Ngoma, T.: World health organization cancer priorities in developing countries. Ann. Oncol. **17**, viii9–viii14 (2006)

48. Nilsson, J.A., Cleveland, J.L.: Myc pathways provoking cell suicide and cancer. Oncogene **22**(56), 9007–9021 (2003). https://doi.org/10.1038/sj.onc.1207261
49. Peltomäki, P.: Deficient DNA mismatch repair: a common etiologic factor for colon cancer. Hum. Mol. Genet. **10**(7), 735–40 (2001)
50. Polak, R., Buitenhuis, M.: The PI3K/PKB signaling module as key regulator of hematopoiesis: implications for therapeutic strategies in leukemia. Blood **119**(4), 911–23 (2012). https://doi.org/10.1182/blood-2011-07-366203
51. Pópulo, H., Lopes, J.M., Soares, P.: The mTOR signalling pathway in human cancer. Int. J. Mol. Sci. **13**(2), 1886–1918 (2012). https://doi.org/10.3390/ijms13021886
52. Prakash, D., Sharma, G. (eds.): Phytochemicals of nutraceutical importance. CABI, Wallingford (2014). https://doi.org/10.1079/9781780643632.0000
53. Roberts, P.J., Der, C.J.: Targeting the Raf-MEK-ERK mitogen-activated protein kinase cascade for the treatment of cancer. Oncogene **26**(22), 3291–3310 (2007). https://doi.org/10.1038/sj.onc.1210422
54. Rochlitz, C.F., Herrmann, R., de Kant, E.: Overexpression and amplification of c-myc during progression of human colorectal cancer. Oncology **53**(6), 448–54 (1996). https://doi.org/10.1159/000227619
55. Rosen, L.S., Ashurst, H.L., Chap, L.: Targeting signal transduction pathways in metastatic breast cancer: a comprehensive review. Oncologist **15**(3), 216–35 (2010). https://doi.org/10.1634/theoncologist.2009-0145
56. Rottenberg, S., Jonkers, J.: MEK inhibition as a strategy for targeting residual breast cancer cells with low DUSP4 expression. Breast Cancer Res. **14**(6), 324 (2012). https://doi.org/10.1186/bcr3327
57. Saini, K.S., et al.: Targeting the PI3K/AKT/mTOR and Raf/MEK/ERK pathways in the treatment of breast cancer. Cancer Treat. Rev. **39**(8), 935–46 (2013). https://doi.org/10.1016/j.ctrv.2013.03.009
58. Srinivasan, D., Plattner, R.: Activation of Abl tyrosine kinases promotes invasion of aggressive breast cancer cells. Can. Res. **66**(11), 5648–5655 (2006). https://doi.org/10.1158/0008-5472.CAN-06-0734
59. Srinivasan, D., Sims, J.T., Plattner, R.: Aggressive breast cancer cells are dependent on activated Abl kinases for proliferation, anchorage-independent growth and survival. Oncogene **27**(8), 1095–1105 (2008). https://doi.org/10.1038/sj.onc.1210714
60. Steelman, L.S., et al.: Roles of the Raf/MEK/ERK and PI3K/PTEN/Akt/mTOR pathways in controlling growth and sensitivity to therapy-implications for cancer and aging. Aging **3**(3), 192–222 (2011). https://doi.org/10.18632/aging.100296
61. Tan, M.C.S., Enriquez, M.L.D., Arcilla, R.G., Noel, M.G.: Determining the apoptotic-inducing property of isothiocyanates extracted from three cultivars of Raphanus sativus Linn. Using the comet assay. J. Appl. Pharm. Sci. **7**(9), 44–51 (2017). https://doi.org/10.7324/JAPS.2017.70906
62. Tsygvintsev, A., Marino, S., Kirschner, D.E.: A mathematical model of gene therapy for the treatment of cancer. In: Mathematical Methods and Models in Biomedicine, pp. 367–385. Springer (2013)
63. Tubeza, P.C.: Breast cancer is the most prevalent in PH — Inquirer News (2012)
64. Wang, Y., et al.: Mutant N-RAS protects colorectal cancer cells from stress-induced apoptosis and contributes to cancer development and progression. Cancer Discov. **3**(3), 294–307 (2013). https://doi.org/10.1158/2159-8290.CD-12-0198
65. Wee, K.B., Aguda, B.D.: Akt versus p53 in a network of oncogenes and tumor suppressor genes regulating cell survival and death. Biophys. J. **91**(3), 857–865 (2006). https://doi.org/10.1529/BIOPHYSJ.105.077693
66. WHO — early diagnosis and screening (2017)
67. Xiong, A., Yang, Z., Shen, Y., Zhou, J., Shen, Q.: Transcription factor STAT3 as a novel molecular target for cancer prevention. Cancers **6**(2), 926–957 (2014). https://doi.org/10.3390/cancers6020926

Bifurcation Analysis of a Tuberculosis Model with the Risk of Re-infection

Fatima Sulayman, Mohd Hafiz Mohd, and Farah Aini Abdullah

Abstract In this paper, a deterministic model for assessing the influences of exogenous re-infection, re-infection on individuals that have been treated and the transmission rate of the outbreak of tuberculosis are examined. The disease-free (DFE) and endemic (EEP) 'equilibria together with the existence criterion of different equilibria are established. The local stability analysis of the DFE and EEP equilibria is also performed. The basic reproduction number is derived by using the next generation matrix and it is found that the disease free equilibrium is stable when $R_0 < 1$, and the model undergoes a transcritical bifurcation at $R_0 = 1$. By using bifurcation analysis, further investigation reveals the existence of several threshold conditions, which trigger some bifurcational changes in dynamics to occur in this epidemiological system. In particular, we observe the emergence of saddle-node and transcritical bifurcations and the interaction between these two bifurcations can shape the overall dynamics of the system.

Keywords Tuberculosis · Basic reproduction number · Stability analysis · Bifurcation analysis · Saddle-node and transcritical bifurcations

1 Introduction

Tuberculosis (TB) is a communicable disease that affects the lungs. Some studies have suggested that up to a third of the world human population has contracted the disease [1]. Tuberculosis is caused by family of bacteria called Mycobacterium Tuberculosis. This airborne infectious disease is a public health challenge worldwide: in USA [2] and European nation [3, 4], and developing countries. In other words, tuberculosis remains a major epidemic that contributes to the morbidity and mortality [5, 6] of both human and animal populations [7]. Usually, when the dis-

F. Sulayman (✉) · M. H. Mohd · F. A. Abdullah
School of Mathematical Sciences, Universiti Sains Malaysia, USM, 11800 George Town, Penang, Malaysia

M. H. Mohd
School of Mathematical Sciences, Faculty of Science and Technology, Universiti Kebangsaan Malaysia, UKM, 43600 Bangi, Selangor, Malaysia

M. H. Mohd et al. (eds.), *Modelling, Simulation and Applications of Complex Systems*, Springer Proceedings in Mathematics & Statistics 359,
https://doi.org/10.1007/978-981-16-2629-6_10

ease is contracted by individuals, it is not immediately obvious. For instance, when a person is infected, he/she may remain infected for years, or even latently-infected for life [8]. It is attested to be highly prevalent in middle-income countries, among other regions [6]. The prevalence of TB, globally, in 2018 was about 1 million cases more than that of HIV/AIDS, and there were up to 1.5 million more deaths [6].

Mostly, TB can be transmitted whether through direct or indirect contact with an infected person [4, 9]. The symptoms of tuberculosis include bad cough that produces blood and/or sputum (phlegm from deep inside the lungs), which can persist for up to three weeks, chest pain, fever, and sweating at night, among others [10]. TB may take from two to three months to become symptomatic, due to long incubation period. The long latent period of MTb bacteria prolongs the start of the active phase of the disease, which further makes it difficult to understand how the disease develops. Another point related to the spread of this disease is the level of exogenous re-infection (see [5, 11]). Already infected Individuals (even in the dormant stage), with low levels of immunity may become newly infected through contact with an infectious individual [12]. Therefore, anyone in the latent phase of TB can move to the active phase due to exogenous re-infection rate [12].

Bandera et al. [13] discussed exogenous re-infection in locations with low prevalence of TB; the authors established that re-infection occurs at a lower rate in developed countries than in high-risk areas. Uys et al. [14] studied a deterministic model of TB with the probability of re-infection component and suggested that the rate of re-infection is a multiple of the rate of first time infection. Other studies have investigated the influences of re-infection or multiple infections (see [11, 15]) as well as on the effects of exogenous re-infection on the transmission dynamics of TB see [17–21]. Feng et al. [16] proposed a deterministic model for TB with exogenous re-infection and submitted that an infectious individual can infect another one with each contact per unit of time.

The present paper is motivated by the study of [12], in which the authors proposed a TB model with SEIT components, and it is later being reduced to SEI model. The present study built on the work of [12], by finding the equilibria and the stability of the equilibria, as well as using simulation for the bifurcation part. We investigate the significant impacts of transmission rate β and exogenous re-infection p on the bifurcational changes in dynamics in this epidemiological system.

This article is presented as follows: a mathematical model of TB is considered in Sect. 2. In Sect. 3 we focus on the existence of steady state and stability of disease free (DFE). In Sect. 4 stability of endemic equilibrium point (EEP) and bifurcation analysis are carried out. The numerical simulation and bifurcation analysis results are reported in Sect. 5. Finally, the discussion and conclusion of the study are given in Sect. 6.

2 Model Formulation

We examined the transmission dynamics of TB by employing a nonlinear ODE system of SEIT type system, which is developed by Kar and Mondal [12]. Our aim is to better understand the interplay between exogenous re-infection and transmission rate. In determining the outbreaks of TB. The whole population $N(\mathrm{t})$is classified into four (4) compartments, S, E, I and T, which respectively denote susceptible, exposed, infected and treated individuals. Kar and Mondal [12] assumed that individuals that have been treated can be re-infected if their immunity is low. This leads to S-E-I-T-E type system to the model transmission dynamics of TB. The SEITE process of TB spreading is shown in Fig. 1.

In this epidemiological system, the susceptible compartment is increased by recruiting individuals, either by immigration or birth, into the population at the constant rate Λ. The term μ is taken to be natural death rate. The exposed compartment becomes infectious and progresses to active infected state at a constant rate κ. Infected individuals also develop active tuberculosis because of exogenous re-infection (can acquire new infection from another infectious individual) at a rate p. Infected individuals are recovered at the rate r. The treated individuals return to the exposed compartment due to low immunity at the rate σ.

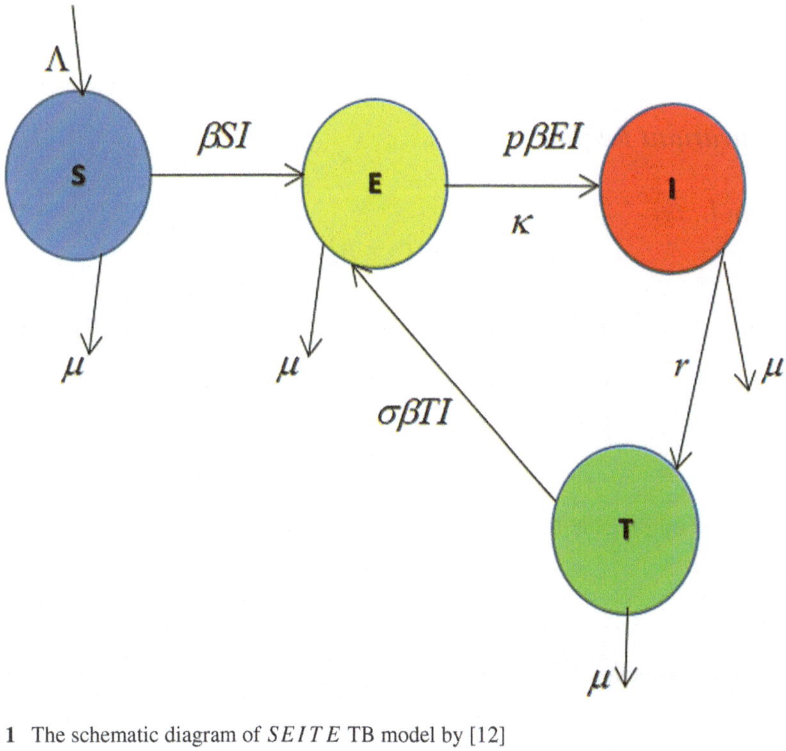

Fig. 1 The schematic diagram of $SEITE$ TB model by [12]

2.1 The Model

The nonlinear model that is considered in this study consists of a system of ordinary differential equations (ODE):

$$\begin{aligned}
\tfrac{dS}{dt} &= \Lambda - \mu S - \beta SI \\
\tfrac{dE}{dt} &= \beta SI - p\beta EI - (\kappa + \mu)\, E + \sigma\beta TI \\
\tfrac{dI}{dt} &= p\beta EI + \kappa E - (\mu + r)\, I \\
\tfrac{dT}{dt} &= rI - \mu T - \sigma\beta TI
\end{aligned} \tag{1}$$

with

$$N = S + E + I + T.$$

3 Mathematical Model Analysis

In this section, we seek to qualitatively study dynamical properties of the TB model (1) by means of invariant and positivity solutions.

3.1 Invariant Region

Theorem 1. *Let*

$$\Phi = \{(S, E, I, T) \in R_+^4 : N \leq \frac{\Lambda}{\mu}\}$$

The region Φ is positively-invariant and attract all positive solutions of the model.

Proof. As $N = S + E + I + T$, adding all the equations of the system (1), we have

$$\frac{dN}{dt} = \Lambda - \mu N \tag{2}$$

By standard comparison theorem in [22] we see that

$$\frac{dN}{dt} \leq \Lambda - \mu N \tag{3}$$

which yield (by method of integration factor).

$$N(t) \leq \frac{\Lambda}{\mu} + \left[N(0) - \frac{\Lambda}{\mu} \right] e^{-\mu t}$$

In particular, if $N(0) \leq \frac{\Lambda}{\mu}$, then $N(t) \leq \frac{\Lambda}{\mu}$. Hence Φ is positively-invariant and an attractor so that no solution path leaves through any boundary of Φ [23].

3.2 *Positivity of Solutions*

In order for the tuberculosis infection model (1) to be epidemiological realistic, it is essential to demonstrate that all the state variables are positive at all times.

Let initial data be

Theorem 2. $\{(S, E, I, T) \geq 0\} \in \Phi$.

Then, the solution set $\{S(t), E(t), I(t), T(t)\}$ of the model system (1) is non-negative for all $t > 0$.

Proof. As in Obasi and Mbah [24] from the non-linear system of model system (1) we take the first equation

$$\frac{dS}{dt} = \Lambda - \mu S - \beta SI$$

$$\frac{dS}{dt} \geq -(\mu + \beta I)S \tag{4}$$

$$\Rightarrow \frac{dS}{S} \geq -(\mu + \beta I)dt. \tag{5}$$

integrating (5) gives

$$S(t) \geq S(0)e^{\int -(\mu+\beta I)dt} \geq 0. \tag{6}$$

In the same way, it can be verified that $E(t) > 0$, $I(t) > 0$ and $T(t) > 0$ for all $t > 0$ [21].

Thus, the disease is uniformly consistent for every positive solution. The above result can also be proved by using the approach in [25].

The system (1) stands for the dynamic transmission of TB in humans; hence, all the related parameters are taken to be positive. As a result, the following positivity and invariance results must hold.

It is assumed that Eq. (2) admits a positive equilibrium N_0 that satisfies $\Lambda = \mu N$ with N_0 represents population equilibrium in the absence of disease. It is also assumed that this equilibrium is asymptotically stable and unique $N_0 > 0$. This implies that the total population will still be in equilibrium while the epidemic is spreading. Therefore, Eq. (1) can be considered if $S + E + I + T = N_0$.

The system of Eqs. (1) is not solvable analytically because of the non-linearities. Instead, we can eliminate the variable T in order to reduce system (1) into three dimensional system. In particular, we set $T = N_0 - S - E - I$ can be written as:

$$\begin{aligned} \tfrac{dS}{dt} &= \mu N_0 - \mu S - \beta SI \\ \tfrac{dE}{dt} &= \beta SI - p\beta EI - (\kappa + \mu) E + \sigma\beta I (N_0 - S - E - I) \\ \tfrac{dI}{dt} &= p\beta EI + \kappa E - (\mu + r) I \end{aligned} \tag{7}$$

3.3 Analysis of Disease-Free Equilibrium (DFE), P_0, and Basic Reproduction Number R_0

The disease-free equilibrium (DFE) state, P_0 is a steady state solution where there is no infection in the community. Disease class can be described as the infected human population. Taking the first equation of (3) with $E = I = 0$ into consideration, we arrive at:

$$\mu N - \mu S = 0$$

i.e.,

$$S = N_0$$

Then, the disease free equilibrium (DFE) state P_0 is given by $(S, E, I) = (N_0, 0, 0)$. Diekmann et al. [26] define basic reproduction number, R_0 as the effective number of secondary infections caused by a primary infected individual. To obtain the basic reproduction number, we employed next generation matrix method by [27], where F is the matrix of the new infection terms and V the matrix of the transition terms. The matrices F and V are determined from the coefficients of E and I, in the second and the last equations of the system (7). Starting with the newly infective classes, the model equations can then be written as:

$$\frac{dE}{dt} = \beta SI - p\beta EI - (\mu + k) E - \sigma\beta I (E + I), \tag{8}$$

$$\frac{dI}{dt} = p\beta EI + kE - (\mu + r) I \tag{9}$$

$$F = \begin{pmatrix} \beta S \\ 0 \end{pmatrix}, V = \begin{pmatrix} (\mu + \kappa) & 0 \\ -k & (\mu + r) \end{pmatrix}$$

Next, the Jacobian matrix of **F** and **V** at the disease free equilibrium $(N_0, 0, 0)$ is obtained to give:

$$F = \begin{bmatrix} 0 & \beta N_0 \\ 0 & 0 \end{bmatrix}, V = \begin{bmatrix} (\mu+\kappa) & 0 \\ -k & (\mu+r) \end{bmatrix}$$

It is always known that for any two by two matrix for example, $A = \begin{bmatrix} a & b \\ c & d \end{bmatrix}$, its inverse can be obtained by:

$$A^{-1} = \frac{1}{ad - bc} \begin{bmatrix} d & -b \\ -c & a \end{bmatrix}. \tag{10}$$

In a similar way, the inverse of V can be obtained as given below:

$$V^{-1} = \frac{1}{(\mu+\kappa)(\mu+r) - 0} \begin{bmatrix} (\mu+r) & 0 \\ k & (\mu+\kappa) \end{bmatrix} = \begin{bmatrix} \frac{1}{(\mu+\kappa)} & 0 \\ \frac{\kappa}{(\mu+\kappa)(\mu+r)} & \frac{1}{(\mu+r)} \end{bmatrix} \tag{11}$$

then, we now compute the product of F and V^{-1}, which becomes,

$$FV^{-1} = \begin{bmatrix} 0 & \beta N_0 \\ 0 & 0 \end{bmatrix} \begin{bmatrix} \frac{1}{(\mu+\kappa)} & 0 \\ \frac{k}{(\mu+\kappa)(\mu+r)} & \frac{1}{(\mu+r)} \end{bmatrix} \tag{12}$$

$$FV^{-1} = \begin{bmatrix} \frac{\beta\kappa N_0}{(\mu+\kappa)(\mu+r)} & \frac{1}{(\mu+r)} \\ 0 & 0 \end{bmatrix}. \tag{13}$$

Basic reproduction number is the spectral radius of FV^{-1} [26, 28]. It must be considered when analyzing any epidemiological model [27]. Thus, R_0 for this epidemiological system is:

$$R_0 = \frac{\beta\kappa N_0}{(\mu+\kappa)(\mu+r)}. \tag{14}$$

4 Stability and Bifurcation Analysis

The Jacobian matrix Σ of the system (7) at endemic equilibrium (S^*, E^*, I^*) is given by

$$\Sigma = \begin{pmatrix} -\mu - \beta I^* & 0 & -\beta S^* \\ -\sigma\beta I^* + \beta I^* & -p\beta I^* - \sigma\beta I^* - (\mu+\kappa) & \beta(1-\sigma)S^* - (p+\sigma)\beta E^* + \sigma\beta(N_0 - 2I^*) \\ 0 & p\beta I^* + \kappa & p\beta E^* - (\mu+r) \end{pmatrix}. \tag{15}$$

For the endemic equilibrium points, setting the right hand sides of the equations in model (7) to zero we have:

$$\begin{aligned}
&\mu N_0 - \mu S^* - \beta S I^* = 0 \\
&\beta(1-\sigma)S^*I^* - (p+\sigma)\beta E^*I^* - (\kappa+\mu)\,E^* + \sigma\beta I^*(N_0 - I^*) = 0 \\
&p\beta E^*I^* + \kappa E^* - (\mu + r)\,I^* = 0
\end{aligned} \tag{16}$$

For endemic equilibrium $P = (S^*, E^*, I^*)$, from first and last equations of (16)

$$S^* = \frac{\mu N_0}{(\mu + \beta I^*)} \tag{17}$$

$$E^* = \frac{(\mu + r)I^*}{(\kappa + p\beta I^*)} \tag{18}$$

Substituting (17) and (18) into Eq. (16), the endemic equilibrium conditions become a cubic polynomial equation of I^* which is given below:

$$f(I) = n_1 I^{*3} + n_2 I^{*2} + n_3 I^* + n_4 = 0 \tag{19}$$

where

$$\begin{aligned}
n_1 &= p\sigma\beta \\
n_2 &= p\sigma N_0(\beta_0 - \beta) \\
n_3 &= (p\mu + \kappa\sigma)N_0\left(\tfrac{\beta_1}{\beta} - 1\right) \\
n_4 &= \mu(\mu + r)(\mu + \kappa)\tfrac{(1-R_0)}{\beta^2}
\end{aligned}$$

where

$$R_0 = \frac{\beta\kappa N_0}{(\mu + \kappa)(\mu + r)}$$

and β_0, β_1 are two threshold parameters as follows:

$$\beta_0 = \tfrac{(p+\sigma)(\mu+r)+\sigma(\kappa+p\mu)}{p\sigma N_0}$$

$$\beta_1 = \tfrac{\mu(\mu+r)(p+\sigma+1)+\kappa(\mu+r+\mu\sigma)}{(p\mu+\kappa\sigma)N_0}$$

It is clear the I^* is the positive real roots of the polynomial (19). Currently, number of possible positive real roots of the cubic polynomial (19), depend on the signs of n_2, n_3 , and , n_4.

4.1 Local Asymptotic Stability (LAS) of Disease Free Equilibrium Point P_0

Theorem 3. *The disease free equilibrium state, P_0 of the system (7) is LAS when $R_0 < 1$ and otherwise unstable.*

Proof. Consider the model system (7)

$$f_1 = \frac{dS}{dt}, f_2 = \frac{dE}{dt}, f_3 = \frac{dI}{dt}, \tag{20}$$

then, the Jacobian of the model system (7) will be given by:

$$\Sigma(P_0) = \begin{pmatrix} \frac{\partial f_1}{\partial S} & \frac{\partial f_1}{\partial E} & \frac{\partial f_1}{\partial I} \\ \frac{\partial f_2}{\partial S} & \frac{\partial f_2}{\partial E} & \frac{\partial f_2}{\partial I} \\ \frac{\partial f_3}{\partial S} & \frac{\partial f_3}{\partial E} & \frac{\partial f_3}{\partial I} \end{pmatrix} \tag{21}$$

at equilibrium point (P_0), the Jacobian becomes;

$$\Sigma(P_0) = \begin{pmatrix} -\mu & 0 & -\beta S \\ \beta I & -(\kappa+\mu) & \beta S + \sigma\beta R \\ 0 & \kappa & -(\mu+r) \end{pmatrix} \tag{22}$$

Evaluating the (22) at the P_0 to ascertain the LAS of the system yield;

$$\Sigma(P_0) = \begin{pmatrix} -\mu & 0 & -\beta N_0 \\ 0 & -(\kappa+\mu) & \beta N_0 \\ 0 & \kappa & -(\mu+r) \end{pmatrix} \tag{23}$$

Calculate the eigenvalues of the Jacobian, evaluated at the equilibrium point (in this case, the P_0) ($\Sigma(P_0)$, thus: The eigenvalue of (23) are given by

$$|\Sigma P_0 - \alpha I| = 0$$

where $\Sigma(P_0)$ represents Jacobian matrix at disease free equilibrium, α is the characteristic equation of the matrix while I is the identity matrix. The characteristics equation of the resulting Jacobian from (23) is given by:

$$\Sigma(P_0) = \begin{vmatrix} -\mu-\alpha & 0 & -\beta N_0 \\ 0 & -(\kappa+\mu)-\alpha & \beta N_0 \\ 0 & \kappa & -(\mu+r)-\alpha \end{vmatrix} = 0. \tag{24}$$

Thus, the characteristics equation of (24) are given by

$$(-\mu-\alpha)\left[-(\mu+\kappa)-\alpha)(-(\mu+r)-\alpha) - \kappa\beta N_0\right] = 0$$

$$(\mu+\alpha)\left[(\mu+\kappa)(\mu+r)+(\kappa+\mu+\mu+r)\alpha+\alpha^2-\kappa\beta N_0\right]=0$$

$$(\mu+\alpha)\left[\alpha^2+(\kappa+2\mu+r)\alpha+(\mu+\kappa)(\mu+r)-\kappa\beta N_0\right]=0$$

$$(\mu+\alpha)\left[\alpha^2+(\kappa+2\mu+r)\alpha+(\mu+\kappa)(\mu+r)\left(1-\frac{\kappa\beta N_0}{(\mu+\kappa)(\mu+r)}\right)\right]=0$$

$$(\mu+\alpha)\left[\alpha^2+(\kappa+2\mu+r)\alpha+(\mu+\kappa)(\mu+r)(1-R_0)\right]=0 \tag{25}$$

From Eq. (25)

$$((\mu+\alpha)=f(\alpha))$$

therefore

$$f(\alpha)=\delta_0\alpha^2+\delta_1\alpha+\delta_2. \tag{26}$$

where

$\delta_0=1$, $\delta_1=(\kappa+2\mu+r)\alpha$, $\delta_2=(\mu+\kappa)(\mu+r)(1-R_0)$.

In order to show the LAS of DFE, we set up conditions that will allow the quadratic equation (26) to have negative real roots. In this case, we apply the Routh-Hurwitz (R-H) Criterion.

Clearly, from Eq. (26), $\delta_0=1>0$, $\delta_1=(\kappa+2\mu+r)\alpha>0$, while $\delta_2=(\mu+\kappa)(\mu+r)(1-R_0)>0$ if and only if $R_0<1$. Therefore, disease–free equilibrium is locally asymptotically stable when $R_0<1$ and unstable when $R_0>1$.

5 Bifurcation Analysis and Numerical Simulation Studies

Bifurcation analysis is an essential tool that enables systematic identification of where dynamics of interest can be located in the parameter space [29]. The important parameters that play crucial roles in the system are varied in order to understand dynamical behaviours of the model. These parameters are:

- Basic Reproduction number, R_0: this term is one of epidemiological value that plays a critical role in epidemics [27, 30–32].
- Exogenous re-infection rate, p: this quantity refers to when a previously infected individual (in dormant stage) in an exposed class acquire new infection from another infectious person, as a result of low level of immunity [12].
- Transmission rate β: this term is the extent of transmission probability of the disease due to contact with infectious individuals [33].

Table 1 Parameter and values used for the model (7)

Parameters	Descriptions	Values	Source
μ	Natural death rate	0.15	[16]
β	Transmission rate	2.6	[12]
N_0	Total population	5	[12]
σ	Reinfection among the treated individuals	0.25–1	[34]
κ	Progression rate	0.02	[12]
r	Recovery rate	2	[35]
p	Exogenous reinfection	0.25–1	[34]

This section presents some findings from our numerical simulation analysis. We also highlight our bifurcation analysis results to demonstrate the bifurcational changes in dynamics through the occurrence of distinct bifurcations. The appearance and stability of different branches of (DFE and EEP) equilibria are analyzed as β and p are changed. Numerical simulation and bifurcation analysis are performed using MATLAB and XPPAUT [39], respectively. Figure 2 illustrates the bifurcation diagram of the system (7) as the basic reproduction number (R_0) varies against the infected population (I). The first (top-left) diagram of Fig. 2 depicts the bifurcation diagram at $\beta = 1.7$. In general, there are several branches of steady states: (i) the upper branch corresponds to stable endemic equilibria, EEP; (ii) the middle branch represents unstable EEP; (iii) the lower branch corresponds disease free equilibria, DFE, which can be stable or unstable depending on the magnitudes of R_0. As in our theoretical analysis section, this R_0 quantity can be calculated using Eq. (14). Our bifurcation analysis results also reveal the occurrence of two threshold quantities, i.e., transcritical (BP, with this bifurcation occurs when $R_0 = 1$) and saddle-node (LP) bifurcations.

When $R_0 > 1$, only EEP is stable and this situation leads to an outbreak of TB disease. Also noticed is that as basic reproduction number decreases and lies below LP point i.e., $R_0 < LP$, DFE is stable in this case; consequently, this situation leads to elimination of the disease. To further understand the transmission dynamics of system (3), we plot some time series diagrams in Fig. 3 as transmission rate β changes. As can be seen in the top left and right diagrams, one of the possibilities when the magnitudes of β are high is that this situation would lead to the persistence of TB disease (i.e., the trajectories converge to EEP) in this epidemiological system. However, as the transmission rate β decreases, this results in small R_0 quantities as shown by the bottom-left (respectively, bottom-right) of Fig. 3 with $\beta = 1.5$ (respectively, $\beta = 1$) and $R_0 = 0.4104$ (respectively, $R_0 = 0.2736$); consequently, the TB disease is eradicated and DFE is stable in a long run.

Furthermore, other interesting dynamics are realized when $LP < R_0 < 1$, whereby alternative steady states phenomenon emerges. In this case, the outcomes of this epidemiological system depend on initial conditions and the unstable EEP (middle branch) acts as a basin boundary separating the two stable (DFE and EEP) equilibria. Hence, the trajectories will converge to either DFE (elimination of disease) or EEP (outbreaks of TB), depending on the initial conditions. For example, an initial condition below the unstable EEP (basin boundary) will converge to DFE. Otherwise, the trajectories will shift and converge to EEP. To demonstrate these possible outcomes, the time series diagrams in Fig. 4 are plotted using distinct initial conditions. It can be seen that when $\beta = 1.7$ (with other parameter values fixed as in Table 1) and $R_0 = 0.4651$, this situation is equivalent to a bifurcation diagram in the top left of Fig. 2 where alternative steady states occur. In this case, the trajectories converge to either EEP (outbreaks of TB: Fig. 4 left) or DFE (elimination of disease: Fig. 4 right), depending on the choice of initial conditions.

A closer investigation of our bifurcation analysis results in Fig. 2 demonstrate that the interaction between saddle-node and transcritical bifurcations leads to contrasting observations in the model (7). As the value of transmission rate β decreases, we observe that the frequency of alternative steady states phenomenon diminishes and eventually eliminated under R_0-variation. This is evident when we examine the fifth (bottom-left) and sixth (bottom-right) diagrams of Fig. 2: the two bifurcation points, namely BP and LP, approach each other and finally coalesce. Consequently, alternative stable states incident disappear and the trajectories would converge to a stable equilibrium of the system; for instance, as $\beta = 1.5$, EEP (respectively, DFE) is stable when $R_0 > 1$ (respectively, $R_0 < 1$) and this situation causes an outbreak (respectively, elimination) of TB disease.

To examine the influence of the force of exogenous reinfection on the dynamical behaviour of the model, we performed one-parameter bifurcation analysis as shown in Fig. 5. These findings demonstrate some bifurcation diagrams of system (7) with **y**-axis representing infected population (I) and **x**-axis corresponding to basic reproduction number (R_0) as parameter p changes.

Similar to our previous observation, there are some branches of steady states where the upper (respectively, middle) branch corresponds to stable (respectively, unstable) EEP. In addition, there occurs lower branch of equilibrium for DFE, which can be stable or unstable, depending on the magnitudes of R_0. The appearances of transcritical (BP) and saddle-node (LP) bifurcations are also observed. As p increases (from top to bottom diagrams), we notice that the alternative stable states region first vanishes and then emerges. The bi-stable region gets wider (i.e., occurs at more values of R_0) the force of exogenous re-infection (p) gets higher (see bottom left and right diagrams); in this case, when $LP < R_0 < 1$, the outcomes of system (7) determined by the alternative stable states phenomenon with the trajectories converging to either DFE (elimination of disease) or EEP (outbreaks of TB), depending on the initial conditions. It is realized that the interaction between these distinct bifurcations can shape the overall dynamics of the epidemiological system (7).

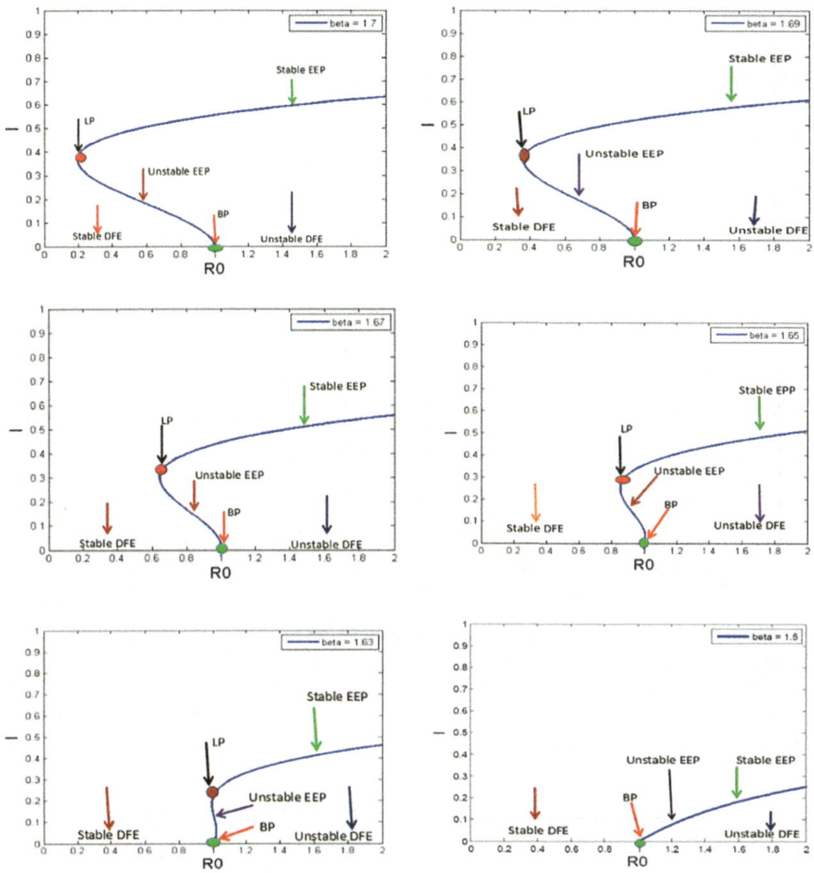

Fig. 2 Bifurcation diagram of system model (7), there is different types of behaviour as the bifurcation parameter β changes from the top left, different values of β are used in clockwise direction respectively: values are $\beta = 1.7$, $\beta = 1.69$, $\beta = 1.67$, $\beta = 1.65$, $\beta = 1.63$, $\beta = 1.61$, while other parameters values are fixed $\mu = 0.15, N_0 = 5, p = 0.5, \kappa = 0.02, \sigma = 0.7, r = 2$. The label LP corresponds to saddle node bifurcation, BP correspond to transcritical bifurcation, EEP corresponds to endemic equilibrium point and DFE correspond to disease free equilbrium. These graphs are computed using Matlab

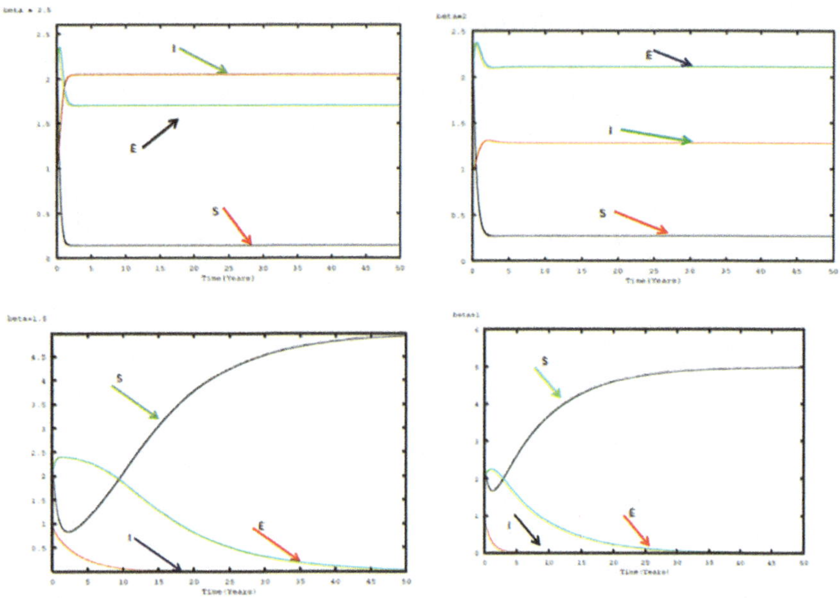

Fig. 3 Simulation results showing the impact of transmission rate (β) on the spread of tuberculosis. The values of transmission rate are set to be $\beta = 2.5$, $\beta = 2$, $\beta = 1.5$, $\beta = 1$, in clockwise direction respectively while other parameters values are $\mu = 0.15$, $N_0 = 5$, p = 0.5 $k = 0.02$, $\sigma = 0.7$, $r = 2$ and the initial conditions S(0) = 2.5, E(0) = 2 and I(0) = 1. These plots are computed by numerical continuation XPPAUT

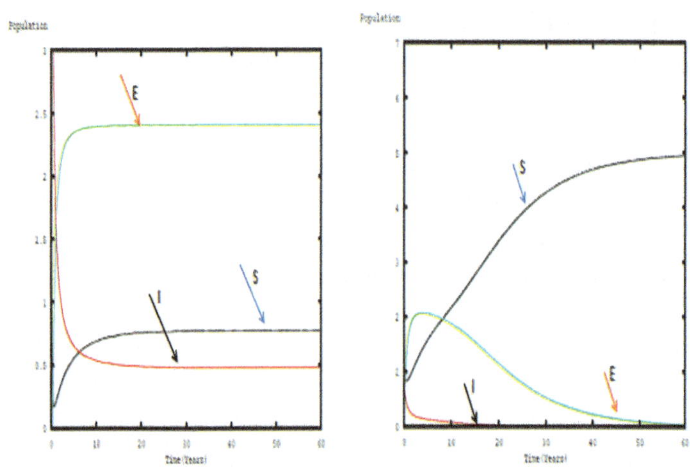

Fig. 4 Time series plots of system (7) showing the endemic equilibrium points (EEP) left and disease free equilibrium (DFE) right with the initial conditions of (S(0), E(0), I(0)) = (3.5, 2.5, 4) and (2.5, 2,1) at the parameter values $\mu = 0.15$, $N_0 = 5$, $\beta = 1.7$, $\sigma = 0.7$, $k = 0.02$, $p = 0.5$, $r = 2$. These plots are computed by numerical continuation XPPAUT

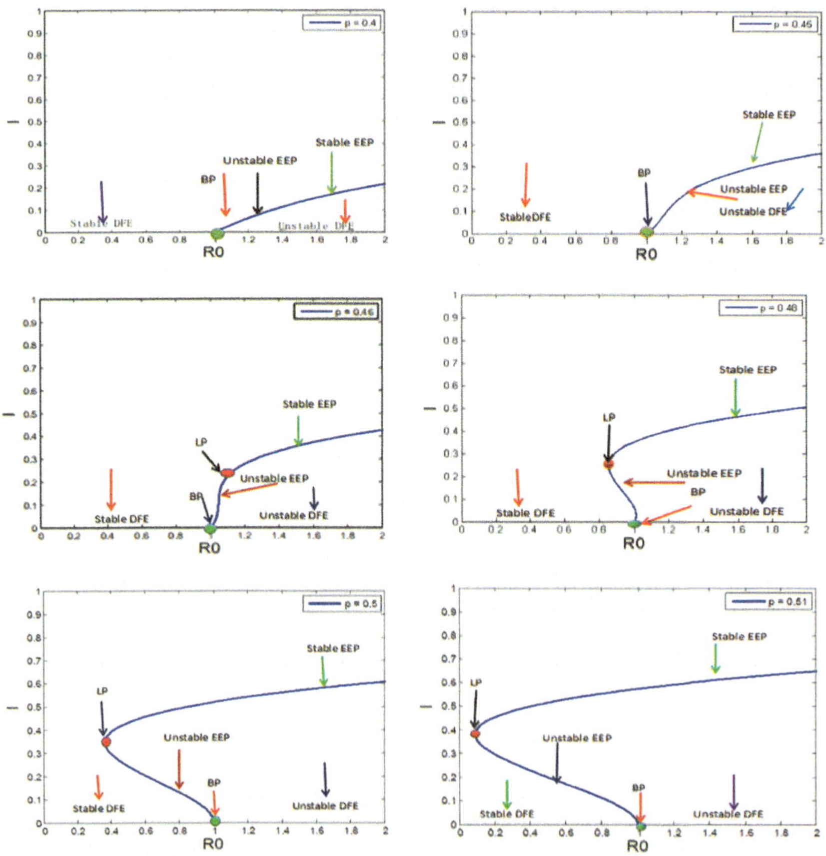

Fig. 5 Bifurcation diagram of system model (7), there is different types of behaviour as the bifurcation parameter p changes from the top left, different values of p are used: values are from 0.4 to 0.51 in clockwise direction respectively while other parameters values are fixed $\mu = 0.15, N_0 = 5$, $\beta = 1.69$, $\kappa = 0.02$, $\sigma = 0.7$, $r = 2$. The label LP corresponds to saddle node bifurcation, BP corresponds to transcritical bifurcation, EEP corresponds to endemic equilibrium point and DFE corresponds to disease free equilibrium. These graphs are computed using Matlab

6 Discussion and Conclusion

In this present paper, the influences of transmission rate, β, and exogenous re-infection, p, on the dynamical behaviors of TB outbreaks have been investigated. It is observed from the simulation and bifurcation analysis results that the threshold quantity known as basic reproduction number R_0 is very critical when determining the persistence and exclusion of the epidemic. By using bifurcation analysis, we observed the occurrences of transcritical and saddle node bifurcations in the system. The interplay between these two local bifurcations shapes the overall dynamics of the system and determines the outbreaks of TB disease. In general, the interaction between saddle-node and transcritical bifurcations is also considered in other biological systems such as Mohd et al. [36–38] and Kooi et al. [40]. We conclude that different epidemiological forces, such as the transmission rate β and exogenous reinfection, p, exert significant effects on the transmission dynamics of TB. We suggest that future research should focus on mitigating their joint impacts on the society.

Acknowledgement This research was supported by the Universiti Sains Malaysia (USM) Fundamental Research Grant Scheme (FRGS) No. 203/PMATHS/6711645 and the Bridging Grant No. 304/PMATHS/6316285 from Research Creativity and Management Office (RCMO) USM.

References

1. Bloom, B.R., Murray, C.J.: Tuberculosis: commentary on a reemergent killer. Science **257**(5073), 1055–1064 (1992)
2. Hill, A.N., Becerra, J.E., Castro, K.G.: Modelling tuberculosis trends in the USA. Epidemiol. Infection **140**(10), 1862–1872 (2012)
3. Abubakar I., Dara, M., Manissero, D., Zumla, A.: Tackling the spread of drug-resistant tuberculosis in Europe. Lancet **379**(9813), e21–e23 (2012)
4. Behr, M.A.: Tuberculosis due to multiple strains: a concern for the patient? A concern for tuberculosis control? (2004)
5. Moghadas, S.M., Alexander, M.E.: Exogenous reinfection and resurgence of tuberculosis: a theoretical framework. J. Biol. Syst. **12**(02), 231–247 (2004)
6. World Health Organization. Global tuberculosis report. *Google Scholar* 214 (2018)
7. Castillo-Chavez, C., Song, B.: Dynamical models of tuberculosis and their applications. Math. Biosci. Eng. **1**(2), 361 (2004)
8. Adebiyi, A.O.: Mathematical modeling of the population dynamics of tuberculosis. University of the Western Cape, South Africa (2016)
9. Richardson, M., et al.: Multiple Mycobacterium tuberculosis strains in early cultures from patients in a high-incidence community setting. J. Clin. Microbiol. **40**(8), 2750–2754 (2002)
10. https://www.cdc.gov/tb/publications/guidelines/default.htm
11. van Rie, A., et al.: Exogenous reinfection as a cause of recurrent tuberculosis after curative treatment. N. Engl. J. Med. **341**(16), 1174–1179 (1999)
12. Kar, T.K., Mondal, P.K.: Global dynamics of a tuberculosis epidemic model and the influence of backward bifurcation. J. Math. Modell. Alg. **11**(4), 433–459 (2012)
13. Bandera, A., et al.: Molecular epidemiology study of exogenous reinfection in an area with a low incidence of tuberculosis. J. Clin. Microbiol. **39**(6), 2213–2218 (2001)
14. Uys, P.W., van Helden, P.D., Hargrove, J.W.: Tuberculosis reinfection rate as a proportion of total infection rate correlates with the logarithm of the incidence rate: a mathematical model. J. R. Soc. Interface **6**(30), 11–15 (2009)
15. Verver, S., et al.: Rate of reinfection tuberculosis after successful treatment is higher than rate of new tuberculosis. Am. J. Respir. Crit. Care Med. **171**(12), 1430–1435 (2005)
16. Feng, Z., Castillo-Chavez, C., Capurro, A.F.: A model for tuberculosis with exogenous reinfection. Theor. Popul. Biol. **57**(3), 235–247 (2000)
17. Cohen, T., Lipsitch, M., Walensky, R.P., Murray, M.: Beneficial and perverse effects of isoniazid preventive therapy for latent tuberculosis infection in HIV-tuberculosis coinfected populations. Proc. Natl. Acad. Sci. **103**(18), 7042–7047 (2006)
18. Gerberry, D.J.: Practical aspects of backward bifurcation in a mathematical model for tuberculosis. J. Theor. Biol. **388**, 15–36 (2016)
19. Singer, B.H., Kirschner, D.E.: Influence of backward bifurcation on interpretation of R_0 in a model of epidemic tuberculosis with reinfection. Math. Biosci. Eng. **1**(1), 81 (2004)

20. Wu, P., Lau, E.H., Cowling, B.J., Leung, C.C., Tam, C.M., Leung, G.M.: The transmission dynamics of tuberculosis in a recently developed Chinese city. PloS one **5**(5), e10468 (2010)
21. Egonmwan, A. Okuonghae, D.: Analysis of a mathematical model for tuberculosis with diagnosis
22. Lakshmikantham, V., Leela, S., Martynyuk, A.A.: Stability analysis of nonlinear systems, pp. 249–275. M. Dekker, New York (1989)
23. Okuonghae D. A mathematical analysis of epidemiological models for infectious diseases, p. 120. University of Benin, Nigeria (2016)
24. Obasi, C., Mbah, G.C.E.: On the stability analysis of a mathematical model of Lassa fever disease dynamics. J. Niger. Soc. Math. Biol. **2**, 135–144 (2019)
25. Sharomi, O.Y., Safi, M.A., Gumel, A.B., Gerberry, D.J.: Exogenous re-infection does not always cause backward bifurcation in TB transmission dynamics. Appl. Math. Comput. **298**, 322–335 (2017)
26. Diekmann, O., Heesterbeek, J.A.P., Metz, J.A.: On the definition and the computation of the basic reproduction ratio R_0 in models for infectious diseases in heterogeneous populations. J. Math. Biol. **28**(4), 365–382 (1990)
27. Van den Driessche, P., Watmough, J.: Further Notes on the Basic Reproduction Number in Mathematical Epidemiology, pp. 159–178. Springer, Heidelberg (2008)
28. Sharomi, O., Podder, C.N., Gumel, A.B., Elbasha, E.H., Watmough, J.: Role of incidence function in vaccine-induced backward bifurcation in some HIV models. Math. Biosci. **210**(2), 436–463 (2007)
29. Sharma, S., Coetzee, E.B., Lowenberg, M.H., Neild, S.A., Krauskopf, B.: Numerical continuation and bifurcation analysis in aircraft design: an industrial perspective. Philos. Trans. Royal Soc. Math. Phys. Eng. Sci. **373**(2051), 2014:0406 (2015)
30. Anderson, R.M., Anderson, B., May, R.M.: Infectious Diseases of Humans: Dynamics and Control. Oxford University Press, Oxford (1992)
31. Hethcote, H.W.: The mathematics of infectious diseases. SIAM Rev. **42**(4), 599–653 (2000)
32. Castillo-Chavez, C., Cooke, K., Huang, W., Levin, S.A.: Results on the dynamics for models for the sexual transmission of the human immuno-deficiency virus (1989)
33. Das, D.K., Kar, T.K.: Dynamical analysis of an age-structured tuberculosis mathematical model with LTBI detectivity. J. Math. Anal. Appl. **492**(1), 124407 (2020)
34. Gomes, M.G.M., et al.: How host heterogeneity governs tuberculosis reinfection? Proc. Royal Soc. B Biol. Sci. **279**(1737), 2473–2478 (2012)
35. Wangari, I.M., Davis, S., Stone, L.: Backward bifurcation in epidemic models: problems arising with aggregated bifurcation parameters. Appl. Math. Model. **40**(2), 1669–1675 (2016)
36. Mohd, M.H., Murray, R., Plank, M.J., Godsoe, W.: Effects of biotic interactions and dispersal on the presence-absence of multiple species. Chaos Solitons Fractals **99**, 185–194 (2017)
37. Mohd, M.H., Murray, R., Plank, M.J., Godsoe, W.: Effects of different dispersal patterns on the presence-absence of multiple species. Commun. Nonlinear Sci. Numer. Simul. **56**, 115–130 (2018)
38. Mohd, M.H.: Numerical bifurcation and stability analyses of partial differential equations with applications to competitive system in ecology. In: SEAMS School on Dynamical Systems and Bifurcation Analysis - Springer Proceedings in Mathematics and Statistics, vol. 295, pp. 117–132 (2019)
39. Omaiye, O.J., Mohd, M.H.: Computational dynamical systems using XPPAUT. In: SEAMS School on Dynamical Systems and Bifurcation Analysis - Springer Proceedings in Mathematics and Statistics, vol. 295, pp. 175–203 (2019)
40. Kooi, B.W., Boer, M.P., Kooijman, S.A.L.M.: Resistance of a food chain to invasion by a top predator. Math. Biosci. **157**(1–2), 217–236 (1999)

Managing Regime Shift in Lake Systems by Modelling and Simulation

Chai Jian Tay, Su Yean Teh, Hock Lye Koh, Mohd Hafiz Mohd, and Zhiming Zhang

Abstract Large, abrupt and persistent change in ecosystem structures and dynamics, known as regime shift, is not always observed in advance. This lack of early warning poses problem for prevention and mitigation measures. Regime shift can be driven by natural processes or anthropogenic activities that push the ecosystem across a threshold. Lake or reservoir eutrophication is an example of regime shift that developed gradually over a long period due to persistent high levels of nutrients in the stagnant system. Driven by excessive nutrient loading, eutrophication involves the abrupt change from a clear-water low algal state (oligotrophic) to a turbid algal-dominated condition (eutrophic). Eutrophication degrades water quality, and impairs the supply of safe drinking water, leading to public health risks and economic losses. Hence, identifying the regime shift threshold for an oligotrophic or eutrophic lake is vital for determining the effective intervention measures or restoration measures, respectively. For this purpose, mathematical models linking algal concentration (μg/L chlorophyll *a*) to phosphorus concentration (μg/L) are formulated for two case studies. The first case study involves a temperate, large, deep and oligotrophic lake known as Lake Fuxian in China while the second examines a tropical, small, shallow and highly eutrophic lake known as Tasik Harapan (TH) in Universiti Sains Malaysia, Penang. Model simulations coupled with bifurcation analysis of the mathematical model identify the regime shift threshold and determine the type of lake response to nutrient loading. Model simulations revealed that the reversible and deep Lake Fuxian would

C. J. Tay · S. Y. Teh (✉) · M. H. Mohd
School of Mathematical Sciences, Universiti Sains Malaysia,
11800 USM, Pulau Pinang, Malaysia
e-mail: syteh@usm.my

H. L. Koh
Jeffrey Sachs Center on Sustainable Development, Sunway University, 47500 Bandar Sunway, Selangor, Malaysia

Z. Zhang
Institute of Ecology and Geobotany, School of Ecology and Environmental Sciences, Yunnan University, Kunming, Yunnan, China

M. H. Mohd et al. (eds.), *Modelling, Simulation and Applications of Complex Systems*,
Springer Proceedings in Mathematics & Statistics 359,
https://doi.org/10.1007/978-981-16-2629-6_11

only become eutrophic (algal concentration ≥ 10 μg/L) by year 2380 if the current external phosphorus loading rate continues to increase linearly by a modest rate of 0.00016 μg/L/d per year. However, Lake Fuxian could abruptly shift from oligotrophic to eutrophic in just three years if there is a large increase in external phosphorus loading beyond the critical threshold of 0.0765 μg/L/d. For the highly eutrophic tropical and shallow TH, model analysis suggests that TH shifted to eutrophic state when the external phosphorus loading exceeded 0.01595 μg/L/d. In its current condition, a large reduction in the phosphorus input through dredging and flushing is needed to restore the irreversible state of TH.

Keywords Water · Bifurcation · Eutrophication · Mathematical model · Tropical lake · Temperate lake

1 Introduction

Increasingly reported in a variety of aquatic systems worldwide, regime shift is characterized by large, abrupt and persistent changes in ecosystem structures and dynamics [1, 2]. Having more than one equilibrium state, an ecosystem may undergo a regime shift, transitioning suddenly from one equilibrium state to another when a critical threshold is crossed. Driven by natural processes or anthropogenic activities that push the system across a threshold [3], regime shift is not always observed in advance. This lack of early warning poses problem for prevention and mitigation measures. The identification of regime shift is a daunting task because it requires a comprehensive knowledge on a clear identification of the alternate regimes and on the behaviour of the systems in these alternate regimes [3, 4]. Significant efforts have been devoted to the assessment of the likelihood of regime shift, involving the application of time series analysis. Further, appropriate management of regime shift entails the detection of early warning signals and the examination of regime shift thresholds [4]. The application of time series analysis and detection of early warning signals of regime shift involve statistical analysis that depends on the availability of large quality time series data. In the study of regime shift thresholds, bifurcation analysis is commonly used to identify the alteration of stability between different regimes when the system exceeds the threshold value. Examples of regime shift in other ecosystems include kelp transition [5], soil salinization [6], coral reef degradation [7] and lake eutrophication [8].

In general, the trophic state of a lake is classified into three states: oligotrophic, mesotrophic and eutrophic [9]. The oligotrophic state is the desired state with low algal concentration and low nutrients. In contrast, the undesired eutrophic state has high algal level and rich nutrient constitution. With intermediate level of algal and nutrient concentrations, mesotrophic is the state between oligotrophic and eutrophic. The two regimes involved in lakes are the oligotrophic and eutrophic states of the water. Regime shift occurs in a lake when its clear water state (oligotrophic) turns into a turbid state (eutrophic) once the nutrient level persistently exceeds a threshold. In this study, the external phosphorus loading threshold is defined as the external phosphorus loading above which eutrophication occurs ($A \geq 10$ μg/L) [10]. The algal concentration, A is reflective of the state of the system while the external phosphorus loading defines the critical threshold that triggers regime shift. Hence, to reduce the risk of regime shift to the undesirable eutrophic state, the external phosphorus loading must be reduced. The undesired shift to eutrophic state can be reversed if the external phosphorus loading is managed and reduced to a point below the critical threshold. But regime shift may occur unexpectedly and is difficult to predict in advance. Hence, restoration efforts to reverse the shift may be difficult or even impossible [10, 11]. Nevertheless, regime shift models have been used in many studies of eutrophication remediation [12–14], in eutrophication management [10, 15] and in early warning system of regime shift in lakes [16, 17]. These studies aim to offer insights on the loading thresholds and geo-environmental conditions that trigger the dramatic shift in the lake from one trophic state to another, allowing early intervention measures for oligotrophic lakes and restoration measures for eutrophic lakes.

With the growing human population, rapid urbanization and accompanying anthropogenic activities, eutrophication problem is expected to intensify due to the excessive nutrients discharged into the water bodies [18]. Eutrophication degrades water quality, poses public health risks and causes economic losses. Hence, there is a need to examine the causes of eutrophication and to provide holistic solution by using regime shift models. This paper is organized into two case studies in two distinct climatic conditions, one for the temperate Lake Fuxian and another for the tropical Tasik Harapan. For each case study, a brief overview of the study site is provided followed by the formulation of the mathematical model that links algal concentration (μg/L chlorophyll a) to phosphorus concentration (μg/L). Model simulations and bifurcation analysis of the mathematical model are then performed to identify the regime shift thresholds and to determine the type of lake responses to nutrient loading for each case study. The model analysis for Lake Fuxian aims (i) to determine the nutrient loading threshold and the geo-environmental conditions that could trigger an undesired regime shift from oligotrophic to eutrophic states, and (ii) to formulate appropriate early intervention measures. For the highly eutrophic Tasik Harapan (TH), the model analysis would provide insights on methodology for reversing the undesired state of the highly eutrophic TH. The results of the regime shift analysis for both lakes are then compared and discussed. This paper ends with conclusions that summarize the findings and provide recommendation for future research.

2 Case Study 1: Temperate Lake Fuxian, China

Located at Yunnan, China (Fig. 1), Lake Fuxian has an area of 211 km^2, a mean depth of 89 m, a maximum depth of 155 m and a water storage of 189×108 m^3 [19]. As a large, oligotrophic lake and the second deepest lake in China, Lake Fuxian

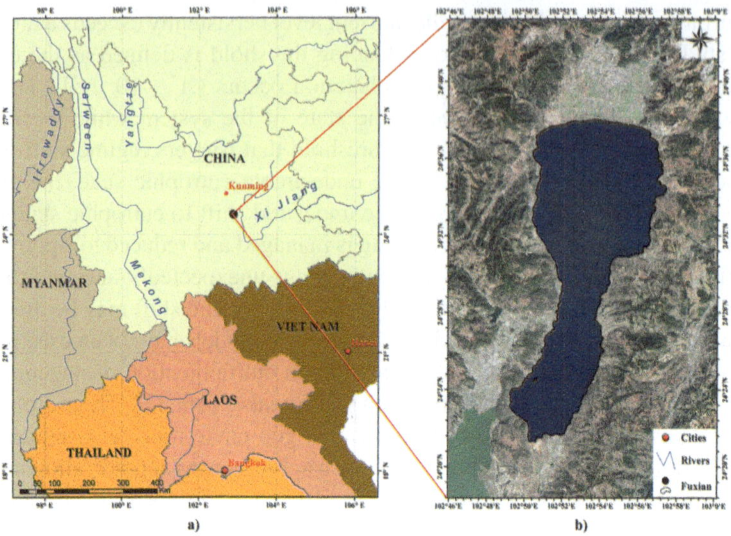

Fig. 1 **a** Map of China with a black dot indicating the location of Lake Fuxian. **b** Map of Lake Fuxian at Yunan, China

contributes 9.16% of freshwater storage in China [20]. It is 31.5 km long and 6.7 km wide on average. More than 20 rivers flow into the lake with the Gehe Watergate as the main inflow and the Haikou River as the only outflow [21]. Limited outflow has the propensity to increase water retention time of Lake Fuxian to 167 years. The lake serves as a major drinking water source and is one of the most important freshwater resources in China. However, it has been reported that the lake's water quality has been deteriorating since 1980s due to pollutants arising from human activities such as agricultural and industry development, urbanization and tourism being discharged into the lake [21].

2.1 Mathematical Model for Lake Fuxian

Phosphorus is commonly considered as the primary driver of eutrophication. Numerous studies have indicated the existence of a strong relationship between algal concentration and nutrient loading [22]. Algal population is a common indicator of eutrophication while phosphorus is the primary driver of eutrophication. Hence, algae and phosphorus are the two most important components of eutrophication model as formulated in Eq. (1) for the temperate Lake Fuxian [8, 23]. The sources of phosphorus input into Lake Fuxian include external phosphorus loading (l_p), excretion associated with zooplankton grazing (egA) and recycling from sediment (r_1). The loss of phosphorus from the lake is due to flushing ($h_1 P$) and uptake by algae for growth ($bA(P - p_a A)/(h_a + P - p_a A)$). Phosphorus is a nutrient limiting factor for algal growth and is represented by the term $(P - p_a A)/(h_a + P - p_a A)$. The loss of algae is due to flushing $(h_1 A)$, grazing by zooplankton (gA) and mortality (sA/z_e).

In this model, the phosphorus recycling term (r_1 in Eq. (1)) suitable for temperate lake is formulated based upon the phosphorus release rate from the sediment into the water column and the duration of anoxia during summer stratification [8]. The term r_1 is governed by the oxygen depletion rate, which is a function of algal concentration (A), hypolimnion thickness (z_h), and hypolimnion temperature (T) [24]. It is assumed that the phosphorus released from the hypolimnion into the epilimnion occurs at the maximal rate, i.e. 14 mg m^{-2} d^{-1} multiplied by the proportion of anoxic days in the hypolimnion during the stratified season, which is reported to be 150 days long [8]. Table 1 displays the definition and unit of the parameters used in this algae-phosphorus model for the temperate Lake Fuxian.

$$
\begin{aligned}
\frac{dA}{dt} &= bA\frac{P-p_a A}{h_a+P-p_a A} - \left(h_1 + g + \frac{s}{z_e}\right)A \\
\frac{dP}{dt} &= l_p + r_1 + egA - bA\frac{P-p_a A}{h_a+P-p_a A} - h_1 P \\
r_1 &= \frac{R}{z_e},\ R = \frac{14}{150} \times \left[150 - \frac{DO\times(50+z_h)}{3.8\times\frac{1.15A^{1.33}}{9+1.15A^{1.33}}\times 2^{\frac{T-4}{10}}+0.12}\right]
\end{aligned}
\tag{1}
$$

2.2 *Curve Fitting and Bifurcation Analysis for Lake Fuxian*

The analysis of regime shift in Lake Fuxian involves several components. First, curve fitting is performed to identify the best set of parameter values that closely reflect the eutrophication dynamic in Lake Fuxian. External phosphorus loading rate, l_p is the most important parameter that governs the dynamics of lake eutrophication [22]. Although some parameter values, such as zooplankton grazing rate and algal sinking rate, can be obtained from literature review, the parameter value for external phosphorus loading rate, l_p should be estimated from curve fitting to reflect the recorded algal concentration in Lake Fuxian. The algal concentration data for the period 1990 to 2014 for Lake Fuxian is obtained from [19], as shown in Fig. 2(a). To the date of submission of this manuscript, only algal data in the period 1990 to 2014 is available in published literature. Hence, we only consider algal data in this period. The algal concentration, A in Lake Fuxian increased significantly in year 2001 from an average concentration of 0.5776 μg/L (1990 to 2000) to 2.1256 μg/L (2001). This sudden increase in algal concentration is attributed to increased sewage discharge into the lake as a result of development of agriculture, industry and urbanization [21]. Overall, the algal concentration data indicates that Lake Fuxian is presently classified as oligotrophic [19]. However, Chen et al. [27] opined that Lake Fuxian is at the risk of becoming increasing polluted, as may be observed from the deteriorating water quality over the course of the past 25 years.

Table 1 Definition and unit of the parameters in the algae-phosphorus model for the temperate Lake Fuxian

Variable/ Parameter	Definition	Unit	Value	Source	Range
A	Algal concentration	μg/L	–	–	–
P	Phosphorus concentration	μg/L	–	–	–
t	Time	d^{-1}	–	–	–
b	Algal growth rate	d^{-1}	0.7	[8]	0–1
h_a	Half saturation constant	μg/L	10	[23]	0–10
p_a	Phosphorus content percentage	–	1	[23]	0–1
h_1	Flushing rate	d^{-1}	1.6×10^{-5}	[21]	0–0.8
g	Zooplankton grazing rate	d^{-1}	0.03	[8]	0–0.91
s	Algal sinking rate	m d^{-1}	0.085	[8]	0–3
l_p	External phosphorus loading rate	μg/L/d	Different scenarios	–	–
DO	Saturated oxygen concentration corresponding to temperature	mg/L	10.3	[25]	–
z_h	Mean of hypolimnion thickness	m	88.55	[26]	–
z_e	Epilimnion thickness	m	27.5	[26]	–
T	Temperature at hypolimnion	°C	14.21	[26]	–
e	Phosphorus excretion associated with grazing	$\mu g/\mu g$	0.65	[8]	0.4–0.8

The algae-phosphorus model in Eq. (1) is used in the curve fitting simulation, with the definition and value of the parameters shown in Table 1. The best fit to the average values of algal concentration (0.5776 μg/L) for the years 1990 to 1999 is obtained with external phosphorus loading rate of $l_p = 0.0038$ μg/L/d. In year 2000, the sudden increase in A is simulated by a significant increase in the loading rate $l_p = 0.0140$ μg/L/d. Subsequently, for the year 2001 and beyond, the increasing trend of algal concentration is simulated by increasing the external phosphorus loading rate linearly by 0.00016 μg/L/d per year from the initial loading of 0.0140 μg/L/d in 2001. Table 2 shows the summary of fitted l_p from year 1990 to 2014 for Lake Fuxian.

With the estimated l_p, bifurcation analysis is then performed by using XPPAUT to determine the type of lake response for Lake Fuxian. Developed by Bard Ermentrout [28], XPPAUT is an open-source numerical tool for simulating, animating, and analysing dynamical systems. Since phosphorus is the main driver of eutrophication, l_p is selected as the bifurcation parameter in the analysis. Figure 2(b) shows the bifurcation diagram of algal concentration, A (μg/L) against external phosphorus loading rate l_p (μg/L/d). Based on Fig. 2(b), $l_p = 0.0765$ μg/L/d is obtained as the external phosphorus loading threshold for Lake Fuxian, where any l_p value equal to or greater

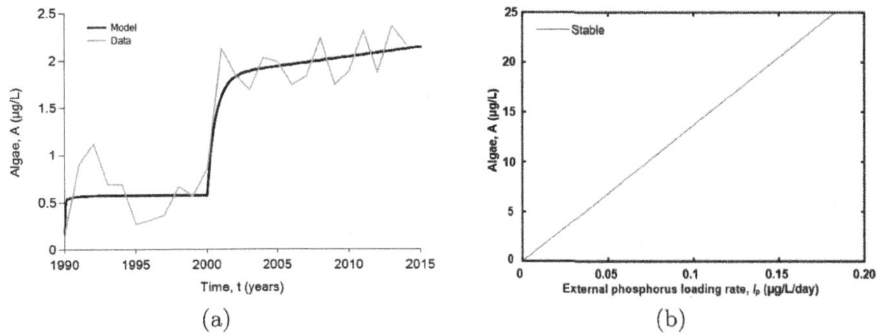

Fig. 2 **a** Curve fitting of algal concentration, A (μg/L) in Lake Fuxian from year 1990 to 2014 [19]. **b** Bifurcation diagram of algal concentration, A (μg/L) against external phosphorus loading rate, l_p (μg/L/d)

Table 2 Summary of fitted external phosphorus loading, l_p (μg/L/d) into Lake Fuxian for years 1990 to 2014

Year	Algal concentration, A (μg/L)	Fitted external phosphorus loading, l_p (μg/L/d)
1990–1999	0.5776 (average value)	0.0038
2000	0.8597	0.0140
2001–2014	1.6894–2.3636	l_p increases constantly by 0.00016 μg/L/d per year

than 0.0765 μg/L/d could result in $A \geq 10$ μg/L (eutrophic state). If l_p lies below this catastrophic tipping point (for example $l_p = 0.07$ μg/L/d or $l_p = 0.065$ μg/L/d), a sudden jump in state would not occur. In Fig. 2(b), no multiple equilibria are obtained and A increases as l_p increases. This is indicative of a reversible behaviour of Lake Fuxian where eutrophication can be reversed by controlling the l_p alone.

2.3 *External Phosphorus Loading Threshold for Lake Fuxian*

The external phosphorus loading threshold is determined here by analysing the estimated external phosphorus loading rate (l_p). Information on eutrophication threshold is vital in early warning analysis with the goal to prevent regime shift in Lake Fuxian. Two scenarios of the numerical simulations are presented. The first scenario assumes that the current loading scenario for years 2001 to 2014 persisted beyond 2014. Hence, the external phosphorus loading rate continues to increase linearly by 0.00016 μg/L/d per year beyond 2014. Simulation result in Fig. 3(a) suggests that if the current modest loading scenario persists, Lake Fuxian would only become

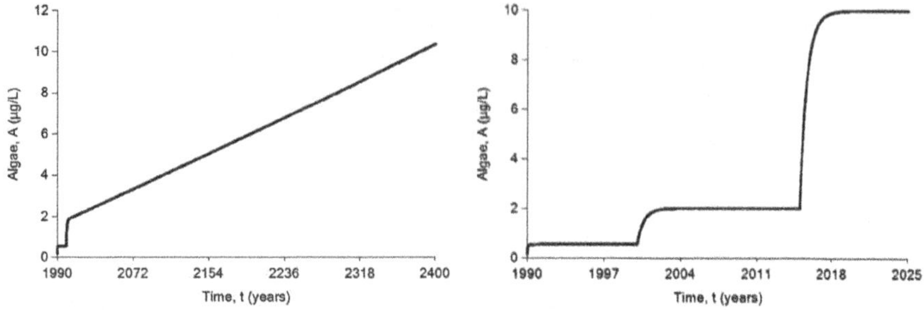

(a) (b)

Fig. 3 **a** Projection of algal concentration, A (μg/L) based on an increment of 0.00016 μg/L/d per year in external phosphorus loading, l_p. **b** Graph of algal concentration, A (μg/L) against time (year) when 2014 s l_p is increased by five times in year 2015

eutrophic by year 2380. Since the lake has large volume ($189 \times 108\,\text{m}^3$) with deep depth (mean 89 m) and long retention time (167 years), Lake Fuxian is resilient to eutrophication and may change more slowly to any direct modest anthropogenic influence [29].

In the second scenario, the level of l_p that would result in a sudden regime shift in Lake Fuxian is examined. Simulation results in Fig. 2(b) indicate that a sudden increase in l_p value to $l_p = 0.0765$ μg/L/d (which is five times the rate in year 2014) in year 2015 could trigger a regime shift in Lake Fuxian in just three years. As shown in Fig. 3(b), the value of A increases to $A = 10$ μg/L in three years after l_p is increased to 0.0765 μg/L/d. This implies that $l_p = 0.0765$ μg/L/d may be identified as the external phosphorus loading threshold that triggers eutrophication in Lake Fuxian. Any value of l_p which is equal to or greater than 0.0765 μg/L/d could result in eutrophication and produce a sudden increase in algal concentration as indicated in Fig. 3(b) for Lake Fuxian. Hence, this threshold value of l_p serves as a guide for the relevant authorities in controlling or limiting the input of nutrients into the lake.

3 Case Study 2: Tropical Tasik Harapan, Malaysia

Located in Universiti Sains Malaysia (USM), Penang, Malaysia, TH (Fig. 4) is a small and shallow tropical lake with a surface area of about $6070\,\text{m}^2$ (1.5 acres), volume of $10{,}000\,\text{m}^3$ and a mean depth of 1.0 to 1.5 m [30]. Constructed for flood mitigation in USM in 1990, the lake water has since turned eutrophic due to excessive algal growth supported by high nutrients, particularly phosphorus. A lake is classified as eutrophic if the algal concentration is equal to or greater than 10 μg/L chl *a* [31]. With a reported high algal concentration occasionally exceeding 300 μg/L chl *a*, TH is highly eutrophic [30].

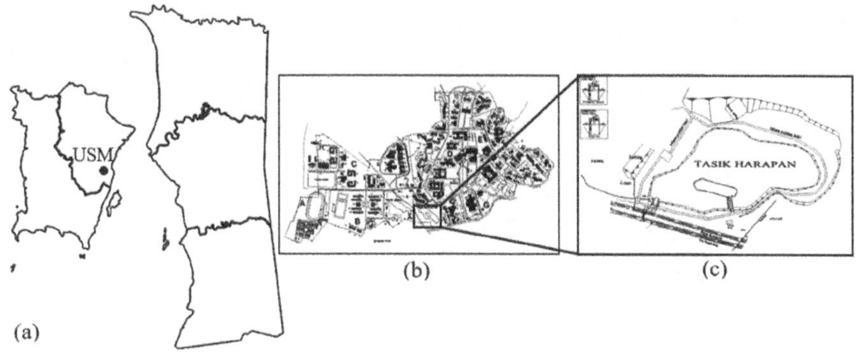

Fig. 4 **a** Location of USM in Penang, Malaysia. **b** Map of USM, Penang, Malaysia. **c** Map of Tasik Harapan (TH) in USM, Penang, Malaysia

3.1 Mathematical Model for Tasik Harapan

In this study, the mathematical model for the tropical TH differs from the model for the temperate Lake Fuxian only with regards to the distinct phosphorus recycling term. The term r_1 in Eq. (1) is formulated for temperate lake involving the concept of proportion of anoxic days during summer stratification. Hence, it is deemed not suitable to represent the phosphorus recycling in a tropical lake with no summer stratification and with little fluctuation in temperature throughout the year. Hence, another phosphorus recycling term appropriate for tropical lake, namely $rP^q/(P^q + n^q)$ is considered. The choice of this formulation is based on three reasons: (i) the term $rP^q/(P^q + n^q)$ is a sigmoid function that could represent the probability of anoxia in lakes [33, 34], (ii) the alternative states of lake can be accommodated by using the term, and (iii) the major recycling mechanism in shallow lake is phosphorus re-suspension which is proportional to the amount of phosphorus available in water [34]. The mathematical model in Eq. (2) is formulated to analyze the regime shift in the tropical TH. Table 3 shows the definition and unit of the parameters used in this algae-phosphorus model for the tropical Tasik Harapan. Important parameter values such as external phosphorus loading rate, algal growth rate and phosphorus recycling rate are estimated from curve fitting in a previous study [35]. Other parameter values are obtained from literature review.

$$\begin{aligned} \frac{dA}{dt} &= bA\frac{P-p_aA}{h_a+P-p_aA} - \left(h_1 + g + \frac{s}{z_e}\right)A \\ \frac{dP}{dt} &= l_p + \frac{rP^q}{P^q+n^q} + egA - bA\frac{P-p_aA}{h_a+P-p_aA} - h_1P \end{aligned} \quad (2)$$

Based on various findings of lake eutrophication studies [34, 37–43], recycling should increase with P following the sigmoid form $rP^q/(P^q + n^q)$ [32]. In particular, the findings of [43] indicates that the probability of anoxia in lakes is a sigmoid function of P and this probability is correlated with the number of days that sedi-

Table 3 Definition and unit of the parameters in the algae-phosphorus model for the tropical TH

Variable/ Parameter	Definition	Unit	Value	Source	Range
A	Algal concentration	μg/L	–	–	–
P	Phosphorus concentration	μg/L	–	–	–
t	Time	d^{-1}	–	–	–
b	Algal growth rate	d^{-1}	0.7	[35]	0–1
h_a	Half saturation constant	μg/L	10	[23]	0–10
p_a	Phosphorus content percentage	–	1	[23]	0–1
h_1	Flushing rate	d^{-1}	0	–	0–0.8
g	Zooplankton grazing rate	d^{-1}	0.03	[8]	0–0.91
s_v	Algal mortality rate	d^{-1}	0.085	[8]	0–0.9
l_p	External phosphorus loading rate	μg/L/d	0.3	[35]	–
r	Phosphorus recycling rate	μg/L/d	0.3	[35]	0–14
n	Half saturation value of recycling function	μg/L	10	[36]	0–10
q	Parameter for steepness of sigmoid function near n	–	20	[32]	0–20
e	Phosphorus excretion associated with grazing	$\mu g/\mu g$	0.65	[8]	0.4–0.8

ment is overlain by anoxic water [33]. Generally, phosphorus recycling depends on oxygen depletion at the sediment-water interface (hypolimnion). When P is high, the recycling rate is near its maximum as the hypolimnion is anoxic almost all the time. At high phosphorus concentration, active algal growth leads to the formation of organic matter in lake sediment. The decomposition of this organic matter depletes oxygen in the hypolimnion, inducing release of phosphorus from sediment to the water column. Conversely, when P is low, the recycling rate is at lower level because the hypolimnion is oxygenated most of the time. At intermediate levels of P, the recycling rate increases rapidly with P.

With the sigmoid function of phosphorus recycling $rP^q/(P^q + n^q)$, the alternative states of lake can be illustrated. To show this, the bifurcation diagram of algal, A against external phosphorus loading rate, l_p with and without sigmoid function in phosphorus recycling term is plotted in Fig. 5, for comparison. As an example, values of $l_p = 0.1$ μg/L/d, $r = 0.1$ μg/L/d and $h = 0.001\ d^{-1}$, together with other parameter values from Table 3 are considered in the comparison. Based on Fig. 5(a), the inclusion of sigmoid function in phosphorus recycling term would induce saddle-node bifurcation, where alternative states of lake exist, i.e., oligotrophic ($A < 2.6$ μg/L) or mesotrophic (2.6 μg/L $\leq A < 10$ μg/L) and eutrophic ($A \geq 10$ μg/L) states. The details of saddle-node bifurcation in lake eutrophication is discussed further in Sect. 3.2. With these alternative states of lake, it is possible to identify l_p threshold

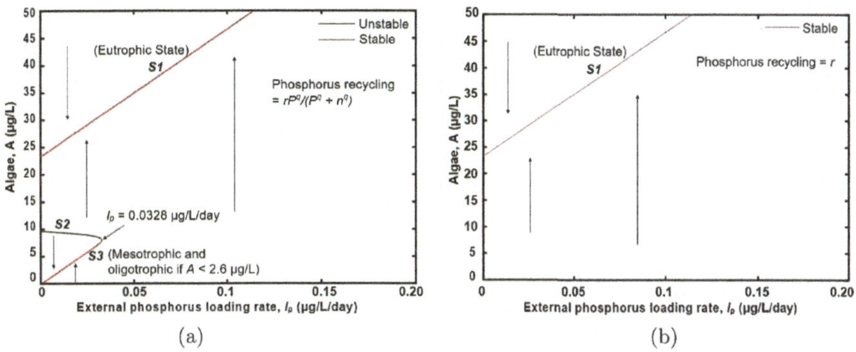

Fig. 5 Bifurcation diagrams of algal concentration, A (μg/L) against external phosphorus loading rate, l_p (μg/L/d) with two different phosphorus recycling term: **a** $rP^q/(P^q + n^q)$ and **b** r

where regime shift in lake system occurs. In contrast, only stable equilibrium exists in Fig. 5(b) when the phosphorus recycling is a *constant* term *without* the sigmoid function. Hence, the sigmoid function in phosphorus recycling $rP^q/(P^q + n^q)$ is considered to illustrate regime shift in lake system.

3.2 *Regime Shift for Tasik Harapan*

Using the parameter values listed in Table 3 and the algae-phosphorus model in Eq. (2), the regime shift threshold for lake eutrophication in TH is determined through bifurcation analysis. This analysis (i) provides information on the threshold value at which the lake became eutrophic, (ii) identifies the lake bifurcation behaviour and (iii) determines the effective restoration measures. Bifurcation analysis is performed by means of XPPAUT using external phosphorus loading rate, l_p as the main bifurcation parameter. Figure 6(a) shows the bifurcation diagram of algal concentration A against external phosphorus loading l_p. From Fig. 6(a), TH shows an irreversible behaviour because the lake remains eutrophic (A = 131 μg/L) even though l_p is reduced to zero (no P inflow). Therefore, other restoration method such as flushing, dredging or bio-manipulation is needed in conjunction with reduction in external loading to further reduce the algal concentration in TH. Bio-manipulation refers to the reduction in algal concentration through an increase in zooplankton grazing [44].

Based on Fig. 6(a), a saddle-node bifurcation is observed, where the unstable steady state S2 collides with the stable steady state S3, resulting in a turning point at l_p = 0.01595 μg/L/d. Three equilibria (two stables and one unstable) exist when 0 μg/L/d < l_p < 0.01595 μg/L/d. When l_p increases, A increases along the stable steady state $S3$ until a certain value of A is reached at the right-side inflection point of equilibrium (l_p = 0.01595 μg/L/d, A = 7.353 μg/L). Further increase in l_p causes the equilibrium to "jump" to another stable steady state S1 and remain in S1, where

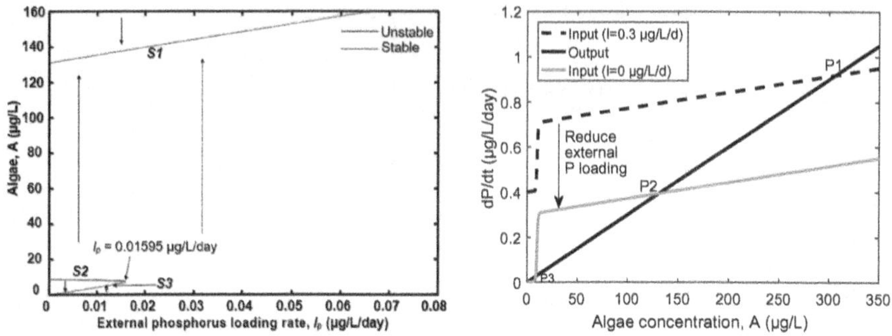

(a) (b)

Fig. 6 **a** Bifurcation diagrams of algal concentration, A (μg/L) against external phosphorus loading rate, l_p (μg/L/d). **b** Restoration of irreversible TH by a large reduction in the phosphorus input

eutrophication occurs. The choice of initial values determines which steady state A will approach, either the oligotrophic S3 or eutrophic S1. For example, if the initial value of A lies between S1 and S2, it will be attracted to the stable steady state S1. The trajectories of A will never approach the unstable steady state S2. An algal concentration A that is below S2 with $l_p < 0.01595$ μg/L/d will approach the stable steady state S3 ($A < 7.353$ μg/L), indicating mesotrophic state of the lake and oligotrophic if $A < 2.6$ μg/L. In Fig. 6(a), it is observed that the basin of attraction for oligotrophic state S3 in TH is small. This implies that TH has a small chance to become oligotrophic. Based on this analysis, the regime shift threshold for the eutrophic TH is determined as $l_p = 0.01595$ μg/L/d.

Since TH shows irreversible behaviour, the lake eutrophication cannot be reversed even by a large reduction in l_p, as demonstrated in Fig. 6(b). The dashed curve represents the phosphorus input before reduction in l_p, the black solid line represents the phosphorus output and the grey solid curve denotes the phosphorus input after reduction in l_p. From Fig. 6(b), P1 is the current state of TH, in which $A = 300$ μg/L and $l_p = 0.3$ μg/L/d. When l_p is reduced to $l_p = 0$ μg/L/d, the steady state of A moves to P2 ($A = 131$ μg/L) which is still eutrophic. The intersection point P3 ($A = 8.702$ μg/L) is an unstable steady state and the system will never approach it. The minimum l_p is not low enough to shift the eutrophic lake to oligotrophic state. This implies that other lake restoration methods such as flushing and sediment dredging coupled with reduction in l_p are essential to significantly improve the eutrophic state of TH.

3.3 Bifurcation Analysis for Tasik Harapan

In order to determine the flushing rate required to restore TH to oligotrophic state ($A < 2.6$ μg/L), a bifurcation diagram of algal concentration, A against flushing rate h_1 is plotted as displayed in Fig. 7(a). As expected, the algal concentration decreases when flushing rate increases. Figure 7(a), three equilibria (two stables and one unstable) exist when $0.018\,d^{-1} < h_1 < 0.02575\ d^{-1}$. Within this range of h_1, the value of A will be attracted to either the stable steady state S4 ($A > 10\ \mu g/L$, eutrophic) or stable steady state S6 (mesotrophic if $2.6\ \mu g/L \leq A < 10\ \mu g/L$ and oligotrophic if $A < 2.6\ \mu g/L$). If A is above S5 (with $0.018\ d^{-1} < h_1 < 0.02575\ d^{-1}$), it approaches the stable steady state S4. An algal concentration lower than S5 (with $0.018\ d^{-1} < h_1 < 0.02575\ d^{-1}$) will decrease and approach the stable steady state S6. The value of A will never approach the unstable steady state S5. From Fig. 7(a), a flushing rate of $h_1 > 0.02575\ d^{-1}$ would restore TH to mesotrophic state ($A < 9.735\ \mu g/L$), while the oligotrophic state of TH can be achieved when $h_1 > 0.042\ d^{-1}$ ($A < 2.6\ \mu g/L$).

Figure 7(b) shows the bifurcation diagram of algal concentration, A ($\mu g/L$) against algal mortality rate s_υ (d^{-1}). Based on Fig. 7(b), two limit points (L1 and L2) and two Hopf points (H1 and H2) exist. At L1 and L2, the stable equilibria collide with the unstable equilibrium and disappear. This is known as the saddle-node bifurcation. Three equilibria (two stables and one unstable) exist in the range $0.04543\ d^{-1} < s_\upsilon < 0.06732\ d^{-1}$. One stable steady state exists at the range $0.06732\ d^{-1} < s_\upsilon < 0.1494\ d^{-1}$. At $s_\upsilon = 0.1494\ d^{-1}$ (H1), Hopf bifurcation occurs. Here, the solution behaviour changes from stable steady state to periodic solutions. An enlarged bifurcation diagram is plotted in Fig. 8(a) to provide a clearer picture of the Hopf bifurcation. In Fig. 8(a), the green and filled circles represent stable limit cycles while the blue and empty circles correspond to unstable limit cycles. At $s_\upsilon = 0.1494\ d^{-1}$ (H1), unstable limit cycle occurs and there is a change in stability at limit point cycle, LPC1 ($s_\upsilon = 0.1481\ d^{-1}$), i.e., from unstable limit cycle to stable limit cycle. The stable limit cycle exists within the range $0.1481\ d^{-1} < s_\upsilon < 0.1560\ d^{-1}$ and loses its stability at LPC2 ($s_\upsilon = 0.1560\ d^{-1}$). Unstable limit cycle is formed and it connects LPC2 with H2 ($s_\upsilon = 0.1548\ d^{-1}$). Based on Fig. 8(a), two ranges of bi-stability are observed, consisting of a stable equilibrium and a stable limit cycle are observed, i.e., LPC1 $< s_\upsilon <$ H1 and H2 $< s_\upsilon <$ LPC2. Subcritical Hopf bifurcation occurs at both H1 and H2, where the unstable limit cycle surrounds a stable equilibrium. As an example, a time series plot of algal and phosphorus concentrations at $s_\upsilon = 0.152\ d^{-1}$ is plotted and displayed in Fig. 8(b). Based on Fig. 8(b), the oscillatory solution behaviour indicates the existence of a stable limit cycle.

4 Discussion

Table 4 displays the differences in lake responses for Lake Fuxian and Tasik Harapan, regarding regime shift thresholds and restoration methods. Lake Fuxian is identified as a reversible lake with regime shift threshold of $l_p = 0.0765\ \mu g/L/d$, while TH is an irreversible lake with a lower value regime shift threshold of $l_p = 0.01595\ \mu g/L/d$. A lower value of regime shift threshold implies that eutrophication ($A > 10\ \mu g/L$) would occur more easily at a lower value of l_p. For a reversible lake, eutrophication can be reversed by controlling l_p alone. However, for the highly eutrophic and irre-

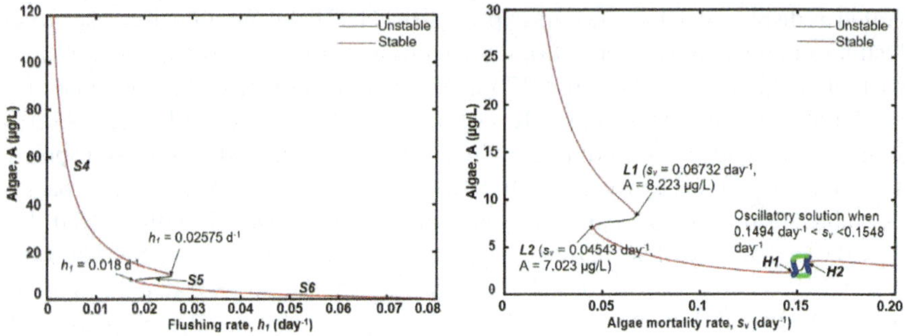

(a) (b)

Fig. 7 Bifurcation diagrams of algal concentration, A (μg/L) against **a** flushing rate, h_1 (d^{-1}) and **b** algal mortality rate, s_v (d^{-1})

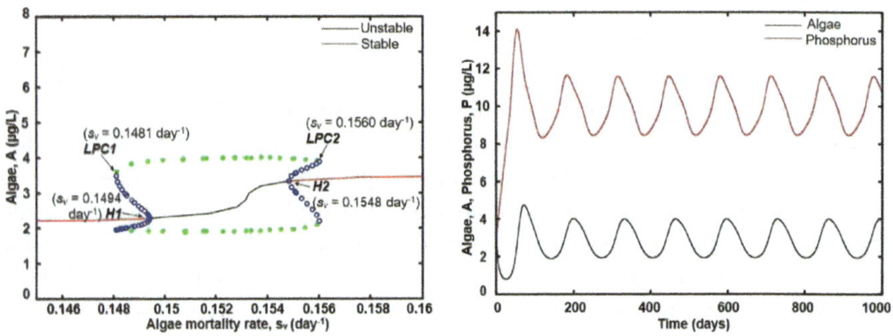

(a) (b)

Fig. 8 **a** Hopf bifurcation occurs as algal mortality rate, s_v (d^{-1}) is varied. **b** Time series plot of algal and phosphorus concentrations at $s_v = 0.152\ d^{-1}$

versible TH, model analysis suggests that a combination of reduction in l_p with other restoration methods such as flushing and sediment dredging is needed to restore the highly eutrophic and irreversible TH. Here, the differences in regime shifts for Lake Fuxian and TH are attributable to two critical factors, i.e., temperature and depth of lake as highlighted by [8].

Lake Fuxian is resilient to eutrophication because of its temperate climate and deep depth. On the other hand, TH is prone to eutrophication due to its tropical higher temperature and shallow depth. Located near the equator, Malaysia has a tropical rainforest climate which is hot and humid throughout the year [45]. Together with high solar irradiance condition, an average temperature of 26–28 °C in Malaysia favours algal growth [46], which is optimal in the temperature range 20–30 °C [46]. A higher lake temperature, as in TH, increases the bacterial activity, stimulates the mineralization of organic matter and reduces oxygen solubility. As a result, more phosphorus is released from sediment into the water column and is readily absorbed

Table 4 Comparison of type of lake response, regime shift threshold and restoration method between Lake Fuxian and Tasik Harapan

Lake	Lake Fuxian (Temperate Lake)	Tasik Harapan (Tropical Lake)
Type of lake response	Reversible	Irreversible
Regime shift threshold	$l_p = 0.0765\ \mu$g/L/d	$l_p = 0.01595\ \mu$g/L/d
Restoration method(s)	Lake eutrophication can be reversed by controlling l_p alone	A combination of reduction in l_p with other restoration methods such as dredging and flushing

for algal growth. It is reported that a higher temperature would cause eutrophication to be more likely to occur and the restoration to be more difficult [8].

On the other hand, the epilimnion and hypolimnion temperatures of Lake Fuxian during summer are in the range of 22.81–25.51 °C and 13.70–14.72 °C, respectively [26]. This lower temperature inhibits algal growth. Further, Lake Fuxian is a deep lake with mean depth of 89 m and maximum depth of 155 m. The phosphorus released from the sediment may not readily reach the epilimnion due to the thick hypolimnion that dilutes the recycled phosphorus [8, 47]. In contrast, the phosphorus released from sediment is more available to the photic zone of the shallow TH (lake mean depth of 1.0 to 1.5 m). Further, wind-induced resuspension of phosphorus is more likely to occur in the shallow TH [48]. Hence, the removal of the lake bottom sediment via dredging, which would stop the release of phosphorus from sediment into the water column, is deemed as the most effective restoration method, especially for the shallow TH [36, 48].

5 Conclusion

The basic mathematical model consisting of algae and phosphorus for examining regime shift in lakes is formulated for the temperate Lake Fuxian and the tropical TH, with distinctly different phosphorus recycling terms. For the temperate Lake Fuxian, the phosphorus recycling term used is the product of maximal phosphorus release rate and proportion of anoxic days. In the study of regime shift in lakes, most of the mathematical models were developed for temperate lakes. In order to overcome this limitation and to examine regime shift in a tropical lake, a sigmoid function of the phosphorus concentration is selected to represent the phosphorus recycling in the tropical TH. This choice is justified based on three reasons: (i) the sigmoid function could represent the probability of anoxia in lakes, (ii) the alternative states of lake can be represented by using the sigmoid function, and (iii) the major recycling mechanism in shallow lake is well represented as phosphorus resuspension is proportional to the amount of phosphorus available in water. Model simulations coupled with bifurcation analysis indicate that the current state of Lake Fuxian is reversible, with estimated regime shift threshold of $l_p = 0.0765$ μg/L/d. Any value of l_p greater than this value could result in lake eutrophication. Two distinct scenarios with distinct phosphorus loading rates l_p are used to estimate the time when Lake Fuxian could become eutrophic. First, a modest l_p is used, i.e. the l_p is increased constantly by 0.00016 μg/L/d per year. With this slow rate of l_p increase, Lake Fuxian is expected to slowly become eutrophic in year 2380 (Fig. 3(a)). Second, when a sharp increase in l_p beyond the critical threshold ($l_p > 0.0765$ μg/L/d) is applied, Lake Fuxian quickly becomes eutrophic in just three years. In contrast, the regime shift threshold for irreversible TH is $l_p = 0.01595$ μg/L/d. This difference in regime shift for Lake Fuxian and TH is attributed to large contrast in the temperature and depth of the lakes. Lake Fuxian is located in an area with temperate climate and has deep depth while TH is located in an area with tropical climate and has shallow depth. Higher temperature favours algal growth. Moreover, the phosphorus released from sediment will be more well-mixed in shallow waters than in deep waters, granting algae an easy access to sediment phosphorus in shallow lake. Hence, Lake Fuxian (regime shift threshold of $l_p = 0.0765$ μg/L/d) is more resilient to eutrophication than TH (regime shift threshold of $l_p = 0.01595$ μg/L/d). Model analysis suggests that for TH, merely reducing the external phosphorus loading to the minimum l_p may not shift the eutrophic lake to oligotrophic state. In its current eutrophic condition, a large reduction in the phosphorus input through dredging and flushing is needed to restore the irreversible state of TH. For future work, the bifurcation dynamics of the model in Eq. (2) will be further investigated.

Acknowledgement Financial support provided by Kementerian Pengajian Tinggi FRGS grant 203.PMATHS. 6711806 is gratefully acknowledged.

References

1. Bannister, W., McGowan, S., Santos-Borja, A.C., Quak, J., Fong, L.S., Mendoza, M., Papa, R.D.S., Taylor, D.: Potential anthropogenic regime shifts in three freshwater lakes in Tropical East Asia. Freshwater Biol. 1–15 (2018)
2. Genkai-Kato, M.: Regime shifts: catastrophic responses of ecosystems to human impacts. Ecol. Res. **22**, 214–219 (2006)
3. Gal, G., Anderson, W.: A novel approach to detecting a regime shift in a lake ecosystem. Methods Ecol. Evol. **1**, 45–52 (2010)
4. Filatova, T., Polhill, J.G., Ewijk, S.V.: Regime shifts in coupled socio-environmental systems: review of modelling challenges and approaches. Environ. Softw. Model. **75**, 333–347 (2016)
5. Filbee-Dexter, K., Scheibling, R.E.: Sea urchin barrens as alternative stable states of collapsed kelp ecosystems. Mar. Ecol. Prog. Ser. **495**, 1–25 (2014)
6. Lal, R.: Restoring soil quality to mitigate soil degradation. Sustain **7**, 5875–5895 (2015)
7. Schmitt, R.J., Holbrook, S.J., Davis, S.L., Brooks, A.J., Adam, T.C.: Experimental support for alternative attractors on coral reefs. P. Natl. Acad. Sci. USA **116**(10), 4372–4781 (2019)
8. Genkai-Kato, M., Carpenter, S.R.: Eutrophication due to phosphorus recycling in relation to lake morphometry, temperature, and macrophytes. Ecology **86**(1), 210–219 (2005)
9. Organisation for Economic Cooperation and Development. Eutrophication of waters: monitoring, assessment and control. OECD (1982)

10. Genkai-Kato, M.: Regime shifts in response to human impacts and management in ecosystems. Kuroshio Sci. **5**(1), 33–37 (2011)
11. Crépin, A., Biggs, R., Polasky, S., Troell, M., Zeeuw, A.D.: Regime shifts and management. Ecol. Econ. **84**, 15–22 (2012)
12. Schallenberg, M., Sorrell, B.: Regime shifts between clear and turbid water in New Zealand lakes: environmental correlates and implications for management and restoration. New Zeal. J. Mar. Fresh. **43**(3), 701–712 (2009)
13. Wang, Y., Wang, W., Wang, Z., Li, G., Liu, Y.: Regime shift in Lake Dianchi (China) during the last 50 years. J. Oceanology Limnol. **36**(4), 1075–1090 (2018)
14. Zhang, K., Dong, X., Yang, X., Kattel, G., Zhao, Y., Wang, R.: Ecological shift and resilience in China's lake systems during the last two centuries. Global Planet. Change **165**, 147–159 (2018)
15. Janssen, A.B.G., van Wijk, D., van Gerven, L.P.A., Bakker, E.S., Brederveld, R.J., DeAngelis, D.L., Janse, J.H., Mooij, W.M.: Success of lake restoration depends on spatial aspects of nutrient loading and hydrology. Sci. Total Environ. **679**, 248–259 (2019)
16. Carpenter, S.R., Cole, J.J., Pace, M.L., Batt, R., Brock, W.A., Cline, T., Coloso, J., Hodgson, J.R., Kitchell, J.F., Seekell, D.A., Smith, L., Weidel, B.: Early warnings of regime shifts: a whole-ecosystem experiment. Science **332**, 1079–1082 (2011)
17. Wang, R., Dearing, J.A., Langdon, P.G., Zhang, E., Yang, X., Dakos, V., Scheffer, M.: Flickering gives early warning signals of a critical transition to a eutrophic lake state. Nature **492**, 419–422 (2012)
18. Stachelek, J., Ford, C., Kincaid, D., King, K., Miller, H., Nagelkirk, R.: The national eutrophication survey: lake characteristics and historical nutrient concentrations. Earth Syst. Sci. Data **10**, 81–86 (2018)
19. Zhou, Q., Zhang, Y., Li, K., Huang, L., Yang, F., Zhou, Y., Chang, J.: Seasonal and spatial distributions of euphotic zone and long-term variations in water transparency in a clear oligotrophic Lake Fuxian. China. J. Environ. Sci. **72**, 185–197 (2018)
20. Wu, X., Deng, K., Ge, P., Wang, Y., Xue, Y.: Investigation on the ecological environment and resource protection management system of Fuxian lake. Adv. Eng. Res. **141**, 290–295 (2017)
21. Li, Y., Gong, Z., Xia, W., Shen, J.: Effects of eutrophication and fish yield on the diatom community in Lake Fuxian, a deep oligotrophic lake in southwest China. Diatom Res. **26**(1), 51–56 (2011)
22. Li, Y.P., Tang, C.Y., Yu, Z.B., Acharya, K.: Correlations between algae and water quality: factors driving eutrophication in Lake Taihu. China. Int. J. Environ. Sci. Technol. **11**, 169–182 (2014)
23. Jones, M.: Using a coupled bio-economic model to find the optimal phosphorus load in Lake Tainter WI. Honors Research Projects 632 (2018)
24. Charlton, M.N.: Hypolimnion oxygen consumption in lakes: discussion of productivity and morphometry effects. Can. J. Fish. Aquat. Sci. **37**, 1531–1539 (1980)
25. Rounds, S.A., Wilde, F.D., Ritz, G.F.: Dissolved oxygen (ver. 3.0): U.S. Geological Survey techniques of water-resources investigations, book 9, chap. A6, sec. 6.2 (2013)
26. Wang, L.J., Yu, H., Niu, Y., Niu, Y., Zhang, Y.L., Liu, Q., Ji, Z.Y.: Distribution characteristics of water temperature and water quality of Fuxian Lake during thermal stratification period in summer. Huang Jing Ke Xue **38**(4), 1384–1392 (2017)
27. Chen, J., Lyu, Y., Zhao, Z., Liu, H., Zhao, H., Li, Z.: Using the multidimensional synthesis methods with non-parameter test, multiple time scales analysis to assess water quality trend and its characteristics over the past 25 years in the Fuxian Lake. China. Sci. Total Environ. **655**, 242–254 (2019)
28. Ermentrout, B.: Simulating, Analyzing, and Animating Dynamical Systems: A Guide to XPPAUT for Researchers and Students. SIAM, Philadelphia (2002)
29. Zhang, Y., Su, Y., Liu, Z., Chen, X., Yu, J., Di, X., Jin, M.: Sediment lipid biomarkers record increased eutrophication in Lake Fuxian (China) during the past 150 years. J. Great Lakes Res. **41**, 30–40 (2015)

30. Teh, S.Y., Koh, H.L., Ismail, A.I.M., Mansor, M.: Determining photosynthesis rate constants in Lake Harapan Penang. Int. Conf. Biomed. **1**, 585–590 (2008)
31. Chapra, S.C.: Surface Water-quality Modeling. McGraw-Hill (1997)
32. Carpenter, S.R., Ludwig, D., Brock, W.A.: Management of eutrophication for lakes subject to potentially irreversible change. Ecol. Appl. **9**(3), 751–771 (1999)
33. Nürnberg, G.K.: Quantifying anoxia in lakes. Limnol. Oceanogr. **40**(6), 1100–1111 (1995)
34. Nürnberg, G.K.: The prediction of internal phosphorus load in lakes with anoxic hypolimnia. Limnol. Oceanogr. **29**(1), 111–124 (1984)
35. Tay, C.J., Teh, S.Y., Koh, H.L.: Eutrophication bifurcation analysis for Tasik Harapan restoration. Int. J. Environ. Sci. Dev. **11**(8), 407–413 (2020)
36. Carpenter, S.R.: Eutrophication of aquatic ecosystems: bistability and soil phosphorus. P. Natl. Acad. Sci. USA **102**, 10002–10005 (2005)
37. Baines, S.B., Pace, M.L.: Sinking fluxes across lakes spanning a trophic gradient: patterns and implications for the fate of planktonic primary production. Can. J. Fish. Aquat. Sci. **51**, 25–36 (1994)
38. Caraco, N.F., Cole, J.J., Likens, G.E.: A cross-system study of phosphorus release from lake sediments. In: Cole, J.J., Lovett, G., Findlay, S. (eds.) Comparative Analyses of Ecosystems, pp. 241–258. Springer-Verlag, New York, USA (1991)
39. del Georgio, P.A., Peters, R.H.: Patterns in planktonic P: R ratios in lakes: influence of lake trophy and dissolved organic carbon. Limnol. Oceanogr. **39**, 772–787 (1994)
40. National Research Council: Restoration of Aquatic Ecosystems: Science, Technology and Public Policy. National Academy Press, Washington, D.C., USA (1992)
41. Sas, H.: Lake Restoration by Reduction of Nutrient Loading: Expectations, Experiences, Extrapolations. Academia Verlag, Richarz, St. Augustin, Germany (1989)
42. Schindler, D.W., Fee, E.J., Rusczynski, T.: Phosphorus input and its consequences for phytoplankton standing crop and production in the experimental lakes area and in similar lakes. J. Fish. Res. Board Can. **35**, 190–196 (1978)
43. Reckhow, K.H.: Empirical lake models for phosphorus: development, applications, limitations, and uncertainty. In: Scavia, D., Robertson, A. (eds.) Perspectives on Lake Ecosystem Modelling, pp. 193–221. Ann Arbor Science, Ann Arbor, Michigan, USA (1979)
44. Jeppesen, E., Søndergaard, M., Lauridsen, T.L., Davidson, T.A., Liu, Z., Mazzeo, N., Trochine, C., Özkan, K., Jensen, H.S., Trolle, D., Starling, F., Lazzaro, X., Johansson, L.S., Bjerring, R., Liboriussen, L., Larsen, S.E., Landkildehus, F., Egemose, S., Meerhoff, M.: Biomanipulation as a restoration tool to combat eutrophication: recent advances and future challenges. Adv. Ecol. Res. **47**, 411–488 (2012)
45. Tang, K.H.D.: Climate change in Malaysia: trends, contributors, impacts, mitigation and adaptations. Sci. Total Environ. **650**, 1858–1871 (2019)
46. Singh, S.P., Singh, P.: Effect of temperature and light on the growth of algae species: a review. Renew. Sust. Energ. Rev. **50**, 431–444 (2015)
47. Welch, E.B., Cooke, G.D.: Internal phosphorus loading in shallow lakes: importance and control. Lake Reserv. Manage. **21**(2), 209–217 (2005)
48. Hickey, C.W., Gibbs, M.M.: Lake sediment phosphorus release management-decision support and risk assessment framework. New. Zeal. J. Mar. Fresh. **43**(3), 819–856 (2009)

Stability Analysis of Magnetohydrodynamic Mixed Convection Flow and Heat Transfer over a Moving Flat Plate in Ferrofluids with Suction and Slip Effects

Norshafira Ramli and Syakila Ahmad

Abstract This paper investigated the magnetohydrodynamic (MHD) mixed convection flow and heat transfer over a moving flat plate in ferrofluids with suction, slip effects and constant heat flux. Using appropriate similarity variables, the governing partial differential equations are reduced to ordinary differential equations before being solved numerically by a shooting method with Maple implementation. The variations of the numerical solutions for the skin friction coefficient and local Nusselt number, as well as the velocity and temperature profiles are obtained for pertinent parameters by considering two types of base fluids (water and kerosene) with three selected ferroparticles (magnetite, cobalt ferrite and manganese-zinc ferrite). It is found that triple solutions exist for both assisting and opposing flows. Also, the kerosene-based magnetite provides the highest heat transfer rate compared to the other two ferroparticles. The stability analysis is then carried out to test the stability of the non-uniqueness solutions by applying bvp4c solver in Matlab. Results from the stability analysis revealed that the first solution is stable and physically realizable while the second and third solutions are unstable and not physically realizable.

Keywords Stability analysis · MHD · Mixed convection · Heat transfer · Moving flat plate · Ferrofluids · Suction · Slip effects

1 Introduction

The research about flow analysis of nanofluids with the interaction of magnetic field has increased enormously. Magnetic nanofluids which are also known as ferrofluids are colloidal suspensions of magnetic nanoparticles with a size range of 5–15 nm in

N. Ramli (✉) · S. Ahmad
School of Mathematical Sciences, Universiti Sains Malaysia, USM, 11800 Penang, Malaysia
e-mail: norshafiraramli@usm.my

M. H. Mohd et al. (eds.), *Modelling, Simulation and Applications of Complex Systems*, Springer Proceedings in Mathematics & Statistics 359,
https://doi.org/10.1007/978-981-16-2629-6_12

diameter scattered in non-conducting base fluid (see [1]). The magnetic nanoparticles which are commonly used include magnetite, cobalt, and ferrite while the base fluids such as water, kerosene, heptane, and hydro-carbons (see [2]). Ferrofluids have significant potential to absorb electromagnetic field in order to enhance the heat transfer since they are attainable to regulate the magnetic force applied to the ferrofluid via a suitable idea of the external magnetic field. The applications of ferrofluids are found in the field of medical, industrial engineering, aeronautical, science and technology (see [3]). Recent studies report that the presence of an external magnetic field produced in a meaningful increment in the thermal conductivity of ferrofluids. Gan Jia Gui et al. [4] investigated the effect of various commercial water-based ferrofluid types on the heat transfer properties under the application of an external magnetic field. They concluded that ferrofluids have higher thermal conductivity than their base fluids and increasing magnetic flux decreases heat transfer enhancement. Doganay et al. [5] developed a review on the effect of external magnetic field on thermal conductivity and viscosity of ferrofluids. The results showed that an addition of the ferroparticles into the base fluid increases the thermal conductivity and viscosity.

Mixed convection flows or combined free and forced convection flows, arise in many transport processes both naturally and in engineering applications. They play an important role, for example, at atmospheric boundary layer flows, solar collectors, heat exchangers and nuclear reactors and in electronic equipment (see [6]). A study towards a better understanding of MHD in the mixed convection flow in ferrofluids promising the heat transfer enhancement in the boundary layer flow. Gul et al. [7] examined the problem of heat transfer in MHD mixed convection flow of ferrofluid along a vertical channel using magnetite (Fe_3O_4) and non-magnetic (Al_2O_3) nanoparticles in water-based fluid. The solutions are obtained using the perturbation method and the spherical shape of nanoparticles were preferred in this study. The results indicate that the thermal conductivity and viscosity of the spherical ferrofluids were raised with the increment of the volume fraction of nanoparticles. However, increasing in temperature led to the decrease in the viscosity of ferrofluids. Apart from that, Rashad [2] also studied the problem of MHD mixed convection flow but he used cobalt ferromagnetic particles in the kerosene-based fluid. This problem was solved by an implicit numerical scheme of finite-difference type. Later on, the effect of a magnetic source on the mixed convection of pulsating ferrofluid flow over a backward-facing step was examined by Selimefendigil et al. [8]. Then, Ramli et al. [3] investigated the stability analysis of MHD mixed convection flow over a moving flat plate in ferrofluids with thermal radiation in heat flux case. More recently, Jamaludin et al. [9] also determined the influence of thermal radiation on MHD mixed convection stagnation-point flow towards a nonlinearly moving sheet that was positioned vertically.

The motion of heat transfer from a surface can be found into two ways, either a stationary or moving fluid and it occurs in many material processing applications such as drawing, extrusion and hot rolling (see [10]). Sakiadis (see [11], [12] and [13]) was the first person who published the papers dealing with boundary layer flow on a continuous moving surface. In his pioneering papers, he pointed out the study of the approximate and exact solutions for the momentum equation in laminar and turbulent flows on a surface moving through a stagnant fluid. In addition, Pop et al. [14] presented the influence of variable viscosity for the flow and heat transfer due to a continuous moving flat plate. They solved the problem by using the Runge-Kutta method and the results have demonstrated that the assumption of constant properties may introduce severe errors in the prediction of surface friction factor and heat transfer rate. Then, Pantokratoras [15] has extended the work by considering the boundary layer flow along an isothermal, continuous moving flat plate on the variable viscosity with temperature. Then, the problems within the moving surface have drawn considerable attention and a good amount of literature has been found in different situations such as [16–22]. On the other hand, the impact of suction and slip effects on MHD forced convection of ferrofluids over a moving flat plate in a convective condition were examined by Ramli et al. [23].

In contrast to the amount of material published on the flow and heat transfer associated with a moving flat plate, there has been lack analysis of the heat transfer for ferrofluids. Therefore, the present work attempts to extend the work by Ramli et al. in [23] considering the MHD mixed convection in ferrofluids with the influence of suction and slip effects. Here, we employed the mathematical nanofluid model suggested by Tiwari and Das [24]. The numerical investigation is carried out for different governing parameters and the results that have been obtained are discussed and plotted graphically.

2 Basic Equations

Consider a steady, two-dimensional, laminar mixed convection boundary layer flow and heat transfer of ferrofluids over a moving permeable vertical flat plate in a variable magnetic field $B(x)$. In formulating this problem, this study considered the water- and kerosene-based ferrofluids containing magnetite, cobalt ferrite and manganese-zinc ferrite ferroparticles. It is assumed that the flow takes place in the region $y > 0$ and is driven by a moving surface with the velocity $u_w = \lambda U_\infty + u_{slip}$, where u_{slip} is the surface slip velocity, λ is a dimensionless constant moving parameter with $\lambda > 0$ for a surface moving out of the origin and $\lambda < 0$ for a surface moving towards the origin and U_∞ denotes the free stream velocity, as shown in Fig. 1. It is also assumed that the uniform heat flux from the surface is q_w, while the temperature of the ambient fluid is T_∞. Based on the Boussinesq approximation, order of magnitude analysis and taking into consideration of the above assumptions, the basic equations of the problem are given by (see [23] and [30])

$$\frac{\partial u}{\partial x} + \frac{\partial v}{\partial y} = 0, \tag{1}$$

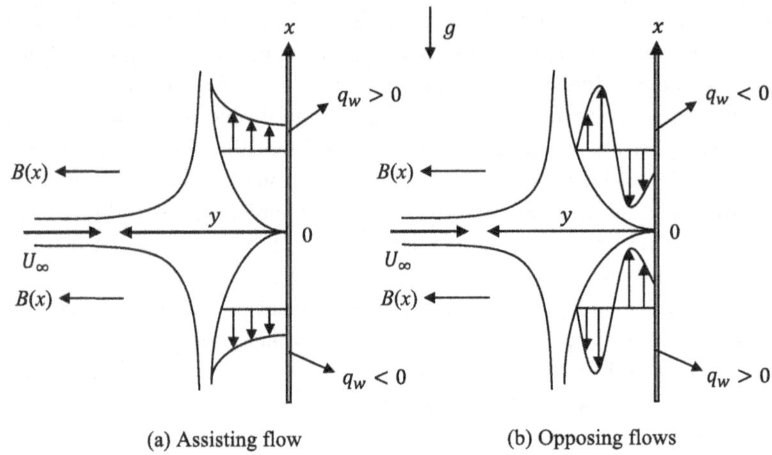

Fig. 1 Physical model and coordinate system

$$u\frac{\partial u}{\partial x} + v\frac{\partial u}{\partial y} = \nu_{ff}\frac{\partial^2 u}{\partial y^2} - \frac{\sigma B^2(x)}{\rho_{ff}}(u - U_\infty) + g\beta_{ff}(T - T_\infty), \tag{2}$$

$$u\frac{\partial T}{\partial x} + v\frac{\partial T}{\partial y} = \alpha_{ff}\frac{\partial^2 T}{\partial y^2}. \tag{3}$$

where x and y are the Cartesian coordinates measured perpendicular and normal to the plate, respectively, u and v are the velocity components along x- and y-axes, respectively, σ is the electric conductivity, B is the total magnetic field, g is the gravitational acceleration, T is the temperature of the ferrofluid and T_∞ is the temperature of the ambient fluid. Here, ν_{ff} is the kinematic viscosity of the ferrofluid, ρ_{ff} is the density of the ferrofluid, β_{ff} is the coefficient of thermal expansion of the ferrofluid and α_{ff} is the thermal diffusivity of the ferrofluid.
Let us present some physical properties of the ferrofluids such that

$$\mu_{ff} = \frac{\mu_f}{(1-\varphi)^{2.5}}, \qquad \rho_{ff} = (1-\varphi)\rho_f + \varphi\rho_s, \qquad \nu_{ff} = \frac{\mu_{ff}}{\rho_{ff}}, \tag{4}$$

where μ_{ff} is the dynamic viscosity of the ferrofluid, μ_f is the dynamic viscosity of the base fluid, φ is the volume fraction of solid particle of the ferrofluid, ρ_f is the density of the base fluid and ρ_s is the density of the solid. Furthermore, the thermal properties are

$$\left(\rho C_p\right)_{ff} = (1-\varphi)\left(\rho C_p\right)_f + \varphi\left(\rho C_p\right)_s,$$
$$\frac{k_{ff}}{k_f} = \frac{k_s + 2k_f - 2\varphi\left(k_f - k_s\right)}{k_s + 2k_f + \varphi\left(k_f - k_s\right)}, \qquad \alpha_{ff} = \frac{k_{ff}}{\left(\rho C_p\right)_{ff}}, \tag{5}$$

where $\left(\rho C_p\right)_{ff}$ is the heat capacity of the ferrofluid, $\left(\rho C_p\right)_f$ is the heat capacity of the base fluid, $\left(\rho C_p\right)_s$ is the heat capacity of the solid, k_{ff} is the thermal conductivity of the ferrofluid, k_f is the thermal conductivity of the base fluid and k_s is the thermal conductivity of the solid.

Then, it is assumed that Eqs. (1)–(3) are subjected to the boundary conditions

$$u = \lambda U_\infty + \Omega \nu_f N \frac{\partial u}{\partial y}, \quad v = v_w, \quad -k_{ff}\frac{\partial T}{\partial y} = q_w \quad \text{at} \quad y = 0,$$
$$u \to U_\infty, \quad T \to T_\infty \quad \text{as} \quad y \to \infty, \tag{6}$$

where Ω is the slip parameter, ν_f is the kinematic viscosity of the base fluid, N is the slip factor velocity and v_w is the mass transfer velocity. Further, we assume that q_w is given by $q_w = q_0 x$ where q_0 is the constant heat flux characteristic with $q_0 > 0$ for the assisting flow and $q_0 < 0$ for the opposing flow, respectively.

In order Eqs. (1)–(3) to have similarity solutions, we have to assume that $B(x) = B_0 x^{-1/2}$, where $B_0 \neq 0$ is the strength of the applied magnetic field and Ω should be constant, we take $\Omega = x^{1/2} D / N\sqrt{U_\infty \nu_f}$ where D is the dimensionless slip parameter. Then, we look for a similarity solution of Eqs. (1)–(3) with boundary conditions (6) by introducing the following transformation

$$\psi(x, y) = \nu_f \sqrt{\mathrm{Re}_x}\, f(\eta), \qquad \theta(\eta) = \frac{T - T_\infty}{q_w x / k_f}\sqrt{\mathrm{Re}_x}, \qquad \eta = \frac{y}{x}\sqrt{\mathrm{Re}_x}, \tag{7}$$

where η is the similarity variable and $\mathrm{Re}_x = U_\infty x / \nu_f$ is the local Reynolds number based on the free stream velocity. Furthermore, $\psi(x, y)$ is the streamline, which represents in the common form as $u = \partial\psi/\partial y$ and $v = -\partial\psi/\partial x$. Thus, we have

$$u = U_\infty f'(\eta), \qquad v = -\nu_f \frac{\sqrt{\mathrm{Re}_x}}{2x}\left[f(\eta) - \eta f'(\eta)\right], \tag{8}$$

where prime ($'$) denotes differentiation with respect to η. It is worth noting that

$$v_w = -\nu_f \frac{\sqrt{\mathrm{Re}_x}}{2x} f(0). \tag{9}$$

We now take $f(0) = S$, where $S\left(= -2x v_w / \nu_f \sqrt{\mathrm{Re}_x}\right)$ is the constant mass transfer parameter with $S > 0$ for suction and $S < 0$ for injection.

Now, substituting Eqs. (7) and (8) into Eqs. (1)–(3) and boundary conditions (6), we have the ordinary differential equations

$$\frac{1}{(1-\varphi)^{2.5}\left(1-\varphi+\varphi\rho_s/\rho_f\right)}f''' + \frac{1}{2}ff'' + \frac{M}{1-\varphi+\varphi\rho_s/\rho_f}\left(1-f'\right) + \omega\theta = 0, \tag{10}$$

$$\frac{k_{ff}/k_f}{1-\varphi+\varphi\left(\rho C_p\right)_s/\left(\rho C_p\right)_f}\frac{1}{\Pr}\theta'' + \frac{1}{2}\left(f\theta' - f'\theta\right) = 0, \tag{11}$$

subject to the boundary conditions

$$f(0) = S, \quad f'(0) = \lambda + Df''(0), \quad \theta'(0) = -\frac{k_f}{k_{ff}},$$
$$f'(\eta) \to 1, \quad \theta(\eta) \to 0 \quad \text{as} \quad \eta \to \infty, \tag{12}$$

where primes (′) denote differentiation with respect to η and $M\left(= \sigma B_0^2/\rho_f U_\infty\right)$ is the magnetic parameter. Moreover, ω is the constant mixed convection, which is defined as

$$\omega = \frac{\text{Gr}}{(\text{Re}_x)^{5/2}}, \tag{13}$$

where $\text{Gr} = g\beta_{ff}\left(xq_w/k_f\right)x^3/\nu_f^2$ is the Grashof number and $\text{Re}_x = U_\infty x/\nu_f$ is the Reynolds number. It is worth mentioning that $\omega > 0$ corresponds to the assisting flow and $\omega < 0$ corresponds to the opposing flows.

Quantities of physical interest in this problem are the skin friction coefficient C_f and the local Nusselt number Nu_x, which are defined as

$$C_f = \frac{\tau_w}{\rho_f U_\infty^2}, \qquad \text{Nu}_x = \frac{xq_w}{k_f\left(T_w - T_\infty\right)}, \tag{14}$$

where τ_w is the skin friction or shear stress along the plate, such that

$$\tau_w = \mu_{ff}\left(\frac{\partial u}{\partial y}\right)_{y=0}, \qquad q_w = -k_{ff}\left(\frac{\partial T}{\partial y}\right)_{y=0}. \tag{15}$$

Using Eqs. (7), (14) and (15), we get

$$\text{Re}_x^{1/2}C_f = \frac{f''(0)}{(1-\varphi)^{2.5}}, \qquad \text{Re}_x^{-1/2}\text{Nu}_x = \frac{1}{\theta(0)}. \tag{16}$$

3 Stability Analysis

Weidman et al. [25] and Roşca and Pop [26] have demonstrated that the lower branch solutions are unstable (not physically realizable), while the upper branch solutions are stable (physically realizable). These characteristics are assessed by looking at the unsteady form of Eqs. (10), (11) and boundary conditions (12). Thus, the new dimensionless time variable $\tau = U_\infty t/x$ is proposed and we have to note that the use of τ is related with an initial value problem. Thus, the unsteady Eqs. (1)–(3) are

$$\frac{\partial u}{\partial t} + u\frac{\partial u}{\partial x} + v\frac{\partial u}{\partial y} = \nu_{ff}\frac{\partial^2 u}{\partial y^2} - \frac{\sigma B^2(x)}{\rho_{ff}}(u - U_\infty) + g\beta_{ff}(T - T_\infty), \tag{17}$$

$$\frac{\partial T}{\partial t} + u\frac{\partial T}{\partial x} + v\frac{\partial T}{\partial y} = \alpha_{ff}\frac{\partial^2 T}{\partial y^2}, \tag{18}$$

where t denotes the time. Now, introducing the new variables of

$$u = U_\infty\frac{\partial f}{\partial \eta}(\eta, \tau), \qquad v = -\frac{\nu_f}{2x}\sqrt{\mathrm{Re}_x}\left[f(\eta, \tau) - \eta\frac{\partial f}{\partial \eta}(\eta, \tau) - 2\tau\frac{\partial f}{\partial \tau}\right],$$
$$\theta(\eta, \tau) = \frac{T - T_\infty}{q_w x/k_f}\sqrt{\mathrm{Re}_x}, \qquad \eta = \frac{y}{x}\sqrt{\mathrm{Re}_x}, \qquad \tau = \frac{U_\infty}{x}t. \tag{19}$$

So, Eqs. (17) and (18) can be written as

$$\frac{1}{(1-\varphi)^{2.5}\left(1 - \varphi + \varphi\rho_s/\rho_f\right)}\frac{\partial^3 f}{\partial \eta^3} + \frac{1}{2}f\frac{\partial^2 f}{\partial \eta^2} + \left(\tau\frac{\partial f}{\partial \eta} - 1\right)\frac{\partial^2 f}{\partial \eta \partial \tau}$$
$$- \tau\frac{\partial f}{\partial \tau}\frac{\partial^2 f}{\partial \eta^2} + \frac{M}{1 - \varphi + \varphi\rho_s/\rho_f}\left(1 - \frac{\partial f}{\partial \eta}\right) + \omega\theta = 0, \tag{20}$$

$$\frac{k_{ff}/k_f}{1 - \varphi + \varphi\left(\rho C_p\right)_s/\left(\rho C_p\right)_f}\frac{1}{\mathrm{Pr}}\frac{\partial^2 \theta}{\partial \eta^2} + \frac{1}{2}\left(f\frac{\partial \theta}{\partial \eta} - \frac{\partial f}{\partial \eta}\theta\right)$$
$$+ \left(\tau\frac{\partial f}{\partial \eta} - 1\right)\frac{\partial \theta}{\partial \tau} - \tau\frac{\partial f}{\partial \tau}\frac{\partial \theta}{\partial \eta} = 0, \tag{21}$$

subjected to the boundary conditions

$$f(0, \tau) = S, \quad \frac{\partial f}{\partial \eta}(0, \tau) = \lambda + D\frac{\partial^2 f}{\partial \eta^2}(0, \tau), \quad \frac{\partial \theta}{\partial \eta}(0, \tau) = -\frac{k_f}{k_{ff}},$$
$$\frac{\partial f}{\partial \eta}(\eta, \tau) \to 1, \quad \theta(\eta, \tau) \to 0 \ \text{ as } \ \eta \to \infty. \tag{22}$$

To examine the stability of the steady flow solution $f(\eta) = f_0(\eta)$ and $\theta(\eta) = \theta_0(\eta)$ satisfying the boundary-value problem of Eqs. (10)–(12), we define (see [25] and [26]),

$$f(\eta, \tau) = f_0(\eta) + e^{-\xi\tau}F(\eta, \tau), \qquad \theta(\eta, \tau) = \theta_0(\eta) + e^{-\xi\tau}G(\eta, \tau), \tag{23}$$

where ζ is an unknown eigenvalue parameter, and $F(\eta, \tau)$ and $G(\eta, \tau)$ are small relative to $f_0(\eta)$ and $\theta_0(\eta)$. Solutions of the eigenvalue problem of Eqs. (20) to (22), give an infinite set of eigenvalues $\zeta_1 < \zeta_2 < \ldots$. It is worth noting that if the smallest eigenvalue is negative, there is an initial growth of disturbances and the flow is unstable. However, there is an initial decay and the flow is stable when ζ_1 is positive.

Thus, substituting (23) into Eqs. (20) and (21), we obtain the linearized problem as follows:

$$\frac{1}{(1-\varphi)^{2.5}\left(1-\varphi+\varphi\rho_s/\rho_f\right)}\frac{\partial^3 F}{\partial\eta^3}+\frac{1}{2}f_0\frac{\partial^2 F}{\partial\eta^2}$$
$$+\left(\zeta-\tau\zeta f_0'-\frac{M}{1-\varphi+\varphi\rho_s/\rho_f}\right)\frac{\partial F}{\partial\eta}+\left(\frac{1}{2}+\tau\zeta\right)f_0''F$$
$$-\tau f_0''\frac{\partial F}{\partial\tau}+\left(\tau f_0'-1\right)\frac{\partial^2 F}{\partial\eta\partial\tau}+\omega G=0, \tag{24}$$

$$\frac{k_{ff}/k_f}{1-\varphi+\varphi\left(\rho C_p\right)_s/\left(\rho C_p\right)_f}\frac{1}{\Pr}\frac{\partial^2 G}{\partial\eta^2}+\frac{1}{2}f_0\frac{\partial G}{\partial\eta}-\left(\frac{1}{2}f_0'+\tau\zeta f_0'-\zeta\right)G$$
$$+\left(\frac{1}{2}+\tau\zeta\right)F\theta_0'-\frac{1}{2}\theta_0\frac{\partial F}{\partial\eta}+\left(\tau f_0'-1\right)\frac{\partial G}{\partial\tau}-\tau\theta_0'\frac{\partial F}{\partial\tau}=0, \tag{25}$$

along with the boundary conditions

$$F(0,\tau)=0,\qquad \frac{\partial F}{\partial\eta}(0,\tau)=D\frac{\partial^2 F}{\partial\eta^2}(0,\tau),\qquad \frac{\partial G}{\partial\eta}(0,\tau)=0,$$
$$\frac{\partial F}{\partial\eta}(\eta,\tau)\to 0,\qquad G(\eta,\tau)\to 0\quad \text{as}\quad \eta\to\infty. \tag{26}$$

As suggested by Weidman et al. [25], we investigate the stability of the steady flow and heat transfer solution $f_0(\eta)$ and $\theta_0(\eta)$ by setting $\tau = 0$, and therefore $F = F_0(\eta)$ and $G = G_0(\eta)$ in Eqs. (24) and (25) to identify the initial growth or decay of the solution (23). To test our numerical process, we have to solve the linear eigenvalue problem and the results is obtained as

$$\frac{1}{(1-\varphi)^{2.5}\left(1-\varphi+\varphi\rho_s/\rho_f\right)}F_0'''+\frac{1}{2}f_0F_0''+\left(\zeta-\frac{M}{1-\varphi+\varphi\rho_s/\rho_f}\right)F_0'$$
$$+\frac{1}{2}f_0''F_0+\omega G_0=0, \tag{27}$$

$$\frac{k_{ff}/k_f}{1-\varphi+\varphi\left(\rho C_p\right)_s/\left(\rho C_p\right)_f}\frac{1}{\Pr}G_0''+\frac{1}{2}f_0G_0'-\left(\frac{1}{2}f_0'-\zeta\right)G_0+\frac{1}{2}\theta_0'F_0$$
$$-\frac{1}{2}\theta_0F_0'=0, \tag{28}$$

along with the boundary conditions

$$\begin{aligned} &F_0(0)=0, \quad F_0'(0)=DF_0''(0), \quad G_0'(0)=0,\\ &F_0'(\eta)\to 0, \quad G_0(\eta)\to 0 \quad \text{as} \quad \eta\to\infty. \end{aligned} \tag{29}$$

It should be mentioned that for particular values of Pr, ω, D, φ, and M, the stability of the steady flow solution $f_0(\eta)$ and $\theta_0(\eta)$ are obtained from the smallest eigenvalue ζ. According to Harris et al. [27], the range of possible eigenvalues can be determined by relaxing a boundary condition on $F_0'(\eta)$ and $G_0(\eta)$. For the present problem, we have to relax the condition that $F_0'(\eta)\to 0$ as $\eta\to\infty$ and for a fixed value of ζ, we solve the system (27)–(29) along with the new boundary condition $F_0''(0)=1$.

4 Results and Discussion

The system of ordinary differential equations (10) and (11) subject to the boundary conditions (12) were solved numerically using the shooting method which is done by a *shootlib* function in Maple software, for the case of uniform heat flux. Note that in this problem, we carry out a study that demonstrates the influences of the mixed convection parameter ω, magnetic parameter M, mass transfer parameter S, moving parameter λ, slip parameter D and volume fraction of solid ferroparticle φ on the dimensionless velocity $f'(\eta)$, temperature $\theta(\eta)$, skin friction coefficient $\mathrm{Re}_x^{1/2}C_f$ and local Nusselt number $\mathrm{Re}_x^{-1/2}\mathrm{Nu}_x$ for the three selected ferroparticles, namely magnetite, cobalt ferrite and manganese-zinc ferrite in water- and kerosene-based fluids. Throughout this research, the value of the Prandtl number was chosen to be 6.2 and 21, for the base fluids water and kerosene, respectively. Also, the volume fraction of solid ferroparticle φ is considered in the range $0\le\varphi\le 0.1$, where $\varphi=0$ represents the pure fluid water or kerosene. This is considered because a sedimentation will arise if we used $\varphi>0.1$. For example, if we take $\varphi=0.2$, then 20% of solid ferroparticle is in the fluid, and heavy particles can no longer be supported whilst the desired results cannot be obtained.

The thermophysical properties of both the base fluids (water and kerosene) and the ferroparticles (magnetite Fe_3O_4, cobalt ferrite $CoFe_2O_4$ and manganese-zinc ferrite Mn-$ZnFe_2O_4$) are given in Table 1. In order to verify the accuracy of the present results, the values of $f''(0)$ are compared with those reported by Cortell [28], Yazdi et al. [29] and Khan et al. [30], as shown in Table 2. The present results are found to be in excellent agreement with the published data and thus gives us some confidence in the present numerical results.

Table 1 Thermophysical properties of base fluids and magnetic nanoparticles (see [30])

Physical properties	Base fluids		Magnetic nanoparticles		
	Water	Kerosene	Fe_3O_4	$CoFe_2O_4$	$Mn\text{-}ZnFe_2O_4$
$\rho\,(kg/m^3)$	997	783	5180	4907	4900
$C_p\,(J/kg.K)$	4179	2090	670	700	800
$k\,(W/m.K)$	0.613	0.15	9.7	3.7	5

Table 2 Comparison of $f''(0)$ for $D = 0, 0.5$ and $M = 0, 1$ when $\varphi = 0$, $S = 0$ and $\lambda = 0$

D	M	[28]	[29]	[30]	Present study
0	0	0.33206	–	0.33206	0.33206
	1	–	1.0440	1.04400	1.04400
0.5	0	–	0.6987	0.69872	0.69872

It is clear that triple solutions (first, second and third solutions) exist for Eqs. (10) and (11) subject to the boundary conditions (12) for both assisting and opposing flows when the mass transfer parameter of suction ($S > 0$) and moving parameter for a surface moving towards the origin ($\lambda < 0$) are applied. The triple solutions were acquired by defining three different initial guesses for the missing values of $f''(0)$ and $\theta(0)$ in which the velocity and temperature profiles satisfy the boundary condition (12) with different patterns. In Figs. 2, 3, 4, 5 and 7–8, the solid, dash and dot lines represent the first, second and third solutions, respectively.

The variation of the skin friction coefficient $\mathrm{Re}_x^{1/2} C_f$ and local Nusselt number $\mathrm{Re}_x^{-1/2}\mathrm{Nu}_x$ with the mixed convection parameter ω for various values of M and D are displayed in Figs. 2, 3, 4, 5. All these figures show that unique solutions exist for Eqs. (10) and (11) with the boundary conditions (12) when $\omega < \omega_{c1}$ and $\omega > \omega_{c2}$, while triple (first, second and third) solutions for $\omega_{c1} \le \omega \le \omega_{c2}$. It is observed that the first, second and third solutions merge with one another at the critical values of the mixed convection parameter ω_{c1} and ω_{c2} for negative (opposing flows) and positive (assisting flow) values of ω, respectively. These solutions merge at the critical points $\omega = \omega_{c1}(< 0)$ and $\omega = \omega_{c2}(> 0)$ where the boundary solutions beyond this point are possible to be obtained since the separation of the boundary layer from the plate does not occur. Then, the specified values of critical point ω_{c1} and ω_{c2} are shown in Table 3. From this table, it indicates that the critical value $|\omega_{c1}|$ increases as the parameters M and D are increasing, suggesting that M and D increase the range of existence of the solutions of the governing equations. Thus, these processes (magnetic and slip) delay the boundary layer separation. However, the opposite behavior can be viewed for the critical value ω_{c2} in which ω_{c2} decreases as the parameters M and D increasing, thus these processes accelerate the boundary layer separation.

Furthermore, it can be seen that Eqs. (10) and (11) with the boundary conditions (12) gives the singularities ω_{si} at some values of $\omega = \omega_{si}$ (< 0) as illustrated in Table 4. The singularity is occured when the flat plate is moving to the differ-

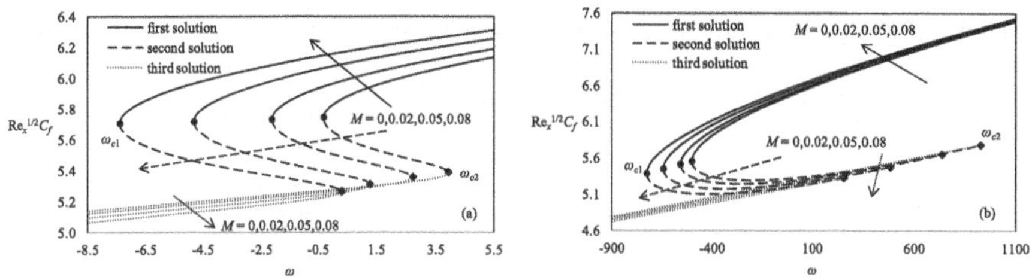

Fig. 2 Variation of $Re_x^{1/2}C_f$ with ω for Fe_3O_4, **a** water-based ferrofluid, Pr = 6.2 and **b** kerosene-based ferrofluid, Pr = 21, when $S = 3$, $\lambda = -7.2$, $D = 1$, $\varphi = 0.1$ and with varying M

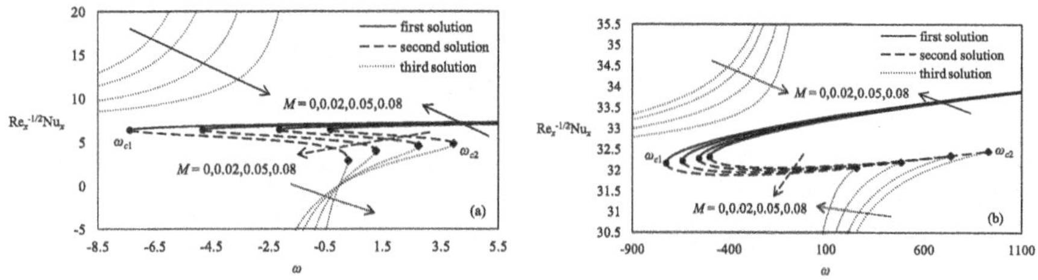

Fig. 3 Variation of $Re_x^{-1/2}Nu_x$ with ω for Fe_3O_4, **a** water-based ferrofluid, Pr = 6.2 and **b** kerosene-based ferrofluid, Pr = 21, when $S = 3$, $\lambda = -7.2$, $D = 1$, $\varphi = 0.1$ and with varying M

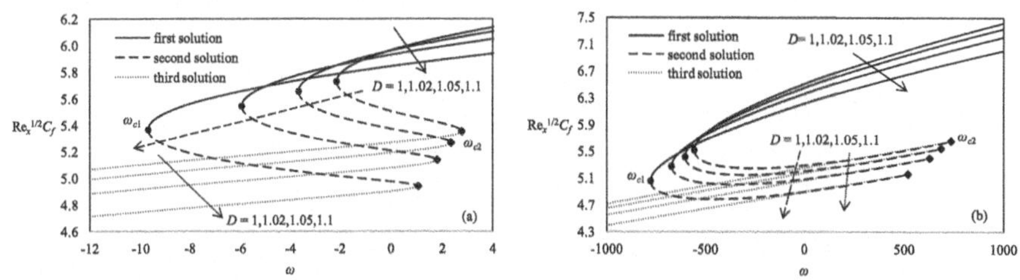

Fig. 4 Variation of $Re_x^{1/2}C_f$ with ω for Fe_3O_4, **a** water-based ferrofluid, Pr = 6.2 and **b** kerosene-based ferrofluid, Pr = 21, when $M = 0.02$, $S = 3$, $\lambda = -7.2$, $\varphi = 0.1$ and with varying D

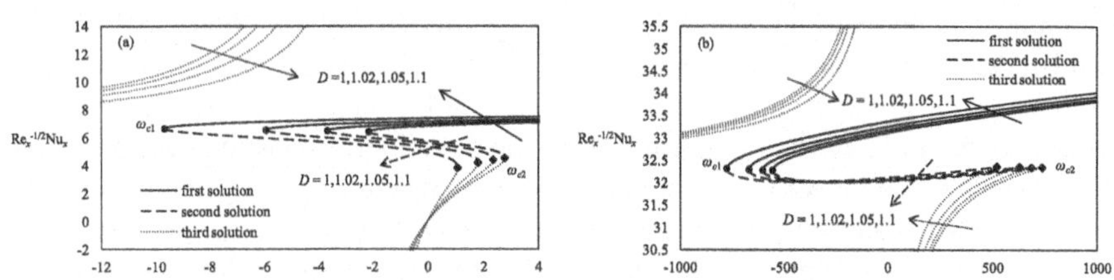

Fig. 5 Variation of $Re_x^{-1/2}Nu_x$ with ω for Fe_3O_4, **a** water-based ferrofluid, Pr = 6.2 and **b** kerosene-based ferrofluid, Pr = 21, when $M = 0.02$, $S = 3$, $\lambda = -7.2$, $\varphi = 0.1$ and with varying D

Table 3 Value of critical point ω_{c1} and ω_{c2} for different figure and parameter in water- and kerosene-based ferrofluids

Figure	Parameter	Water-based ferrofluid (a)		Kerosene-based ferrofluid (b)	
		ω_{c1}	ω_{c2}	ω_{c1}	ω_{c2}
2 and 3	$M = 0$	−0.3054	3.9341	−499.4804	929.8413
	$M = 0.02$	−2.1292	2.7352	−555.5109	738.3931
	$M = 0.05$	−4.7983	1.2815	−639.4782	483.6249
	$M = 0.08$	−7.3968	0.3199	−723.3299	257.8729
4 and 5	$D = 1$	−2.1262	2.7727	−555.5109	738.3931
	$D = 1.02$	−3.6604	2.3475	−602.4757	690.4983
	$D = 1.05$	−5.9483	1.7917	−669.8351	627.6642
	$D = 1.1$	−9.7020	1.0615	−774.5518	515.5914

Table 4 Singularity point of $\mathrm{Re}_x^{-1/2}\mathrm{Nu}_x$ for different figures and parameters with water- and kerosene-based ferrofluids

Figure	Parameter	Singularity point (ω_{si})	
		Water-based ferrofluid	Kerosene-based ferrofluid
3	$M = 0$	$-3.81 < \omega_{si} < -3.80$	$-20.51 < \omega_{si} < -20.50$
	$M = 0.02$	$-3.17 < \omega_{si} < -3.16$	$-17.13 < \omega_{si} < -17.12$
	$M = 0.05$	$-2.22 < \omega_{si} < -2.21$	$-11.99 < \omega_{si} < -11.98$
	$M = 0.08$	$-1.27 < \omega_{si} < -1.26$	$-6.79 < \omega_{si} < -6.78$
5	$D = 1$	$-3.23 < \omega_{si} < -3.22$	$-17.13 < \omega_{si} < -17.12$
	$D = 1.02$	$-2.97 < \omega_{si} < -2.96$	$-15.98 < \omega_{si} < -15.97$
	$D = 1.05$	$-2.61 < \omega_{si} < -2.60$	$-14.33 < \omega_{si} < -14.32$
	$D = 1.1$	$-2.04 < \omega_{si} < -2.03$	$-11.81 < \omega_{si} < -11.80$

ent direction of the free stream. Then, we continued the discussion of the effects of parameters M and D as illustrated in Figs. 2, 3, 4, 5. For the case of stable solution, it is noticed that the skin friction coefficient $\mathrm{Re}_x^{1/2}C_f$ and local Nusselt number $\mathrm{Re}_x^{-1/2}\mathrm{Nu}_x$ increase monotonically with the magnetic parameter M. This happens because large values of M are responsible to enhance the Lorentz force within boundary layer region which reverse the flow in the opposite direction. Then, the opposite behavior is noticed from Figs. 4 and 5 where $\mathrm{Re}_x^{1/2}C_f$ tends to decrease and $\mathrm{Re}_x^{-1/2}\mathrm{Nu}_x$ increases with the slip parameter D. Physically, the increase of the slip parameter D indicates the increase of the slip velocity in x-direction, and thus reduces the adhesion force between the fluid and the sheet surface, which causes the shear stress on the x-direction to decrease. Hence, the velocity slip effect delays the boundary layer separation.

In order to ascertain which of the three solutions are stable and unstable, the stability analysis is performed by obtaining the smallest eigenvalues of each solutions using `bvp4c` Matlab solver. This analysis is accomplished by solving the

Table 5 Smallest eigenvalues ζ for Fe_3O_4, $CoFe_2O_4$ and Mn-$ZnFe_2O_4$ ferroparticles at several values of ω (< 0, opposing flow), with various values of M, when $S = 3, \lambda = -7, D = 1, \varphi = 0.1$ and $\Pr = 6.2$ (water-based ferrofluid)

Ferroparticle	M	ω	ζ		
			First solution	Second solution	Third solution
Fe_3O_4	0.02	−5.52	0.0697	−0.0388	−0.3560
		−5.60	0.0566	−0.0262	−0.3575
		−5.65	0.0458	−0.0157	−0.3584
		−5.67	0.0403	−0.0103	−0.3588
	0.05	−8.06	0.0953	−0.0488	−0.3994
		−8.10	0.0918	−0.0455	−0.3999
		−8.20	0.0787	−0.0329	−0.4016
		−8.34	0.0519	−0.0069	−0.4040
	0.08	−10.70	0.1025	−0.0412	−0.4487
		−10.78	0.0923	−0.0314	−0.4499
		−10.85	0.0819	−0.0212	−0.4510
		−10.93	0.0665	−0.0062	−0.4523
$CoFe_2O_4$	0.02	−3.00	0.0660	−0.0499	−0.3042
		−3.10	0.0531	−0.0376	−0.3060
		−3.20	0.0348	−0.0200	−0.3079
		−3.25	0.0170	−0.0025	0.3088
	0.05	−6.11	0.0769	−0.0460	−0.3455
		−6.20	0.0656	−0.0353	−0.3470
		−6.30	0.0492	−0.0194	−0.3487
		−6.32	0.0449	−0.0152	−0.3490
	0.08	−9.20	0.0795	−0.0339	−0.3927
		−9.25	0.0727	−0.0273	−0.3934
		−9.30	0.0648	−0.0196	−0.3941
		−9.32	0.0612	−0.0161	−0.3944
Mn-$ZnFe_2O_4$	0.02	−3.00	0.0512	−0.0360	−0.3048
		−3.06	0.0407	−0.0259	−0.3059
		−3.10	0.0314	−0.0168	−0.3067
		−3.14	0.0145	−0.0002	−0.3074
	0.05	−6.11	0.0583	−0.0282	−0.3478
		−6.13	0.0549	−0.0249	−0.3481
		−6.20	0.0398	−0.0102	−0.3493
		−6.21	0.0368	−0.0073	−0.3495
	0.08	−9.10	0.0705	−0.0250	−0.3948
		−9.15	0.0621	−0.0168	−0.3955
		−9.20	0.0513	−0.0063	−0.3963
		−9.22	0.0457	−0.0007	−0.3966

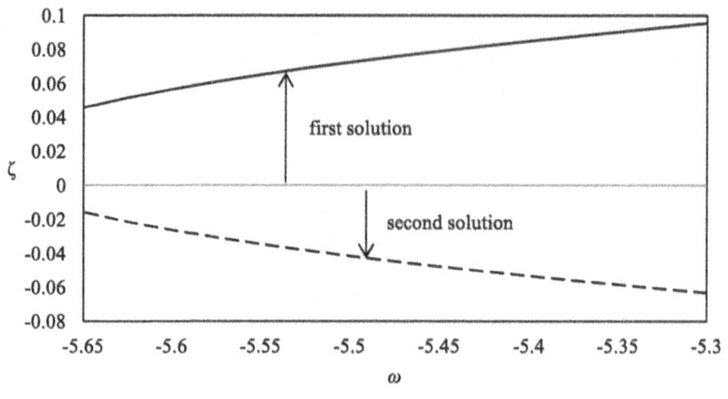

Fig. 6 Plot of smallest eigenvalues ζ as a function of ω

linear eigenvalue problem of Eqs. (27) and (28) subject to the boundary conditions (29) to determine which solution is stable and physically reliazable. If the smallest eigenvalue is positive, then the solution is stable and physically realizable while if the smallest eigenvalue is negative, then the solution is unstable and not physically realizable. Table 5 shows the smallest eigenvalues ζ for Fe_3O_4, $CoFe_2O_4$ and Mn-$ZnFe_2O_4$ ferroparticles at several values of ω (> 0, assisting flow), with various values of M when $S = 3$, $\lambda = -7$, $D = 1$, $\varphi = 0.1$ and $\Pr = 6.2$ for water-based ferrofluid. From this table, we notice that, as we approach the critical value ω_c, the smallest eigenvalue ζ approaches 0 for both first and second solutions, whereas the third solution acts in the opposite manner.

Then, let us discuss the trend for the third solution by looking at Fig. 3. From this figure, we can see that the third solution is separated into two distinct regions. This happened when the singularities occur in which a point is not define or in exceptional set where it fails to be well-behaved in some particular values. In the stability analysis, we rather choose the critical values based on the first and second solutions since the first one is assumed as a stable solution. Figure 3 also illustrates that the third solution is separated from the graph (if we take one value of the critical point) and this is the reason why the smallest eigenvalues behave in the opposite trend. Also, Fig. 6 shows a plot of the smallest eigenvalues ζ as a function of the mixed convection parameter, ω. It is clearly shown from Fig. 6 that ζ is positive for the first solution and negative for the second solution. For Table 5, it is found that the first solution is positive while the second and third solutions are positive. Besides, from Eq. (23), we can see that the results will converge to steady flow solution when ζ is positive. Thus, we conclude that of the triple steady flow solutions, the first solution is linearly stable while those on the second and third solutions are linearly unstable. This is because there is an initial decay and the flow is stable for the first solution.

Tables 6 and 7 illustrate the variations of the skin friction coefficient $\mathrm{Re}_x^{1/2} C_f$ and local Nusselt number $\mathrm{Re}_x^{-1/2} \mathrm{Nu}_x$, respectively, as a function of the assisting and opposing flow parameter ω for the three selected ferroparticles in water- and kerosene-based ferrofluids by considering the first solution is stable. It is shown that

Table 6 Variation of the skin friction coefficient $\mathrm{Re}_x^{1/2}C_f$ and the local Nusselt number $\mathrm{Re}_x^{-1/2}\mathrm{Nu}_x$ with ω (> 0, assisting flow) for different ferroparticles with water- and kerosene-based ferrofluids when $\varphi = 0.1$, $M = 0.02$, $\lambda = -7$, $S = 3$ and $D = 1$

Ferroparticle	ω	$\mathrm{Re}_x^{1/2}C_f$		$\mathrm{Re}_x^{-1/2}\mathrm{Nu}_x$	
		Water-based ferrofluid	Kerosene-based ferrofluid	Water-based ferrofluid	Kerosene-based ferrofluid
Fe_3O_4	0.1	5.9406	6.3882	7.1225	33.2283
	0.5	5.9546	6.3886	7.1403	33.2286
	1	5.9714	6.3892	7.1615	33.2291
$CoFe_2O_4$	0.1	5.8060	6.2965	7.0350	33.1189
	0.5	5.8230	6.2970	7.0576	33.1192
	1	5.8432	6.2975	7.0838	33.1197
Mn-$ZnFe_2O_4$	0.1	5.8020	6.2941	7.0926	34.0357
	0.5	5.8193	6.2945	7.1162	34.0361
	1	5.8397	6.2950	7.1437	34.0365

Table 7 Variation of the skin friction coefficient $\mathrm{Re}_x^{1/2}C_f$ and the local Nusselt number $\mathrm{Re}_x^{-1/2}\mathrm{Nu}_x$ with ω (< 0, opposing flow) for different ferroparticles with water- and kerosene-based ferrofluids when $\varphi = 0.1$, $M = 0.02$, $\lambda = -7$, $S = 3$ and $D = 1$

Ferroparticle	ω	$\mathrm{Re}_x^{1/2}C_f$		$\mathrm{Re}_x^{-1/2}\mathrm{Nu}_x$	
		Water-based ferrofluid	Kerosene-based ferrofluid	Water-based ferrofluid	Kerosene-based ferrofluid
Fe_3O_4	−0.1	5.9335	6.3880	7.1133	33.2281
	−0.5	5.9189	6.3876	7.0943	33.2278
	−1	5.8998	6.3871	7.0693	33.2274
$CoFe_2O_4$	−0.1	5.7971	6.2963	7.0231	33.1187
	−0.5	5.7785	6.2959	6.9979	33.1184
	−1	5.7534	6.2954	6.9632	33.1180
Mn-$ZnFe_2O_4$	−0.1	5.7929	6.2939	7.0801	34.0356
	−0.5	5.7740	6.2935	7.0536	34.0353
	−1	5.7483	6.2929	7.0169	34.0349

the skin friction coefficient $\mathrm{Re}_x^{1/2}C_f$ and local Nusselt number $\mathrm{Re}_x^{-1/2}\mathrm{Nu}_x$ decrease with the increment of the mixed convection parameter ω in water- and kerosene-based ferrofluids. It is interesting to note that kerosene-based ferrofluid has higher local Nusselt number compared to water-based ferrofluid. This is due to higher Prandtl number of kerosene which make the thermal boundary layer thickness for kerosene become smaller.

The dimensionless velocity $f'(\eta)$ and temperature $\theta(\eta)$ profiles for different values of M and ω for Fe_3O_4 in both water- and kerosene-based ferrofluids are depicted

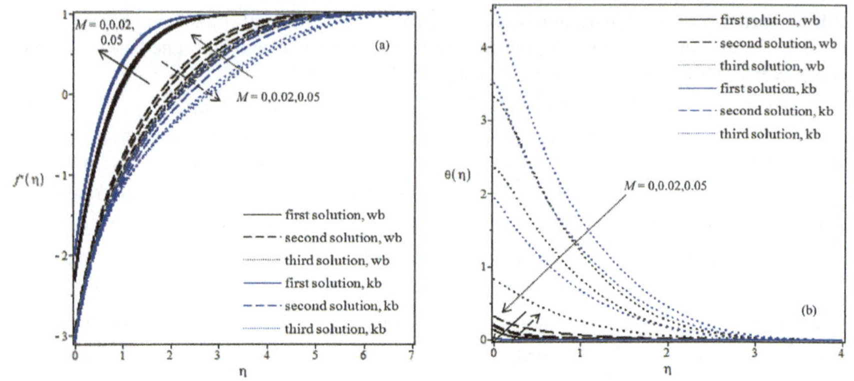

Fig. 7 Dimensionless **a** velocity profiles $f'(\eta)$ and **b** temperature profiles $\theta(\eta)$, for several values of M, Fe_3O_4, when $\omega = 0.1$ (> 0, assisting flow), $\varphi = 0.1$, $S = 3$, $\lambda = -7.2$ and $D = 1$

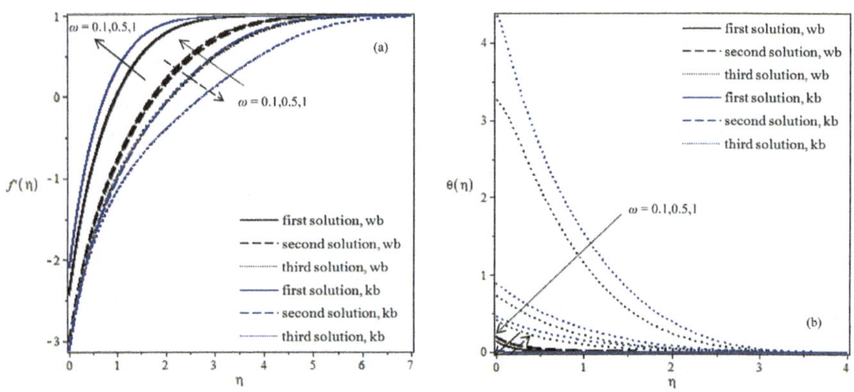

Fig. 8 Dimensionless **a** velocity profiles $f'(\eta)$ and **b** temperature profiles $\theta(\eta)$, for several values of ω (> 0, assisting flow), Fe_3O_4, when $M = 0.02$, $\varphi = 0.1$, $S = 3$, $\lambda = -7$ and $D = 1$

in Figs. 7 and 8, respectively. It is seen that the velocity and thermal boundary-layer thicknesses decrease in the first and third solutions while increase in the second solution with the increase of M and ω. Further, it is obviously noticed that the boundary layer thickness for the first solution is thinner than the second and third solutions. The velocity and thermal boundary-layer thicknesses are seem to be smaller for each kerosene-based ferrofluid due to the higher Prandtl number of kerosene. Thus, the far field boundary condition (12) are satisfied asymptotically and thus support the validity of the present numerical results.

5 Conclusions

The mixed convection on MHD flow and heat transfer over a moving flat plate in ferrofluids with uniform heat flux, slip effects and suction has been studied in this research. The governing boundary layer equations were solved numerically for both assisting and opposing flow regimes using the shooting method with the aid of *shootlib* function in Maple software. From the analysis of results, the following conclusions can be drawn:

- Triple (first, second and third) solutions exist for the assisting ($\omega > 0$) and opposing ($\omega < 0$) flow cases with $\omega_{c1} \leq \omega \leq \omega_{c2}$.
- The critical values ω_{c1} and ω_{c2} decrease for the skin friction coefficient and local Nusselt number when the magnetic and slip parameters are increasing.
- The stability analysis has revealed that the first solution is stable and physically realizable, while the second and third solutions are unstable and not physically realizable.
- It is found that on the first solution, the value of skin friction coefficient $\mathrm{Re}_x^{1/2} C_f$ increases with M but decreases with the increase of D. However, the local Nusselt number $\mathrm{Re}_x^{-1/2} \mathrm{Nu}_x$ increases with the increase of magnetic parameter M and slip parameter D.
- It is interesting to note that kerosene-based ferrofluid has higher skin friction and heat transfer rate compared to the water-based ferrofluid. The heat transfer of ferrofluids in the presence of magnetic fields could occur due to the changes in ferrofluids thermophysical properties such as thermal conductivity and heat capacity.
- Magnetite ferroparticle in the kerosene-based ferrofluid gives the highest heat transfer rate compared to the other two ferroparticles, which are cobalt ferrite and manganese-zinc ferrite.

References

1. Sheikholeslami, M., Rashidi, M.: Effect of space dependent magnetic field on free convection of Fe3O4-water nanofluid. J. Taiwan Inst. Chem. Eng. **56**, 6–15 (2015)
2. Rashad, A.M.: Impact of anisotropic slip on transient three dimensional MHD flow of ferrofluid over an inclined radiate stretching surface. J. Egyptian Math. Soc. **25**(2), 230–237 (2017)
3. Ramli, N., Ahmad, S.: Stability analysis of MHD mixed convection flow over a moving flat plate in ferrofluids with thermal radiation, suction and second-order slip effects: heat flux case. In: AIP Conference Proceedings, vol. 2184, 060036 (2019)
4. Gan Jia Gui, N., Stanley, C., Nguyen, N.-T., Rosengarten, G.: Ferrofluids for heat transfer enhancement under an external magnetic field. Int. J. Heat Mass Transf. **123**, 110–121 (2018)
5. Doganay, S., Alsangur, R., Turgut, A.: Effect of external magnetic field on thermal conductivity and viscosity of magnetic nanofluids: a review. Mater. Res. Express **6**, 112003 (2019)
6. Jha, B.K., Aina, B.: Role of suction/injection on steady fully developed mixed convection flow in a vertical parallel plate microchannel. Ain Shams Eng. J. **9**(4), 747–755 (2018)
7. Gul, A., Khan, I., Shafie, S., Khalid, A., Khan, A.: Heat transfer in MHD mixed convection flow of a ferrofluid along a vertical channel. PloS One **10**(11), e0141213 (2015)

8. Selimefendigil, F., Oztop, H.F., Chamka, A.J.: Mixed convection of pulsating ferrofluid flow over a backward-facing step. Iran J. Sci. Technol. Trans. Mech Eng. **43**, 593–612 (2019)
9. Jamaludin, A., Naganthran, K., Nazar, R., Pop, I.: Thermal radiation and MHD effects in the mixed convection flow of Fe3O4-water ferrofluid towards a nonlinearly moving surface. Processes **8**(1), 95 (2020)
10. Jaluria, Y.: Transport from continuously moving materials undergoing thermal processing. Ann. Rev. Heat Transf. **4**, 187–245 (1992)
11. Sakiadis, B.C.: Boundary-layer behavior on continuous solid surfaces: I. Boundary-layer equations for two-dimensional and axisymmetric flow. AIChE J. **7**, 26–28 (1961a)
12. Sakiadis, B.C.: Boundary-layer behavior on continuous solid surfaces: II. The boundary layer on a continuous flat surface. AIChE J. **7**, 221–225 (1961b)
13. Sakiadis, B.C.: Boundary-layer behavior on continuous solid surfaces: III. The boundary layer on a continuous cylindrical surface. AIChE J. **7**, 467–472 (1961c)
14. Pop, I., Gorla, R.S.R., Rashidi, M.: The effect of variable viscosity on flow and heat transfer to a continuous moving flat plate. Int. J. Eng. Sci. **30**(1), 1–6 (1992)
15. Pantokratoras, A.: Further results on the variable viscosity on flow and heat transfer to a continuous moving flat plate. Int. J. Eng. Sci. **42**, 1891–1896 (2004)
16. Makinde, O.D.: On MHD heat and mass transfer over a moving vertical plate with a convective surface boundary condition. Can. J. Chem. Eng. **88**, 983–990 (2010)
17. Mustafa, M., Hayat, T., Pop, I., Aziz, A.: Unsteady boundary layer flow of a Casson fluid due to an impulsively started moving flat plate. Heat Transf. Asian Res. **40**, 563–576 (2011)
18. Motsumi, T.G., Makinde, O.D.: Effects of thermal radiation and viscous dissipation on boundary layer flow of nanofluids over a permeable moving flat plate. Physica Scripta **86**, 045003 (2012)
19. Uddin, M.J., Khan, W.A., Ismail, A.I.M.: Similarity solution of double diffusion free convective flow over a moving vertical flat plate with convective boundary condition. Ain Shams Eng. J. **6**(3), 1105–1112 (2015)
20. Roşca, A.V., Uddin, M.J., Pop, I.: Boundary layer flow over a moving vertical flat plate with convective thermal boundary condition. Bull. Mal. Math. Sci. Soc. **39**(4), 1287–1306 (2016)
21. Khazayinejad, M., Hatami, M., Jing, D., Khaki, M., Domairry, G.: Boundary layer flow analysis of a nanofluid past a porous moving semi-infinite flat plate by optimal collocation method. Powder Technol. **301**, 34–43 (2016)
22. Hafidzuddin, E.H., Nazar, R., Arifin, N.M., Pop, I.: Effects of anisotropic slip on three-dimensional stagnation-point flow past a permeable moving surface. Eur. J. Mech. B Fluids **65**, 515–521 (2017)
23. Ramli, N., Ahmad, S., Pop, I.: MHD forced convection flow and heat transfer of ferrofluids over a moving flat plate with uniform heat flux and second-order slip effects. Scientia Iranica B **25**(4), 2186–2197 (2018)
24. Tiwari, R.K., Das, M.K.: Heat transfer augmentation in a two-sided lid- driven differentially heated square cavity utilizing nanofluids. Int. J. Heat Mass Transf. **50**, 2002–2018 (2007)
25. Weidman, P.D., Kubitschek, D.G., Davis, A.M.J.: The effect of transpiration on self-similar boundary layer flow over moving surfaces. Int. J. Eng. Sci. **44**, 730–737 (2006)
26. Roşca, A.V., Pop, I.: Flow and heat transfer over a vertical permeable stretching/shrinking sheet with a second order slip. Int. J. Heat Mass Transf. **60**, 355–364 (2013)
27. Harris, S.D., Ingham, D.B., Pop, I.: Mixed convection boundary-layer flow near the stagnation point on a vertical surface in a porous medium: Brinkman Model with slip. Transp. Porous Media **77**, 267–285 (2009)
28. Cortell, R.: Numerical solutions of the classical Blasius flat-plate problem. Appl. Math. Comput. **170**, 706–710 (2005)
29. Yazdi, M.H., Abdullah, S., Hashim, I., Sopian, K.: Effects of viscous dissipation on the slip MHD flow and heat transfer past a permeable surface with convective boundary conditions. Energies **4**, 2273–2294 (2011)
30. Khan, W.A., Khan, Z.H., Haq, R.U.: Flow and heat transfer of ferrofluids over a flat plate with uniform heat flux. Eur. Phys. J. Plus **130**(86), 1–10 (2015)

Data Science and Optimization of Complex Systems

Modeling and Analysis of the Dengue Activity in Baguio City Using Two-Mode and One-Mode Networks

Rosel R. Oryan, Joel M. Addawe, and Donnabel Tubera-Panes

Abstract In this study, the two-mode network is used to model and investigate a given set of data regarding the dengue cases that have been reported across the 128 *barangays* of Baguio City, Philippines for the years 2010–2018. Three different structure perspectives are used: time-location, time, and location-age, from which respective models were patterned from. The study proposes variants of the Newman and weighted Newman projection method to aid in preserving connectivity information and is compared with Newman, and weighted Newman. Measures of *global clustering coefficient* and *two-mode degree* are applied upon the two-mode models while *strength*, *closeness*, and *betweenness* measures are used in the analysis of the different one-mode projections. These models have shown that *barangays* labeled 98, 85, and 80 have the most concentrated and repeated dengue activity based from the *barangay*-month model. For the month-year model, July to September and February to April were shown to be the peak and hollow seasons of dengue respectively. From the age-*barangay* and age-district models, age-group of 17–22 years were shown to be the most common among dengue patients.

Keywords Bimodal · Network modelling · Dengue

1 Introduction

The Dengue Virus is transmitted to humans through the bites of an infected female mosquito mainly by the aedes genus of mosquitoes. It is not contagious however the virus can be spread to uninfected mosquitoes through blood contact (ingestion) when a virus carrier is bitten. It takes the virus 4–10 days to incubate before the

R. R. Oryan · J. M. Addawe (✉)
Department of Mathematics and Computer Science, College of Science,
University of the Philippines Baguio, Baguio, Philippines
e-mail: jmaddawe@up.edu.ph
URL: https://web.upb.edu.ph/jmaddawe

D. Tubera-Panes
City Health Services Office, Baguio, Philippines

M. H. Mohd et al. (eds.), *Modelling, Simulation and Applications of Complex Systems*,
Springer Proceedings in Mathematics & Statistics 359,
https://doi.org/10.1007/978-981-16-2629-6_13

manifestation of Dengue Fever. The symptoms last for 2–7 days and may include high fever around 40°C, pain behind the eyes, muscle and joint pains, nausea, vomiting, and rashes. Dengue fever seldom causes deaths but it may develop into the more potentially deadly Severe Dengue or Dengue Hemorrhagic Fever which may cause severe bleeding and organ impairment [1, 2]. *Aedes Aegypti* mosquitoes are the main dengue virus vectors. They breed in enclosed environments, usually indoors, making them more likely to inhabit these places which are less affected by shifts in climate and thus allowing for a longer life span. A secondary vector is the *Aedes Albopictus* mosquito which is a forest dweller that has successfully adapted into human environments. This species produces eggs that can endure very dry conditions and survive many months without water. Both species are widely spread throughout the world [2, 3].

Dengue is more prevalent in tropical and subtropical countries and this includes the Philippines. The World Health Organization (WHO) identified that 2016 experienced large dengue outbreaks worldwide wherein the Philippines reported at least 176,411 cases [1, 2]. Different mathematical models were already formulated to describe the transmission of dengue virus. In Baguio City, the summer capital of the Philippines that is located at the northern part of Luzon, the City Health Services Office reported a 400% increase with 1,875 dengue cases recorded in the first six months of 2016 compared to the 239 cases of the first six months of 2015 [4]. The study of [5] presented a deterministic approach and [6, 7] presented stochastic approaches in the mathematical modeling of dengue dynamics using data from Baguio City, Philippines.

To further contribute to the understanding of dengue dynamics in the Philippines, we use the concept of two-mode network model to analyse the dengue incidence data from 2010 to 2018 in Baguio City, Philippines. Networks illustrate relationships and interactions between objects such that the *nodes* of a network represent objects and its *links* represent connections or correlation between objects [8]. One-mode networks (unimodal networks) then are networks with one set of nodes with a set of links connecting these nodes together while two-mode (bimodal) networks have two distinct set of nodes with links that connect nodes from the first set to the second.

One way of analysing networks is through measures that examine connections between nodes such as clustering, another is through centrality measures such as degree, betweenness, and closeness, to examine node positions in the graph. The studies of [9–11] have proposed a generalized clustering measure for weighted networks as well as degree (strength), closeness, and betweenness for weighted networks. Furthermore, the study of [9] has also proposed a clustering measure specifically for two-mode networks.

Two-mode networks have been used to investigate different relationships between entities to create realistic models from real-world datasets [12–18]. The studies by [9, 12, 13, 19] have highlighted the main disadvantages of two-mode network that lacks appropriate analytical tools hence the need for its conversion to unimodal networks using the method of projection before evaluation is performed, thus may result to loss of some information. The studies done by [20, 21] have compared different ways of transforming weights from two-mode networks to projected links (such as the binary, sum, Newman, and weighted Newman methods) in order to preserve lost information.

In this paper, we model and analyse the dynamics on how dengue spreads by constructing a bimodal network and use the data on dengue incidence in Baguio City Philippines. Using the actual dataset, we construct a two-mode network and project it into a one-mode network and perform analysis using Newman, weighted-Newman and proposed variants of the weighted-Newman projection methods.

2 Methodology

In this study, we considered recorded data that contains monthly dengue cases from the 128 different *barangays*, being the smallest political unit in the City of Baguio, Philippines. The dataset consists of raw data on the dengue reports filed by hospitals around Baguio City from January 2010 to May 2018. The dengue data were coded and only the set containing information mainly on the date of admission, age of patient, and address were extracted. The date of admission consists of a specified month and year, age of patient, while the address only indicates the *barangay* of the patient's residence (or the *barangay* it is closest to). From the three types of information namely: location, time (given in month or year), and patient age, the *barangay* is then grouped into different districts. Also, for certain locations that are considered rare (*barangay* which are not specified under the 128 official list) are combined with one of the closest *barangay* listed in the area. In this study, we choose a combination in which a type of information will be arranged according to importance. In particular, *barangay* will be prioritized over month, month over year, and age over *barangay*.

The given dataset can be viewed through varied perspectives from which specific models can be derived based on the different combinations of data points. Two-mode network models similar to the study of [12] are then constructed. In this study, we only consider using a time-location perspective, time perspective, and location-age perspective that are represented by bimodal graphs M with the general form of $M = \{V_{primary}, V_{secondary}, E\}$ where

- $V_{primary}$ is the set of vertices representing data points and the focus of analysis
- $V_{secondary}$ is the set of vertices secondary to $V_{primary}$
- E is the set of edges connecting a vertex u to v with its weight being the incidence of dengue, u with respect to v where $u \in V_{primary}$ and $v \in V_{secondary}$.

Given the different levels of importance, we enumerate the two-mode model specifications in Table 1.

In addition to the four two-mode models, a one-mode district connectivity model would be used to check which districts are likely to influence the dengue activity in the city. This model is represented by the simple graph $M_5 = \{V_{district}, E_{border}\}$ where links $(u, v) \in E_{border}$ exists if some district u shares a border (geographically adjacent) with another district v where $u, v \in V_{district}$.

Table 1 Two-mode models

Perspective	Model	Model representation
Location-time	*barangay*-month	$M_1 = \{V_{barangay}, V_{month}, E_1\}$
Time	month-year	$M_2 = \{V_{month}, V_{year}, E_2\}$
Age-location	age-*barangay*	$M_3 = \{V_{age}, V_{barangay}, E_3\}$
	age-district	$M_4 = \{V_{age}, V_{district}, E_4\}$

2.1 Projections

In order to appropriately apply certain centrality measures to examine the models in their respective perspectives, the models are to be projected in the form of a one-mode network.

A two-mode network has two distinct set of nodes and the projection of a bimodal to a unimodal network involves choosing one set of nodes and removing the other while trying to maintain the respective connections (for convention, primary nodes are retained and secondary nodes are discarded).

Suppose the two-mode network is given by $G = (V_{primary}, V_{secondary}, E)$, the resulting one mode network is denoted by $G_{projected} = (V_{primary}, E_{projected})$ where the link $(i, j) \in E_{projected}$, can only be established given that the links (i, p) and (j, p) exist such that $p \in V_{secondary}$ and $i, j \in V_{primary}$.

Projecting two-mode to one-mode means correlating primary vertices to one another while removing the secondary vertices that exist between these primary vertices. In other words, we are trying to relate two primary nodes i to j by considering the connections between them. For simplicity, we consider a portion of the actual dataset to visualize the one-mode network produced by the projection of a complex two-mode network. In Fig. 1, we consider humans with 0–12 years of age in the first district with six barangays.

To carry-over certain information (such as weights and co-occurrence) as weights w_{ij} into the projected link (i, j), methods of weight transformation as summarized in Table 2 will be implemented.

To illustrate, consider p is a node joining nodes i and j. In Newman method, $N_p - 1$ means considering all the connections of secondary node p to primary nodes other than its established connection to node i. We then assume that these connections are weighted. Hence to carry over these weights, instead of using N_p (degree of p) we use the strength of p or the sum of all weighted connections of p. Since we are considering all of p's connections other than it's connection to node i, we subtract the value of the edge/connection of node i to p, hence we come up with strength of p or $\left(\sum_q w_{p,q}\right) - w_{i,p}$. Since we are trying to connect i to j via node p, then we must consider the connection of p to j. Returning to $1/(N_p - 1)$, as a whole, it means considering the value of a certain connection in relation to the rest of node p's connections. If we let this connection become the edge from node p to j, then it would make sense since it means finding the significance of the connection from p

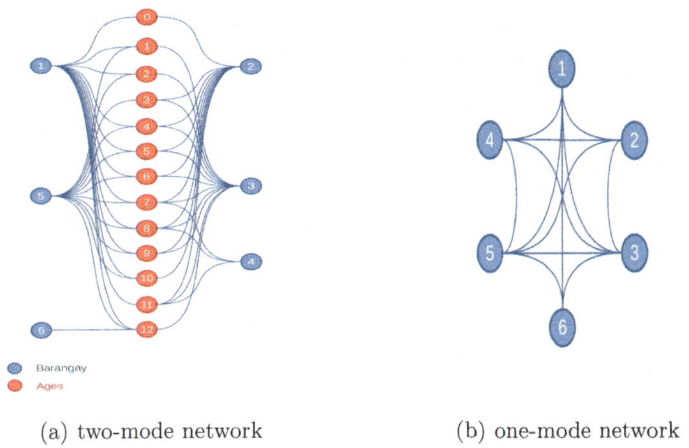

Fig. 1 Projecting two-mode network to one-mode for the first district having six *barangays* with dengue cases aged 0–12 years

to j in comparison to all other connections other than i to p. This relationship can be written as the First-variant method.

The Binary method considers co-occurrences, Sum method considers both weight and co-occurence, Newman considers neighborhood connectivity, and weighted-Newman method generalizes the Newman method while considering edge weights.

However, the Newman methods do not consider specific connections. While Newman methods identify the number of neighbors attached to secondary nodes, it does not differentiate between neighbors. Hence, we propose some other weighted projection methods that aim to carry-over neighbor identity.

From the given two-mode network, its projected one-mode network would contain edge (i, j) with weight $w_{i,j}$ given by

Table 2 Projection methods

Method	$w_{i,j}$
Binary	$\sum_p 1$
Sum [9]	$\sum_p w_{i,p}$
Newman [22]	$\sum_p \frac{1}{N_p - 1}$
Weighted-Newman [9]	$\sum_p \frac{w_{i,p}}{N_p - 1}$
Proposed method	
First-variant	$\sum_p \frac{w_{p,j}}{\left(\sum_q w_{p,q}\right) - w_{i,p}}$
Second-variant	$\sum_p \frac{w_{i,p} \cdot w_{p,j}}{\left(\sum_q w_{p,q}\right) - w_{i,p}}$

$$w_{i,j} = \sum_{p} \frac{w_{p,j}}{\left(\sum_q w_{p,q}\right) - w_{i,p}} \quad \text{(First-variant)} \tag{1}$$

where $\sum_q w_{p,q}$ is the strength of node p, $w_{p,q}$ is the weight of the edge (p, q), $w_{i,p}$ is the weight of edge (i, p), and $w_{p,j}$ is the weight of the edge (p, j) such that $i, j, q \in V_{primary}$, $i \neq j$, and $p \in V_{secondary}$.

Equation (1) is then modified to consider the general weighted neighborhood to get

$$w_{i,j} = \sum_{p} \frac{w_{i,p} \cdot w_{p,j}}{\left(\sum_q w_{p,q}\right) - w_{i,p}} \quad \text{(Second-variant)} \tag{2}$$

2.2 Measures

The projected models are evaluated using different measures to determine whether specific trends exist throughout the selected set of data or if certain nodes in each model hold any significant differences from the rest.

Table 3 Measures for one-mode network

Method	Equation	Description
Degree and strength [23]	$deg(v) = n$ (no. of links) $str(v) = \sum_n w_{v,n}$	Degree of a vertex v. Strength of vertex v is the sum of the weights w of all edges connected to v
Closeness [24]	$C_C(v) = \left[\sum_i^N d(v, i)\right]^{-1}$	N is the total number of vertices in G and $d(v, i)$ is the distance of the geodesic (shortest path) from node v to any other node $i \in V$, $i \neq v$
Betweenness [22, 24, 25]	$C_B(i) = \sum_j^N \sum_k^N \frac{g_{jk}(i)}{g_{jk}}$	N is the number of vertices in the graph, g_{jk} is the total number of geodesics from vertex j to vertex k, and $g_{jk}(i)$ is the number of geodesics from j to k that passes through vertex i, $j \neq k$
GCC [9, 11]	$CC_{global}(G) = \frac{3\|\tau_\Delta\|}{\|\tau\|}$	Global clustering coefficient (GCC) used in [11] and two-mode GCC in [9], where $\|\tau\|$-number set of triplets, $\|\tau_\Delta\|$-number of set of closed triplets

As the projections are in the form of one-mode networks, then given any one-mode network represented by the graph $G = \{V, E\}$ with vertices V and edges E, the following measures in Table 3 are used.

3 Results and Discussion

Each model are projected using the methods of Newman, weighted-Newman, and the proposed First-variant and Second-variant to have a one-mode network. The networks were then evaluated using centrality measures of degree, strength, closeness, betweenness, and global clustering coefficient (GCC). The four main models (*barangay*-month, month-year, age-*barangay*, and age-district) were then evaluated. In this study, GCC and degree measures were used to evaluate the original two-mode network while strength, closeness, and betweenness measures were used on the one-mode network projections. Binary and Sum methods assign weights by accumulating connections. Newman method allows the weights to be distributed across links based on secondary node connections, that is, the Newman method only considers the number of connections as secondary node to other primary nodes. The proposed projection method distributes weights across projected links based on strength of connections where stronger connections are allotted more weight, weaker connections are given less.

3.1 Two-Mode Network

Due to the complexity of the two-mode network, we only consider the degree measure and GCC to evaluate the network for the different models.

3.1.1 Degree Measure

The degree measure of a network is a feature of counting the number of links. For a two-mode network, counting the links can be easily be carried to the projected one-mode network. After converting a two-mode network to one-mode using the different methods of projection, the generated one-mode network will result in having the same one-mode degrees. This result is due to each generated one-mode network having the same connections, but only with different magnitudes or weights. Since the degree is not weighted, it remains constant for each projected one-mode network, thus we consider the two-mode network. Based on our simulations, no significant changes are seen from the computed degree measures between two-mode and one-mode, thus we omit to include the table of generated values.

Table 4 Two-mode global clustering coefficients

Models	Binary	Arithmetic	Geometric	Minimum	Maximum	Avg
barangay-month	0.99765	0.99805	0.99805	0.99791	0.99799	0.99793
month-year	1.0	1.0	1.0	1.0	1.0	1.0
age-*barangay*	0.99288	0.99280	0.99355	0.99392	0.99169	0.99297
age-district	0.98448	0.98577	0.98585	0.98495	0.98562	0.98533

3.1.2 Global Clustering Coefficient

From the different approaches (Binary, Arithmetic Mean, Geometric, Maximum, Minimum) in computing GCC as shown in Table 4, the average GCC values that are very close to 1 are: *barangay*-month model (0.99793), month-year model (1.0), age-*barangay* model (0.99297), and age-district model (0.98533). All show the highly interconnected nature of the dataset. The GCC of the month-year model having all values equal to one indicates that a dengue case is reported in each month throughout the years in the dataset.

3.2 One-Mode Network

In most networks, it is preferable to analyze networks in their original form. Due to limited number of methods that exist for analysing two-mode networks, we transform a two-mode network into a one-mode network by projection to analyze these networks. For the projected one-mode network, we use strength, closeness and betweenness measures to analyse the different models.

3.2.1 M1: *Barangay*-Month Model

Newman and First-variant strengths shown in Fig. 2(a) and (c) respectively achieved the total number of links of each primary node from the original two-mode form. This reveals the frequency of being affected by dengue which it was observed that *barangay* 98 has been affected the most repeatedly (has reported cases during 85 out of 105 months), followed by *barangays* 85 (73 months) and 80 (72 months).

The weighted-Newman and Second-variant strengths shown in Fig. 2(b) and (d) respectively achieved the total sum of weights of each primary node from the original two-mode network form. This shows that the total dengue incidence in each *barangay* throughout the dataset timeframe from which it was observed is that *barangay* 98 has reported the highest number of Dengue incidence with 667 cases followed by *barangay* 80 with 406 cases. In the *barangay*-month model, we note that *barangays*

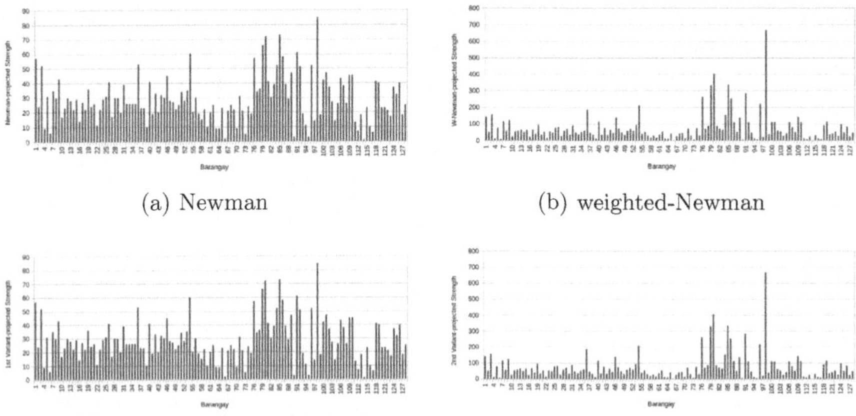

(a) Newman (b) weighted-Newman

(c) First-variant (d) Second-variant

Fig. 2 Values for measure of strength on the one-mode network generated using (a) Newman, (b) weighted-Newman, (c) First-variant and (d) Second-variant projections for the *barangay*-month Model

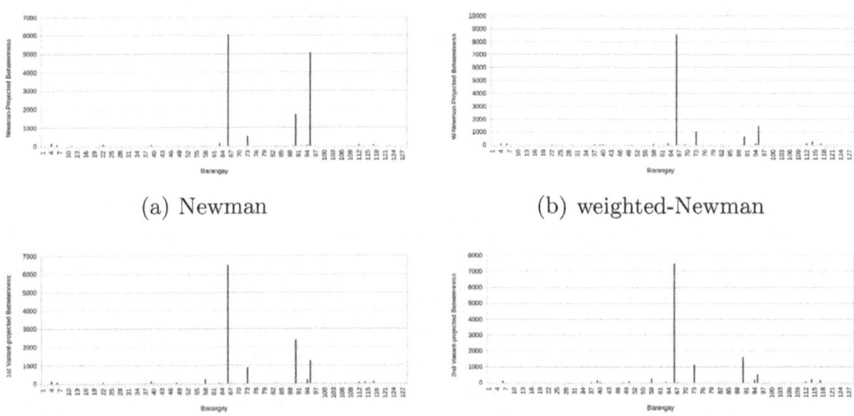

(a) Newman (b) weighted-Newman

(c) First-variant (d) Second-variant

Fig. 3 Betweenness measures for one-mode network generated using (a) Newman, (b) weighted-Newman, (c) First-variant and (d) Second-variant projections for the *barangay*-month Model

98, 85, 80, and 79 have the highest weighted-Newman and Second-variant strengths of 667, 336, 406, and 331 respectively.

A combination of measures can give significant meanings. For instance, nodes with a high closeness measure value mean that these nodes have one of the most number but also the weakest connections. Figures 3(a)–(d) show that nodes with high betweenness imply that the graph converges towards that node which can be due to having the same conditions. In Fig. 2, we noted that *barangays* 98, 85, 80, and 79 with the highest weighted-Newman and Second-variant strengths of 667, 336, 406, and

Table 5 Computed measure values from the one-mode network generated by projection methods applied on the *barangay*-month model

Projection methods	Strength		Closeness		Betweenness	
	Brgy	Value	Brgy	Value	Brgy	Value
Newman	98	85	66	0.43510	66	6059.18
	85	73	95	0.38584	95	5036.88
	80	72	90	0.37910	90	1715.87
	79	66	27	0.34894	73	529.688
Weighted-Newman	98	667	66	0.37815	66	8608.02
	80	406	90	0.29556	95	1472.65
	85	336	58	0.28776	73	1043.52
	79	331	97	0.28322	90	648.199
First-variant	98	85	66	0.686529	66	6485.47
	85	73	95	0.664427	95	2384.12
	80	72	90	0.652798	90	1231.72
	79	66	112	0.617615	73	863.657
Second-variant	98	667	66	0.573983	66	7477.05
	80	667	90	0.516665	90	1604.04
	85	667	111	0.509235	73	1126.8
	79	667	58	0.508047	95	534.056

331 respectively have two-mode network degree measures of 85, 73, 72, and 66, and weighted-Newman closeness and betweenness values that are among the lowest, with weighted-Newman closeness of 0.09100 for *barangays* 98, 85, and 80, and 0.06507 for *barangay* 79. Furthermore, computed betweenness of 3.24045 for *barangay* 98 and zero for *barangays* 85, 80, and 79. Based on the indicated measures, these *barangays* observed to be the most frequently affected by dengue and when affected, the number of cases are among the highest. Second-variant closeness, however, has *barangays* 66, 90, 111, 58, and 112 as its highest with values of 0.57398, 0.51667, 0.50924, 0.50805, and 0.50238 respectively which means that these *barangays* are locations with the weakest but with the highest number of connections.

To show results of our simulations, a summary is shown in Table 5. The first four from the computed measures for strength, closeness and betweenness are included in the list.

3.2.2 M2: Month-Year Model

The impact of these distributions are apparently observed from the closeness measures given in Fig. 4 and betweenness measures given in Fig. 5. Both measures deal with shortest path. For instance, Binary, Sum, Newman, and weighted-Newman betweenness values are zero, which means that based on the projected weights from

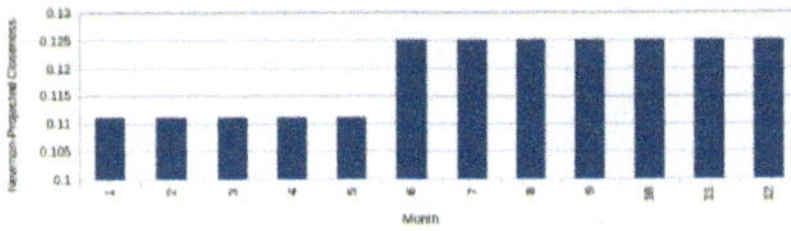

(a) Newman-projected closeness.

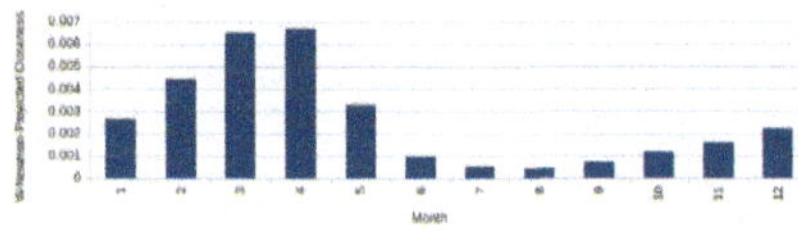

(b) Weighted-Newman-projected closeness

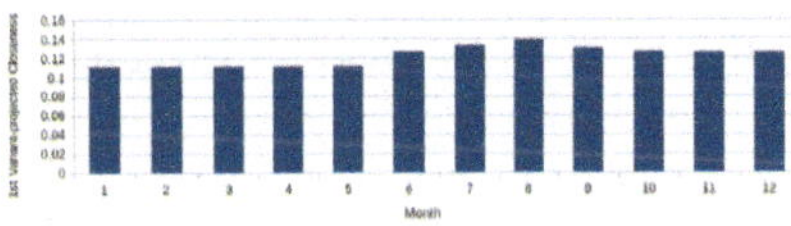

(c) First-variant-projected closeness.

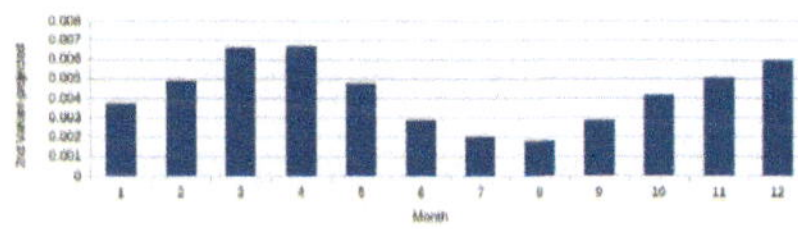

(d) Second-variant-projected closeness

Fig. 4 Computed closeness measures of the one-mode network generated by the month-year Model

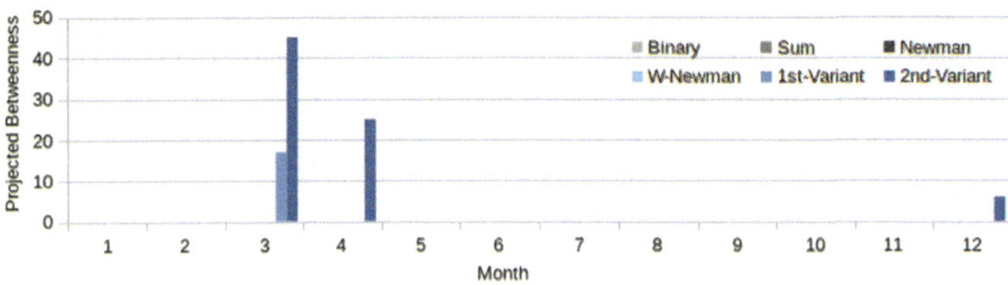

Fig. 5 Betweenness measures for the month-year Model: Binary, Sum, Newman, weighted-Newman, First-variant and Second-variant

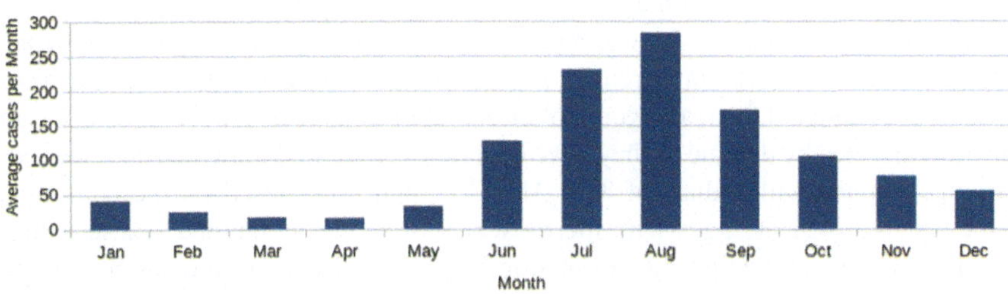

Fig. 6 Average case per month

these methods, the shortest path between each node is only the direct link connecting the nodes. With the proposed methods, March was found to be with betweenness measure value of 17 from the First-variant projection method while March, April, and December with betweenness values of 45, 25, and 6 respectively for the Second-variant.

The distribution of dengue activity throughout the year in Baguio City is not constant as shown in Fig. 6. The year can be divided into quarters where the first quarter would be from January to March, the second quarter from April to June, third quarter from July to September, and the fourth quarter from October to December.

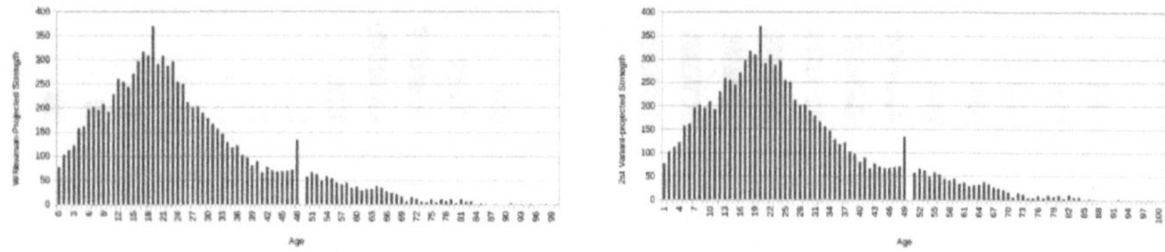

(a) age-*barangay* (weighted-Newman) (b) age-*barangay* (Second-variant).

Fig. 7 Strengths of age-*barangay* model

In the case of strength, we can instead take the ACPM through strength measures.[1] The lowest point of dengue activity is during April (with an ACPM of 16) and from there, steadily increases (by 102% from April to May, then by 279% from May to June) and enters the months of July to September when activity is highest (as indicated by its weighted-Newman or second-variant strength). These months can then be interpreted to be the peak season of dengue.

3.2.3 M3: Age-*Barangay* Model

In general, most recorded dengue cases in Baguio involve people with ages between 12–25 years. Computations show the highest weighted-Newman and Second-variant strengths above 45 from the age-*barangay* model shown in Fig. 7(a)–(b). The dataset does not contain patients with ages 49, 87, 89, 90, 94, 96, and 99.

From the age-*barangay* model, an age-group with the highest weighted-Newman strength and second-variant strength means that it is observed to have the most patients accumulated from a large number of barangays. Ages 19, 17, 18, and 21 are the most affected by dengue as they have the highest weighted-Newman and Second-variant strengths with values of 370, 317, 309, and 309. Based on the computed values, ages 17–21 are ages with the most risk of acquiring dengue. These values may be dependent on which age group dominates the majority of the population. To show results of our simulations, a summary is shown in Table 6. The first four from the computed measures for strength, closeness and betweenness are included in the list.

[1]ACPM. Average-case-per-month can be obtained by dividing weighted-Newman strengths over Newman strengths or Second-variant over First-variant strengths. It may also be approximated with sum over binary strengths.

Table 6 First four measure values generated from the one-mode network for the age-barangay model

Projection methods	Strength		Closeness		Betweenness	
	Age	Value	Age	Value	Age	Value
Newman	19	108	86	0.60319	86	5145.63
	21	100	93	0.48393	93	319.267
	17	94	88	0.44647	67	144.839
	16	93	100	0.44647	100	114.376
Weighted-Newman	19	370	86	0.60319	97	927.747
	17	317	93	0.47759	86	850.113
	21	309	88	0.43747	95	755.0
	18	309	100	0.43747	93	635.585
First-variant	19	108	17	1.02359	86	2190.01
	21	100	6	1.01860	93	906.412
	17	94	24	1.01704	100	863.568
	16	93	47	1.01472	91	293.941
Second-variant	19	370	86	0.99396	86	932.659
	17	317	69	0.97746	93	796.468
	21	309	74	0.95953	100	710.205
	18	309	73	0.95864	95	528.0

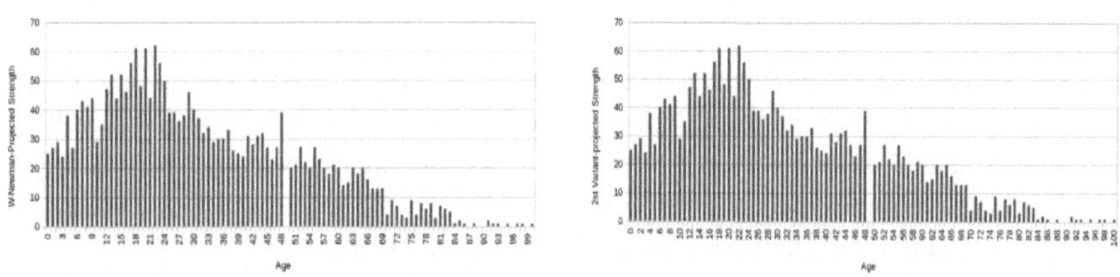

(a) age-district (weighted-Newman) (b) age-district (Second-variant)

Fig. 8 Strengths of age-district model

3.2.4 M4: Age-District Model

The 128 different *barangays* in the City of Baguio is grouped into different districts. Each district can have between 5 and 7 barangays as indicated in Table 8. Most recorded dengue cases in Baguio involve people with ages between 12–25 years. Computations show the highest weighted-Newman and Second-variant strengths above 45, and strength (incidence) above 240 from the age-district model shown in Fig. 8(a)–(b).

From the age-district model, an age-group with the highest strengths, such as ages 22, 18, and 20 are age-groups observed to have the most patients from majority of the districts.

Table 7 First four measure values generated from the one-mode network for the age-district model

Projection methods	Strength		Closeness		Betweenness	
	Age	Value	Age	Value	Age	Value
Newman	22	20	93	0.69489	93	5214.39
	21	20	86	0.68224	86	1314.54
	20	20	98	0.64276	98	137.067
	19	20	97	0.64247	84	108.0
Weighted-Newman	22	62	93	0.69489	86	2528.74
	18	61	86	0.68224	93	2344.97
	20	61	98	0.64276	98	776.476
	17	56	97	0.64247	97	525.633
First-variant	22	20	92	0.97689	93	2976.41
	21	20	62	0.88941	98	1262.73
	20	20	65	0.88606	100	1140.27
	19	20	63	0.88464	86	772.104
Second-variant	22	62	93	0.97689	93	1693.2
	18	61	65	0.88606	86	1271.44
	20	61	54	0.87896	98	1176.27
	17	56	98	0.87492	100	934.956

The trend of incline and decline in the number of patients across age-groups might possibly be attributed to the risk of contracting dengue. Based on the computed values, we can see that 17–22 are ages with the most risk of acquiring dengue. Results of our simulations is summarized in Table 7. The first four from the computed measures for strength, closeness and betweenness are included in the list.

3.2.5 M5: District Connectivity

The district connectivity model illustrates the geographical layout of the districts through a one-mode network represented by $M_5 = \{V_{district}, E_{border}\}$ where

- $V_{district}$ is the set of districts within Baguio City,
- E_{border} is the set of edges between $V_{district}$ vertices,

such that an edge $(u, v) \in E_{border}$ exists if the vertices $u, v \in V_{district}$ are geographically adjacent. Two distinct districts are considered connected if they share a border. Figure 9 illustrates the district connectivity network that contains the different *barangays*.

The information derived from the analysis about districts can supplement the results of the four perspectives. It aims to provide a clearer understanding of incidence distribution across districts and to possibly support previous inferences. A summary

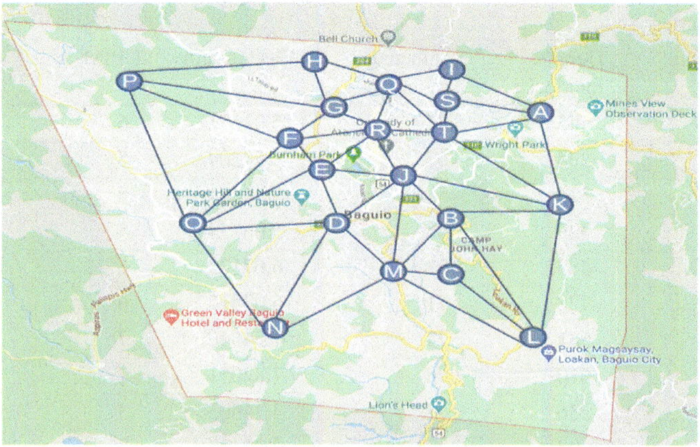

District	Barangay
A	1-6
B	7-11
C	12-17
D	18-24
E	25-32
F	33-39
G	41-47
H	48-55
I	56-62
J	63-67
K	69-73
L	74-79
M	80-84
N	85-90
O	91-97
P	98-103
Q	104-109
R	110-117
S	118-123
T	124-128

Fig. 9 District connectivity model on the actual location of twenty districts in the City of Baguio

on the coded districts located in Baguio City is given in Table 8, where the case-per-infected-month (CPIM) is the approximated number of cases during months when each district is affected by dengue and is rounded down from cases per month.

Based on the district-connectivity model, Fig. 10 shows the locations with the greatest one-mode degree. The values are 7 for district J and 6 for M, Q, R, and T of which districts J, R, and T were found to be with the highest closeness values of 0.02941, 0.02857, and 0.02778 that are observed to be the centers, and with district J having the highest betweenness of 40.472, which is seen to be the most accessible from all other districts.

Degree results indicate that Districts M, N, O and P are the most frequently affected by the disease with dengue activity spanning 89, 89, 81, and 92 months (links) respectively. In addition, strength results show that Districts M and P as well as N have had the most dengue incidents with 1111, 1040, and 892 cases in that order. Based on the additional district summaries, Districts M, N, and P have shown the greatest number of cases-per-month during months with dengue activity, with usually 12, 10, and 11 patients respectively during infected-months. Therefore, these three districts are shown to be the most dengue-active districts and during its infected-months, the number of cases is high. This one-mode model is supplemented with additional data summaries obtained from simple averaging and summations, resulting to obtaining the number of infected-months, total incidence of dengue, and CPIM of each district. From this, district J was found to be with the least dengue incidents of only 93 cases and is the least affected with 42 months of dengue activity, resulting to having the least CPIM of only 2 cases-per-month.

Table 8 District summaries

District	Barangay	Dengue incidence	Infected-months	Case/month	CPIM
A	1–6	445	74	6.01351	6
B	7–11	374	67	5.58209	5
C	12–17	324	57	5.68421	5
D	18–24	329	65	5.06154	5
E	25–32	487	73	6.67123	6
F	33–39	428	71	6.02817	6
G	41–47	581	77	7.54546	7
H	48–55	610	78	7.82051	7
I	56–62	236	63	3.74603	3
J	63–67	93	42	2.21429	2
K	69–73	208	55	3.78182	3
L	74–79	525	69	7.60870	7
M	80–84	1111	89	12.48315	12
N	85–90	892	89	10.02247	10
O	91–97	697	81	8.60494	8
P	98–103	1040	92	11.30435	11
Q	104–109	451	71	6.35211	6
R	110–117	210	57	3.68421	3
S	118–123	362	69	5.24638	5
T	124–128	299	61	4.90164	4

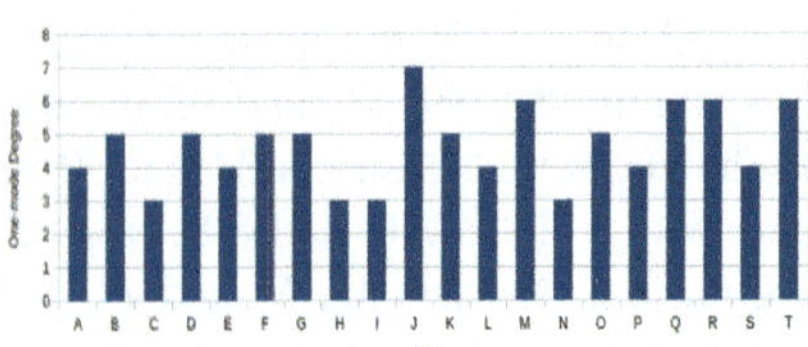

(a) degree

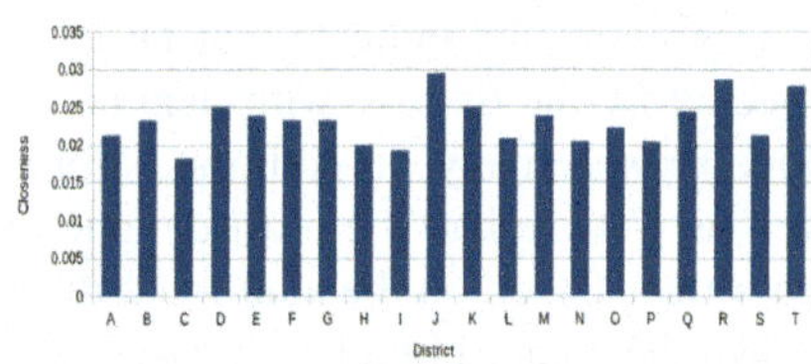

(b) closeness

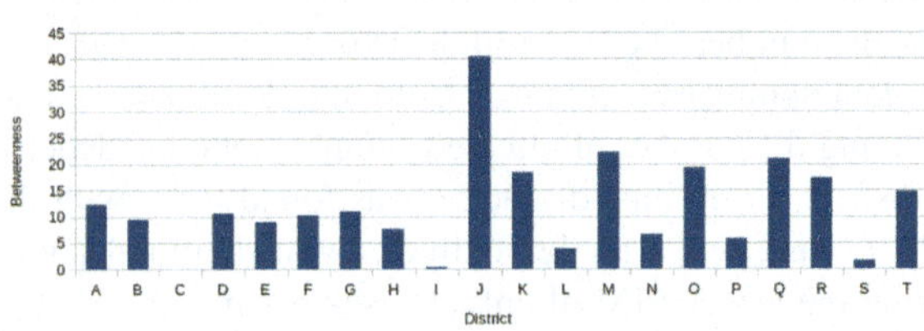

(c) betweenness

Fig. 10 Computed measures for district connectivity model

4 Conclusions and Recommendations

As demonstrated from the summarised data on the district connectivity model, information such as the total infected-months, incidence, and CPIM can be obtained through simpler procedures rather than network modeling and measuring. This information is obtained from bimodal network models by first storing data in the model and measuring, even requiring transformations (projection) to extract information from the model using an appropriate measure.

However, two-mode networks provide a more flexible container for information. Flexible, since it can be viewed from the two sets of nodes. From the discussion, our *barangay*-month model has focused on the *barangay* set of data points. The same model can be viewed using the month data points as we apply network measures and projections upon the secondary set of nodes. Information that can be gained from the two-mode network is only limited by the transformations applied upon it, and thus different projections can produce different information when measured. That is, through the *barangay*-month model, infected-months can be obtained from two-mode degree, Newman strengths and First-variant strengths, and incidence from weighted-Newman and Second-variant strengths.

Two-mode networks allow straightforward mapping between segregated data points and its projection to one-mode collapses certain information such as the summation of weights into a unimodal link from individual bimodal links or the removal of multiple links. One indicator on the loss of data is that the two-mode model, once transformed, cannot be returned to its original state. That is, when projecting using the different methods of weight transformations, co-occurent nodes and the links tied to these node are combined, resulting to merging of data.

Among the binary, sum, Newman, and weighted-Newman methods, the Newman and weighted-Newman carries over the most information (co-occurrence and neighborhood connectivity), while in the proposed First-variant and Second-variant models, both provide a proportioned division between neighbors as it considers neighborhood weights. From the age-*barangay* and age-district models, an age-group with the highest weighted-Newman strength and second-variant strength measures show similar age-group.

Different perspectives can translate to varied interpretation of results, that is, certain results can be directly used for interpretation while some require further external evaluations before a concrete conclusion can be made. To illustrate, the weighted-Newman strength values of the *barangay*-month model and month-year model result to the number of dengue cases per *barangay* for the *barangay*-month and dengue cases per month for the month-year model. The *barangay*-month model allows the use of the stated strength values as a dengue activity indicator for each *barangay*.

Acknowledgements The authors would like to thank the Baguio City Health Services Office for the data used in the paper gathered through the Philippine Integrated Disease Surveillance and Response program of DOH.

References

1. Dengue severe dengue frequently asked questions. World Health Organization (2017). http://www.who.int/denguecontrol/faq/en/index5.html
2. Dengue and severe dengue. World Health Organization (2018). http://www.who.int/en/news-room/fact-sheets/detail/dengue-and-severe-dengue
3. The Mosquito. World Health Organization (2018). http://www.who.int/denguecontrol/mosquito/en/
4. Official website of the city government of Baguio: Baguio dengue cases increase by 400 percent (2006). http://www.baguio.gov.ph/content/baguio-dengue-cases-increase-400-percent. Accessed Aug 2017
5. Libatique, C.P., Pajimola, A.J., Addawe, J.M.: Bifurcation analysis of dengue transmission model in Baguio City, Philippines. In: AIP Conference Proceedings, vol. 1905, p. 030023 (2017). https://doi.org/10.1063/1.5012169
6. Magsakay, C.B., De Vera, N.U., Libatique, C.P., Addawe, R.C., Addawe, J.M.: Treatment on outliers in UBJ-SARIMA models for forecasting dengue cases on age groups not eligible for vaccination in Baguio City, Philippines. In: AIP Conference Proceedings, vol. 1905, p. 050028 (2017).https://doi.org/10.1063/1.5012247
7. Addawe, R.C., Addawe, J.M., Magadia, J.C.: Optimization of seasonal ARIMA models using differential evolution - simulated annealing (DESA) algorithm in forecasting dengue cases in Baguio City. In: AIP Conference Proceedings, vol. 1776, p. 090021 (2016).https://doi.org/10.1063/1.4965385
8. Wasserman, S., Faust, K.: Social Network Analysis: Methods and Applications. Cambridge University Press, Cambridge (1994)
9. Opsahl, T.: Triadic closure in two-mode networks: redefining the global and local clustering coefficients. Soc. Netw. **35**(2), 159–167 (2013). https://doi.org/10.1016/j.socnet.2011.07.001
10. Opsahl, T., Agneessens, F., Skvoretz, J.: Node centrality in weighted networks: generalizing degree and shortest paths. Soc. Netw. **32**(3), 245–251 (2010). https://doi.org/10.1016/j.socnet.2010.03.006
11. Opsahl, T., Panzarasa, P.: Clustering in weighted networks. Soc. Netw. **31**(2), 155–163 (2009). https://doi.org/10.1016/j.socnet.2009.02.002
12. Malik, H.A.M., Mahesar, A.W., Abid, F., Wagas, A., Wahiddin, M.R.: Two-mode network modeling and analysis of dengue epidemic behavior in Gombak, Malaysia. Appl. Math. Model. **43**, 207–220 (2017). https://doi.org/10.1016/j.apm.2016.10.060
13. Padròn, B., Nogales, M., Traveset, A.: Alternative approaches of transforming bimodal into unimodal mutualistic networks. The usefulness of preserving weighted information. Basic Appl. Ecol. **12**(8), 713–721 (2011). https://doi.org/10.1016/j.baae.2011.09.004
14. Niekamp, A., Mercken, L.A.G., Hoebe, C.J.P.A., Dukers-Muijrers, N.H.T.M.: A sexual affiliation network of swingers, heterosexuals practicing risk behaviours that potentiate the spread of sexually transmitted infections: a two-mode approach. Soc. Netw. **35**(2), 223–236 (2013). https://doi.org/10.1016/j.socnet.2013.02.006
15. Snijders, T.A.B., Lomi, A., Torl, V.J.: A model for the multiplex dynamics of two-mode and one-mode networks, with an application to employment preference, friendship, and advice. Soc. Netw. **35**(2), 265–276 (2013). https://doi.org/10.1016/j.socnet.2012.05.005
16. Aksoy, S., Kolda, T.G., Pinar, A.: Measuring and modeling bipartite graphs with community structure. J. Complex Netw. **5**, 581–603 (2017)
17. Everett, M.G.: Centrality and the dual-projection approach for two-mode social network data. Methodol. Innov. **9** (2016). https://doi.org/10.1177/2059799116630662
18. Broccatelli, C., Everett, M., Koskinen, J.: Temporal dynamics in covert networks. Methodol. Innov. **9** (2016). https://doi.org/10.1177/2059799115622766
19. Malinick, T.E., Tindall, D.B., Diani, M.: Network centrality and social movement media coverage: a two-mode network analytic approach. Soc. Netw. **35**(2), 148–158 (2013). https://doi.org/10.1016/j.socnet.2011.10.005

20. Saracco, F., Straka, M.J., Di Clemente, R., Gabrielli, A., Caldarelli, G., Squartini, T.: Inferring monopartite projections of bipartite networks: an entropy-based approach. New J. Phys. **19**(5), 053022 (2017). http://stacks.iop.org/1367-2630/19/i=5/a=053022
21. Neal, Z.: The backbone of bipartite projections: inferring relationships from co-authorship, co-sponsorship, co-attendance and other co-behaviors. Soc. Netw. **39**, 84–97 (2014). https://doi.org/10.1016/j.socnet.2014.06.001
22. Newman, M.E.J.: Scientific collaboration networks. II. Shortest paths, weighted networks, and centrality. Phys. Rev. E **64**(1), 016132016132 (2001). https://doi.org/10.1103/PhysRevE.64.016132
23. Newman, M.E.J.: Analysis of weighted networks. Phys. Rev. E **70**(5), 056131 (2004). https://doi.org/10.1103/PhysRevE.70.056131
24. Freeman, L.C.: Centrality in social networks conceptual clarification. Soc. Netw. **1**(3), 215–239 (1978). https://doi.org/10.1016/0378-8733(78)90021-7
25. Brandes, U.: A faster algorithm for betweenness centrality. J. Math. Sociol. **25**(2), 163–177 (2001). https://doi.org/10.1080/0022250X.2001.9990249

Spatio-Temporal Distribution of Dengue Infections in Baguio City, Philippines

Louie Ville A. Balino, Kenneth S. Caasi, and Rizavel C. Addawe

Abstract This paper investigates the clustering of dengue fever in Baguio City using spatial analysis. The Moran's index and Average Nearest Neighbor for the extent of spatial autocorrelation are employed. The Local Moran's *I* is utilized for the cluster map which identifies hotspots of dengue occurrence. The map was constructed by overlaying hotspots for the years 2014, 2016, and 2018. The months of April and July are the representatives for each year.

Keywords Spatio-temporal analysis · Moran's *I* · Average nearest neighbor

1 Introduction

Dengue is an infectious disease caused by dengue viruses transmitted by the *Aedes Aegypti* and *Aedes Albopictus* [1]. This disease is considered to be a major health threat in subtropical and tropical countries around the world, with a 30-fold increase in incidence over the last 50 years. In 2012, the World Health Organization (WHO) ranked dengue as the most important mosquito-borne viral disease in the world due to the global burden it brings [2]. It also became the most widespread vector-borne disease of humans in the second half of the twentieth century with estimates of about forty (40) percent of the world's population at risk [3, 4].

Philippines, a dengue-endemic country, recorded 138,000 cases of dengue fever in the first three quarters of 2018, which represents a twenty-one percent (21%) increase compared to the number of cases in the same period of 2017 as per the Department of Health (DOH) [5]. Meanwhile, the Cordillera Administrative Region (CAR) experienced a hundred percent increase in the incidences for the first ten months last 2018 [6].

Baguio City, a highly urbanized city in CAR, documented a high number of dengue occurrences. The recorded dengue fever cases sharply rose in the year 2016

L. V. A. Balino (✉) · K. S. Caasi · R. C. Addawe
Department of Mathematics and Computer Science, College of Science,
University of the Philippines Baguio, 2600 Baguio, Philippines
e-mail: labalino@up.edu.ph

M. H. Mohd et al. (eds.), *Modelling, Simulation and Applications of Complex Systems*,
Springer Proceedings in Mathematics & Statistics 359,
https://doi.org/10.1007/978-981-16-2629-6_14

by five-folds, putting the city into an outbreak level [7]. Researcher from the Baguio City Epidemiology and Surveillance Unit, reported one hundred forty-three (143) dengue cases in Baguio City from January to May in 2018, an increase of almost forty-six percent (45.92%) from ninety-eight (98) cases recorded during the same period in the year 2017 [8].

These alarming numbers caught the attention of the Philippine government and its different sectors. There are ongoing initiatives locally and nationally that aim to increase dengue fever awareness. These include dengue prevention control programs of DOH such as surveillance, case management, diagnosis, integrated vector management, outbreak response, health promotion, and research [9]. The DOH-CAR and the Philippine Red Cross–Baguio City Chapter urge the residents to observe the 4S strategy against dengue for the said prevention control [10].

Due to the demand to scrutinize the epidemiology of dengue fever, this study intends to explore the spatial distribution of its extent and magnitude in Baguio City. Epidemiologists commonly use the ratio of case numbers at a given time to a past case to determine increase in case occurrences of diseases in the form of mean or median [11]. However, since the incidence of dengue fever varies from one place to another, the spatial and time components must also be taken into consideration [12]. Disease mapping is the first step in the spatial descriptive analysis [13]. Disease maps provide an overview of the extent and magnitude of the public health problem. The results of the general and specific clustering [14] of the disease may be of importance to the health authorities. General clustering includes an analysis of the overall clustering trend of disease incidence in the study region and is accompanied by an evaluation of the global spatial autocorrelation in which the exact location of clusters is not examined. On the other hand, specific clustering uses specific disease-clustering methods to examine the exact location of the clusters.

There are researches conducted in several countries that utilized spatial analysis to map the incidence of vector-borne diseases. A study in Indonesia [15] used Moran's I statistics and Local Indicator of Spatial Association (LISA) (see the study of Anselin in [18] for further description) to assess the global autocorrelation of dengue incidence and to detect local spatial autocorrelation, respectively. The findings indicated that dengue cases in Makassar, Indonesia were spatially random for all five years. Moreover, results showed that the central location of dengue fever transmission was Biringkanaya whereas Manggala, Makassar, and Tamalanrea were potentially susceptible to dengue fever infection. Currently, the National Capital Region (NCR) is the only region in the Philippines that has been subjected to spatial distribution analysis of dengue fever. Prior to the study, Moran's I and Kulldorff's spatial scan (SaTScan) statistics were employed to identify clusters of dengue incidence. Analyses identified significant dengue incidence clusters and revealed that NCR's northern cities, such as Caloocan, Malabon, Navotas, and Valenzuela, exhibited high spatial autocorrelation [16]. These studies motivated the adaptation of spatial analysis to analyze the distribution of dengue incidence in Baguio City.

The outcome of this research may provide insights for the health office in determining hotspots and also serve as a basis for which they can properly allocate resources for the prevention of possible outbreaks and treatment of the dengue fever.

Furthermore, it may be used in future studies and programs with regards to risk factors and vector management of the respective disease.

2 Methodology

2.1 Area of Study

Baguio City is located at latitude $16°24'59.99N$ and longitude $120°35'59.99E$ and is situated in the Province of Benguet. The area of the city is 57.49 km^2 enclosed in the perimeter of 30.98 km. The developed portion of the city corresponds to the plateau that rises to an elevation of 1,400 m. Most of it lies in the northern half of the city. The city is landlocked within the province of Benguet, thus bounding it on all sides by its different municipalities on the North by the capital town of La Trinidad, on the East by Itogon, and to the South and West by Tuba. With City Hall as a reference point, it extends 8.2 km from East to West and 7.2 km from North to South. Baguio City has twenty (20) administrative districts and consists of 129 barangays in total [17]. Barangay is the smallest unit of the community.

2.2 Data Collection

The records of dengue cases were provided by the Baguio City Health Services Office. Dengue incidence, expressed in the number of cases per month and per week, were recorded over a nine-year period from 2010 to 2018 and sorted to each of the 129 barangays of the city. The data was also categorized according to age and sex per barangay. Only cases with positive results who reside in Baguio City constitute the subjects of the study. Population data for the years 2010 and 2015 were obtained from the Philippine Statistics Authority (PSA).

2.3 Global and Local Index

There are two common global spatial autocorrelation indices, namely Moran's I and Geary's C. Global Moran's I statistics is the most frequently used global spatial autocorrelation indicator which was first introduced by Patrick Moran in 1948 [20]. It examines whether a phenomenon is clustered or not. The interpretation of Moran's I is as follows: a value close to 0 indicates randomness, whereas a positive (negative) value indicates positive (negative) autocorrelation. Its inference is based on a null hypothesis of spatial randomness.

The LISA by Anselin [18], is a localized form of the Moran's index. It focuses on the heterogeneity of spatial association of the clusters over the whole area of study. It offers visualization by significance and cluster maps.

As given in Chen [19], if the spatial weight matrix W is symmetric with zero diagonal and normalized in form, then we have

$$I = z^T W z \tag{1}$$

where $z = (x - \mu)/\sigma$ for μ as the mean incidence and σ the standard deviation. The local Moran's index at position i is defined as

$$I_i = diag(zz^T W)_i. \tag{2}$$

The global and local index is related by the equation

$$I = \sum_i I_i. \tag{3}$$

2.4 Average Nearest Neighbor (ANN)

The ANN is a tool used in determining if the incidence is clustered or not. It is given by the following:

$$ANN = \frac{\frac{1}{n}\sum_{i=1}^{n} d_i}{\frac{0.5}{\sqrt{n/A}}} = \frac{D_O}{D_E}; \tag{4}$$

$$z_{ANN} = \frac{D_O - D_E}{SE} \tag{5}$$

where

$$SE = \frac{0.26136}{\sqrt{n^2/A}}. \tag{6}$$

Here, d_i is the distance of the object i to its nearest neighbor, A as the total area of study and n is the number of objects. The extent of clustering is described as: 0 for clustered, 1 as random and 2.15 for uniformly distributed. The z-score is given above by z_{ANN}. The z-score of ANN is interpreted as follows: if z_{ANN} is greater than 1.96, then this indicates that the data points is distributed while if z_{ANN} is less than -1.96, there is a clustering of the data points. The 1.96 value is based on a 95% confidence level.

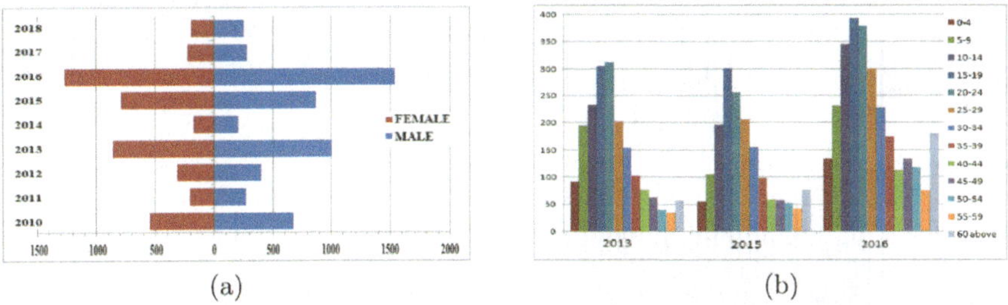

Fig. 1 Annual dengue cases **a** from 2010–2018 according to sex and, **b** according to age group for 2013, 2015 and 2016.

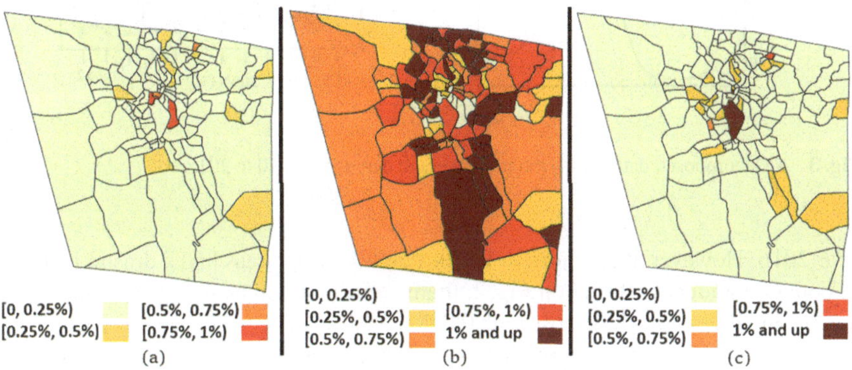

Fig. 2 Geographic maps for dengue incidence rates: **a** 2014, **b** 2016, and **c** 2018.

3 Results and Discussion

3.1 Descriptive Analysis

A total of 10,068 dengue cases were reported in Baguio City from January 2010 to December 2018. Majority of the cases (54.66%) were male as in Fig. 1(a) and most of it belonged to the age groups either 15–19 or 20–24 given in Fig. 1(b). The highest number of dengue cases recorded was in the year 2016 followed by years 2013 and 2015 whereas 2014 recorded the lowest. In the year 2016, an outbreak was declared in the city. In general, the actual trend in the number of dengue cases fluctuates.

As mentioned earlier, the lowest and highest number of dengue incidences were recorded in 2014 and 2016, respectively. These lead to the construction of geographical maps in these years, with 2018 added, as shown in Fig. 2. In the year 2014, the business areas such as Session Road and Rizal Monument had the highest rate of dengue incidence with 0.75 to 1% of their respective population and is followed by a mountainous area (Ambiong) with incidence rate ranging from 0.5 to 0.75%. The dengue incidence rate in Session Road and Rizal Monument area declined in the

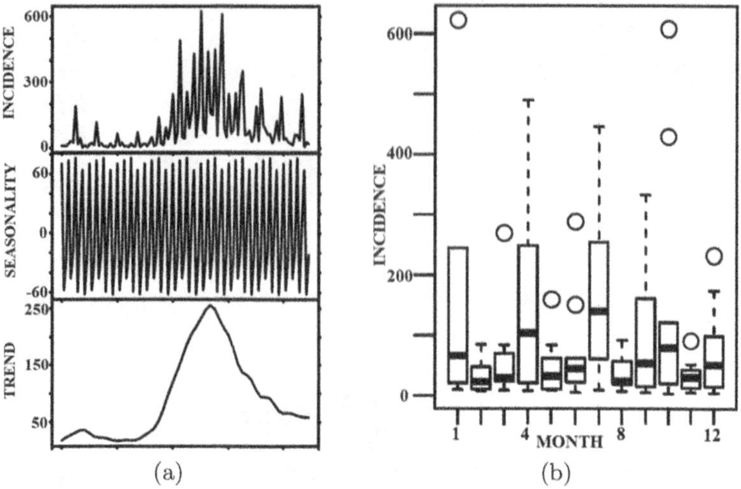

Fig. 3 **a** Decomposed data, and **b** monthly average data from 2010 to 2018.

year 2016. However, the majority of their neighbouring areas had a drastic increase in dengue incidence. Based on Fig. 2, there were more areas with 0.5% and above dengue incidence rate than those with 0.5% and below. The incidence rates decreased to as low as 0 to 0.25% in some locations in the year 2018. On the other hand, the number of cases in some other business areas such as Legarda-Burnham-Kisad and Kayang-Hilltop either had increased or retained high dengue incidences with at least 1% of their respective population.

3.2 *Temporal-Spatial Analysis*

As seen from Fig. 3, seasonality of the monthly number of dengue cases is detected by decomposing the time series into three components. The decomposed monthly series was obtained using the *decomp*() function in R. Note that Fig. 3 shows that the dengue incidence reaches peak values in April and July for the given study period. These months will be used as representatives of each year for the results of ANN and Global Moran's index.

Baguio City has 129 barangays, so the size of the contiguity matrix is 129 by 129. Here, we adopt the rook's case in which two places are considered as neighbors if they share a common borderline. Barangay Happy Hollow, located near the outskirts of Baguio City, has the greatest number of neighbors. On another note, Kias, Mines View Park, and Alfonso Tabora have the least number of neighbors. The values of Global Moran's I and its corresponding z-score computed for a row-standardized spatial weight matrix based on first-order contiguity are given in Table 1.

Table 1 Global Moran's *I* for the month of April and July.

April	2010	2011	2012	2013	2014	2015	2016	2017	2018
Moran's *I* (z-score)	−0.0461 −0.7073	−0.0538 1.1387	−0.0223 −0.2680	0.0637 1.3190	0.1300 2.5451	−0.0232 −0.2851	−0.0228 −0.2783	0.1144 2.2571	−0.0185 −0.1980
July	2010	2011	2012	2013	2014	2015	2016	2017	2018
Moran's *I* (z-score)	0.0539 1.1392	0.0084 0.3008	0.0228 0.5663	0.0390 0.8639	0.0012 0.1673	−0.0242 −0.3022	−0.0402 −0.5979	−0.0126 −0.0887	−0.0137 −0.1094

In this study, areas with positive and significant Local Moran's index with z-score greater than 1.96 were considered as hotspots. There is a spatial pattern (clustering of hotspots) in the study area based on the computed z-scores of Moran's index in April 2015 and April 2017. There are also several years where there is a spatial autocorrelation but not statistically significant.

We utilized ANN to consider the distance analysis of dengue incidence in Baguio City. The ANN ratio and z-score are given as shown in Table 2 below. Results show that most of the incidence are dispersed for July. Thus, there is less than 5% likelihood that this dispersed pattern could be the result of random chance. On the other hand, the years 2011, 2012, and 2014 shows spatial clustering for the month of April. One observes that the result of ANN and Moran's I coincide for April 2014 though the ANN ratio is not statistically significant since its z-score is not less than -1.96. The difference in the result is possibly due to the use of neighbourhood with rook's case for Moran's I and average distance for ANN.

We note here that autocorrelation index is sensitive to the type of neighborhood matrix [21]. This means there is a possibility that other neighbourhood matrix may provide a result that coincides with ANN.

Moreover, the ANN describes the distribution of dengue cases while the Moran's index checks the autocorrelation between spatial location and an attribute value of points. For Moran's I, barangays are point particles. The ANN uses location representation (longitude and latitude chosen from PhilAtlas) and then computes the distance from one point (barangay) to another.

An example of difference in the result is found in [22] where the ANN resulted in a dispersed pattern while the Moran's I claimed a clustered pattern of dengue outbreak distribution in Selangor, Malaysia. This also occured in a study of spread of Mountain Pine Bettle infestation [23] and spatial distribution of lakes [24].

Table 2 Average Nearest Neighbor for April and July.

April	2010	2011	2012	2013	2014	2015	2016	2017	2018
ANN (z-score)	1.3749 1.8979	0.8013 −1.0055	0.8492 −0.5768	1.0888 0.7408	0.7635 −1.279	0.9997 −0.001	1.0582 0.6872	1.0917 0.3510	1.0632 0.3825
July	2010	2011	2012	2013	2014	2015	2016	2017	2018
ANN (z-score)	1.2072 3.2702	0.9672 −0.3813	1.1683 2.6756	1.1224 2.2339	1.0626 0.6341	1.1366 2.0586	1.2041 4.0956	1.1231 1.6662	1.2090 2.6220

Table 3 Number of barangays per category.

Label	Count
(0, 0, 0)	77
(1, 0, 0)	11
(0, 1, 0)	17
(0, 0, 1)	15
(1, 0, 1)	1
(1, 1, 0)	4
(0, 1, 1)	5

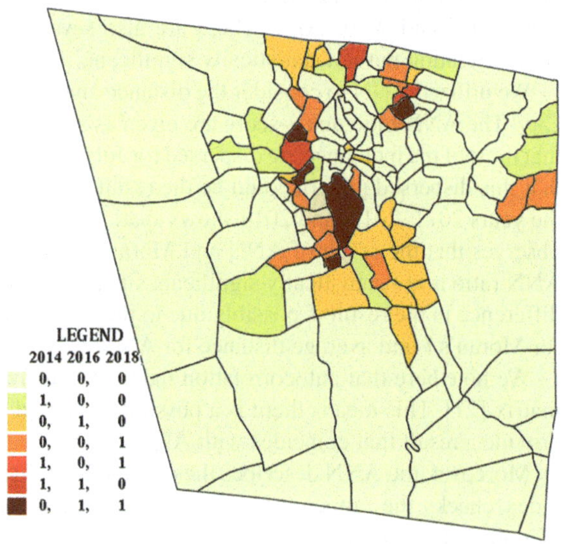

Fig. 4 Cluster map (2014, 2016, 2018)

We will now deal with location assessment to provide a more informative analysis of dengue cases [18]. For the same reason in considering the years 2014, 2016, and 2018 in constructing the maps in Fig. 2, another map of Baguio City was constructed as shown in Fig. 4. It was made by overlaying the temporal hotspots in 2014, 2016, and 2018, respectively. In this study, areas with positive and significant Local Moran's index, that is with z-score greater than 1.96, were considered as hotspots. Similar to Hu et al. [25], the following findings may provide an efficient way of dengue supervision.

If the barangay is a hotspot in the given year, then we denote it as 1, otherwise 0. Observe that groups for the classification of hotspots were categorized into seven. By overlapping hotspots for those three years, the area of the study may be categorized into eight. However, since no barangay was classified as a hotspot for three years, the "1, 1, 1" category was omitted. The number of locations that fall under each category is stipulated in Table 3. The label "0, 0, 0" indicates that the barangay has never been a hotspot for the three years. Baguio General Hospital compound is the only barangay labelled as “1, 0, 1”. Campo Filipino, Aurora Hill Proper, Aurora Hill North Central, and Happy Homes-Lucban fall under the “1, 1, 0” category. On the other hand, the five barangays under “0, 1, 1” category include Legarda-Burnham-Kisad, Poliwes, MRR-Queen of Peace, Andres Bonifacio (Lower Bokawkan), and Modern Site-West. These barangays possess one or more of the following characteristics: densely populated, tourist spot, central business area, nearby schools, universities, and market place. As mentioned in [15], hotspots can possibly cause neighbouring areas to be prone to higher dengue cases.

4 Conclusion

The spatial clustering test was done using Moran’s index and ANN. The representation of year by month was due to the seasonality of dengue. Though there were similarities of results from the two tests, there are more differences that were due to how each method uses neighbourhoods. It is evident that ten barangays are hotspot for two years.

The local government should consider monitoring neighbouring barangays of hotspots, especially to those determined majority of cases. This study can help prevent or contain possible outbreaks shortly. For an insight in forecasting monthly dengue cases in each barangay in Baguio City, the ARIMA forecasting model can be considered for future research.

Acknowledgement The authors gratefully acknowledge the Baguio City Epidemiology and Surveillance Unit and Dr. Donnabel Tubera-Panes for the raw data of dengue incidence.

References

1. Bäck, A.T., Lundkvist, A.: Dengue viruses-an overview. Infect. Ecol. Epidemiol. **3**(1), 19839 (2013). https://doi.org/10.3402/iee.v3i0.19839
2. World Health Organization: Global strategy for dengue prevention and control 2012–2020. World Health Organization, Geneva (2012). http://www.who.int/denguecontrol/9789241504034/en/
3. Stephenson, J.R.: The problem with dengue. Trans. Roy. Soc. Trop. Med. Hyg. **99**, 643–646 (2005)
4. World Health Organization: Comprehensive guidelines for prevention and control of dengue and dengue hemorrhagic fever. Revised and expanded edition. WHO Regional Office for South-East Asia (2011). https://apps.who.int/iris/handle/10665/204894
5. De Vera A.: Dengue cases up. Manila Bull. (2018). https://news.mb.com.ph/2018/10/24/dengue-cases-up/
6. Comanda, Z.: Cordillera dengue cases up 100%. Manila Bull. (2018). https://news.mb.com.ph/2018/11/21/cordillera-dengue-cases-up-100/
7. Guieb, D.: Baguio records highest dengue cases. SunStar Baguio (2016). https://www.sunstar.com.ph/article/87549

8. Dar, C.C.: Baguio health office cautions public of dengue, rainy day illnesses. Philippine Information Agency (2018). https://pia.gov.ph/news/articles/1010062
9. Department of Health Philippines: Dengue prevention and control program. https://www.doh.gov.ph/national-dengue-prevention-and-control-program
10. SunStar: Red Cross renews drive on dengue prevention. SunStar Baguio (2018). https://www.sunstar.com.ph/article/1750767
11. Nakhapakorn, K., Jirakajohnkool, S.: Temporal and spatial autocorrelation statistics of dengue fever. Dengue Bull. **30**, 177 (2006). WHO Regional Office for South-East Asia. https://apps.who.int/iris/bitstream/handle/10665/170259/db2006v30p177.pdf
12. Er, A.C., Rosli, M.H., Asmahani, A., Naim, M.M., Harsuzilawati, M.: Spatial mapping of dengue incidence: a case study in Hulu Langat District, Selangor. Malaysia. Int. J. Geol. Environ. Eng. **4**(7), 251–255 (2010)
13. Lai, P., So, F., Chan, K.: Spatial Epidemiological Approaches in Disease Mapping and Analysis. CRC Press, Boca Raton (2008)
14. Rezaeian, M., Dunn, G., St Leger, S., Appleby, L.: Geographical epidemiology, spatial analysis and geographical information systems: a multidisciplinary glossary. J. Epidemiol. Community Health **61**(2), 98–102 (2007)
15. Aswi, A., Sukarna, S.: Spatial distribution analysis of dengue incidence in Makassar, Indonesia. Int. J. Appl. Math. Stat. **53**(6), 202–208 (2015)
16. Pangilinan, M.A.P., Gonzales, D.P.G., Leong, R.N.F., Co, F.F.: Spatial analysis of the distribution of reported dengue incidence in the National Capital Region, Philippines. Acta Med. Philipp. **51**(2), 126–132 (2017)
17. City Planning and Development Office: Comprehensive land use plan: Baguio City, 2013–2023, vol. 1. City Government of Baguio (2013)
18. Anselin, L.: Local indicators of spatial association-LISA. Geogr. Anal. **27**(2), 93–115 (1995)
19. Chen, Y.: New approaches for calculating Moran's index of spatial autocorrelation. PLoS ONE **8**(7), e68336 (2013). https://doi.org/10.1371/journal.pone.0068336
20. Moran, P.A.: Notes on continuous stochastic phenomena. Biometrika **37**(1), 17–23 (1950). https://doi.org/10.2307/2332142
21. Ding, Y., Fotheringham, S.: The integration of spatial analysis and GIS. Cornput. Environ. Urban Syst. **16**, 3–19 (1992)
22. Abd Latif, Z., Mohamad, M.: Mapping of dengue outbreak distribution using spatial statistics and geographical information system, pp. 1–6 (2015). In proceedings
23. Fox, E., Balram, S., Dragicevic, S., Roberts, A.: Spatial analysis of high resolution aerial photographs to analyze the spread of mountain pine beetle infestations. J. Sustain. Dev. **5**(9), 106–129 (2012)
24. Taylor, S.: Pattern detectors. In: Introduction to GIS Using ArcView GIS: Statistical Analysis. Western Oregon University. https://people.wou.edu/~taylors/g492/avstat3.pdf
25. Hu, W., Clements, A., Williams, G., Tong, S.: Spatial analysis of notified dengue fever infections. Epidemiol. Infect. **139**(3), 391–399 (2011)

Spatio-Temporal Analysis of Measles Cases in Baguio City, Philippines from 2010–2018

Rizavel Addawe, Gervy Marie Angeles, and Maricar Balolong

Abstract We study the spatio-temporal autocorrelation of the reported measles cases in Baguio City, Philippines from 2010–2018 by utilizing Moran's I statistic. Using cluster analysis, we construct a disease map by overlaying hot spots for the years 2010, 2014, and 2018. We reject the null hypothesis of spatial randomness based on clustering of the identified hot spots. The number of measles cases in the city is therefore represented by a positive spatial autocorrelation. We also assess the fit of a time-series Susceptible-Infected-Recovered (TSIR) model in `R` in the `tsiR` package for the analysis of infectious disease time-series data.

Keywords Spatio-temporal analysis · Spatial autocorrelation · Measles · Moran's index · TSIR model

1 Introduction

A major public health concern in recent years is an acute viral infection called measles [1]. Measles is a highly contagious disease caused by a virus that replicates in the nose and throat of an infected individual. The virus is transmitted by direct contact with infectious droplets or by airborne spread. Despite worldwide efforts toward measles eradication, the provisional data [2] from the World Health Organization (WHO) reported outbreak declarations from multiple countries of all WHO regions. Among the countries with the largest measles outbreaks is the Philippines [2, 3]. In 2014, a total of 58,010 measles cases was reported in [4], which is nine times higher compared to the reported cases in 2013. Poor implementation of strategies against the spread of disease was pinpointed as a cause of measles reemergence [5].

R. Addawe (✉) · G. M. Angeles · M. Balolong
University of the Philippines Baguio, Governor Pack Road, 2600 Baguio City, Philippines
e-mail: rcaddawe@up.edu.ph

G. M. Angeles
e-mail: gmangeles1@up.edu.ph

M. Balolong
e-mail: mpbalolong2@up.edu.ph

M. H. Mohd et al. (eds.), *Modelling, Simulation and Applications of Complex Systems*, Springer Proceedings in Mathematics & Statistics 359,
https://doi.org/10.1007/978-981-16-2629-6_15

More recently, measles cases rose by about 550% above the normal average from 2017 to 2018 [6].

In this work, we examine the spatial and temporal autocorrelation of reported measles cases from 2010 to 2018 in Baguio City, a mountainous area in Northern Luzon, Philippines. The number of cases is grouped according to *barangays*, the smallest administrative divisions in the Philippines. In particular, Baguio City is composed of 129 barangays. Because the city features a cool climate relative to lowland areas, travelers from different regions go back and forth to the city throughout the year. However, most attractions are clustered in a relatively small area, causing predominance of the transmission of viral infections. As a consequence, the city becomes vulnerable to disease outbreaks (see e.g. measles outbreak [7]). Despite the alarming situation, the authors were not able to find research works about measles in the city (with the exception of the epidemiology of measles cases during 2010–2012 [8]) dealing with localized spatio-temporal patterns of cases. For this reason, the authors hope that the results of this study would help the local government in providing scientific information to control and prevent measles outbreaks, and in general, to provide disease surveillance.

The spatio-temporal analysis of measles cases is carried out using Moran's I statistic [9, 10]. We compute the global Moran's I for each year to obtain a general assessment of the spatio-temporal patterns of cases. A further investigation requires localizing the statistic at the barangay level. This is achieved by utilizing local Moran's I. Statistically significant values are identified and interpreted, and are then used to assemble a disease map by overlaying 2010, 2014, and 2018 hot spots. Based on clustering of cases, we reject the null hypothesis of spatial independence and randomness of measles cases among barangays.

We also present a dynamical systems approach in modeling an infectious disease using time series analysis. First described in [11], the time-series Susceptible-Infected-Recovered (TSIR) model is a discrete time analogue of the Susceptible-Infected-Recovered (SIR) type epidemic models. It reconstructs the susceptible compartment to allow fitting of time-series data. We use a simple discrete-time model for the susceptibles in the context of measles [11, 12]. Several analysis of the TSIR model with measles as a case study are available in the literature (see e.g. [13–16]). In the present study, we only attempt to fit the TSIR model (1) in `tsiR`,[1] an open-source software package in the `R` programming language. This allows us to infer parameters, such as epidemic saturation, from the cases data.

This paper is organized as follows: In the next section, we study the spatial and temporal autocorrelation of measles cases in Baguio City. In Sect. 3, we present a TSIR model and its fit in the `tsiR` package. We summarize the results and give recommendations for future work in Sect. 4.

[1]Codes are publicly available on GitHub (www.github.com/adbecker/tsiR).

2 Spatio-Temporal Analysis Using Moran's *I*

2.1 Data and Demographics

The number of reported measles cases is obtained from Baguio City Health Services Office (HSO). We plot the measles cases for each year from 2010 to 2018 according to age groups (see Fig. 1), and according to sex (see Fig. 2). A significant peak of measles cases is observed in 2014, dominated by age groups 1–4 and 15–19, both of which are mostly males. These demographics are consistent with the report of the 2013 measles outbreak [7] in the Cordillera region, a group of provinces surrounding Baguio City. As pointed out in [5], the resurgence of measles in 2013–2014 reflects the decrease in coverage of measles elimination strategies.

2.2 Testing Global Autocorrelation

In Global Moran's *I* statistic, the null hypothesis states that measles cases in Baguio City are spatially independent and are randomly distributed among barangays. We compute for Global Moran's *I* using the formula [9, 10]:

$$I = \frac{n}{\sum_{i=1}^{n}\sum_{j=1}^{n} w_{ij}} \frac{\sum_{i=1}^{n}\sum_{j=1}^{n} w_{ij}(x_i - \overline{x})(x_j - \overline{x})}{\sum_{i=1}^{n}(x_i - \overline{x})^2},$$

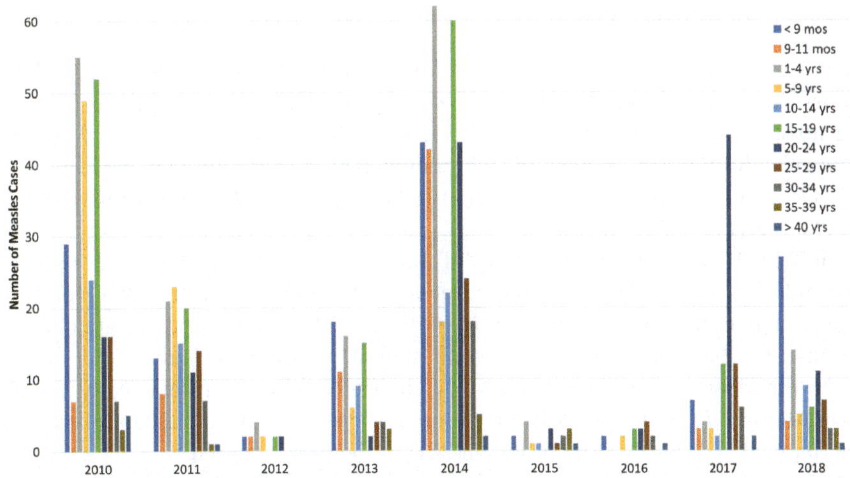

Fig. 1 Measles cases from years 2010 to 2018 according to age groups

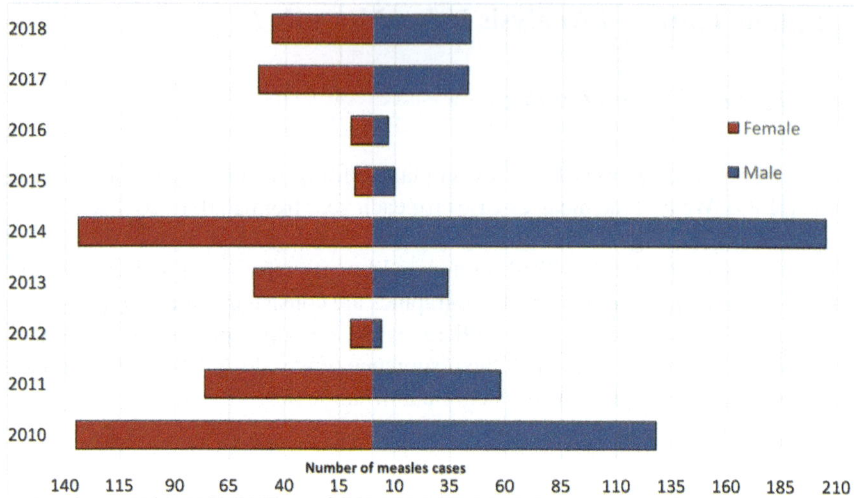

Fig. 2 Measles cases from years 2010 to 2018 according to sex

where x_i, x_j represent the number of cases in a particular barangay i, j, respectively. The sample size $n = 129$ since the city is composed of 129 barangays. The mean of the variables x_i for $i = 1, \dots, n$ is denoted by $\overline{x}$. For determining the relationship between locations i and j, we use rook weights w_{ij} defined as

$$w_{ij} = \begin{cases} 1 & \text{if locations } i \text{ and } j \text{ share a boundary,} \\ 0 & \text{otherwise.} \end{cases}$$

Values of I range between -1 and +1. Values near $+1$ indicate clustering, while values near -1 indicate dispersion. If $I = 0$, then the pattern is random, meaning no spatial autocorrelation exists. To test whether the Global Moran's I is significantly different from random, we compute its z-score through the equation $z = (I - E(I))/(\sqrt{\text{var}(I)})$, where $E(I)$ is the expected value of I with variance $\text{var}(I) = E(I^2) - E(I)^2$.

In Table 1, we provide a summary of the computed Global Moran's I, its corresponding z-value, and p-value based on a 5% level of significance. We pick two representative months for each year. In particular, March and February are chosen because they have the highest and second highest, respectively, number of cases in each year. Observe that none of the values in March returned a statistically significant Moran's I. On the other hand, we notice a positive spatial autocorrelation in February 2014. Existence of clusters in 2014 is therefore inferred.

Table 1 Global Moran's I of log rate of measles incidence (March and February)

Year	Global Moran's I (March)	z-score	p-value	Global Moran's I (February)	z-score	p-value
2010	−0.010215	2.016254	0.538769	0.004878	1.016254	0.303597
2011	−0.053890	14.016254	0.969020	−0.028014	13.016254	0.793418
2012	−0.003597	26.016254	0.432196	0.002697	25.016254	0.335146
2013	−0.012421	38.016254	0.574041	0	37.016254	0.375819
2014	0.026302	50.016254	0.083495	0.079666	49.016254	0.000197
2015	−0.021461	62.016254	0.709835	−0.016016	61.016254	0.630173
2016	0	74.016254	0.375819	0	73.0162542	0.375819
2017	−0.016380	86.016254	0.635723	−0.026854	85.016254	0.779747
2018	−0.007411	98.016254	0.493516	−0.009738	97.016254	0.531092

2.3 *Disease Map of the City*

While Global Moran's I gives an overall assessment of the spatial pattern of measles cases, we would like to further examine which barangays are similar to or different from other barangays in their neighborhood. This is done by employing Local Moran's I_i, or Local Indicator Spatial Association (LISA) [9]. In contrast to Global Moran's I, each location i receives its own I_i value, as well as its own z-value, expected value of I_i, and variance of I_i. Local Moran's I_i is computed using

$$I_i = (x_i - \overline{x}) \sum_{j=1}^{n} w_{ij}(x_j - \overline{x}), \quad \text{for each location } i.$$

A positive value of I_i indicates that barangay i is spatially clustered with barangays of similar high or low measles cases. A negative value means that a barangay is surrounded by neighboring barangays with different values. To test the significance of I_i, we calculate its z-score according to the equation $z_i = (I_i - \overline{x})/(\sigma(I_i))$, where $\sigma(I_i)$ is the standard deviation of I_i.

A hot spot is defined as a barangay with positive and statistically significant (based on a 5% level of confidence) local Moran's I_i. We construct a disease map (see Fig. 3) of the city by overlaying hot spots of measles cases in 2010, 2014, and 2018. We label each barangay according to eight categories: If a barangay is a hot spot for a particular year, then we label it with 1, otherwise, 0. Hence, we assign "1,1,1" to a barangay if it is a hot spot for the years 2010, 2014 and 2018. If a barangay is a hot spot for the years 2010 and 2014 but not in 2018, we label it as "1,1,0". Thus "0,0,0" means that the barangay is not a hot spot for each of the three years. The number of barangays per label is listed in Table 2. It is worth noting that the barangays labeled "1,1,1" are *Lower Magsaysay* and *Teodora Alonzo*. Each of these barangays possesses one or more of the following: populous area, high population density, and proximity to schools and market places.

Table 2 Number of barangays per label

Label	Count
0,0,0	39
1,0,0	6
0,1,0	49
0,0,1	10
1,0,1	2
1,1,0	10
0,1,1	11
1,1,1	2

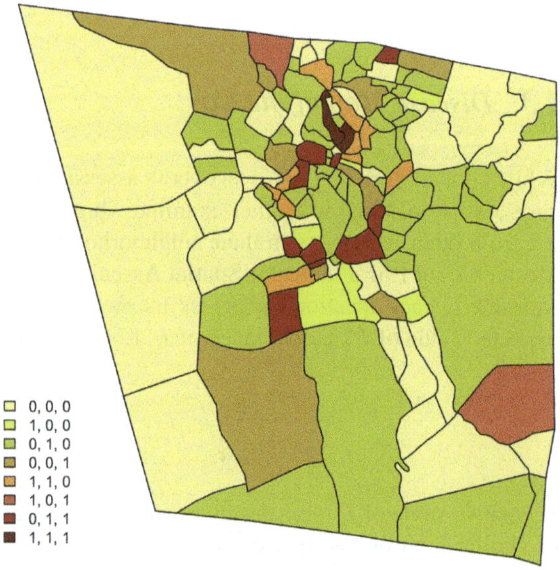

Fig. 3 Disease map for 2010, 2014, 2018

3 Fitting a TSIR Model in the `tsiR` Package

We reconstruct the susceptible compartment of the SIR models by employing a simple discrete-time model [11, 12]. In particular, the probability that a susceptible individual S at time t will become infected at time $t+1$ is given by

$$\begin{cases} S_{t+1} = B_{t+1} - S_t - I_{t+1}, \\ E[I_{t+1}] = \beta_{t+1} S_t I_t^{\alpha}, \end{cases} \tag{1}$$

where we assume that infected individuals I are infectious for a unit time before being removed from the dynamics. Here, B denotes number of births, and $E[\cdot]$ means expected value. The parameter α describes epidemic saturation, and a correction for shifting from continuous to discrete time [14, 17]. On the other hand, β_{t+1} is the contact rate between susceptible and infected individuals at time $t + 1$.

3.1 `tsiR` *Implementation*

A data frame with *time*, *cases*, *births*, and *pop* (population size) columns is constructed from gathered data. Since the number of births and the population size are on different time scales compared to the reported cases, we interpolate the data to the proper infectious period timescale. At each point in time, the column *births* is the number of births that occur within the infectious period (IP) in weeks. Typically, IP = 2 for measles, however, in this work we use IP = 4. We justify this choice by the fact that Baguio City is relatively small compared to the region(s) considered in [12], so the number of cases is also relatively low. The main function of the `tsiR` package is `runtsir`. It reconstructs the susceptible dynamics, estimates parameters, and runs forward simulation under the fitted parameters. A comprehensive description of the function and its arguments can be found in [12].

The output generated by the following code is plotted in Fig. 4. The file `baguio.csv` is created using the Baguio shape file available at the website http://philgis.org/city-and-capital-page/baguio-city.

```
BaguioMeas <- read.csv(file="baguio.csv", header=TRUE, sep=",")
BaguioRes <- runtsir(data=BaguioMeas,IP=4,
                     xreg='cumcases', regtype='gaussian',
                     alpha=NULL, sbar=NULL,
                     family='gaussian', link='identity',
                     method='negbin', nsim=100,
                     sigmamax=0.5)
plotres(BaguioRes)
```

Forward simulation (with number of simulations `nsim` = 100) yields the computed mean value $\overline{\beta} = 0.0012$ for the contact rate, which is typical. On the other hand, we obtain $\alpha = 0.39$, which is significantly different from the commonly used value $\alpha = 0.97$. In Fig. 4, measles cases (blue) are plotted against the TSIR simulations (brown). Observe that the random stochastic simulations (brown) are mostly below the data (blue). We speculate that the under-reporting of cases, ubiquitous in developing regions, is the main proponent of this issue. Moreover, in the numerical experiments we conducted, the reporting rate falls below 1% even when we decrease `sigmamax`, the inverse kernel width for Gaussian regression.

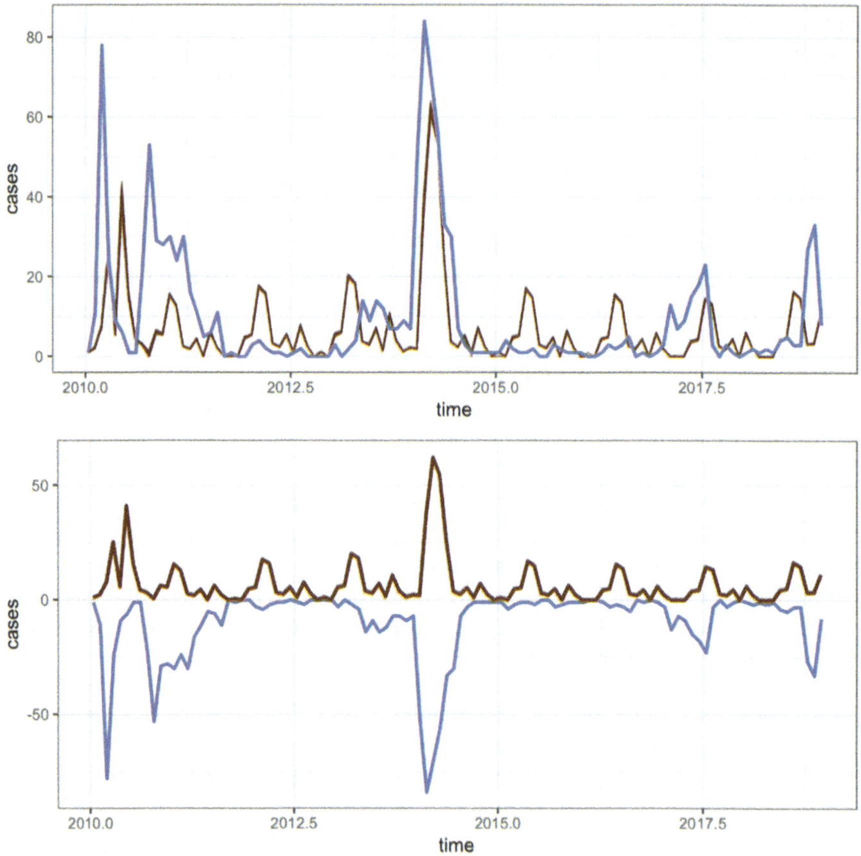

Fig. 4 Data (blue) against 10 randomly chosen stochastic simulations (brown) and the (inverse) data against mean of the simulations with confidence intervals

4 Conclusions and Recommendations

In this work, we examined the spatial and temporal autocorrelation of measles cases in Baguio City in 2010–2018. Clustering of measles cases was observed in February 2014. Therefore, we reject the null hypothesis of spatial randomness. The number of measles cases is distributed nonrandomly, suggestive of a positive spatial autocorrelation. We further studied the spatio-temporal dynamics of measles cases in the city by considering its smallest administrative units, called *barangays*, as spatial locations. The results revealed that two barangays, namely, *Lower Magsaysay* and *Teodora Alonzo*, are hot spots of measles cases in 2010, 2014, and 2018. Furthermore, we fitted a TSIR model, a synthesis of time-series data and continuous-time infectious disease models, in the `R` package `tsiR`. The obtained simulations were considerably different from the data. We conjecture that this is due to a rather high under-reporting of measles cases in the city.

The results we obtained are only applicable at the barangay level. This means that they are subject to aggregation bias, so we emphasize that we do not conclude anything about individual-level behavior. We therefore propose to consider the socio-economic, environmental, and other factors associated with the identified hot spots in this study. On the other hand, the TSIR model we employed makes several assumptions that may not be realistic, e.g. fixed infectious period, and a linear relationship between cumulative cases and births. We propose to utilize other TSIR models reviewed in [16, Section 23.4.2]. We also point out that under-reporting can be incorporated to system (1) (see e.g. [11] [16, Section 22]), which may lead to a better model fit.

Acknowledgments The authors would like to thank Dr. Donna Tubera–Panes, Medical Officer IV at the Office of Baguio Health Department, and the anonymous reviewers for their invaluable suggestions and comments.

References

1. Abad, C.L., Safdar, N.: The reemergence of measles. Curr. Infect. Dis. Rep. **17**(12), 51–58 (2015). https://doi.org/10.1007/s11908-015-0506-5
2. World Health Organization (WHO). New measles surveillance data from WHO. https://www.who.int/immunization/newsroom/new-measles-data-august-2019/en/
3. United Nations Children's Fund (UNICEF). Measles explained: What's behind the recent outbreaks? Reported cases of measles rose by 200 percent from 2018 to 2019. https://www.unicef.org/stories/measles-explained-whats-behind-recent-outbreaks
4. Herriman, R.: Philippines measles outbreak 2014: 58,010 cases, 110 deaths (2015)
5. Takashima, Y., Schluter, W.W., Mariano, K.M.L., Diorditsa, S., de Quiroz Castro, M., Ou, A.C., Ducusin, M.J.U., Garcia, L.C., Elfa, D.C., Dabbagh, A., Rota, P.: Progress toward measles elimination—Philippines, 1998–2014. MMWR **64**(13), 357–362 (2015)
6. International Federation of Red Cross and Red Crescent Societies. Emergency Plan of Action (EPoA) Philippines: Measles Outbreak (2019)
7. Ching, P.K., Zapanta, M.J., De Los Reyes, V.C., Tayag, E., Magpantay, R.: Investigation of a measles outbreak in Cordillera. Northern Philippines. WPSAR **7**(3), 1–5 (2016). https://doi.org/10.5365/WPSAR.2015.6.4.007
8. Gonzales Jr., A., Tubera-Panes, D.L., Purificacion, S.: Population-based study of measles and vaccination coverage in Baguio City, Philippines. Pediatr. Infect. Dis. Soc. Philippines J. **16**(1), 28–35 (2015)
9. Anselin, L.: Local indicators of spatial association-LISA. Geogr. Anal. **27**(2), 93–115 (1995). https://doi.org/10.1111/j.1538-4632.1995.tb00338.x
10. Chen, Y.: New approaches for calculating moran's index of spatial autocorrelation. PLoS ONE **8**(7), e68336 (2013). https://doi.org/10.1371/journal.pone.0068336
11. Finkenstädt, B.F., Grenfell, B.T.: Time series modelling of childhood diseases: a dynamical systems approach. J. Royal Stat. Soc. Ser. C (Appl. Stat.) **49**(2), 187–205 (2000)
12. Becker, A.D., Grenfell, B.T.: tsiR: an R package for time-series Susceptible-Infected-Recovered models of epidemics. PLoS ONE **12**(9), e0185528 (2017). https://doi.org/10.1371/journal.pone.0185528

13. Bjørnstad, O.N., Finkenstädt, B.F., Grenfell, B.T.: Dynamics of measles epidemics: estimating scaling of transmission rates using a time series SIR model. Ecol. Monographs **72**(2), 169–184 (2002)
14. Glass, K., Xia, Y., Grenfell, B.T.: Interpreting time-series analyses for continuous-time biological models-measles as a case study. J. Theor. Biol. **223**(1), 19–25 (2003). https://doi.org/10.1016/s0022-5193(03)00031-6
15. Grenfell, B.T., Bjørnstad, O.N., Finkenstädt, B.F.: Dynamics of measles epidemics: scaling noise, determinism, and predictability with the TSIR model. Ecol. Monographs **72**(2), 185–202 (2002)
16. Held, L., Hens, N., D O'Neill, P., Wallinga, J. (Eds.): Handbook of infectious disease data analysis. CRC Press (2019). https://doi.org/10.1201/9781315222912
17. He, D., Dushoff, J., Day, T., Ma, J., Earn, D.J.: Mechanistic modelling of the three waves of the 1918 influenza pandemic. Theor. Ecol. **4**(2), 283–288 (2011)

Spatiotemporal Analysis of Typhoid Cases in Baguio City, Philippines

Nheil Ignacio, Roberto Liwag, and Rizavel Addawe

Abstract We explore the surveillance data of typhoid fever from 2010 to 2018 in Baguio City, Philippines. The time series model exhibits the behavior of the monthly incidence rates in the 129 barangays. Frequency counts according to sex and age groups were fitted to Poisson processes. Spatial autocorrelation using global and local Moran's index was employed to identify clusters of the disease among the barangays. Moreover, the associations of typhoid fever with environmental factors such as precipitate, humidity, and temperature were verified.

Keywords Spatiotemporal · Typhoid fever · SARIMA

1 Introduction

Typhoid fever is a systemic disease caused by the gram-negative *Salmonella enterica* serovar Typhi (S. Typhi), known as enteric fevers, together with *Salmonella enterica* serovars Paratyphi (S. Paratyphi) A, B, and C. While S. Paratypi C has been reported to infect a variety of animals, S. Typhi and S. Paratyphi A and B infections are limited to humans and they are transmitted, via the fecal-oral route, through ingestion of contaminated water or food, or person-to-person contact [1]. There are more than 10.9 million cases worldwide occurring in the year 2017 alone which resulted in more than 116,800 deaths [2].

A unified prevention approach involving vaccines, improvements in water and sanitation infrastructure, proper hygiene, improved diagnostics, and appropriate antibiotics makes typhoid fever preventable [3]. Social inequity is evident in the control of this disease. On one hand, high-income countries have lesser incidence due to increased access to safe water and food, better sanitation and hygiene, improved surveillance and diagnostics, and access to required antibiotics making typhoid fever a disease of the past [4, 5]. On the other hand, this disease is still endemic in low- and middle-income countries due to poor housing, poor sanitation, and personal hygiene,

N. Ignacio (✉) · R. Liwag · R. Addawe
Department of Mathematics and Computer Science, College of Science, University of the Philippines Baguio, 2600 Baguio, Philippines
e-mail: njignacio1@up.edu.ph

M. H. Mohd et al. (eds.), *Modelling, Simulation and Applications of Complex Systems*, Springer Proceedings in Mathematics & Statistics 359,
https://doi.org/10.1007/978-981-16-2629-6_16

contaminated water and food, and limited treatment options hence this disease is essentially a disease of the poor [5–7].

Saharan Africa tops the list with 7.2 million cases expected followed by South Asia with 3.6 million cases, North Africa and the Middle East with 2.6 million, and Southeast Asia (including East Asia, and Oceania) with 2.21 million cases [8]. Although the number of cases is seen to decrease through the years, the emergence of multi-drug resistance typhoid and its spread to other regions, because of increasing international travel, is becoming a serious concern in the fight against the disease [7, 9, 10].

The Philippines is one of the countries where typhoid fever is endemic in big cities and their surroundings and even outside these major urban centers. Moreover, multi-drug resistant S. Typhi strains have been detected which aggravates the burden of the disease [11, 12]. This disease has moved up in the ranking of causes of premature deaths in the country from the year 1990 to 2010 [13].

Based on a report of the City Government of Baguio's official website, in the year 2018, there are no typhoid-related deaths compared to other years. However, there are a total of 15% increase of typhoid cases in the Cordillera Administrative Region (CAR) [14]. The office of the regional epidemiology and surveillance unit (RESU) is active in monitoring the number of patients having symptoms of typhoid and in identifying the clustering of typhoid fever cases region-wide. In this research, we will describe the spatial clusters of the disease based on historical data. Furthermore, the results are of particular importance to continuous surveillance and the reduction of typhoid incidence.

2 Materials and Method

2.1 Study Area

Baguio City, Benguet is part of CAR in the Philippines. It is composed of 129 barangays—smallest political unit in the Philippines—which is the basis for the demographics and geographic distribution of this study. With a total land area of 57.5 square kilometers and according to the Philippine Statistics Authority (PSA), the total census of population is roughly 345,366 in the year 2015 [15]. The city is known as the Summer Capital of the Philippines because of its annual low temperature. Based on Philippine Atmospheric, Geophysical and Astronomical Services Administration (PAGASA), from the year 2010 to 2018, the average temperature is 19 °C per year, with lowest recorded at 7.3 °C and highest recorded at 30 °C. In addition, precipitation or the amount of rainfall per year ranges from 2,860 mm to 5,490 mm [16].

2.2 Collection and Management

The data used in this study was requested from the Baguio City - Health Service Office (HSO) which collects and maintains surveillance data of infectious diseases in the area. For privacy considerations, the personal identifiers for the patients were deleted. The data was categorized according to the following: year, morbidity week and month, age, sex, and barangay. It comprises of patients who were diagnosed of typhoid fever through blood culture, TUBEX, Typhi DOT, and Widal tests by their attending physicians. Only patients with positive results who reside in Baguio City constitute the sample of the study within the year 2010 to 2018.

The record of the city's population in May 2010 and April 2015 were taken from the census data of the PSA. Exponential growth model were used to estimate the population for each month from 2010 to 2018. That is

$$A(t, j) = A(0, j)k_j^t, \tag{1}$$

where $A(t, j)$ is the population at month t and barangay j. Let $t = 5$ denotes May 2010 and $t = 64$ denotes April 2015. Since the initial data given is $A(5, j)$, we can manipulate the growth model as follows

$$A(t, j) = A(0, j)k_j^t = A(0, j)k_j^5 \cdot k_j^{t-5} = A(5, j)k_j^{t-5}. \tag{2}$$

The next census of population declares the population at time $t = 64$, we can solve for k_j. That is,

$$k_j = \exp\left[\frac{\ln A(64, j) - \ln A(5, j)}{64 - 5}\right]. \tag{3}$$

This was done to prevent negative population estimates.

2.3 Descriptive Analysis

A Baguio City shapefile map was obtained online from PhilGIS.org from which the city map is generated [17]. The number of typhoid cases were analyzed per geographic distribution by the total number of cases from the year 2010 to 2018. The monthly incidence rates per 1000 individual from each barangay, $IR(t, j)$ is computed by

$$IR(t, j) = \frac{C(t, j)}{A(t, j)} \times 1000, \tag{4}$$

where $C(t, j)$ is the number of cases at month t and barangay j.

The cases were classified by age group based on the Philippines' Department of Health (DOH) reports on typhoid cases [25]. In addition, the ratio of the number of male to number of female who acquired the disease is also computed and analyzed. From the monthly incidence data, the average monthly rate was computed which was used to create a Poisson distribution model that describes the probability of acquiring the disease.

2.4 Spatial Autocorrelation

Typhoid cases autocorrelation were computed using global Moran's Index for Baguio City using the approach in [18]. Suppose there are n barangays in the city. Let t be fixed, $x_t := [x_{tj}]$ is a column vector with entries $x_{tj} := IR(t, j)$ for barangay $j = 1, 2, \ldots, n$, $z_t := [z_{tj}]$ is the standardized vector of x_t and $W := [w_{ij}]$ is the unitary spatial weights adjacency matrix. Then the global Moran's Index at month t is given by

$$I(t) = z_t^T W z_t. \tag{5}$$

The neighboring of the adjacency matrix is based on queen contiguity where in the two barangays are neighbors if and only if they shared at least one point. This was used to determine if there were clustering of high (low) typhoid fever incidences during the study period, years 2010 to 2018.

Meanwhile, local Moran's Index or the local indicator of spatial autocorrelation (LISA) was used to compute for the autocorrelation for each barangay in Baguio City and its neighboring barangays. Using the new approach, the local Moran's Index of barangay b at month t, $I_b(t)$, is given by

$$I_b(t) = (z_t z_t^T W)_{bb}, \tag{6}$$

that is the bth row and b^{th} column of the square matrix

$$z_t z_t^T W = \begin{bmatrix} z_{t1} \sum_{j=1}^{n} w_{1j} z_{tj} & z_{t1} \sum_{j=1}^{n} w_{2j} z_{tj} & \cdots & z_{t1} \sum_{j=1}^{n} w_{nj} z_{tj} \\ z_{t2} \sum_{j=1}^{n} w_{1j} z_{tj} & z_{t2} \sum_{j=1}^{n} w_{2j} z_{tj} & \cdots & z_{t2} \sum_{j=1}^{n} w_{nj} z_{tj} \\ \vdots & \vdots & \ddots & \vdots \\ z_{tn} \sum_{j=1}^{n} w_{1j} z_{tj} & z_{tn} \sum_{j=1}^{n} w_{2j} z_{tj} & \cdots & z_{tn} \sum_{j=1}^{n} w_{nj} z_{tj} \end{bmatrix}.$$

The local Moran's Index was used to identify which barangays are of significant incidence rate and where the clusters are. The statistical significance of Moran's index is measured based on p-value less than 0.05.

2.5 Time-Series Analysis

The univariate Box-Jenkins autoregressive integrated moving average (UBJ- ARIMA) [19] modeling method which served as a tool for the formulation of appropriate models only applies to stationary series, a series with constant mean, variance, and autocorrelation function (ACF) over time. The Augmented Dickey-Fuller (ADF) test is a useful test for verifying the stationarity of the data. In this test, a stationary series must give a p-value of less than 0.05. Otherwise, differencing or (and) logarithmic transformation of the data may be necessary. ARIMA modeling consists of three essential steps: identification, estimation, and diagnostic checking.

In the model identification and estimation, the estimated ACF and partial autocorrelation functions (PACF) are used as guides in choosing one or more seasonal or nonseasonal models that best fit the stationary series. The ARIMA (p, d, q) comprises of three parameters: (p) for the autoregressive (AR), (d) for the number of differencing induced, and (q) for the moving average (MA). When the series has temporal dependence, additional parameters are included to account for seasonality. The multiplicative seasonal ARIMA $(p, d, q)(P, D, Q)_s$ model has additional parameters: P for the seasonal AR, D for the seasonal differencing, and Q for the seasonal MA. Using the backward shift notational device $By_t = y_{t-1}$ when working with time series lags, the equation is given by

$$\Phi(B^s)\phi(B)(1-B)^d(1-B^s)^D Y_t = \Theta(B^s)\theta(B)\varepsilon_t \tag{7}$$

where $\Phi(B^s)$ is the seasonal AR operator, $\phi(B)$ is the regular AR operator, $(1-B)^d$ is the regular difference operator, $(1-B^s)^D$ is the seasonal difference operator, Y_t is the data series, $\Theta(B^s)$ is the seasonal MA operator, $\theta(B)$ is the regular MA operator, and ε_t is the series of white noise.

The seasonal part of the model consists of terms that are similar to the non-seasonal components of the model, but involve backshift of the seasonal period. An $ARIMA(1, 1, 1)(1, 1, 1)_{12}$ model (without a constant) for example is for monthly data ($s = 12$) and can be written as

$$(1-\Phi_1 B^{12})(1-\phi_1 B)(1-B)(1-B^{12})Y_t = (1+\Theta_1 B^{12})(1+\theta_1 B)\varepsilon_t. \tag{8}$$

With a total number of observations T, the first $T/4$ lags are examined for significantly large ACF of the resulting residuals. To test the adequacy of the model, the root mean square error (RMSE) and the mean absolute square error (MASE) were computed. In addition, to measure the relative quality of the models, the Akaike Information Criterion (AIC) and the Bayesian Information Criterion (BIC) were used. The AIC measures the amount of information lost when a particular model is used to predict future values of a given data set, while the BIC ensures that the model adheres to the principle of parsimony. The model with relatively smaller AIC and BIC must also satisfy the conditions of stationarity and invertibility of its AR and MA coefficients, respectively.

The forecasting ability of the tentative model is verified by testing if the autocorrelations of its residuals are statistically independent of each other. Using the Ljung-Box test, a model with statistically uncorrelated residuals must output a p-value of greater than 0.05. Also, a normality test of these residuals is needed. For further evaluation of the model, the predicted and actual values are compared [23, 24].

3 Results

3.1 Demographic Characteristics

From the data gathered from the Baguio City-HSO, there were 4,499 suspected typhoid cases from the year 2010 to 2018. Upon filtering the data, there were only 2,240 patients positive to have typhoid fever from Baguio City. Susceptibility to gender was almost equal having 1,117 males and 1,123 females as shown in Fig. 1. Children aged five to fourteen (5 to 14) have the highest number of cases, 630 individuals. Moreover, children of age four are most vulnerable to the disease since it recorded the highest incidence having a total of 105 cases.

Three big barangays recorded more than 100 cases from the year 2010 to 2018. These barangays are *Irisan* having 169, *Loakan Proper* having 125, and *Camp 7* with 110. The year 2015 has the highest recorded number of cases (374 infected individuals), with the highest recorded in August with 105 infected individuals.

As shown in Fig. 2, the monthly average number of cases is about twenty-one (21), the highest number of cases was recorded in August with about thirty (30). There are roughly about sixteen (16) cases every April and December. Moreover,

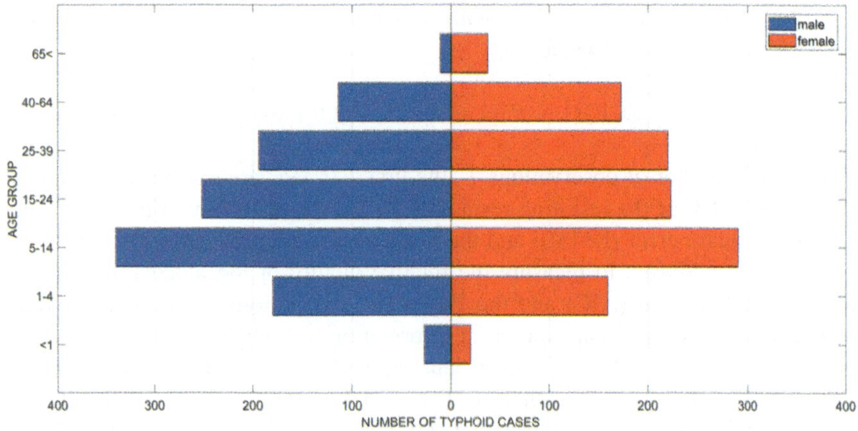

Fig. 1 Total incidence (2010–2018) of typhoid fever according to sex and age group

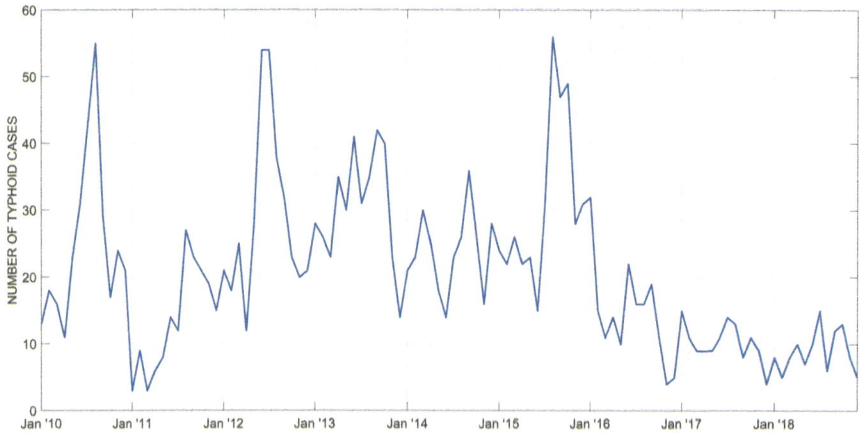

Fig. 2 Monthly counts of typhoid fever from 2010 to 2018

the daily average incidence per 1000 individuals is approximately 0.682. Using the Poisson distribution shown in Fig. 3(a), the probability of having n number of cases daily is given by

$$P(N = n) = e^{-\lambda} \frac{\lambda^n}{n!}, \tag{9}$$

where λ is the daily average case which in this case 0.682 as shown in Fig. 3(b). For example, the probability of 4 typhoid cases to occur within a day is

$$P(N = 4) = e^{-\lambda} \frac{\lambda^4}{4!} \approx 0.35\%. \tag{10}$$

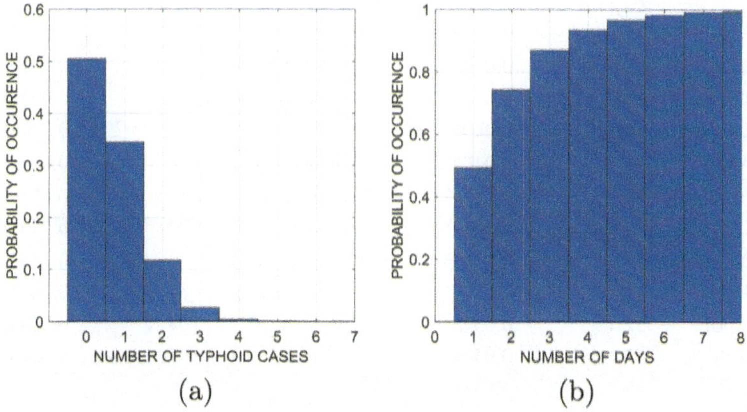

Fig. 3 (a) Poisson Distribution, and (b) Poisson Process with $\lambda = 0.682$

Using the Poisson process, with the same λ, the probability of waiting less than or equal to d days for a typhoid case to occur is given by

$$P(D \leq d) = 1 - e^{-0.682d}. \tag{11}$$

For example, the probability of a typhoid case to occur within two days is

$$P(D \leq 2) = 1 - e^{-\lambda \cdot 2} \approx 74.44\%. \tag{12}$$

3.2 Spatial Autocorrelation

Based on the global Moran's Index, there are 39 months with positive $I(t)$. Using normalization and randomization approaches to compute for z-score and p-value [20], 23 values are significant with $p < 0.05$. These positive Moran's Index values are shown in Table 1. The months t where $I(t) > 0$ with $p < 0.01$ arranged from lowest p-value to highest p-value were $t = 11, 81, 45, 61, 84, 70, 104$.

Now, we define the *hotspot* as the barangay with statistically significant $I_j(t)$ with high incidence rate. That means, for a fixed month t, if the p-value of the z-score of $I_j(t)$ is of threshold say $p < 0.05$, then barangay j has significant typhoid cases. Moreover, if barangay j, $j = 1, 2, \ldots, 129$, has high incidence rate, then j is a hotspot. Combining the significance of global Moran's Index and local Moran's Index, we can identify the *high-high* clusters. That is, for a fixed month t, if $I(t)$ is positive and significant, then there exist neighbor barangays $j \neq k$, $j, k = 1, 2, \ldots 129$

Table 1 Values of Moran's Index in a particular year and month t where ** and * indicate p-values $< .01$ and < 0.05, respectively

Date	t	Moran's $I(t)$	Date	t	Moran's $I(t)$
2010 November	11	0.2357**	2014 August	56	0.0533*
2011 August	20	0.0395*	2015 January	61	0.1119*
2011 December	24	0.0582**	2015 April	64	0.0635*
2012 December	36	0.0406*	2015 June	66	0.0648*
2013 May	41	0.0469**	2015 October	70	0.0847*
2013 September	45	0.1133**	2016 February	74	0.0536*
2013 October	46	0.0649**	2016 July	79	0.0494*
2013 November	47	0.0309*	2016 September	81	0.1507*
2013 December	48	0.0526**	2016 December	84	0.1061*
2014 April	52	0.0393*	2018 August	104	0.0838*
2014 May	53	0.0374*	2018 September	105	0.0529*
2014 July	55	0.0348*			

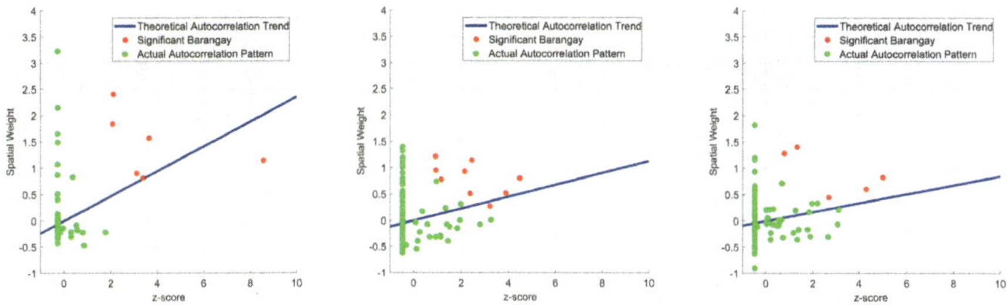

(a) (b) (c)

Fig. 4 Scatter plots of standardized (z-scores) Moran's Index values at (a) month $t = 11$, (b) month $t = 45$, and (c) month $t = 70$

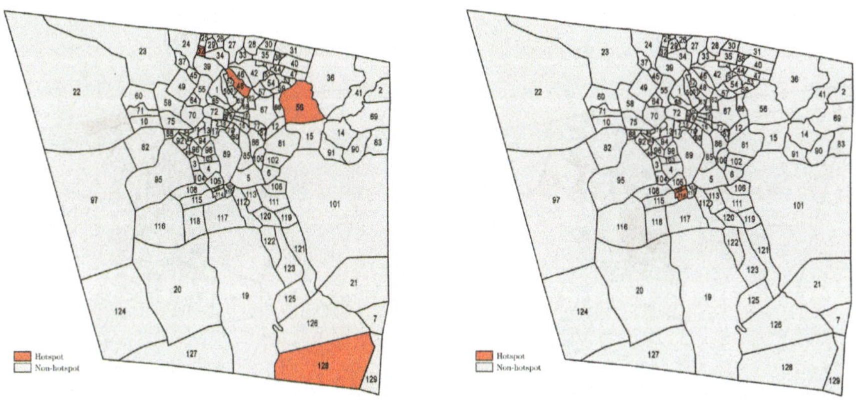

(a) (b)

Fig. 5 Barangays having significant high incidence rate at (a) $t = 7$ and (b) $t = 84$

such that $I_j(t)$ and $I_k(t)$ are statistically significant. Furthermore, if j and k are hotspots, then j and k are in *high-high* cluster.

Looking at the Moran's scatter plot in Fig. 4, we can observe that $I(t)$ is the slope of the regression line for the z-scores and all the hotspots are in the *high-high* cluster.

With the results, the months having greater than or equal to five hotspots were $t = 11, 36, 45, 48, 52, 53, 70$. Notice that for $t = 7$, $I(7)$ is not statistically significant but there are only four hotspots on this month which means that there is no sufficient evidence that there are *high-high* clusters. However, for $t = 84$, $I(84)$ is positive and statistically significant but there are only two hotspots for this month which means that these two barangays are neighbors. To verify this, we look at the scatter plots and their corresponding map in Fig. 5.

The maps on Fig. 6 show the history of barangays having significant high incidence rate at $t = 35, 36$, at $t = 55, 56$, at $t = 69, 70$, and at $t = 104, 105$. Labels "0,0" indicates no hotspot at both time, "1,0" indicates a hotspot at the first given times,

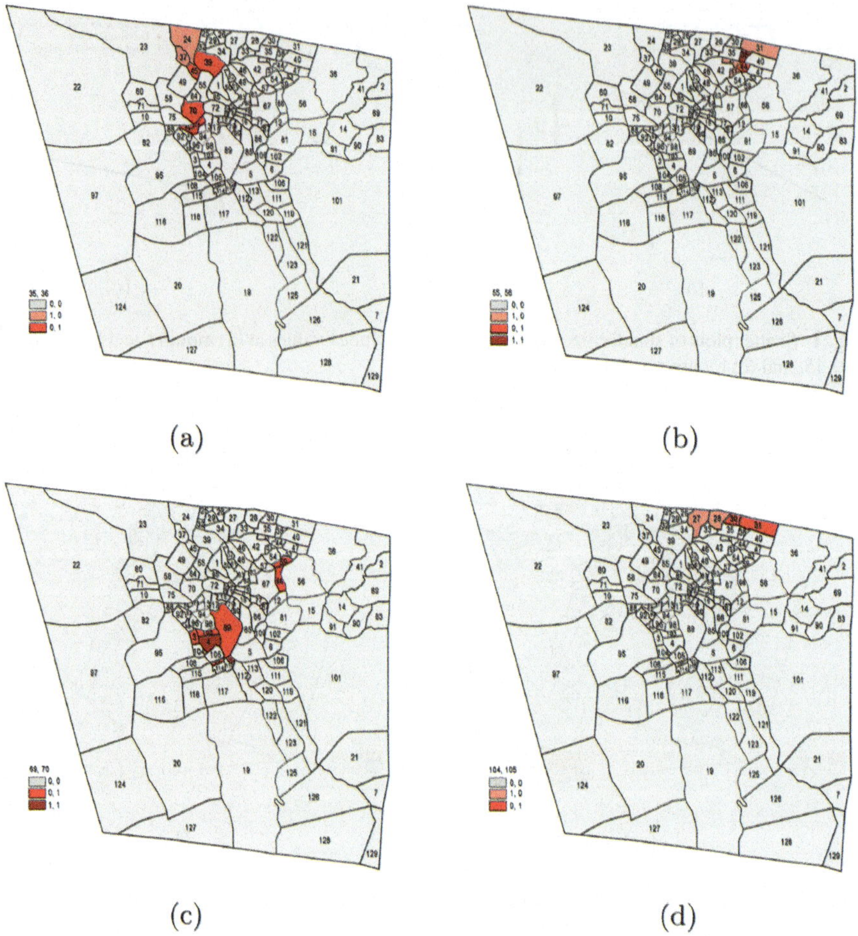

Fig. 6 Barangays having significant high incidence rate at (a) $t = 35, 36$, (b) $t = 55, 56$, (c) $t = 69, 70$, and (d) $t = 104, 105$

"0,1" indicates a hotspot at the second given time, and "1,1" indicates hotspots at both times. In Fig. 6(a), we can see that the hotspots are from barangays $j = 24, 37$ and after a month, the neighboring barangays $j = 39, 45$ also become hotspots. Similarly, for Fig. 6(b), observe that barangays $j = 31, 38, 43, 51$ are hotspots at time $t = 55$, and after a month, the neighboring barangay $j = 44$ became a hotspot while barangay $j = 38$ remained a hotspot. Now, in Fig. 6(c), the hotspot at time $t = 69$ is just barangay $j = 4$, but observing the hotspots at time $t = 70$, the barangay $j = 4$ was still a hotspot including some of its neighboring barangays $j = 3, 89, 103$ and the neighbors of barangay $j = 89$ which are barangays $j = 107, 109$. Lastly, in Fig. 6(d), the hotspots originated from barangays $j = 27, 28$ at time $t = 104$ and after a month, these barangays were no longer hotspots but the neighboring barangay

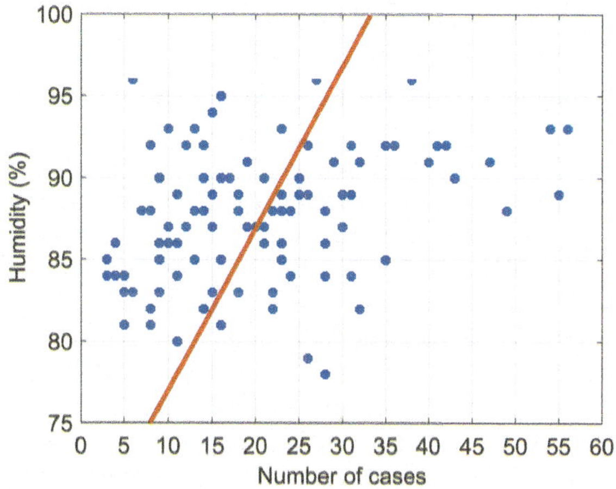

Fig. 7 Scatter plot of number of typhoid cases versus humidity, with a correlation value $r = 0.3363$ which is significant at $\alpha = .05$, and the regression line $C = 1.0138 H - 68.0819$

$j = 30$ along with its neighbor $j = 31$ were hotspots at time $t = 105$. There are no hotspots at time $t = 106$ and no significant evidences that there are clusterings at time $t = 106, 107, 108$ which are the last months of year 2018.

This study also include an investigation of the association of the number of monthly typhoid cases, C, and the variables are humidity (%), precipitate (mm), and monthly mean maximum temperature, mean minimum temperature, highest and lowest temperatures (°C) in Baguio City. As expected, the temperature variables are highly correlated with each other. Similarly, precipitate and relative humidity are also strongly correlated. Using the backward elimination method, the regression model included three independent variables, mean minimum temperature, lowest temperature, and relative humidity with $R^2 = 0.1415$. Continuing with the backward elimination, a model with relative humidity as the only significant variable for typhoid cases is a better fit with a p-value less than 0.05. Meanwhile, using forward selection method, relative humidity has the lowest p-value of 0.0004 with $R^2 = 0.1131$. For any additional variable, there shows no improvement in the model. The scatter plot with the regression line in Fig. 7 is given by $C = 1.0138H - 68.0819$, where H is the measured relative humidity for the month. The number of typhoid cases is significantly associated with relative humidity, with $r = 0.3363$. However, the small value of R^2 implies that humidity explains a very small amount of variability in the number of typhoid cases. Some other latent variables may better explain the variability of the number of typhoid fever cases in the city. These variables may include tourist arrivals, drinking water sources, housing and sanitation facilities, antibiotic treatments, and other environmental factors [3, 4]. These explanatory variables were not included in this study.

3.3 Time Series Analysis

A stationary time series is needed in order to create ARIMA model. The training data projects a seasonal pattern and gradual drop to zero in its ACF as seen in Fig. 8(a) and (b). This is confirmed by the ADF test conducted on the data with p-value >0.05. It implies that there is a need to take the first differences of the series.

After first differencing of this series, a stationary series is obtained with ADF test p-value < 0.01. The ACF and PACF of the stationary series both show negative autocorrelations at lag $= 2$ which gives the following candidate models for ARIMA(p, d, q) taking into account the first differences $d = 1$: ARIMA(0, 1, 0), ARIMA(0, 1, 2), ARIMA(2, 1, 2), and ARIMA(2, 1, 0).

The seasonal pattern of the training data is an indication that a seasonal ARIMA is the appropriate model. With the temporal dependence of typhoid data from a previous study [21] and a strong long-high-low incidence pattern for each year, seasonal differencing is considered giving $D = 1$ in the seasonal ARIMA$(p, d, q)(P, D, Q)_{12}$. The need for seasonal AR (SAR) and seasonal MA (SMA) components are verified through a fit of the training data to the seasonal ARIMA(0, 0, 0)(0, 1, 0) using the $Arima()$ function of the *forecast* package. The ACF of the residuals of the model shows negative autocorrelation at $lag = 12$ indicating that an SMA term is needed. Hence, $Q = 1$. This updates the candidate models to the models in Table 2.

These models were fitted using the $Arima()$ function and the AIC and BIC of each were noted as shown in Table 2. The model with the lowest AIC and BIC is

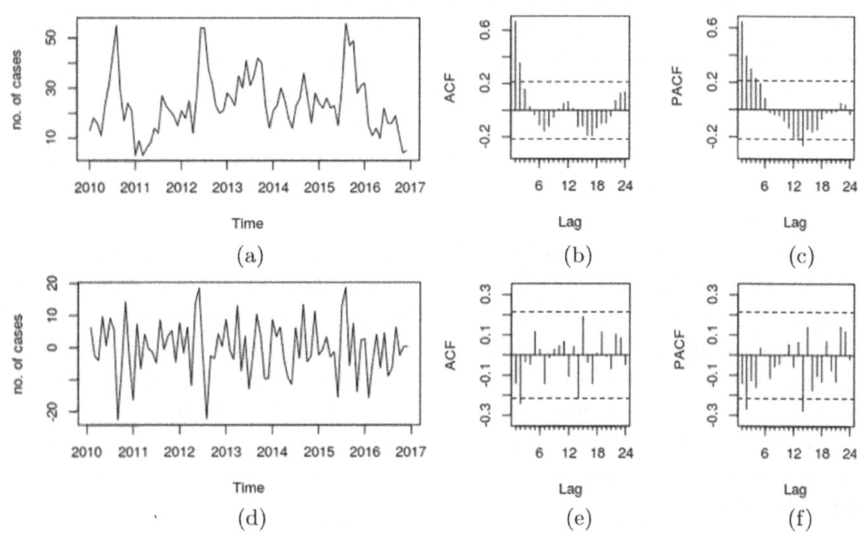

Fig. 8 (a) 2010–2016 series, (b) ACF of 2010–2016 series, (c) PACF of 2010–2016 series, (d) series after first differencing, (e) ACF of series after first differencing, and (f) PACF of series after first differencing

Table 2 Comparison of the AIC and BIC of the different models

Model	AIC	BIC
SARIMA(0, 1, 2)(0, 1, 1)$_{12}$	538.71	547.76
SARIMA(2, 1, 2)(0, 1, 1)$_{12}$	541.16	554.73
SARIMA(2, 1, 0)(0, 1, 1)$_{12}$	541.36	550.41
SARIMA(0, 1, 0)(0, 1, 1)$_{12}$	543.74	548.27

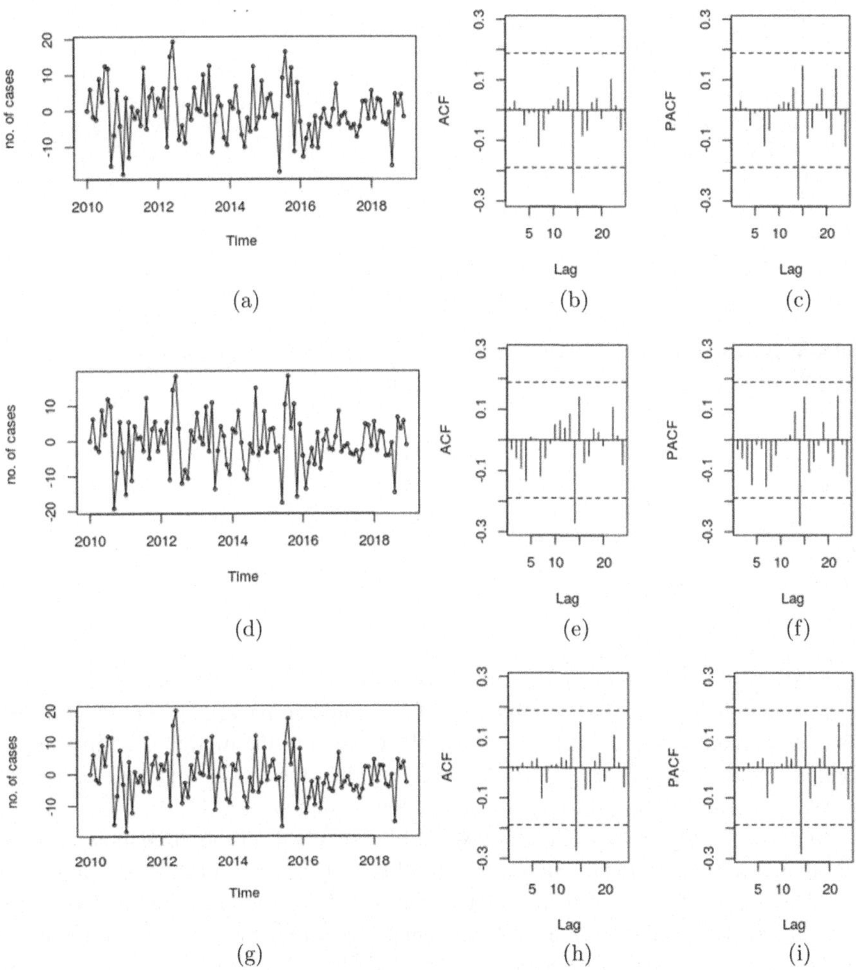

Fig. 9 Plots of (a) residuals, (b) ACF and (c) PACF of SARIMA(0, 1, 2)(0, 1, 1)$_{12}$; (d) residuals, (e) ACF and (f) PACF of SARIMA (2, 1, 0)(0, 1, 1)$_{12}$; and (g) residuals, (h) ACF and (i) PACF of SARIMA (2, 1, 2)(0, 1, 1)$_{12}$ models for the monthly incidence of typhoid fever

Table 3 Comparison of the RMSE, MAE and MASE of the different models

Model	RMSE	MAE	MASE
SARIMA$(0, 1, 2)(0, 1, 1)_{12}$	7.247018	5.630830	0.9214558
SARIMA$(2, 1, 0)(0, 1, 1)_{12}$	7.426731	5.775393	0.9451129
SARIMA$(2, 1, 2)(0, 1, 1)_{12}$	7.222412	5.635906	0.9222865

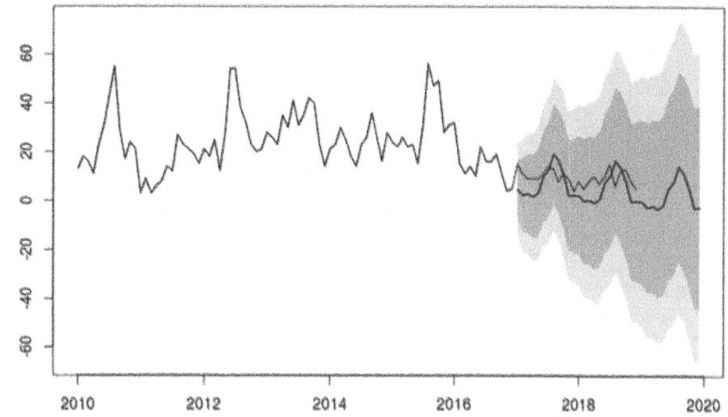

Fig. 10 Three-year forecast of SARIMA$(0, 1, 2)(0, 1, 1)_{12}$ in blue overlayed by the test data. The light gray color is the 95% prediction interval and the dark gray is the 80% prediction interval

Table 4 Forecasted number of typhoid incidence for peak months in three years

Month	2017	2018	2019
June	11	8	5
July	13	11	8
August	20	17	15
September	17	14	12
October	11	9	6

SARIMA$(0, 1, 2)(0, 1, 1)_{12}$. This model is used to forecast the next three years, $h = 36$ using the $forecast()$ function of the same R package [22]. Within this forecast range, the test data from 2017–2018 is fitted to the model as an in-sample forecast data. The accuracy test of the model against the test data is done using the $accuracy()$ function and gives a mean absolute scaled error of 0.38.

As shown in Fig. 9, the ACF and PACF of the residuals of the three models show that there are no significant spikes at first 12 lags. It implies that the residual ACF and PACF is within the 95% confidence interval. If we observe the three models, it is sufficient to compare the root mean square error (RMSE), mean absolute error (MAE), and the mean absolute scaled error (MASE). We show these measures of errors in Table 3.

Since the data is seasonal, we chose the model with the lowest MASE. Hence, the best fit model is SARIMA$(0, 1, 2)(0, 1, 1)_{12}$. It is remarkable that the number of incidence in the forecast data is still high for the year 2019 as seen in Table 4. The plot of the forecast is shown in Fig. 10.

4 Conclusion

The disease surveillance data taken from the Baguio City- HSO showed that there were about 2,240 positive cases of typhoid fever in the city from the year 2010 to 2018. Most of whom are children aged four to fourteen. The three most populated barangays (*Irisan*, *Loakan Proper* and *Camp 7*) recorded the highest number of cases among the 129 barangays in Baguio City. The highest number of cases was recorded in the year 2015 with a peak in August when the humidity is relatively high and has been shown to have positive associations with typhoid incidence. On average, there are around twenty-one monthly recorded cases. This indicates that the city is still far from being typhoid-free. The forecast data for 2019 still shows a high incidence in the peak months.

The results revealed that clusters of typhoid fever with hotspots during the end of the wet season and the beginning of the cold season, November 2010, September 2013, November 2015, and September 2016. Hence, typhoid fever is still endemic to the city as affirmed twenty years back [11]. With the environmental factors associated with the disease and the weather conditions during the months where hotspots are found, the fight to eradicate the city of this preventable disease is far from over. Continuing and better surveillance of typhoid fever is in order and associations of the disease should be further discovered and investigated especially to identified hotspot barangays from 2010 to 2018 and the months when these clusters occurred.

Acknowledgments The authors are expressing their gratitude to Dra. Donnabel Tubera-Panes of Baguio City-HSO for the data and the discussion of surveillance and epidemiology. Also, special thanks to Irineo Dones Jr. RMT and Eduardo Manalo Jr. MClAud for the detailed discussions and clarifications about the topic.

References

1. Dolecek, C.: Typhoid fever and other enteric fevers. In: Infectious Disease. 4th Edition (2017) https://doi.org/10.1016/B978-0-323-04579-7.00109-X
2. Stanaway, D.: The global burden of typhoid and paratyphoid fevers: a systematic analysis for the global burden of disease study 2017. Lancet Infect. Dis. **19**(4), 369–381 (2019). https://doi.org/10.1016/s1473-3099(18)30685-6
3. Coalition Against Typhoid.: Typhoid and Related Illnesses (2018). http://www.coalitionagainsttyphoid.org/why-typhoid/typhoid-related-illnesses/
4. Theiss-Nyland, K., Meiring, J., Laurens, M.: The importance of knowing more: why we need ongoing studies, research, and data about typhoid, 8 March 2019. http://www.coalitionagainsttyphoid.org/the-importance-of-knowing-more-why-we-need-ongoing-studies-research-and-data-about-typhoid/
5. Crump, J.: Progress in typhoid epidemiology. In: 10th International Conference on Typhoid and Other Invasive Salmonellosis, Uganda (2017)
6. Hussein Gassem, M., Dolmans, W.M.V., Keuter M., Djokomoljanto R.: Poor food hygiene and housing as risk factors for typhoid fever in Semarang, Indonesia. Tropical Med. Int. Health **6**(6), 0-00 (2001)

7. Zaidi, A.: Looking to the future: control of typhoid in the next decade. In 10th International Conference on Typhoid and Other Invasive Salmonellosis, Uganda (2017)
8. Antillon, M.: The burden of typhoid fever in low- and middle-income countries: a meta-regression approach. PLOS Neglected Trop. Dis. **11**(2) (2017) https://doi.org/10.1371/journal.pntd.0005376
9. Pang, T., Bhutta, Z.A., Finlay, B.B., Altwegg, M.: Typhoid fever and other salmonellosis: a continuing challenge. Trends Microbiol. **3**(7), 253–255 (1995). https://doi.org/10.1016/s0966-842x(00)88937-4
10. Choe, Y.J., Choe, S.A., Cho, S.: Importation of travel-related infectious diseases is increasing in South Korea: an analysis of salmonellosis, shigellosis, malaria, and dengue surveillance data. Travel Med. Infect. Dis. **19**, 22–27 (2017). https://doi.org/10.1016/j.tmaid.2017.09.003
11. Abucejo, P.E.: Blood culture confirmed typhoid fever in a provincial hospital in the Philippines. Southeast Asian J. Trop. Med. Public Health **32**(3), 531–536 (2001)
12. Dayrit, M.M.: Epidemiology of typhoid fever in the Philippines. Phil. J. Microbiol. Infect. Dis. **21**(2), 36–48 (1992)
13. Institute for Health Metrics and Evaluation.: Global Burden of Diseases, Injuries, and Risk Factors Study 2010. http://www.healthmetricsandevaluation.org
14. The City Government of Baguio. http://www.baguio.gov.ph
15. Philippine Statistics Authority. http://www.psa.gov.ph
16. Philippine Atmospheric, Geophysical and Astronomical Services Administration. http://bagong.pagasa.dost.gov.ph
17. Philippine GIS Data Clearing House. http://philgis.org
18. Chen, Y.: New approaches for calculating moran's index of spatial autocorrelation. PLoS ONE **8**(7) (2013). https://doi.org/10.1371/journal.pone.0068336
19. Box, G.P.E., Jenkins, G.M., Reinsel, G.C., Ljung, G.M.: Time Series Analysis: Forecasting and Control (Wiley Series in Probability and Statistics), 5th edn. John Wiley & Sons Inc., Hoboken, New Jersey (2016)
20. Bivand, R.S., Wong, D.W.S.: Comparing implementations of global and local indicators of spatial association. TEST **27**(3), 716–748 (2018). https://doi.org/10.1007/s11749-018-0599-x
21. Cawiding, O., Natividad, G., Bato, C., Addawe, R.: Forecasting typhoid fever incidence in the cordillera administrative region in the philippines using seasonal ARIMA models. AIP Conf. Proc. **1905**, (2017)
22. R Core Team. R: A language and environment for statistical computing. R Foundation for Statistical Computing, Vienna, Austria (2013). http://www.R-project.org/
23. Hyndman, R.J., Khandakar, Y.: Automatic time series forecasting: the forecast package for R. J. Stat. Softw. **26**(3), 1–22 (2008)
24. Hyndman, R.J., Koehler, A.B.: Another look at measures of forecast accuracy. Int. J. Forecast. **22**(4), 679–688 (2006)
25. Department of Health. http://www.doh.gov.ph

Chaotic Time Series Prediction Using Random Fourier Feature Kernel Least Mean Square Algorithm with Adaptive Kernel Size

Noor A. Ahmad and Shazia Javed

Abstract The random Fourier feature kernel least mean square (RFF-KLMS) algorithm provides a finite dimensional approximation to the kernel least mean square algorithm with radially symmetric Gaussian kernel. RFF-KLMS was introduced to curb the continuously growing radial basis function (RBF) network which prohibits online application of KLMS. RFF-KLMS assumes a fixed kernel size and the application of the method in nonlinear online regression can be quite tedious because it is not always obvious which kernel size to choose for a particular problem. In this paper, we incorporate a stochastic gradient approach in RFF-KLMS to update the kernel size. The efficacy of the new approach is demonstrated in the online prediction of time series generated from two different chaotic systems. In both examples, the RFF-KLMS algorithm with adaptive kernel size demonstrates very good tracking ability.

Keywords Kernel methods · Kernel least mean square · Random fourier features · Chaotic time series prediction · Machine learning

1 Introduction

Modeling of processes in complex systems can be divided into two basic approaches; i) the study of mathematical models which tries to capture the most important qualitative features of the complex systems behavior, ii) the reconstruction of the underlying structure of a nonlinear dynamical system from measured data with use of methods

The authors would like to acknowledge the financial support from Universiti Sains Malaysia through Research University Grant (RUI) (acc. No. 1001/PMATHS/8011040).

N. A. Ahmad (✉)
School of Mathematical Sciences, Universiti Sains Malaysia, 11800 Penang, Malaysia
e-mail: nooratinah@usm.my

S. Javed
Lahore College for Women University, Lahore, Punjab, Pakistan
e-mail: shazia.javed@lcwu.edu.pk

M. H. Mohd et al. (eds.), *Modelling, Simulation and Applications of Complex Systems*, Springer Proceedings in Mathematics & Statistics 359,
https://doi.org/10.1007/978-981-16-2629-6_17

of mathematical statistics, statistical learning, data mining and so on. The second approach is more popularly known as time-delay embedding (also known as the Taken's delay embedding theorem [1]). The theorem states that the dynamics of a system (i.e. the attractor) can be reconstructed from vectors of time-shifted (shift T) states of single variable with nested dimension N

$$\mathbf{x}(t) = (x(t), x(t+T), ..., x(t+[N-1]T))^T,$$

where $N > 2d_A + 1$ and d_A is the dimension of the attractor. Takens embedding theorem lays the foundation for nonlinear time series analysis that allows for the reconstruction of complete system dynamics using a single time series [1]. Although the initial motivation of the theorem was to look for chaotic behaviour in experimental systems, the potential use of the method in a broader range of signal processing activities was soon recognized [2, 3].

Recent developments in chaotic time series prediction witness an increasing number of effort in online time series prediction [4–11]. Online prediction is a sequential or adaptive learning process where the underlying pattern representations from time series data are extracted in a sequential manner. When new data arrive at any time, they are observed and learned by the system. In addition, as soon as the learning procedure is completed, the observations are discarded, without having the necessity to store too much historical information. The online learning setting is significantly in contrast with offline learning, where the learning process has access to the entire data, and is able to go through the data multiple times to train a model, leading to higher accuracy than their online counterparts. Nonlinear online learning [12–14] has drawn a considerable amount of attention as it captures nonlinearity in the data which cannot be effectively modeled in a linear online learning method, and usually achieves better accuracy. One group of nonlinear online learning is based on kernels, that is, kernel-based online learning.

Kernel adaptive learning is a class of kernel-based online learning which are derived in Reproducing Kernel Hilbert Space (RKHS) [15]. In contrast to nonlinear approximators such as the polynomial state-space models [16] and neural networks [10], kernel based learning inherits the convex optimization of its linear counterparts. Typical algorithms from the kernel adaptive filtering (KAF) family include Kernel Least Mean Square (KLMS) [17], Kernel Recursive Least Square [18], Kernel Affine Projection Algorithms [19], Extended Kernel Recursive Least Squares [20]. KLMS is by far the simplest to implement and has been proven to be computationally efficient. In KLMS (and KAFs in general), the solution is given in terms of a linear expansion of kernel functions (centered at the current input data). This linear expansion grows with each incoming data rendering its application prohibitive both in terms of memory as well as computational resources. The centers that make up the linear expansion of solution constitutes the so-called dictionary. Sparsification methods [21, 22] are commonly used to keep the dictionary sufficiently small, however they too require significant computational resources.

A more recent trend in curbing the ever growing structure of the KAFs algorithm is by using approximation methods such as the Nystrom method [23] and the random

Fourier features (RFF) [24–27]. While the Nystrom based method is data-dependent, the RFF based methods are drawn from a distribution that is randomly independent from the training data hence providing a good solution in non-stationary circumstances. The RFF-KLMS algorithm [24, 25] can be seen as a finite-dimensional approximation of the conventional KLMS algorithm, in which the kernel function is approximated by a finite-dimensional inner products. The RFF-KLMS algorithm has been proven to provide good approximation of the Gaussian kernel due to its symmetry and the shift-invariant property [25, 26]. The normalized Gaussian kernel is

$$\kappa(\mathbf{x}(i), \mathbf{x}(j)) = e^{-\|\mathbf{x}(i)-\mathbf{x}(j)\|^2/2\sigma^2}, \tag{1}$$

where $\sigma > 0$ is the kernel size. In both KLMS and RFF-KLMS, the kernel size is treated as a free parameter which is often manually set. The choice of kernel size can be very different from one data set to another, therefore to choose an appropriate kernel size, one may have to resort to empirical approach. Although there are various methods to choose kernel size in batch learning such as using cross validation [28–30], penalizing functions [31] and plug-in methods [31, 32], these methods are computationally intensive and unsuitable for online kernel learning.

In this paper, we introduce a variant of the RFF-KLMS algorithm which allows the kernel size to be adapted using a stochastic gradient method. This technique for adapting kernel size is similar to the technique used in [33, 34]. The effectiveness of our method is demonstrated through the online prediction of chaotic time series generated by two dynamical systems, 1) the Lorenz system, and, 2) the chaotic system proposed by Zhang et al. [36]. The rest of this paper is structured as follows: In Sect. 2 we give a description of nonlinear regression in RKHS, and extend the discussion to online regression via KLMS and RFF-KLMS algorithms in Sect. 3. In Sect. 4, we provide an outline of the RFF-KLMS algorithm with adaptive kernel size, followed by example applications in Sect. 5. Finally, we present the conclusion in Sect. 6.

2 Nonlinear Regression in RKHS

Assume we have a discrete system generating a time series at a single time step forward of the form $\{x(1), x(2), x(3), \ldots, x(n), \ldots\}$, a general nonlinear prediction model of the time series is of the form [1, 2, 37, 38]

$$x(n+1) \approx y(n) = F(\mathbf{x}(n)) \in V,$$

where $\mathbf{x}(n) = [x(n), x(n-1), \ldots, x(n-N+1)]^T$, V is a vector space which we hope to contain the attractor that explains the long-term dynamics of the time series, and $N > 2d_A + 1$ where N is the dimension of V and d_A is the dimension of the attractor. If the basis functions of V is known, $\phi_1(.), \phi_2(.), \ldots, \phi_N(.)$ say, then we may write F as a linearly separable function in terms of the basis functions so that

$$y(n) = \Phi(\mathbf{x}(n))^T \mathbf{w} = w_1\phi_1(\mathbf{x}(n)) + w_2\phi_2(\mathbf{x}(n)) + \cdots + w_N\phi_N(\mathbf{x}(n)).$$

The least squares approach to determining the parameters $\mathbf{w} = [w_1, w_2, \ldots, w_N]^T$ requires the minimization of a loss function of the form

$$f(\mathbf{w}) = \sum_{i=1}^{n} \left| x(i+1) - \mathbf{w}^T \Phi(\mathbf{x}(i)) \right|^2. \tag{2}$$

This problem is just a standard nonlinear regression and this is the approach used in [16, 39, 40] where the basis functions are assumed to be from a class of universal approximators. The difficulty in this approach is that the state space V is most likely infinite dimensional which means N has to be very large to achieve sufficiently accurate prediction.

Alternatively one can avoid working with Φ directly by transforming vectors in the state space construction into the so-called Reproducing Kernel Hilbert space (RKHS). Let $\mathbf{x}(1), \mathbf{x}(2), \ldots, \mathbf{x}(n) \in \mathbb{R}^N$ and let the RKHS be $\mathcal{H}$. The similarity between the elements in $\mathcal{H}$ is measured using its associated inner product $(.,.)_{\mathcal{H}}$ and it is computed via a kernel function $\kappa : \mathbb{R}^N \times \mathbb{R}^N \to \mathbb{R}$ such that $(\mathbf{x}_i, \mathbf{x}_j) \to \kappa(\mathbf{x}_i, \mathbf{x}_j)$. For positive definite kernel functions, we can ensure that for all $\mathbf{x}, \mathbf{x}' \in \mathbb{R}^N$,

$$(\Phi(\mathbf{x}), \Phi(\mathbf{x}'))_{\mathcal{H}} = \kappa(\mathbf{x}, \mathbf{x}'). \tag{3}$$

The property in (3) is called the 'kernel trick' [41]. In RKHS, the optimum prediction model, if determined using the least squares approach, is a minimization problem of the form

$$\min_{\mathbf{w} \in \mathbb{R}^N} \sum_{i=1}^{n} |x(i+1) - (\mathbf{w}, \Phi(\mathbf{x}(i))|^2. \tag{4}$$

The loss function in (4) can be written as

$$f(\mathbf{w}) = \|\hat{\mathbf{x}} - \mathbf{K}\mathbf{w}\|^2 = \hat{\mathbf{x}}^T\hat{\mathbf{x}} - 2\hat{\mathbf{x}}^T\mathbf{K}\mathbf{w} + \mathbf{w}^T\mathbf{K}^T\mathbf{K}\mathbf{w}, \tag{5}$$

where

$$\hat{\mathbf{x}} = [x(2), x(3), \ldots, x(n+1)]^T, \qquad \mathbf{K} = [\Phi(\mathbf{x}(1), \Phi(\mathbf{x}(2)), \ldots, \Phi(\mathbf{x}(n))]^T.$$

Matrix $\mathbf{G} = \mathbf{K}^T\mathbf{K}$ has entries $\mathbf{G}_{ij} = (\Phi(\mathbf{x}(i)), \Phi(\mathbf{x}(i))_{\mathcal{H}} = \kappa(\mathbf{x}(i), \mathbf{x}(j))$ which is positive definite if the kernel function $\kappa(.,.)$ is positive definite. As a consequence, the loss function in (5) is a convex function and the problem in (4) is a convex minimization. To guarantee a convex minimization problem, in the rest of the paper, we adopt the Gaussian kernel in (1).

3 Online Regression in the RKHS via Kernel Least Mean Square Algorithm

In the online scenario, data is collected one at a time in a sequential manner and only a limited set of the most current data is stored while older data are discarded. In this situation, due to the incomplete knowledge of the data set, the loss function (5) needs to be estimated in order to determine the regression parameters. One way to perform online regression is to use the instantaneous approximation of (5), where at any particular time n the current prediction of the time series is of the form $(\Phi(\mathbf{x}(n)), \mathbf{w})_{\mathcal{H}}$ and the estimated loss function is given by

$$f_{inst}(\mathbf{w}) = (e(n))^2 = (x(n+1) - (\Phi(\mathbf{x}(n)), \mathbf{w})_{\mathcal{H}})^2. \tag{6}$$

A gradient based minimization of (6) searches for the optimum parameter vector $\mathbf{w}$ along the negative instantaneous gradient direction which is given by

$$-\nabla_{\mathbf{w}} f(\mathbf{w}) = 2e(n)\Phi(\mathbf{x}(n)). \tag{7}$$

The update equation for $\mathbf{w}$ is then

$$\mathbf{w}(n+1) = \mathbf{w}(n) + \mu e(n)\Phi(\mathbf{x}(n)). \tag{8}$$

Assuming $\mathbf{w}(0) = 0$, it can be shown that [15]

$$\mathbf{w}(n) = \mu \sum_{i=0}^{n-1} e(i)\Phi(\mathbf{x}(i)). \tag{9}$$

As a result, the current prediction can be update as follows:

$$\begin{aligned} y(n) = (\Phi(\mathbf{x}(n)), \mathbf{w}(n))_H &= \mu \sum_{i=0}^{n} e(i)(\Phi(\mathbf{x}(n)), \Phi(\mathbf{x}(i)))_{\mathcal{H}} \\ &= \mu \sum_{i=0}^{n} e(i)\kappa(\mathbf{x}(n), \mathbf{x}(i)). \end{aligned} \tag{10}$$

It is clear from (10) that the current prediction can be computed using only knowledge of the kernel function.

Next, we describe two online gradient based algorithms which attempt to minimize the instantaneous estimated loss function (6).

3.1 The KLMS Algorithm

A straightforward implementation of (7)–(10) leads to the Kernel Least Mean Square (KLMS) algorithm. The basic sequential rule for KLMS is as follows:

Input: Training samples $\{(x(n), d(n))\}$, step-size μ, kernel function $\kappa(.,.)$, set $y(0) = 0$
For $n = 1, 2, \ldots$
$\hat{y}(n) = \sum_{i=0}^{n-1} e(i)\kappa(\mathbf{x}(n), \mathbf{x}(i))$
$e(n) = d(n) - y(n-1)$
$y(n) = \hat{y}(n) + \mu e(n)\kappa(\mathbf{x}(n), .)$

where $\hat{y}(n)$ is the apriori prediction. The apriori prediction is an ever-growing sum; the size of the sum grows with each update and it relies on the entire dictionary $\{\mathbf{x}(1), \mathbf{x}(2), \ldots, \mathbf{x}(n-1)\}$. The ever-growing dictionary size results in an increase in computational resources and memory, thus making the application of KLMS rather prohibitive.

3.2 Random Fourier Feature KLMS (RFF-KLMS)

The prediction in the KLMS algorithm is achieved by first mapping the state vector $\mathbf{x}(n)$ to an infinite dimensional RKHS $\mathcal{H}$, using an implicit map $\phi(\mathbf{x}(n))$, and with the help of the kernel trick, computes the prediction after n data updates as a linear expansion (10) which grows indefinitely as n increases. To overcome the growing sum problem, Rahimi and Recht [24] proposed mapping the state-space vector $\mathbf{x}(n)$ onto a finite dimensional Euclidean space using a randomized map $\Theta : \mathbb{R}^N \to \mathbb{R}^D$. The Bochner's theorem [42] guarantees that the Fourier transform of a positive definite, appropriately scaled, shift-invariant kernel is a probability density, $p(\theta)$ say, such that

$$\begin{aligned} k(\mathbf{x}(i) - \mathbf{x}(j)) &= k(\mathbf{x}(i), \mathbf{x}(j)) \\ &= \int_{R^N} p(\theta) e^{j\theta^T(\mathbf{x}(i)-\mathbf{x}(j))} d\theta = E[\Theta(\mathbf{x}(i))^H \Theta(\mathbf{x}(j))], \end{aligned} \quad (11)$$

where, the last equality in (11) is obtained by defining $\Theta(\mathbf{x}) = e^{j\theta^T\mathbf{x}}$ (H is the conjugate transpose). According to [43], given (11), the D-dimensional random Fourier feature (RFF) of the state-space vector $\mathbf{x}(n)$ that can approximate $\kappa(.,.)$ with L_2 error less than $O(1/\sqrt{D})$ is given by

$$\Theta(\mathbf{x}) = [\psi_1(\mathbf{x}), \psi_2(\mathbf{x}), \ldots, \psi_D(\mathbf{x})]^T, \quad (12)$$

where $\psi_i(\mathbf{x}) = \sqrt{2}\cos(\theta_i^T\mathbf{x} + b_i)$, $i = 1, 2, \ldots, D$, with $\theta_i \in R^N$. In other words, $\Theta(\mathbf{x}(i))^H\Theta(\mathbf{x}(j))$ is an unbiased estimate of $\kappa(\mathbf{x}(i), \mathbf{x}(j))$ when θ is drawn from p. Here we exploit the symmetric property of $\kappa(.,.)$ in which case $\Theta(\mathbf{x}(n))$ can be expressed using real-valued cosine bases. For approximating a Gaussian kernel of size σ given in (1), θ_i is drawn from the Gaussian distribution with zero-mean and covariance matrix $\frac{1}{\sigma^2}\mathbf{I}_N$ with $\mathbf{I}_N$ being an $N \times N$ identity matrix, and, b_i is uniformly sampled from $[0, 2\pi]$ [26, 44].

With the finite dimensional map $\Theta(.)$ defined by (12), the prediction at time n is just $\Theta(\mathbf{x}(n))^T\mathbf{w}$, and the instantaneous loss function is then

$$F_{inst}(\mathbf{w}) = (e(n))^2 = (x(n+1) - \Theta(\mathbf{x}(n))^T\mathbf{w})^2. \quad (13)$$

It follows that, the negative instantaneous gradient direction with respect to $\mathbf{w}$ is $-\nabla_{\mathbf{w}}F(\mathbf{w}) = 2e(n)\Theta(\mathbf{x}(n))$ which results in an update equation for $\mathbf{w}$ in the form

$$\mathbf{w}(n+1) = \mathbf{w}(n) + \mu e(n)\Theta(\mathbf{x}(n)). \quad (14)$$

The update equation in (14) results in the RFF-KLMS algorithm and it requires a computational complexity of a fixed linear order which is $O(D)$. Mean square convergence of RFF-KLMS is presented in [45].

4 RFF-KLMS with Adaptive Kernel Size

In the KLMS and the RFF-KLMS algorithm described in this paper, the kernel function is defined in the form given in (1). This definition requires knowledge of a pre-determined parameter which is the kernel size σ. In the large sample size regime, the asymptotic properties of the mean square approximation are independent of σ, i.e., the choice of σ does not affect the convergence of KLMS and RFF-KLMS. However, σ does affect the dynamics of the algorithm. In particular, in the transient stage when the sample size is small, an optimal kernel size is important to speed up the convergence to the neighbourhood of the optimal solution.

A method for adjusting σ in a sequential optimization framework is proposed in [33] for the KLMS algorithm. An update equation for σ is derived from the minimization of the instantaneous loss function (6) and optimizing it along the negative gradient direction (with respect to σ), i.e.,

$$\sigma_{n+1} = \sigma_n - \rho\frac{\partial f_{inst}}{\partial \sigma_n},$$

where σ_n is the kernel size at time n and ρ is a step-size parameter. The resulting update equation is given by

$$\sigma_{n+1} = \sigma_n + \rho e(n-1)e(n)\|\mathbf{x}(n-1) - \mathbf{x}(n)\|_2^2\frac{\kappa_{\sigma_n}(\mathbf{x}(n-1), \mathbf{x}(n))}{\sigma_n^3}. \quad (15)$$

Table 1 The RFF-KLMS algorithm with adaptive kernel size

Input:

1. Sequential input of time series data $x(1), x(2), x(3), \ldots$
2. State-space dimension N; step-size parameter μ; step-size parameter ρ; random Fourier dimension D.

Output:

1. The one-step ahead predictions $y(1), y(2), y(3), \ldots$
2. The N-dimensional weight vector $\mathbf{w}$; the optimum kernel size σ_{opt}.

Initialization:

1. Initial weight vector $\mathbf{w}(0)$;
2. Sample $\theta_i^{(0)} \in \mathbb{R}^N, i = 1, 2, \ldots, D$ from Gaussian distribution with zero-mean and covariance matrix I_N;
3. Sample $b_i, i = 1, 2, \ldots, D$ uniformly from $[0, 2\pi]$.

For $n = 1, 2, \ldots$

(1) Update the state-space vector: $\mathbf{x}(n) = [x(n), x(n-1), \ldots, x(n-N+1)]$

(2) With $\psi_i(\mathbf{x}(n)) = \sqrt{2}\cos(\theta_i^{(n-1)T}\mathbf{x}(n) + b_i), i = 1, 2, \ldots, D$, compute the random Fourier feature vector:

$$\Theta(\mathbf{x}(n)) = [\psi_1(\mathbf{x}(n)), \psi_2(\mathbf{x}(n)), \ldots, \psi_D(\mathbf{x}(n))]^T .$$

(3) Compute the apriori error $e(n) = x(n+1) - \Theta(\mathbf{x}(n))^T\mathbf{w}(n-1)$.

(4) Update the weight vector: $\mathbf{w}(n) = \mathbf{w}(n-1) + \mu e(n)\Theta(\mathbf{x}(n))$.

(5) Compute current prediction: $y(n) = \Theta(\mathbf{x}(n))^T\mathbf{w}(n)$.

(6) Update σ:

$$\sigma_n = \sigma_{n-1} + \rho e(n-1)e(n)\|\mathbf{x}(n-1) - \mathbf{x}(n)\|_2^2 \frac{\exp(\|\mathbf{x}(n-1)-\mathbf{x}(n)\|^2/2\sigma_{n-1}^2)}{\sigma_{n-1}^3}.$$

(7) Sample $\theta_i^{(n)} \in \mathbb{R}^N, i = 1, 2, \ldots, D$ from Gaussian distribution with zero-mean and covariance matrix $\frac{1}{\sigma_n^2} I_N$;

EndFor.

Since RFF-KLMS is an approximation of KLMS in a D dimensional space, the update equation in (15) can also be used to adapt the kernel size in the RFF-KLMS algorithm. In RFF-KLMS, the kernel size determines the probability density function $p(\theta)$ from which θ_i $(i = 1, 2, \ldots, N)$ is drawn. Thus, adjusting σ also means adjusting θ. To incorporate this adjustment in the algorithm, we initialize θ_i with an N-dimensional vector drawn from the Gaussian distribution with zero-mean and covariance matrix I_N. At each update at time n, a new vector $\theta_i^{(n)}$ is used in place of θ_i, where $\theta_i^{(n)}$ is drawn from the Gaussian distribution with zero-mean and covariance matrix $\frac{1}{\sigma_n^2} I_N$. The complete algorithm is summarized in Table 1.

5 Simulation Examples

In this section, we present several examples to illustrates the performance and application of RFF-KLMS algorithm with adaptive kernel size in short-term prediction of chaotic time series. Two chaotic systems are used: i) the Lorenz system [35], and, ii) the chaotic system proposed by Zhang et al. [36].

5.1 Example 1: Lorenz Chaotic System

Consider the Lorenz oscillator whose state equations are

$$\begin{aligned}\frac{dx}{dt} &= -\beta x + yz \\ \frac{dy}{dt} &= \delta(z - y) \\ \frac{dz}{dt} &= -xy + \lambda y - z\end{aligned} \tag{16}$$

where parameters are set as $\beta = 8/3$, $\delta = 10$ and $\lambda = 28$. Sample time series of x, y and z are given in Fig. 1. The goal is to predict $x(t)$, $y(t)$ and $z(t)$ using the previous eight consecutive samples. The well-known Lorenz attractor is observable in 3-dimensional space (i.e. $d_A = 3$), therefore it is reasonable to choose $N = 8 > 2d_A + 1$.

Mean Squared Error (MSE) Performance. First, we compare the performance of RFF-KLMS with fixed kernel sizes ($\sigma = 1, 2, 5$) and RFF-KLMS with adaptive kernel size. The parameter values used in this experiment are as follows: $\mu = 0.5$, $\rho = 0.005$ and $D = 500$.

For each kernel size, 20 independent (Monte Carlo) simulations are run with different segments of the time series. In each segment, 1000 samples are used for training and a further 100 samples are used for testing. To evaluate the convergence behaviour of the algorithm, the mean squared error (MSE) is used which is defined as MSE = $(\sum_i^{n_{test}} \hat{e}(i)^2)/n_{test}$, where n_{test} is the length of test data. The ith test error $\hat{e}(i)$ is computed based on the test data using the learned weight vector $\mathbf{w}(n)$ and σ_n at that iteration, i.e., $\hat{e}(i) = x_{test}(i+1) - \Theta_{\sigma_n}(\mathbf{x}_{test}(i))^T \mathbf{w}(n)$. All simulation results are averaged over 20 Monte Carlo runs.

Convergence of MSE for each kernel size is shown in Fig. 2(left). It can be seen that RFF-KLMS achieves the best performance in terms of convergence as well as achieving the minimum steady-state MSE. The evolution of the value of kernel size during the learning process is shown in Fig. 2(right) where it is observed that the steady-state (optimum) kernel size is about 3.2. The steady-state MSE for the training and testing are listed in Table 2. The relative sizes of training and testing MSE

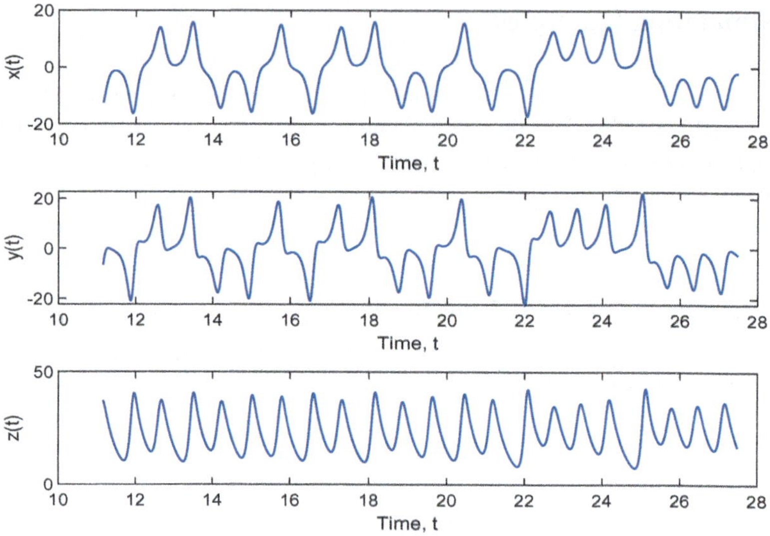

Fig. 1 Time series of the three states in the Lorenz system

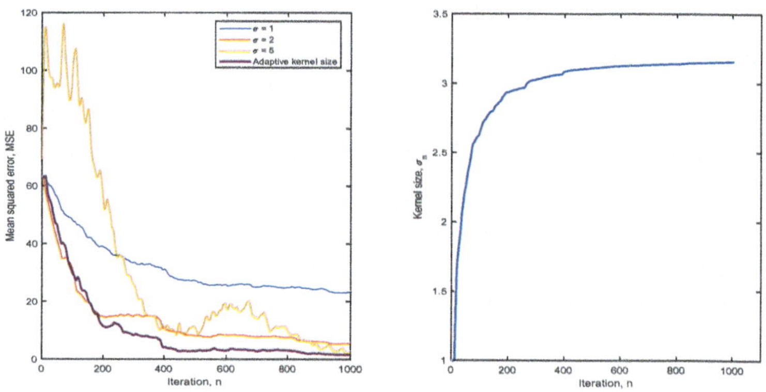

Fig. 2 (Left) Comparison of MSE performance of RFF-KLMS between fixed kernel sizes ($\sigma = 1, 2, 5$) and adaptive kernel size. (Right) Evolution of adaptive kernel size in the course of iteration

are comparable between all kernel sizes which shows that tendency to overlearn is comparable between all kernel sizes.

Prediction. Here we present the time series and the associated Lorenz attractor. The predicted Lorenz attractor is constructed based on the values of the predictions at steady-state.

The Constructed Time Series. Due to space limitation, only the learned prediction of state variable x is presented and this is shown in Fig. 3. The prediction is compared for different values of the kernel size. As expected, the predicted model trained using

Table 2 Steady-state MSE

Kernel size	Training MSE	Testing MSE
$\sigma = 1$	16.7317	21.2890
$\sigma = 2$	0.8227	1.8943
$\sigma = 5$	0.2401	2.8094
Adaptive kernel size	0.1062	0.9333

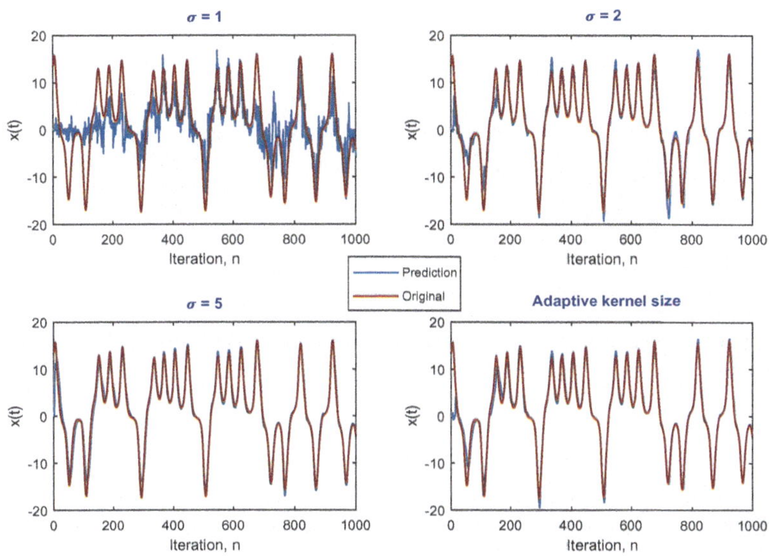

Fig. 3 Predicted time series for state variable x during training

RFF-KLMS with adaptive kernel size is able to capture the dynamics of the time series optimally at steady-state.

Construction of the Lorenz Attractor. Next we reconstruct the Lorenz attractor using the steady-state prediction of state variables x, y and z. For ease of comparison, we choose to present the views of the predicted Lorenz attractor in the 2D planes (x, y) and (x, z) respectively. In Fig. 4, the plots on the left depicts the phase reconstruction using our trained model while the plots on the right depicts the actual Lorentz attractor. It is clearly observed that RFF-KLMS has successfully captured the overall structure of the attractor.

Predicting Sudden Change in Dynamics. In online prediction it is important to analyze the ability of an algorithm to track a time series which is subjected to sudden changes. In order to do so, we have created two time series, $(x^{(1)}, y^{(1)}, z^{(1)})$ and $(x^{(2)}, y^{(2)}, z^{(2)})$, each of which is generated from Lorenz systems with two different values of the parameter λ (the other two parameters are fixed, i.e., $\beta = 10/3$ and $\delta = 10$). The time series $(x^{(1)}, y^{(1)}, z^{(1)})$ is generated with $\lambda = 28$ while the time

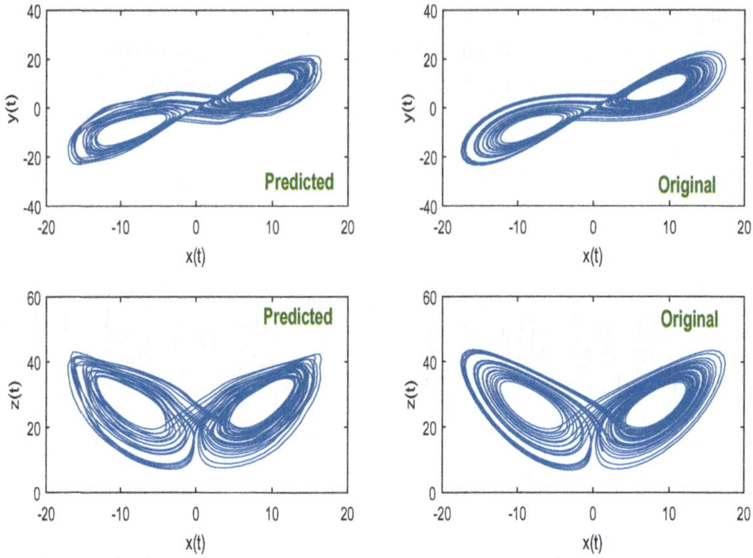

Fig. 4 Lorenz attractor in (x, y) plane (top) and in the (x, z) plane (bottom): Predicted (left), original (right)

series $(x^{(2)}, y^{(2)}, z^{(2)})$ is generated with $\lambda = 15$. These two values results in two completely different dynamics; $\lambda = 28$ results in time series that lie in an invariant manifold associated with the Lorenz chaotic attractor while $\lambda = 15$ results in an invariant manifold associated with a stable focus.

Tracking of Time Series. Only the tracking of the time series of state variable x is shown for the purpose of illustration and this is found in Fig. 5 (Top). The sudden change in the time series occurs at sample $n = 5861$. The corresponding evolution of kernel size σ_n during the tracking process in shown in Fig. 5 (Bottom). It is clearly seen in Fig. 5 that, as the RFF-KLMS algorithm tracks the optimum kernel size the predicted time series becomes more similar to the actual time series. Moreover, once the optimum kernel size is found, minimal adaptation is needed to detect the change in dynamics.

Tracking of Invariant Manifolds. The invariant manifolds are reconstructed from the predicted time series x, y and z. The views of the predicted invariant manifolds in the 2D planes (x, y) and (x, z) are presented in Figs. 6 and 7 respectively. In both Figs. 6 and 7, the top figure depicts the predicted manifolds while the bottom figure depicts the actual observed manifolds. It is evident from these figures that, regardless of the sudden change in λ at $n = 5861$, the overall structure of the manifolds before and after the change have clearly been captured well. This results confirm the extremely good tracking ability of the RFF-KLMS algorithm with adaptive kernel size.

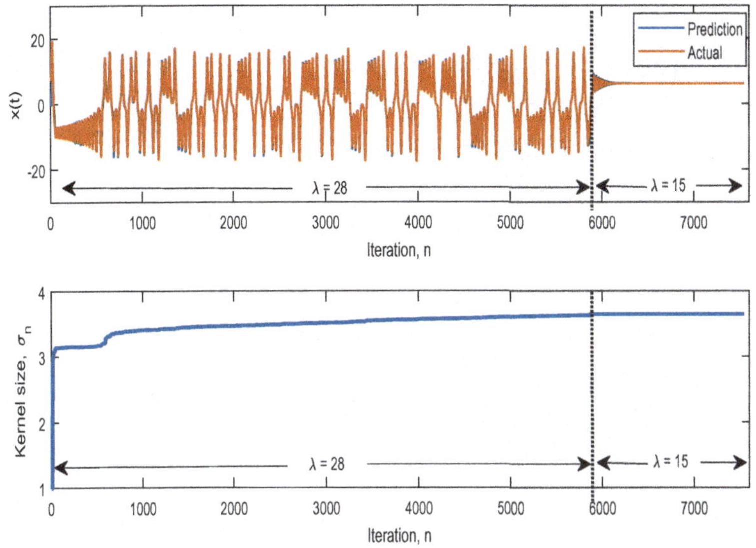

Fig. 5 Tracking sudden changes in the time series: Sudden change occurs at sample $n = 5861$

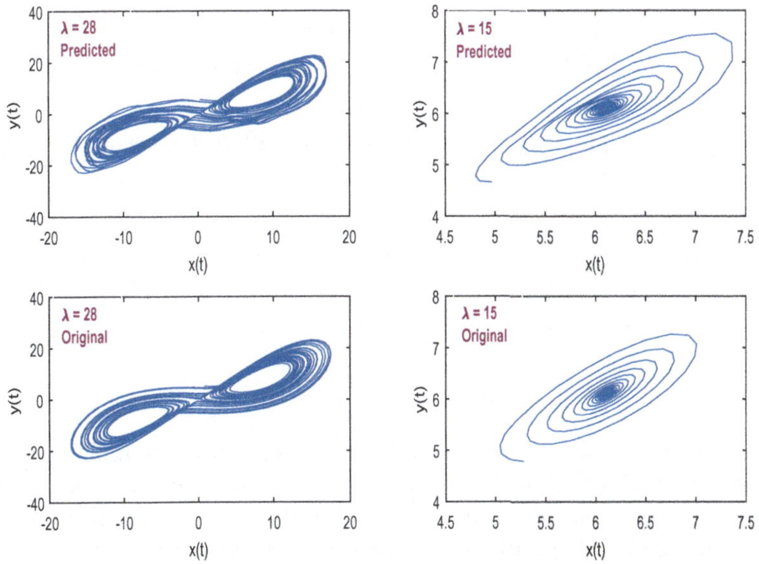

Fig. 6 Lorenz system: Invariant manifold in (x, y) plane. Predicted (top), original (bottom)

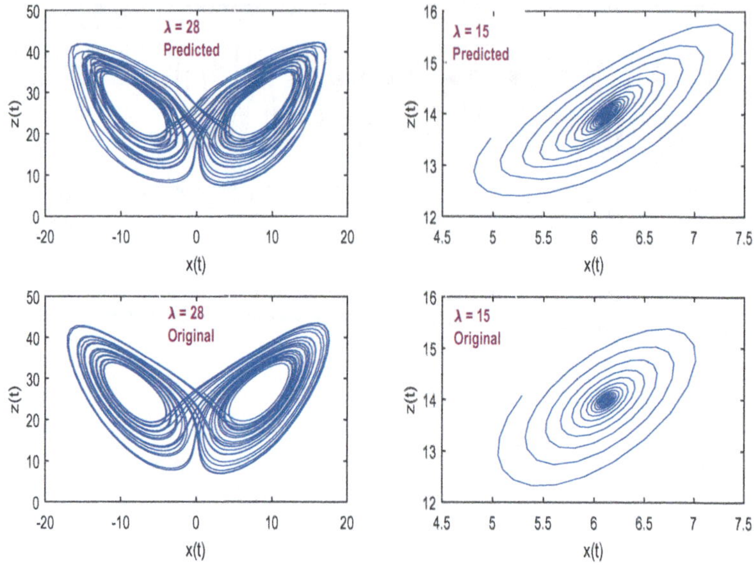

Fig. 7 Lorenz system: Invariant manifold in (x, z) plane. Predicted (top), original (bottom)

5.2 *Example 2: Chaotic System by Zhang et al. [36]*

The chaotic system proposed in [36] is a system of differential equations in term of state variables x, y and z given by

$$\begin{aligned}\frac{dx}{dt} &= -ax + by - yz \\ \frac{dy}{dt} &= x + xz \\ \frac{dz}{dt} &= cz + y^2\end{aligned} \tag{17}$$

where parameters are set as $a = 10, b = 28$ and $c = 6$. In this section, all experiments are conducted using noisy time series where the time series of x, y and z are corrupted by zero-mean Gaussian noise with variance 0.01, 0.1 and 1, with equivalent signal-to-noise ratio (SNR) of 40.8 dB, 30.8 dB and 20.8 dB respectively.

Prediction. Prediction of the chaotic attractor of (17) is done for the three SNR values: 40.8 dB, 30.8 dB and 20.8 dB. The predicted chaotic attractor is constructed based on the values of the time series predictions at steady-state.

The Constructed Time Series. For the purpose of illustration, only the prediction of state variable x is presented and this is shown in Fig. 8. The prediction is compared for different SNR values. Here it is observed that the decrease in SNR value

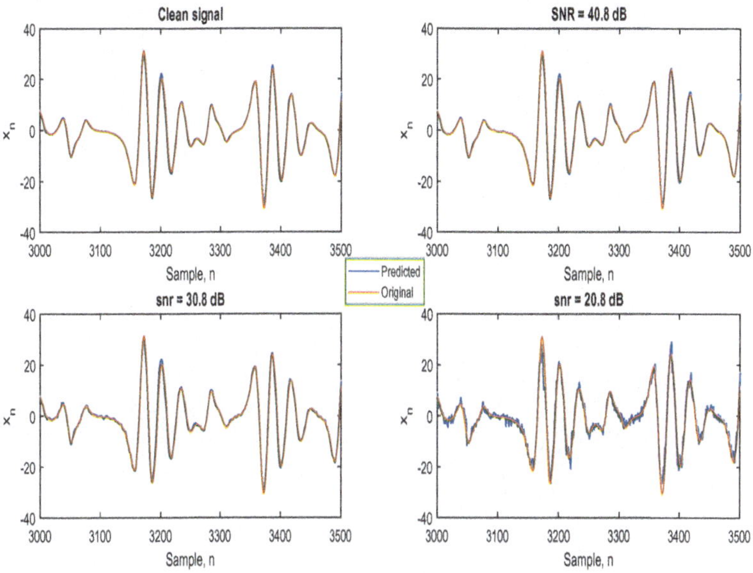

Fig. 8 Predicted time series for state variable x during training

Table 3 Steady-state MSE for different SNR values

SNR (dB)	MSE for x (dB)	MSE for y (dB)	MSE for z (dB)
Clean	–2.451	0.376	1.340
40.8	–1.288	1.352	1.864
30.8	–0.523	1.556	4.449
20.8	7.672	5.896	6.372

(i.e. increase in noise strength) does have some effect on the accuracy of predictions. Table 3 provides a list of the steady-state training MSE for the three different values of SNR from which one can see the increase in steady-state MSE as SNR decreases. This gives a quantitative measure of the decrease in performance as noise strength increases.

Construction of the Chaotic Attractor. The choice of the parameters $a = 10, b = 28$ and $c = 6$ is associated with a chaotic attractor [36]. Using the steady-state prediction of state variables x, y and z, the projections of the attractor in the 2D planes (x, y) and (x, z) are reconstructed and for all the SNR values used in the experiments, it is observed that RFF-KLMS is able to capture the overall structure of the attractor. However the accuracy of the trajectories tend to reduce as SNR value decreases. As an illustration, the reconstruction of the chaotic attractor in the (x, y) plane is shown in Fig. 9 for the three SNR values.

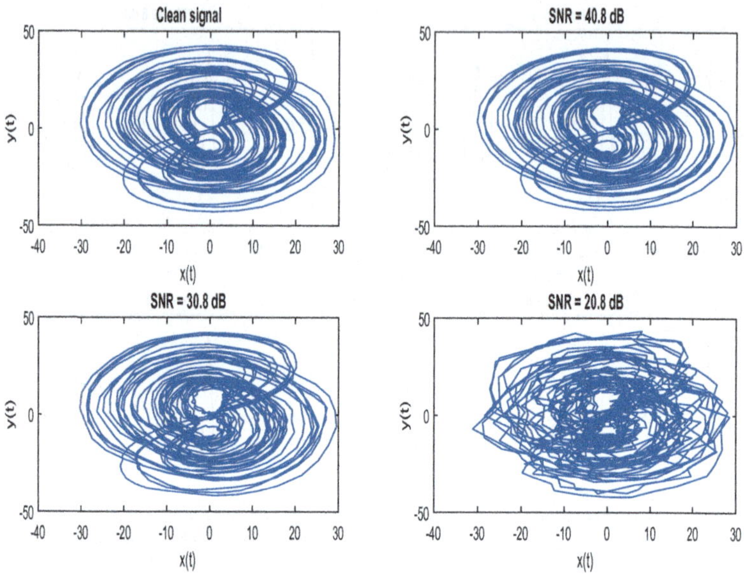

Fig. 9 Reconstruction of the chaotic attractor in the (x, y) plane for different levels of noise strength

Predicting Sudden Change in Dynamics. To investigate the capibility of the RFF-KLMS algorithm with adaptive kernel size in predicting change in dynamics, we have created two time series, $(x^{(1)}, y^{(1)}, z^{(1)})$ and $(x^{(2)}, y^{(2)}, z^{(2)})$, each of which is generated from (17) with two different values of the parameter c (the other two parameters are fixed, i.e., $a = 10$ and $b = 28$). The time series $(x^{(1)}, y^{(1)}, z^{(1)})$ is generated with $c = 6$ while the time series $(x^{(2)}, y^{(2)}, z^{(2)})$ is generated with $c = 158$. These two values results in two completely different dynamics; $c = 6$ results in time series that lie in an invariant manifold associated with the chaotic attractor while $c = 158$ results in an invariant manifold associated with a stable limit cycle [36]. To study the capability of RFF-KLMS in predicting the change in dynamics in the presence of noise, zero-mean Gaussian noise with variances 0.01 (SNR = 40.8 dB), 0.1 (SNR = 30.8 dB) and 1 (SNR = 20.8 dB) is added to both time series.

Tracking of Time Series. Only the tracking of the time series of state variable x is shown for the purpose of illustration. The sudden change in the time series occurs at sample $n = 6197$. The steady-state time series obtained with clean input signal x is compared with the steady-state time series obtained for input signals having SNR 40.8 dB, 30.8 dB and 20.8 dB respectively and the result is shown in Fig. 10. It is clearly seen in Fig. 10 that the sudden change in dynamics is successfully detected by RFF-KLMS algorithm with adaptive kernel size and its ability does not appear to be affected much by the presence of noise. Although some accuracy is loss in the predicted time series when SNR = 20.8 dB, but the overall behaviour of the time series is quite apparent.

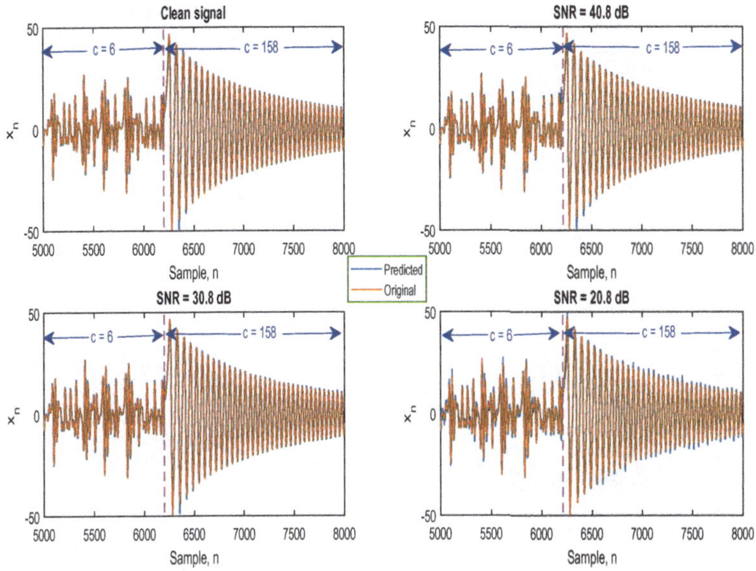

Fig. 10 Comparison of the predicted time series for the state variable x with and without noise: Sudden change in the time series occurs at $n = 6197$.

Tracking of Invariant Manifolds. The invariant manifolds are reconstructed from the predicted time series x, y and z obtained for input signals having SNR 40.8 dB, 30.8 dB and 20.8 dB respectively and the results are compared with the invariant manifold obtained from clean input signal. The views of the predicted invariant manifolds in the 2D planes (x, y) and (x, z) are shown in Figs. 11 and 12 respectively for the most severe case (i.e. SNR = 20.8 dB). In both Figs. 11 and 12, the top figure depicts the predicted manifolds for $c = 6$ while the bottom figure depicts the predicted manifolds for $c = 158$. It is evident from these figures that, regardless of the sudden change in c at $n = 6197$, the overall structure of the manifolds before and after the change have clearly been captured well. The presence of noise somewhat affects the accuracy of trajectories within the manifold, but the presence of the invariant structures are still evident. This results further confirm the robustness of the RFF-KLMS algorithm with adaptive kernel size.

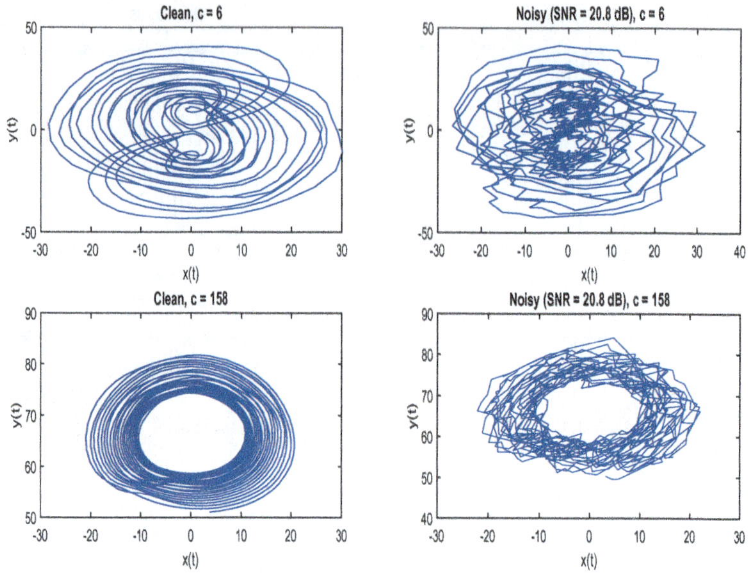

Fig. 11 Invariant manifold in (x, y) plane: c = 6 (top), c = 158 (bottom), clean input signal (left), noisy input signal (right)

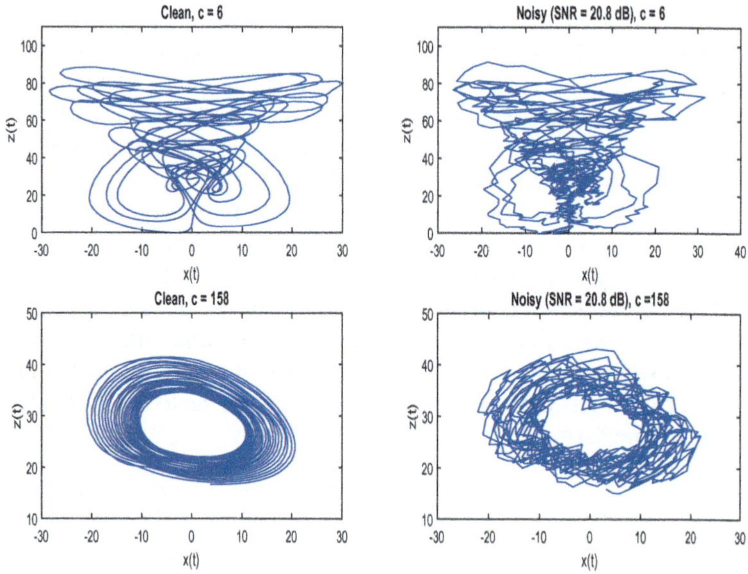

Fig. 12 Invariant manifold in (x, z) plane: c = 6 (top), c = 158 (bottom), clean input signal (left), noisy input signal (right)

6 Conclusion

The kernel size in RFF-KLMS determines the probability density function from which the random features are drawn. In other words, it determines the finite dimensional subspace defines by the feature maps in RFF-KLMS. Quality of approximation by RFF-KLMS is directly determined by this subspace. Therefore, for optimal finite-dimensional mapping, optimal kernel size is needed.

In this paper, we described a procedure for optimizing the kernel size sequentially using a stochastic gradient descent approach. Empirical studies on online prediction of chaotic time series highlights the tracking capability of RFF-KLMS with adaptive kernel size. It not only tracks the time series and the ensuing dynamics of the system well, but the algorithm is also capable of predicting sudden change in dynamics. MSE performance of RFF-KLMS algorithm with adaptive kernel size is also more superior than the MSE of RFF-KLMS with fixed kernel size. The experimental results also highlight the robustness of the algorithm with respect to noise in the input signal. Although some loss in accuracy is observed in the presence of noise, the algorithm can still capture the overall steady-state dynamics of a system.

An immediate future direction of this study is to explore other aspects of the algorithm, for example, combining kernel size adaptation with stepsize adaptation. It is also interesting to investigate the application of RFF-KLMS with adaptive kernel size to look into the possibility of predicting other aspects of system dynamics such as the Lyapunov exponent.

References

1. Takens, F.: Detecting strange attractors in turbulence. Lecture Notes in Math. **898**, 366–381 (1981)
2. Abarbanel, H.D.I.: Analysis of Observed Chaotic Data. Springer-Verlag, New York, Institute for Nonlinear Science (1996)
3. Huke, J.P., Muldoon, M.R.: Embedding and time series analysis. Math. Today **51**(3), 120–123 (2015)
4. Zhang, S., Han, M., Xu, M.: Chaotic time series online prediction based on improved kernel adaptive filter. In: 2018 International Joint Conference on Neural Networks (IJCNN), pp. 1–6. IEEE, Rio de Janeiro (2018)
5. Han, M., Zhang, S., Xu, M., Qiu, T., Wang, N.: Multivariate chaotic time series online prediction based on improved kernel recursive least Squares Algorithm. IEEE Trans. Cybern. **49**(4), 1160–1172 (2019)
6. Lu, L., Zhao, H., Chen, B.: Time series prediction using kernel adaptive filter with least mean absolute third loss function. Nonlinear Dyn. **90**, 999–1013 (2017)
7. Garcia-Vega, S., Zeng, X.-J., Keane, J.: Stock rice prediction using kernel adaptive filtering within a stock market interdependence Approach. Available at SSRN (2018). https://doi.org/10.2139/ssrn.3306250
8. Georga E.I., Principe, J.C., Polyzos, D., Fotiadis, D. I.: Non-linear dynamic modeling of glucose in type 1 diabetes with kernel adaptive filters. In: 2016 38th Annual International Conference of the IEEE Engineering in Medicine and Biology Society (EMBC), pp. 5897–5900, IEEE, Orlando, FL (2016)

9. Ouala, S., Nguyen, D., Drumetz, L., Chapron,B., Pascual, A., Collard, F., Gaultier, L., Fablet, R.: Learning latent dynamics for partially-observed chaotic systems. arXiv preprint, arXiv:1907.02452 (2020)
10. Yin, L., He, Y., Dong, X., Lu, Z.: Adaptive chaotic prediction algorithm of RBF neural network filtering model based on phase Space Reconstruction. J. Comput. **8**(6), 1449–1455 (2013)
11. Feng, T., Yang, S., Han, F.: Chaotic time series prediction using wavelet transform and multi-model hybrid method. J. Vibroeng. **21**(7), 1983–1999 (2019)
12. Kivinen, J., Smola, A., Williamson, R.: Online learning with kernels. In: Advances in Neural Information Processing Systems 14, pp. 785–793. MIT Press (2002)
13. Comminiello, D., Principe, J.C.: Adaptive Learning Methods for Nonlinear System Modeling. Elsevier (2018)
14. Chi, M., He, H., Zhang, W.: Nonlinear online classification algorithm with probability. J. Mach. Learn. Res. **20**, 33–46 (2011)
15. Prıncipe, J.C., Liu, W., Haykin, S.: Kernel Adaptive Filtering: A Comprehensive Introduction. **57**. John Wiley & Sons (2011)
16. Paduart, J., Lauwers, L., Swevers, J., Smolders, K., Schoukens, J., Pintelon, R.: Identification of nonlinear systems using Polynomial Nonlinear State Space models. Automatica **46**(4), 647–656 (2010)
17. Liu, W., Pokharel, P.P., Principe, J.C.: The kernel least-mean-square algorithm. IEEE Trans. Signal Process. **56**(2), 543–554 (2008)
18. Engel, Y., Mannor, S., Meir, R.: The kernel recursive least-squares algorithm. IEEE Trans. Signal Process. **52**(8), 2275–2285 (2004)
19. Liu, W., Prıncipe, J.: Kernel affine projection algorithms. EURASIP J. Adv. Signal Process. **2008**, 1–12 (2008)
20. Liu, W., Park, I., Wang, Y., Prıncipe, J.C.: Extended kernel recursive least squares algorithm. IEEE Trans. Signal Process. **57**(10), 3801–3814 (2009)
21. Platt, J.: A resource-allocating network for function interpolation. Neural Comput. **3**(2), 213–225 (1991)
22. Liu, W., Park, I., Prıncipe, J.C.: An information theoretic approach of designing sparse kernel adaptive filters. IEEE Trans. Neural Netw. **20**(12), 1950–1961 (2009)
23. Wang, S., Wang, W., Dang, L., Jiang, Y.: Kernel least mean square based on the Nystrom method. Circuits Syst. Signal Process. **38**, 3133–3151 (2019)
24. Rahimi, A., Recht, B.: Random features for large-scale kernel machines. In: Proceedings of the 21th Annual Conference on Neural Information Processing Systems (ACNIPS), pp. 1177–1184, Vancouver, BC, Canada (2007)
25. Singh, A., Ahuja, N., Moulin, P.: Online learning with kernels: overcoming the growing sum problem. In: Proceedings of the 2012 IEEE International Workshop on Machine Learning for Signal Process (MLSP), pp. 1–6, Santander, Spain (2012)
26. Bouboulis, P., Pougkakiotis, S., Theodoridis, S.: Efficient KLMS and KRLS algorithms: a random fourier feature perspective. In: 2016 IEEE Statistical Signal Processing Workshop (SSP), pp. 1–5, Palma de Mallorca (2016)
27. Xiong, K., Wang, S.: The online random fourier features conjugate gradient algorithm. IEEE Signal Process. Lett. **26**(5), 740–744 (2019)
28. Racine, J.: An efficient cross-validation algorithm for window width selection for nonparametric kernel regression. Commun. Stat. Simul. Comput. **22**(4), 1107–1114 (1993)
29. Cawley, G.C., Talbot, N.L.: Efficient leave-one-out cross-validation of kernel fischer discriminant classifiers. Pattern Recogn. **36**(11), 2585–2592 (2003)
30. An, S., Liu, W., Venkatesh, S.: Fast cross-validation algorithms for least squares support vector machine and kernel ridge regression. Pattern Recogn. **40**(8), 2154–2162 (2007)
31. Hardle, W.: Applied Nonparametric Regression. Volume 5. Cambridge Univ Press (1990)
32. Herrmann, E.: Local bandwidth choice in kernel regression estimation. J. Comput. Graph. Stat. **6**(1), 35–54 (1997)
33. Chen, B., Liang, J., Zheng, N., Príncipe, J.C.: Kernel least mean square with adaptive kernel size. Neurocomput. **191**, 95–106 (2016)

34. Garcia-Vega, S., Zeng, X.-J., Keane, J.: Learning from data streams using kernel least-mean-square with multiple kernel-sizes and adaptive step-size. Neurocomput. **339**, 105–115 (2019)
35. Lorenz, E.N.: Deterministic aperiodic flow. J. Atmos. Sci. **20**, 130 (1963)
36. Zhang, X., Zhu, H., Yao, H.: Analysis of a new three-dimensional chaotic system. Nonlinear Dyn. **67**, 335–343 (2012)
37. Pelikán, E.: Tutorial: forecasting of processes in complex systems for real-world problems. Neural Netw. World **24**, 567–589 (2014)
38. Bolt, E.M.: Regularized forecasting of chaotic dynamical systems. Chaos, Solitons & Fractals. **94**, 8–15 (2017)
39. Kazem, A., Sharifi, E., Hussain, F.K., Saberi, M., Hussain, O.K.: Support vector regression with chaos-based firefly algorithm for stock market price forecasting. Appl. Soft Comput. **13**(2), 947–958 (2013)
40. Brunton, S.L., Proctor, J.L., Kutz, J.N.: Sparse identification of nonlinear dynamics. Proc. National Acad. Sci. **113**(15), 3932–3937 (2016)
41. Scholkopf, B., Smola, A.J.: Learning with Kernels Support Vector Machines, Regularization, Optimization, and Beyond. MIT Press, Cambridge, MA, USA (2001)
42. Reed, M., Simon, B.: Methods of Modern Mathematical Physics II: Fourier Analysis. Academic Press, Self-Adjointness (1975)
43. Jones, L.K.: A simple lemma on greedy approximation in Hilbert space and convergence rates for projection pursuit regression and neural network training. Ann. Stat. **20**(1), 608–613 (1992)
44. Liu, Y., Sun, C., Jiang, S.: A kernel least mean square algorithm based on randomized feature networks. Appl. Sci. **8**, 458 (2018)
45. Dong, J., Zheng, Y., Chen, B.: A unified framework of random feature KLMS algorithms and convergence analysis. In: 2018 International Joint Conference on Neural Networks (IJCNN), pp. 1–8, IEEE, Rio de Janeiro (2018)

A Robust DEA Model to Handle the Uncertainty in Production Trade-Offs

Rokhsaneh Yousef Zehi and Noor Saifurina Nana Khurizan

Abstract Weight restriction based on production trade-offs as simultaneous changes of inputs and outputs is considered as a weight restriction approach that preserve the technological meaning of efficiency. However, estimating the production trade-offs depends on the expert's judgments and the methodologies that will be applied, hence it may not be always possible to translate and estimate these production trade-offs into exact and precise data. In this article, robust optimization is applied to give a robust counterpart for the weight restriction DEA model which in some of the production trade-offs are subjected to uncertainty. Using the proposed robust weight restriction model (robust WR-TO), it is possible to evaluate efficiency of DMUs in the presence of uncertain production trade-offs to achieve more reliable solutions. The proposed robust counterpart is given using a combination of box and polyhedral uncertainty set. Moreover, an interval weight restriction model (interval WR-TO) is modified to handle the uncertainty in production trade-offs, therefore a comparison between the two proposed model is provided. A real case study of granted research projects is given to validate the proposed models and illustrate the potential application of the proposed robust weight restriction model. The results revealed that the proposed robust WR-TO model gives more reliable effiicency scores and also provides a better ranking of the DMUs compared to the conventional WR-TO model and interval WR-TO model.

Keywords Data Envelopment Analysis · Weight restriction · Production trade-offs · Data uncertainty · Robust optimization

1 Introduction

Data Envelopment Analysis (DEA) is a popular non-parametric technique for the assessment of efficiency of a set of homogeneous decision making units(DMUs) with the same set of inputs and outputs. DEA pioneered by Farrell [1], who

R. Y. Zehi · N. S. N. Khurizan (✉)
School of Mathematical Sciences, Universiti Sains Malaysia, 11800 Pulau Pinang, Malaysia
e-mail: saifurina@usm.my

M. H. Mohd et al. (eds.), *Modelling, Simulation and Applications of Complex Systems*, Springer Proceedings in Mathematics & Statistics 359,
https://doi.org/10.1007/978-981-16-2629-6_18

proposed a non-parametric frontier analysis for solving a linear programing to measure productive efficiency. Later the first DEA model which is called CCR model was developed by Charnes et al. [2]. Since introducing the first DEA model, there has been a massive growth in the theory and application of DEA.

DEA models can be presented either in the form of an envelopment model or a multiplier model. In fact, multiplier model is the dual form of envelopment model. Each one of these two forms, provide a different aspect of interpretation of radial efficiency of the DMU under evaluation (DMU_O) (Cooper et al. [3]). Envelopment form suggest the technological meaning of efficiency for DMU_O as a possible radial improvement factor for its inputs or outputs. Indeed, the DMU_O will be benchmarked against the efficient frontier of the CRS or VRS technology. In the multiplier model a variable weight is associated with each input and output. By varying these weights the efficiency of DMU_O can be determined by maximizing the ratio of weighted sum of its outputs to the weighted sum of its inputs. The optimal weight assessed by the multiplier model, put DMU_O in the best light compared to all other DMUs, because the model is free to set its weight to attempt to reach the efficient frontier. The flexibility in selecting weights permits the DMUs to choose very small weight and even zero for some inputs and outputs which is not acceptable (Thanassoulis et al. [4]). In many cases this flexibility cause lack of discrimination. To overcome this problem and restrict the flexibility of inputs and outputs weights, several models have been presented which are called weight restriction models. These models improve the discrimination power of DEA models. In order to incorporate value judgment and weight restriction into DEA models various methods and models have been presented, such as absolute weight restriction model (Dyson and Thanassoulis,[5]), assurance regions models of type I (Thompson et al. [6]), assurance regions type II (Thompson et al. [7]) and common weights models (Roll et al. [9]). A very detailed classification is given in Allen et al. [8] and Thanassoulis et al. [4]. In the latter group of weight restriction models, weight restrictions will be added into the DEA models in the form of additional constraints on the weights of inputs and outputs in the multiplier model, which leads to the expansion of the production technology (Allen et al. [8]; Roll et al. [9]). Due this expansion the under evaluation DMU will be benchmarked against all units in the technology; the observed units and the new units which have been generated by weight restriction constraints. Therefore, the efficiency measure of DMU_O cannot be interpreted as a realistic improvement input or output factor. To tackle this issue in the weight restriction models, Podinovski [10] introduced the concept of production trade-offs as simultaneous changes to the inputs and outputs which naturally exist in any real production technology. Production trade-offs preserve the technological meaning and radial nature of efficiency and any efficient target obtained in the expanded technology by production trade-off is feasible and producible. The concept of production trade–offs have been used in construction of weight restriction in different real life application of DEA models. Podinovski and Husain [11], evaluated the efficiency of university department using this concept, Khalili et al. [12] applied this concept to assess the efficiency of secondary schools. More related studies can be seen in Amado and Santos [13], Santos et al. [14], Atici and Podinovski [15].

As the production trade-offs naturally exist in any real world technology, they usually will be expressed linguistically and translating them into exact and precise data may not be always possible. So, one of the challenges related to applying production trade-offs is the uncertainty in production trade-offs in real-life applications. In the conventional DEA models, it is assumed that all data are positive and exact values, while in the real world applications it is not always possible to have precise and exact data for the parameters involved in evaluating DMUs. Therefore, by ignoring the uncertainty and perturbation in parameters, the results from DEA models will not be reliable. To overcome this problem in the DEA models, various approaches and models have been presented such as chance constraint DEA models (Land et al. [16]; Olesen and Petersen [17]; Cooper et al. [18]), fuzzy DEA (Sengupta [19]) and interval DEA (Cooper et al. [20]; Despotis and Smirlis, [21]). Another popular approach to deal with uncertainty in data which is widely used recently is robust optimization. The first robust approach was proposed by Soyster [22], and it has been extended by many researchers such as Ben Tal and Nemiroski [23–25] and Bertsimas and Sim [26, 27]. Ben Tal and Nemiroski [25] have proposed modelling the uncertainty in parameters based on ellipsoidal uncertainty set. They showed that a small perturbation in data may lead to the infeasibility of the nominal solution. Bertsimas and Sim [27] used a polyhedral uncertainty set for modeling uncertainty. They argued that the other robust optimization approaches might be very conservative, meaning that they accept a suboptimal solution for the nominal problem in order to ensure that the solution remains feasible. Therefore, they proposed an approach which the conservative level of solution is adjustable and the solution will be more reliable.

Recently many robust DEA models have been developed based upon the robust optimization approaches. Robust DEA models have been introduced to model the uncertainty and perturbation in data in DEA models in order to improve the robustness of efficiency scores and raking of DMUs, so a small perturbation in data cannot change the results. Sadjadi and Omrani [28] were the first authors who proposed a robust DEA model for the performance assessment of electricity distribution companies in Iran, where outputs are considered as uncertain. In recent years the robust DEA field has been developing and various robust DEA models encompassing different uncertainty sets have been introduced. Salahi et al. [29] studied Russell measure and enhanced Russell measure and the robust counterparts are given under interval and ellipsoidal uncertainty sets. Aghayi and Raayatpanah [30] proposed a robust DEA model to measure the overall profit efficiency of DMUs where input and outputs are subjected to uncertainty. Toloo and Mensah [31] introduced a reduced robust approach with non-negative decision variables and the new approach has been applied to introduced a reduced robust DEA model. Salahi et al. [32] proposed a new robust common-weights DEA model, which in the input and output parameters are subjected to uncertainty. A very detailed review on robust DEA field is given in Peykani et al. [33].

The proposed robust DEA models in the literature generally considered that the inputs and/or outputs are subjected to uncertainty, while the other parameters such as the weights assigned to inputs and outputs also can be subjected to uncertainty. In this article we consider the uncertainty in weight restriction model, which in the weight restrictions are constructed based on production trade-offs. A modified robust DEA model is developed to consider the uncertainty in production trade-offs. Moreover, an interval DEA model is modified that can be applied to handle the uncertainty in production trade- off. Furthermore, we will provide a comparison between the two proposed methods that are able to cope with uncertainty in production trade-offs. In addition, we consider a case study of 35 granted research project to validate our proposed model. The obtain results point out that applying robust WR-TO model improve the robustness of efficiency score and gives a more discriminative ranking of DMUs.

The rest of this paper is organized as follows: Sect. 2 provides a background of weight restriction model based on production trade-offs and robust optimization. In Sect. 3, an equivalent robust counterpart for weight restriction model with uncertainty in production trade-offs is provided. A modified interval DEA model for weight restriction model with uncertainty in production trade-offs is given in Sect. 4. Analysis of results is given is Sect. 5 and finally, conclusion is presented.

2 Background

2.1 *Weight Restriction Based on Production Trade-Offs*

Construction of additional weight restriction in weight restriction models, usually is based on value judgment, monetary values or observed importance of an inputs or outputs. By incorporating such weight restriction, the technological manning of efficiency cannot be clear. Adding the dual from of those weight restriction into the envelopment model cause to the expansion of the production technology and due to this expansion DMU_O will be projected on the efficient frontier of the expanded technology and it will be compared and benchmarked against all units in the technology; the observed units and the new units which have been generated by weight restriction constraints. Therefore, the efficiency measure of DMU_O cannot be interpreted as a realistic improvement input or output factor (Podinovski [10]).

Podinovski [10] introduced the concept of production trade-offs as simultaneous changes to the inputs and outputs which naturally exist in any real production technology. Instead of incorporating the production trade- offs into the multiplier model, the trade-offs will be added to the envelopment form of DEA model. However, each production trade-offs can be converted to an equivalent weight restriction which can be used in the multiplier model. Production trade-offs expand the technology in a controlled way because trade-offs are assumed to be technologically possible, as a result the efficient frontier in the expanded technology will be realistic. Therefore, the technological meaning and radial nature of efficiency will be preserved and any efficient target obtained in the expanded technology by production trade-offs is feasible and producible.

Following Podinovski [10], each production trade-off can be can be expressed as a pair of vectors $(P_t, Q_t); t = 1, 2, \ldots, k$ which vector P_t represents inputs changes and vector Q_t shows output changes. Assume there are n DMUs, which each of them produce a nonzero output vector $Y_j = (y_{1j}, \ldots, y_{sj})^t$, using a nonzero input vector $X_j = (x_{1j}, \ldots, x_{mj})^t$, where the superscript "t" indicates the transpose of a vector. Here, the symbol "$\geq$" indicates that at least one component of X_j or Y_j is positive while the remaining inputs and outputs are non-negative. Thus the envelopment form of CCR model, when the production trade-offs are incorporated in the model in order to assess the relative efficiency of $DMU_O(O \in 1, \ldots, n)$ is given as follow:

$$\begin{aligned} &\min \theta \\ &s.t. \ \textstyle\sum_{j=1}^{n} \lambda_j x_{ij} + \sum_{j=1}^{n} \pi_k p_{ik} \leq \theta x_{io}, && i = 1, \ldots, m, \\ &\quad\ \textstyle\sum_{j=1}^{n} \lambda_j y_{rj} + \sum_{j=1}^{n} \pi_k q_{rk} \geq y_{ro}, && k = 1, \ldots, K, \\ &\quad\ \lambda_j \geq 0, && j = 1, \ldots, n. \end{aligned} \tag{1}$$

In order to clarify the concept of production trade-offs consider assessing the efficiency of university departments. Consider staff numbers as input (x_1) and undergraduate, master and Ph.D. students as outputs (y_1, y_2, y_3). All university departments are willing to have the adequate student number in ratio with their staff numbers. It is a tangible production trade-off to assume teaching a master student needs more resources than teaching an undergraduate student, but by no more than a factor of 2. It means that by reducing the number of undergraduate student by 2 units, the adequate resource will be available to increase the number of master student by 1 unit. This trade-off can be notated as $(p_1, q_1, q_2, q_3) = (0, -2, 1, 0)$. Each production trade-off can be stated in the form of a weight restriction as below:

$$U^T Q_t - V^T p_t \leq 0. \tag{2}$$

Production trade-offs can be added in the multiplier model as an additional constraint on inputs and output weights in the form of (2). The multiplier model in such case can be formulated as the following mathematical programming:

$$\begin{aligned} &(WR - TO) \max \textstyle\sum_{r=1}^{s} u_r y_{ro} \\ &s.t. \quad \textstyle\sum_{i=1}^{m} v_i x_{io} = 1, && i = 1, 2, \ldots, m, \\ &\quad\ \textstyle\sum_{r=1}^{s} u_r y_{rj} - \sum_{i=1}^{m} v_i x_{ij} \leq 0, && j = 1, 2, \ldots, n, \\ &\quad\ \textstyle\sum_{r=1}^{s} u_r q_{rt} - \sum_{i=1}^{m} v_i p_{it} \leq 0, && t = 1, 2, \ldots, k, \\ &\quad\ u_r \geq 0, \\ &\quad\ v_i \geq 0. \end{aligned} \tag{3}$$

2.2 Robust Optimization

Robust optimization approaches can be applied to deal with the uncertainty and perturbation in parameters. The first robust approach was proposed by Soyster [22], and it has been extended by many researchers such as Ben Tal and Nemiroski [23–25] and Bertsimas and Sim [26, 27]. Ben Tal and Nemiroski [25] have proposed modelling the uncertainty in parameters based on ellipsoidal uncertainty set. They showed that a small perturbation in data may lead to the infeasibility of the nominal solution. Bertsimas and Sim [27] used a polyhedral uncertainty set for modeling uncertainty. They argued that the other robust optimization approaches might be very conservative, meaning that they accept a suboptimal solution for the nominal problem in order to ensure that the solution remains feasible. Therefore, they proposed an approach that the conservative level of solution is adjustable and the solution will be more reliable. In the following the robust optimization approach by Bertsimas and Sim will be presented.

To present Bertsimas and Sim's approach, consider the following linear programming model:

$$\begin{aligned} &\max\ c'x \\ &s.t.\ \ Ax \geq b, \\ &\qquad\ x \in X. \end{aligned} \tag{4}$$

Without loss of generality we assume that A which is the coefficient matrix is under uncertainty and X is a polyhedron. Consider a particular row i; $(i = 1, \ldots, m)$ of matrix A and let J_i be the set of coefficient in the row i which are under uncertainty assumption. The components in J_i can be modeled as a symmetric and bounded variable which take values according to an unknown symmetric distribution in the interval $[a_{ij} - \hat{a}_{ij}, a_{ij} + \hat{a}_{ij}]$ (a_{ij} is the nominal mean value for $\tilde{a}_{ij}$). The scale deviation from nominal value of a_{ij} is defined as $\eta_{ij} = (\tilde{a}_{ij} - a_{ij})/\hat{a}_{ij}$ which take value in the interval $[-1, 1]$. For each constraint the parameter Γ_i, not necessarily integer is introduced that take value in the interval $[0, |J_i|]$. In fact, the parameter Γ_i adjust the robustness of the model against the level of conservatism of the solution. If $\Gamma_i = 0$, there is no protection against uncertainty and if $\Gamma_i = |J_i|$ the solution is fully protected against uncertainty.

The following linear optimization model which in only a subset of coefficients are allowed to change, is the robust counterpart of (4) that was introduced by Bertsimas and Sim [27].

$$
\begin{array}{lll}
max\ c'x & & \\
s.t. & \sum_j a_{ij} + z_i\Gamma_i + \sum_{j\in J_i} p_{ij} \le b_i, & \forall i, \\
& z_i + p_{ij} \ge \hat{a}_{ij} y_j, & \forall i,\ j \in J_i \\
& -yj \le x_j \le y_j, & \forall j \\
& l \le x_j \le u, & \forall j \\
& p_{ij} \ge 0 & \forall i \in J_I, \\
& z_i \ge 0, & \forall i, \\
& y_j \ge 0, & \forall j.
\end{array} \tag{5}
$$

3 Robust WR-TO Model

As mentioned before production trade-offs naturally exist in any real world technology, they usually will be expressed linguistically and translating them into exact and precise data may not be always possible. For example, in a university we may have a following feasible trade-off:

One teaching post is replaced by one research position. Then no department should lose more than about 20 undergraduate students and we should expect the increase of publication by at least about 0.3 papers a year.

As we mentioned, some of trade-offs between inputs and outputs cannot be express precisely. Hence, we can use robust optimization to improve model (3) while we are considering the perturbation in production trade-offs between inputs and outputs.

Matrix Q includes the trade-offs in outputs and matrix P includes the trade-offs in inputs.

$$
Q = \begin{bmatrix} q_{11} & \cdots & q_{s1} \\ \vdots & \ddots & \vdots \\ q_{1k} & \cdots & q_{sk} \end{bmatrix} \ and \quad P = \begin{bmatrix} p_{11} & \cdots & p_{m1} \\ \vdots & \ddots & \vdots \\ p_{1k} & \cdots & p_{mk} \end{bmatrix},
$$

$Q_t = (q_{1t}, q_{2t}, ..., q_{st})$; $P_t = (p_{1t}, p_{2t}, ..., p_{mt})$.

Suppose that we have uncertainty in t-th constraint of model (3) which is related to production trade-offs and Based upon Bertsimas and Sim's robust optimization method we suppose that the components in this constraint are under uncertainty.

Let $J_t{}^x$ be the set of coefficients $(p_{it}; i \in J_t{}^x)$, and $J_t{}^y$ be the set of coefficients $(q_{rt}; r \in J_t{}^y)$ that are subject to uncertainty. $\tilde{p}_{it}; i \in J_t^x(J_t^x = \{t \setminus \hat{p}_{it} \neq 0\})$ and $\tilde{q}_{rt}; r \in J_t^y(J_t^y = \{t \setminus \hat{q}_{rt} \neq 0\}$ take values according to an unknown but symmetric distribution with a mean value equal to the nominal value p_{it}, q_{rt} in the interval $\tilde{p}_{it} \in [p_{it} - \hat{p}_{it}, p_{it} + \hat{p}_{it}]$ and $\tilde{q}_{rt} \in [q_{rt} - \hat{q}_{rt}, q_{rt} + \hat{q}_{rt}]$ respectively. For each "t" we introduce parameter Γ_t^x ,Γ_t^y which take values in $[0, |J_t^x|]$, $[0, |J_t^y|]$ respectively. By varying parameters Γ_t^x, Γ_t^y, we are able to adjust the robustness of model against the level of conservatism of solution. In other words, up to $\lfloor \Gamma_t^x \rfloor$, $\lfloor \Gamma_t^y \rfloor$ of coefficients are allowed to change and one coefficient p_{id} or q_{rd} change by $(\Gamma_t^x - \lfloor \Gamma_t^x \rfloor)\hat{p}_{id}$ and $(\Gamma_t^y - \lfloor \Gamma_t^y \rfloor)\hat{q}_{rd}$, respectively. Actually, parameters Γ_t^x, Γ_t^y provide flexibility in order to adjust the level of conservatism of the solution. If we consider $\Gamma_t^x = 0$, $\Gamma_t^y = 0$ there is no protection against the uncertainty and the robust counterpart model will be equivalent to the original model, if $\Gamma_t^x = |J_t^x|$, $\Gamma_t^y = |J_t^y|$ the t-th constraint is fully protected against any uncertainty. Therefore, the robust counterpart of model (3) (when production trade-offs are under uncertainty) based on Bertsimas and Sim robust optimization method will be as follow:

$$\begin{aligned}
\max\ & \textstyle\sum_{r=1}^{s} u_r y_{ro} & \\
s.t.\ & \textstyle\sum_{i=1}^{m} v_i x_{io} = 1, & i = 1, 2, \ldots, m, \\
& \textstyle\sum_{r=1}^{s} u_r y_{ro} - \sum_{i=1}^{m} v_i x_{io} \le 0, & j = 1, 2, \ldots, n, \\
& \textstyle\sum_{r=1}^{s} u_r q_{rt} + \beta_i(v, \Gamma_t^x) - \sum_{i=1}^{m} v_i p_{it} + \beta_r(u, \Gamma_t^y) \le 0, & t = 1, \ldots, k, \\
& -z_r \le u_r \le z_r, & \\
& -w_i \le v_i \le w_i, &
\end{aligned} \tag{6}$$

where

$$\beta_i(v, \Gamma_t^x) = \max_{\{s_t^y \cup \{d_t^y\} | s_t^y \subseteq J_t^y; |s_t^y| = \lfloor \Gamma_t^y \rfloor, d_t^y \in J_t^y \setminus s_t^y\}} \left\{ \sum_{r \in s_t^y} \hat{q}_{rt} z_r + (\Gamma_t^y - \lfloor \Gamma_t^y \rfloor) \hat{q}_{rd_t^y} z_d \right\}$$

and

$$\beta_r(u, \Gamma_t^y) = \max_{\{s_t^x \cup \{d_t^x\} | s_t^x \subseteq J_t^x; |s_t^x| = \lfloor \Gamma_t^x \rfloor, d_t^x \in J_t^x \setminus s_t^x\}} \left\{ \sum_{i \in s_t^x\}} \hat{p}_{it} w_i + (\Gamma_t^x - \lfloor \Gamma_t^x \rfloor) \hat{p}_{id_t^x} w_d \right\}.$$

If Γ_t^x and Γ_t^y are chosen as integers, the t-th constraint is protected by

$$\beta_i(v, \Gamma_t^x) = \max_{\{s_t^x | s_t^x \subseteq J_t^x; |s_t^x| = \lfloor \Gamma_t^x \rfloor\}} \left\{ \sum_{i \in s_t^x} \hat{p}_{it} w_i \right\}$$

and

$$\beta_r(u, \Gamma_t^y) = \max_{\{s_t^y | s_t^y \subseteq J_t^y; |s_t^y| = \lfloor \Gamma_t^y \rfloor\}} \left\{ \sum_{r \in s_t^y} \hat{q}_{rt} z_r \right\}.$$

If $\Gamma_t^x = 0$, the t-th constraint is protected by $\beta_i(v, \Gamma_t^x)$.
If $\Gamma_t^y = 0$, the t-th constraint is protected by $\beta_r(u, \Gamma_t^y)$.
And if $\Gamma_t^y = \Gamma_t^x = 0$, the constraints are equivalent to the nominal problem.

Hence by varying $\Gamma_t^x \in [0, |J_t^x|]$ and $\Gamma_t^y \in [0, |J_t^y|]$, we have the flexibility of adjusting the robustness of the method against the level of conservatism of the solution. Model (6) is a nonlinear model, therefore we need to transform it to a linear model, so it will be suitable to solve it using any DEA-solver package.

Proposition 1. *If vectors of u^* and v^* are given, the protection functions of t-th constraint,*

$$\beta_r(u,\Gamma_t^y)=\max_{\{s_t^y\cup\{d_t^y\}|s_t^y\subseteq J_t^y;|s_t^y|=\lfloor\Gamma_t^y\rfloor,d_t^y\in J_t^y\setminus s_t^y\}}\left\{\sum\nolimits_{r\in s_t^y}\hat{q}_{rt}|u_r^*|+(\Gamma_t^y-\lfloor\Gamma_t^y\rfloor)\hat{q}_{rd_t^y}|u_r^*|\right\},$$

$$\beta_i(v,\Gamma_t^x)=\max_{\{s_t^x\cup\{d_t^x\}|s_t^x\subseteq J_t^x;|s_t^x|=\lfloor\Gamma_t^x\rfloor,d_t^x\in J_t^x\setminus s_t^x\}}\left\{\sum\nolimits_{i\in s_t^x}\hat{p}_{it}|v_i^*|+(\Gamma_t^x-\lfloor\Gamma_t^x\rfloor)\hat{p}_{id_t^x}|v_i^*|\right\},$$

Are equal to the objective function of the following linear optimization problems respectively.

$$\begin{array}{lll}\beta_r(u^*,\Gamma_t^y)=\max & \sum_{r\in J_t^y}\hat{q}_{rt}|u_r^*|\gamma_{rt} & \\ s.t. & \sum_{r\in J_t^y}\gamma_{rt}\le\Gamma_t^y, & \\ & 0\le\gamma_{rt}\le 1, & \forall r\in J_t^y,\end{array}\tag{7}$$

$$\begin{array}{lll}\beta_i(v^*,\Gamma_t^x)=\max & \sum_{i\in J_t^x}\hat{p}_{it}|v_i^*|\lambda_{it} & \\ s.t. & \sum_{i\in J_t^x}\lambda_{it}\le\Gamma_t^x, & \\ & 0\le\lambda_{it}\le 1, & \forall i\in J_t^x.\end{array}\tag{8}$$

Proof. It is obvious that the optimal value of model (7) consists of $\lfloor\Gamma_t^y\rfloor$ variables at 1 and one variable at $(\Gamma_t^y-\lfloor\Gamma_t^y\rfloor)$ and this is equivalent to the section of subset $\{s_t^y\cup\{d_t^y\}\mid s_t^y\subseteq J_t^y;|s_t^y|=\lfloor\Gamma_t^y\rfloor,d_t^y\in J_t^y\setminus s_t^y\}$ with corresponding cost function $\{\sum_{r\in s_t^y}\hat{q}_{rt}|u_r^*|+(\Gamma_t^y-\lfloor\Gamma_t^y\rfloor)\hat{q}_{rd_t^y}|u_r^*|\}$. Similarly the optimal value of model (8) consists of $\lfloor\Gamma_t^x\rfloor$ variables at 1 and one variable at $(\Gamma_t^x-\lfloor\Gamma_t^x\rfloor)$ and this is equivalent to the section of subset $\{s_t^x\cup\{d_t^x\}\mid s_t^x\subseteq J_t^x;|s_t^x|=\lfloor\Gamma_t^x\rfloor,d_t^x\in J_t^x\setminus s_t^x\}$ with corresponding cost function $\{\sum_{i\in s_t^x}\hat{p}_{it}|v_i^*|+(\Gamma_t^x-\lfloor\Gamma_t^x\rfloor)\hat{p}_{id_t^x}|v_i^*|\}$. In order to transform model (6) to a linear optimization model let consider z_r^y and α_{rt} be the dual variables corresponding to the first and second constraints in model (7) respectively. Therefore, the dual of model (7) is as follows:

$$\begin{array}{lll}\min & \sum_{r\in J_t^y}\alpha_{rt}+z_r^y\Gamma_t^y & \\ s.t. & z_r^y+\alpha_{rt}\ge\hat{q}_{rt}|u_r^*|, & \forall r\in J_t^y,\\ & z_r^y\ge 0, & \\ & \alpha_{rt}\ge 0, & \end{array}\tag{9}$$

Similarly let consider z_i^x and β_{rt} be the dual variables corresponding to the first and second constraints in model (8) respectively . Therefore, the dual of model (8) is as follows:

$$\begin{array}{lll}\min & \sum_{i\in J_t^x}\beta_{it}+z_i^x\Gamma_t^x & \\ s.t. & z_i^x+\beta_{it}\ge\hat{p}_{it}|v_i^*|, & \forall i\in J_t^x,\\ & z_i^x\ge 0, & \\ & \beta_{it}\ge 0, & \end{array}\tag{10}$$

Using proposition 1, we have that $\beta_r(u^*, \Gamma_t^y)$ and $\beta_i(v^*, \Gamma_t^x)$ are equal to the objective function values of model (7) and model (8) respectively. By substituting to model (6) it is obtained that model (6) is equivalent to the following linear optimization.

$$
\begin{array}{lll}
Robust(WR-TO)\max & \sum_{r=1}^{s} u_r y_{ro} & \\
s.t. & \sum_{i=1}^{m} v_i x_{io} = 1, & i = 1, 2, ..., m, \\
 & \sum_{r=1}^{s} u_r y_{rj} - \sum_{i=1}^{m} v_i x_{ij} \leq 0, & j = 1, 2, ..., n, \\
 & \sum_{r=1}^{s} u_r q_{rt} - \sum_{i=1}^{m} v_i p_{it} \leq 0, & t = 1, 2, ..., k, \\
 & \sum_{r=1}^{s} u_r q_{rt} - \sum_{i=1}^{m} v_i p_{it} + z_i^x \Gamma_t^x + z_r^y \Gamma_t^y & \\
 & + \sum_{i\in J_t^x} \beta_{it} + \sum_{r\in J_t^y} \alpha_{rt} \leq 0, & \forall i \in J_t^x and\ r \in J_t^y, \\
 & z_r^y + \alpha_{rt} \geq \hat{q}_{rt} u_r', & \\
 & z_i^x + \beta_{it} \geq \hat{p}_{it} v_i', & \\
 & -u_r' \leq u_r \leq u_r', & \\
 & -v_i' \leq v_i \leq v_i'. &
\end{array}
\tag{11}
$$

Robust WR-TO model can handle the uncertainty in production trade-offs data without requiring to identify the data distribution function and it can be solved using any linear programming software. In order to apply the above model, we need to consider $\hat{p}_{it} = \varepsilon_i p_{it}$, where ε_i is the value of perturbation for the i-th input trade-offs in the t-th weight restriction constraint and $\hat{q}_{rt} = \varepsilon_r q_{rt}$ where ε_r is the value of perturbation for the r-th output trade-off in the t-th weight restriction constraint. Another important task to be able to implement Robust WR-TO model is the choice of the value of Γ_t^x, Γ_t^y. In order to protect t-th constraint against perturbation it is sufficient to choose Γ_t^{xy} according to the following relation:

$$\Gamma_t^{xy} = \Gamma_t^x + \Gamma_t^y = 1 + \phi^{-1}(1 - \varepsilon_{i,r})\sqrt{n},$$

where ϕ is the cumulative distribution function of the standard Gaussian variable and is the source of uncertainty for each constraint. For example if $n = 200$ and the violation probability of the t-th constraint be less than 1% then Γ_t^{xy} must be at least $33.9200 \approx 17\%$ of the uncertain parameters taking their worst-case value.

4 Handling Uncertainty in Production Trade-Offs Using Interval DEA Approach

In this section we use the interval DEA approach proposed by Smirlis et al. [34] to handle the uncertainty in production trade-offs in WR-TO model. As mentioned before, production trade-offs naturally exist in any real world technology and will be set by decision makers and they usually will be expressed linguistically and translating them into exact and precise data may not be always possible. We assume the exact level of some trade-offs ($i.e.\ q_{rm}, p_{im}$) is unknown, but it is known that their level lie within the upper and lower bounds represented by the intervals $q_{rm} \in$

$[q^l_{rm}, q^u_{rm}]$ and $p_{im} \in [p^l_{im}, p^u_{im}]$. In order to incorporate this type of data in WR-TO model, first we apply the following transformation for variables q_{rm}, p_{im}:

$$q_{rm} = \lambda_r q^u_{rm} + (1-\lambda_r)q^l_{rm} = q^l_{rm} + \lambda_r(q^u_{rm} - q^l_{rm}), \qquad 0 \le \lambda_r \le 1,$$

$$p_{im} = \beta_i p^u_{im} + (1-\beta_i)p^l_{im} = p^l_{im} + \beta_i(p^u_{im} - p^l_{im}), \qquad 0 \le \beta_i \le 1.$$

According to these transformations WR-TO model will be modified as the following mathematical programming:

$$\begin{array}{lll}
\max & \sum_{r=1}^{s} u_r y_{ro} & \\
s.t. & \sum_{i=1}^{m} v_i x_{io} = 1, & i = 1, 2, ..., m, \\
& \sum_{r=1}^{s} u_r y_{rj} - \sum_{i=1}^{m} v_i x_{ij} \le 0, & j = 1, 2, ..., n, \\
& \sum_{r=1}^{s} u_r q_{rt} - \sum_{i=1}^{m} v_i p_{it} \le 0, & t = 1, 2, ..., k \setminus m, \\
& \sum_{r=1}^{s} u_r q^l_{rm} + u_r\lambda_r(q^u_{rm} - q^l_{rm}) & \\
& - \sum_{i=1}^{m} v_i p^l_{im} + v_i\beta_i(p^u_{im} - p^l_{im}) \le 0, & t = m, \\
& u_r \ge 0, & \\
& v_i \ge 0. &
\end{array} \tag{12}$$

Since there are the products of variables $(u_r\lambda_r, v_i\beta_i)$ in model (12), this model is a non-linear model. In order to transform it to a linear programming, we replace these products with new variables $u_r\lambda_r = \gamma_r$; where $0 \le \gamma_r \le u_r$ and $v_i\beta_i = \delta_i$; where $0 \le \delta_i \le v_i$. By substituting these new variables, model (12) is transformed to the following linear programing:

$$\begin{array}{lll}
Interval(WR-TO) \max & \sum_{r=1}^{s} u_r y_{ro} & \\
s.t. & \sum_{i=1}^{m} v_i x_{io} = 1, & i = 1, 2, ..., m, \\
& \sum_{r=1}^{s} u_r y_{rj} - \sum_{i=1}^{m} v_i x_{ij} \le 0, & j = 1, 2, ..., n, \\
& \sum_{r=1}^{s} u_r q_{rt} - \sum_{i=1}^{m} v_i p_{it} \le 0, & t = 1, 2, ..., k \setminus m, \\
& \sum_{r=1}^{s} u_r q^l_{rm} + \gamma_r(q^u_{rm} - q^l_{rm}) & \\
& - \sum_{i=1}^{m} v_i p^l_{im} + \delta_i(p^u_{im} - p^l_{im}) \le 0, & t = m, \\
& 0 \le \gamma_r \le u_r, & \\
& 0 \le \delta_i \le v_i, & \\
& u_r \ge 0, & \\
& v_i \ge 0. &
\end{array} \tag{13}$$

In model (13) if the upper and lower bounds are all zero, Interval WR-TO model will be reduced to WR-TO model. Interval WR-TO model can handle imprecise production trade-offs when decision makers cannot come up with an agreement to set an exact value for the trade-off between two variables in the production. As an example consider assessing the efficiency of farms, driven data and analyses from different regions may show that any farm can produce between 1.2 and 1.7 tons of wheat instead of 1 ton of barley, without claiming additional resources. Hence, in such cases Interval WR-TO model can be applied to handle this type of uncertainty in data.

5 Analysis of Results

This section demonstrates the applicability of the proposed model in this article using data from 35 granted research projects in engineering discipline from universiti Sains Malaysia. In order to analyze the efficiency of these granted research projects the number of researchers (x_1), the amount of grants allocated to complete the research (x_2), the duration token to complete the research (x_3) are considered as inputs. In assessing the efficiency of granted research project it is widely agreed that the most significant research output is publication. In our study output variables included international journal publications (y_1), ISI publications (y_2), local journal publications (y_3), international conference publications (y_4), local journal publications (y_5) and other publications (y_6). A descriptive statistic for data sets is given in Table 1.

In the process of evaluation, it may not seems to be fair to compare the projects under a single rule and assumption. For example, all researchers are agreed that publishing and participating in an international conference surely needs more resources than publishing and participating in a local conference. Hence, by considering the production trade-offs, it is possible to tackle the problem of non-homogeneity in outputs. As the multiplier DEA model determine the potential improvement of the publications by fully utilizing the inputs, the following production trade-offs are considered to improve the weight distribution and discrimination power of the model.

Table 1 Descriptive statistics for data sets

Data set		Mean	Standard deviation	Minimum	Maximum
Inputs					
	Researchers	2.23	0.88	1	4
	Grants	20602.86	7923.66	6674	39257
	Duration	24.46	2.077	24	36
outputs					
	International journal publications	1.49	1.67	0	6
	Local journal publications	0.2	0.47	0	2
	International conference publications	2.03	2.39	0	9
	Local conference publications	0.91	1.17	0	4
	Other publications	0.03	0.17	0	1

Assumption 1. Publishing an international conference paper needs more resources than a local conference paper however by no more than a factor of 5.

$$(p_{11}, p_{21}, p_{31}, q_{11}, q_{21}, q_{31}, q_{41}, q_{51}, q_{61}) = (0, 0, 0, 0, 0, 0, 1, -5, 0)$$

Assumption 2. If the number of researchers is increased by 2, it should be possible to increase the number of publications by at least 0.25.

$$(p_{12}, p_{22}, p_{32}, q_{12}, q_{22}, q_{32}, q_{42}, q_{52}, q_{62}) = (2, 0, 0, 0.25, 0.25, 0.25, 0.25, 0.25, 0.25)$$

Assumption 3. Publishing an international journal paper requires no more than the twice the amount of resources to publish a local journal paper.

$$(p_{13}, p_{23}, p_{33}, q_{13}, q_{23}, q_{33}, q_{43}, q_{53}, q_{63}) = (0, 0, 0, 1, 0, -2, 0, 0, 0)$$

Assumption 4. The increase of the amount of grants by 15000 is sufficient to increase the number of researchers by 2 and increase the number of publications by 0.5.

$$(p_{14}, p_{24}, p_{34}, q_{14}, q_{24}, q_{34}, q_{44}, q_{54}, q_{64}) = (2, 15000, 0, 0.5, 0.5, 0.5, 0.5, 0.5, 0.5)$$

Assumption 5. The reduction of intentional conference and local conference papers by 1, release the sufficient resources to publish at least one international journal paper.

$$(p_{15}, p_{25}, p_{35}, q_{15}, q_{25}, q_{35}, q_{45}, q_{55}, q_{65}) = (0, 0, 0, 1, 0, 0, -1, -1, 0,)$$

Assumption 6. Publishing an ISI paper does not necessarily require more resourses than an international journal paper.

$$(p_{16}, p_{26}, p_{36}, q_{16}, q_{26}, q_{36}, q_{46}, q_{56}, q_{66}) = (0, 0, 0, -1, 1, 0, 0, 0, 0)$$

Each of these production trade-off can be stated in the form of a weight restriction $(\sum_{r=1}^{s} u_r q_{rt} - \sum_{i=1}^{m} v_i p_{it} \leq 0)$ as below:

$$\begin{array}{ll}
1.\ u_4 - 5u_5 \leq 0 & t = 1 \\
2.\ 0.25(u_1 + u_2 + u_3 + u_4 + u_5 + u_6) - 2v_1 \leq 0 & t = 2 \\
3.\ u_1 - 2u_3 \leq 0 & t = 3 \\
4.\ 0.5(u_1 + u_2 + u_3 + u_4 + u_5 + u_6) - 2v_1 - 15000v_2 \leq 0 & t = 4 \\
5.\ u_1 - u_4 - u_5 \leq 0 & t = 5 \\
6.\ -u_1 + u_2 \leq 0 & t = 6
\end{array}$$

Note that more complex trade-offs that links more inputs and outputs can be considered and by adding more complex trade-offs a better weight distribution and efficiency discrimination can be expected.

The result from the multiplier CCR model and WR-TO model which shows the efficiency scores, input-output weights and ranking of the projects are reported in Tables 2 and 3 respectively. The result shows that the discrimination power of the CCR model is much lower than the WR-TO model as 31.43% of projects are reported as efficient projects by applying the classical CCR model in comparison with the WR-TO model which only 11.43% of the project are reported as efficient projects. Moreover, implementing WR-TO model gives a better efficiency score discrimination compared to the classical CCR model. Furthermore, the result for the inputs and outputs weights shows that the WR-TO model distributed the value of inputs-output weights more evenly than those obtained by CCR model as the number of zeros for input-output weights in results is less than the CCR model.

Assigning the production trade-offs depends on the expert's judgments and the methodologies that they have applied. Hence it may not be always possible to translate and estimate these production trade-offs into exact and precise data, and in some cases not all experts may agree on some of these judgments and it is preferred to consider a small perturbation in some or all data. In this case we assume that in trade-off $t = 1$, q_{15} is under uncertainty and let $\epsilon = 0.2$ for it and also in trade of $t = 4$, p_{24} is under uncertainty with $\epsilon = 0.2$. Considering these production trade-offs and uncertainties, the robust WR-TO model is implemented to determine the efficiency scores of the projects. The efficiency score and ranking of projects by applying the robust WR-TO model and interval WR-TO model are presented in Table 4. It is evident that in case of the robust WR-TO model the efficiency scores are decreased in comparison with the other models. A comparison of efficiency scores obtained from CCR, WR-TO, Robust WR-TO and Interval WR-TO model is presented in Fig. 1. In evaluating projects under uncertainty it should be noted that by considering the uncertainty in data the feasibility of the optimal solution can be heavily effected even by a small perturbation of the data. It this study, even though only 2 components are considered under uncertainty, the efficiency scores and the ranking of the projects have been effected. The result of the robust WR-TO model are reported for the full protection as there are only 2 uncertain components, $\Gamma_t^x = 1, \Gamma_t^y = 1$.

The result from the interval WR-TO in comparison with the robust WR-TO model shows that the interval WR-TO model tends to overestimate efficiency scores and it can be less discriminative than the robust WR-TO model. Moreover, the robust WR-TO model with the same uncertain components and $\epsilon = 0.1$ is also solved and the result are shown in Table 4. The result shows that when the level of uncertainty increases the efficiency scores will decrease. Overall, implementing the production trade-offs gives the opportunity to the projects to be evaluated more fairly and the robust WR-TO model is a better option for considering the uncertainty in production trade-offs, especially when the number of parameters that are under uncertainty increases and there is a choice to adjust the level of conservatism.

Table 2 Result from CCR model.

Project	Efficiency score	u_1	u_2	u_3	u_4	u_5	u_6	v_1	v_2	v_3
1	0.2257	0	0	0	0.2257	0	0	0	0.0001	0
2	1	0	0	0.3698	0.077	0.041	0	0	0.00006	0
3	0.250	0	0	0	0	0.250	0	0	0	0.042
4	0.9605	0.1867	0	0	0.067	0	0	0.2151	0.00003	0
5	1	0.074	0	0	0.056	0.1296	0	0	0	0.042
6	0.2979	0.1277	0	0	0.085	0	0	0.021	0	0.039
7	0.3333	0.3333	0	0	0	0	0	0.3333	0	0.028
8	0.8824	0.2350	0	0	0	0.059	0	0.294	0	0.017
9	0.1551	0	0	0	0.1551	0	0	0.1376	0.00005	0
10	0.8571	0	0	0	0.1429	0	0	0.1429	0	0.030
11	0.2489	0	0.2489	0	0	0	0	0.021	0.00001	0.031
12	0.3907	0	0	0	0.1954	0	0	0.1732	0.00006	0
13	0.250	0	0	0	0	0.250	0	0	0	0.042
14	1	0	0.1579	0.4474	0.026	0	0	0.079	0	0.029
15	1	0	0.22220	0	0	0.1111	0	0.1111	0	0.032
16	0.7143	0	0	0.50	0.071	0	0	0.071	0	0.036
17	0.667	0.3333	0	0	0	0	0.3333	0	0.028	0
18	1	0.2917	0	0	0.063	0	0	0.3125	0	0.029
19	1	0	0	0	0	0	1	0	0	0.042
20	0.7143	0	0	0	0.1429	0	0	0.1429	0	0.030
21	0.7030	0	0	0	0.1347	0.082	0	0.00006	0	0
22	1	0.1486	0	0	0	0.037	0	0.1993	0.00001	0
23	0.5499	0	0.2750	0	0	0	0	0.088	0.00005	0
24	1	0.1250	0.1250	0	0	0	0	0.1250	0	0.031
25	0.1281	0	0	0	0.1281	0	0	0.1136	0.00004	0
26	0.6565	0	0	0	0	0.6565	0	0.0002	0	0
27	0.4444	0	0	0	0	0.2222	0	0	0	0.037
28	0.4162	0	0	0	0.1387	0	0	0.1387	0	0.029
29	0.4000	0.200	0	0	0	0	0	0.200	0	0.017
30	1	0	0	0.500	0	0	0	0.271	0.00001	0
31	0.6250	0	0	0.500	0	0.125	0	0	0	0.042
32	0.6000	0.200	0	0	0	0	0	0.200	0	0.017
33	1	0.1423	0	0.4284	0	0	0	0	0.000007	0.032
34	0.2941	0.2353	0	0	0	0.059	0	0.294	0	0.017
35	1	0	0.1875	0	0	0.063	0	0.125	0	0.021

Table 3 Result from WR-TO model.

Project	Efficiency score	u_1	u_2	u_3	u_4	u_5	u_6	v_1	v_2	v_3
1	0.2181	0	0	0	0.2181	0.044	0	0.033	0.00001	0
2	1	0	0	0	0.1471	0.029	0	0.1765	0	0.034
3	0.250	0	0	0	0	0.250	0	0.250	0	0.031
4	0.9268	0.1057	0	0.053	0.095	0.019	0	0.071	0.00005	0
5	1	0.011	0.011	0.005	0.1107	0.053	0	0.024	0.00005	0
6	0.2753	0.10	0	0.050	0.088	0.018	0	0.037	0.00003	0.014
7	0.1538	0.1538	0	0.077	0	0.1538	0	0.3846	0	0.026
8	0.7485	0.1248	0	0.062	0	0.1248	0	0.1292	0.00002	0.011
9	0.1457	0	0	0	0.1457	0.029	0	0.0.087	0.00005	0
10	0.7864	0	0	0	0.1311	0.026	0	0.095	0.000008	0.027
11	0.1760	0.088	0.088	0.044	0	0.088	0	0.1791	0.00002	0.044
12	0.3839	0	0	0	0.1745	0.035	0	0.1040	0.00006	0
13	0.2311	0	0	0	0	0.2311	0	0.029	0.00003	0.018
14	1	0	0	0.2566	0.083	0.017	0	0.044	0.00004	0.002
15	0.9097	0.1011	0.1011	0.051	0	0.1011	0	0.2058	0.00002	0.009
16	0.5444	0	0	0.2965	0.083	0.017	0	0.049	0.00005	0
17	0.5333	0.1333	0.1333	0.067	0	0.1333	0	0.4667	0	0.022
18	0.8405	0.1096	0.1096	0.055	0.091	0.018	0	0.2987	0.00004	0
19	0.5959	0	0	0	0.1402	0.028	0.2875	0.083	0.00005	0
20	0.7073	0	0	0	0.1310	0.026	0	0.095	0.000008	0.027
21	0.6911	0	0	0	0.1392	0.067	0	0.026	0.00006	0
22	1	0.077	0.077	0.038	0.064	0.013	0	0.034	0.00006	0
23	0.4375	0.1094	0.1094	0.055	0	0.1094	0	0.2351	0.00003	0
24	0.8345	0.1043	0.1043	0.0520	0	0.1043	0	0.2123	0.00002	0.009
25	0.1245	0	0	0	0.1245	0.025	0	0.074	0.00004	0
26	0.6248	0	0	0	0	0.6248	0	0.078	0.0001	0
27	0.4320	0	0	0	0	0.2160	0	0.027	0.00003	0.016
28	0.3644	0	0	0	0.1215	0.024	0	0.1457	0	0.028
29	0.1739	0.087	0	0.043	0	0.087	0	0.2174	0	0.014
30	0.9778	0	0	0.4889	0	0	0	0.061	0.00006	0
31	0.3333	0	0	0.3333	0	0	0	0.3333	0	0
32	0.2773	0.092	0	0.046	0	0.092	0	0.096	0.00002	0.008
33	0.5172	0.069	0.069	0.034	0	0.069	0	0.2414	0	0.011
34	0.2222	0.1111	0	0.056	0	0.1111	0	0.2778	0	0.019
35	0.6499	0.059	0.059	0.030	0	0.060	0	0.1203	0.00001	0.005

Table 4 Comparison of the efficiency scores and ranking from different models.

Project	Efficiency score (CCR)	Ranking	Efficiency score (WR-TO)	Ranking	Efficiency score (Robust WR-TO) $\epsilon = 0.1$	Ranking	Efficiency score (Robust WR-TO) $\epsilon = 0.2$	Ranking	Interval (WR-TO) $\epsilon = 0.2$	Ranking
1	0.2257	21	0.2181	27	0.2181	27	0.2181	27	0.2183	26
2	1	1	1	1	1	1	1	1	1	1
3	0.250	19	0.250	24	0.250	24	0.250	24	0.250	23
4	0.9605	2	0.9268	3	0.9268	2	0.9268	2	0.9299	2
5	1	1	1	1	1	1	1	1	1	1
6	0.2979	17	0.2753	23	0.2747	22	0.2738	22	0.2771	22
7	0.3333	16	0.1538	30	0.1538	30	0.1538	30	0.1538	29
8	0.8824	3	0.7485	8	0.7424	8	0.7352	8	0.7582	7
9	0.1551	22	0.1457	31	0.1457	31	0.1457	31	0.1472	30
10	0.8571	4	0.7864	7	0.7860	7	0.7853	7	0.7979	6
11	0.2489	20	0.1760	28	0.1757	28	0.1753	28	0.1765	27
12	0.3907	15	0.3839	19	0.3839	19	0.3839	19	0.3849	18
13	0.250	19	0.2311	25	0.2299	25	0.2284	25	0.2329	24
14	1	1	1	1	1	1	1	1	1	1
15	1	1	0.9097	4	0.9048	3	0.8989	3	0.9175	3
16	0.7143	5	0.5444	14	0.5210	15	0.50	16	0.5915	13
17	0.6667	7	0.5333	15	0.5333	14	0.5333	15	0.5333	14
18	1	1	0.8405	5	0.8405	5	0.8405	4	0.8414	4
19	1	1	0.5960	13	0.5735	13	0.5732	13	0.6713	11
20	0.7143	5	0.7073	9	0.7069	9	0.7062	9	0.7089	8
21	0.7030	6	0.6911	10	0.6911	10	0.6911	10	0.6911	9
22	1	1	1	1	1	1	1	1	1	1
23	0.5499	11	0.4375	17	0.4375	17	0.4375	17	0.4375	16
24	1	1	0.8345	6	0.8323	6	0.8298	5	0.8378	5
25	0.1281	23	0.1245	32	0.1245	32	0.1245	32	0.1251	31
26	0.6565	8	0.6248	12	0.6248	12	0.6248	11	0.6248	12
27	0.4444	12	0.4320	18	0.4318	18	0.4315	18	0.4323	17
28	0.4162	13	0.3644	20	0.3644	20	0.3644	20	0.3721	19
29	0.4000	14	0.1739	29	0.1739	29	0.1739	29	0.1739	28
30	1	1	0.9778	2	0.8964	4	0.8120	6	1	1
31	0.6250	9	0.3333	21	0.3333	21	0.3333	21	0.3333	20
32	0.6000	10	0.2773	22	0.2709	23	0.2635	23	0.2878	21
33	1	1	0.5172	16	0.5172	16	0.5172	14	0.5172	15
34	0.2941	18	0.2222	25	0.2222	26	0.2222	26	0.2222	25
35	1	1	0.6499	11	o.6357	11	0.6192	12	0.6732	10

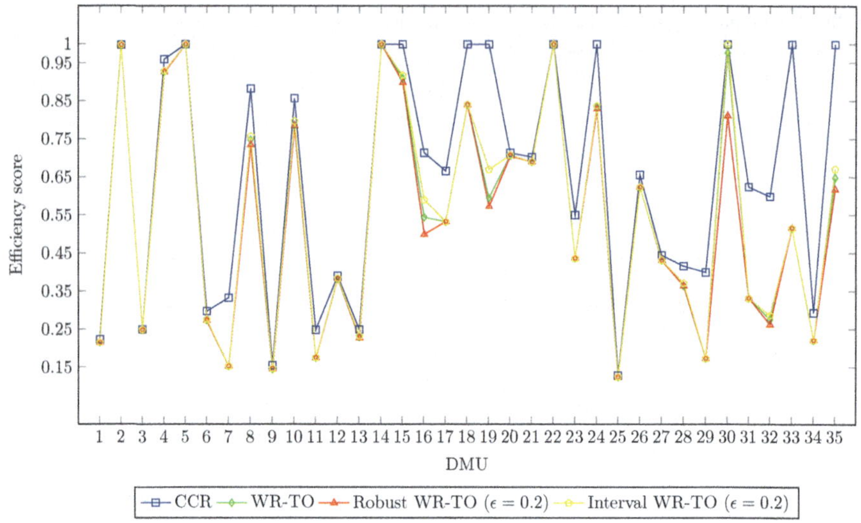

Fig. 1 Comparison of efficiency scores

6 Conclusion

Robust optimization approach has been widely applied in DEA in order to study the problems involving data uncertainty. In the robust DEA literature, usually inputs and/or outputs parameters are considered under uncertainty, however in some other DEA models such as weight restriction models, there are other parameters that might be subjected to uncertainty. In this paper a modified robust weight restriction model based on Bertsimas and Sim's approach [27] where the production trade-offs are subjected to uncertainty is proposed. Moreover, an interval weight restriction model is expanded to handle the uncertainty in production-trade-offs. The proposed models are verified on a case study of granted research projects where the production trade-offs have been considered in the evaluation process in order to tackle the problem of low discrimination and non-homogeneity of outputs. The result indicates that perturbation in data even in one or two parameters can affect the efficiency scores and ranking of DMUs and by increasing the level of perturbation, the efficiency scores will be decreased. Moreover, it is shown that the interval weight restriction model tends to overestimate the efficiency scores and it not an efficient model to be used in cases with a large number of parameters subjected to uncertainty. This paper can be used in real world-applications where the inputs and/or outputs are non-homogenous and experts or decision makers cannot come up with an agreement on the production trade-offs or the weight assigned to inputs and/or outputs to provide a more appropriate and fair evaluation process. Applying robust optimization approaches to the other DEA models, such as models with integer valued data or non-discretionary factors are interesting topics for future researches.

References

1. Farrel, M.J.: The measurement of productive efficiency. J. Royal Stat. Soc. Series A. Part **3** (1957). https://doi.org/10.2307/2343100
2. Charnes, A., Cooper, W.W., Rhodes, E.: Measuring the efficiency of decision making units. Eur. J. Oper. Res. **2**(6), 429–444 (1978). https://doi.org/10.1016/0377-2217(78)90138-8
3. Cooper, W.W., Seiford, L.M., Tone, K.: Data envelopment analysis. Handbook on data envelopment analysis, pp. 1–40. Springer, Boston, MA (2000). https://doi.org/10.1007/b105307
4. Thanassoulis, E., Portela, M.C., Allen, R.: Incorporating value judgments in DEA. In: Handbook on data envelopment analysis, pp. 99–138. Springer, Boston, MA (2004). https://doi.org/10.1007/1-4020-7798-X_4
5. Dyson, R.G., Thanassoulis, E.: (1988). Reducing weight flexibility in data envelopment analysis. J. Oper. Res. Soc. **39**(6), 563–576 (1988). https://doi.org/10.1057/jors.1988.96
6. Thompson, R.G., Singleton Jr, F.D., Thrall, R.M., Smith, B.A.: Comparative site evaluations for locating a high-energy physics lab in Texas. Interfaces **16**(6), 35–49 (1986). https://doi.org/10.1287/inte.16.6.35
7. Thompson, R.G., Langemeier, L.N., Lee, C.T., Lee, E., Thrall, R.M.: The role of multiplier bounds in efficiency analysis with application to Kansas farming. J. Econ. **46**(1–2), 93–108 (1990). https://doi.org/10.1016/0304-4076(90)90049-Y
8. Allen, R., Athanassopoulos, A., Dyson, R., et al.: Weights restrictions and value judgements in Data Envelopment Analysis: evolution, development and future directions. Ann. Oper. Res. **73**, 13–34 (1997). https://doi.org/10.1023/A:1018968909638
9. Roll, Y., Cook, W.D., Golany, B.: Controlling factor weights in data envelopment analysis. IIE Trans. **23**(1), 2–9 (1991). https://doi.org/10.1080/07408179108963835
10. Podinovski, V.V.: Production trade-offs and weight restrictions in data envelopment analysis. J. Oper. Res. Soc. **55**(12), 1311–1322 (2004). https://doi.org/10.1057/palgrave.jors.2601794
11. Podinovski, V.V., Husain, W.R.W.: The hybrid returns-to-scale model and its extension by production trade-offs: an application to the efficiency assessment of public universities in Malaysia. Ann. Oper. Res. **250**(1), 65–84 (2017). https://doi.org/10.1007/s10479-015-1854-0
12. Khalili, M., Camanho, A.S., Portela, M.C.A.S., Alirezaee, M.R.: The measurement of relative efficiency using data envelopment analysis with assurance regions that link inputs and outputs. Eur. J. Oper. Res. **203**(3), 761–770 (2010). https://doi.org/10.1016/j.ejor.2009.09.002
13. Amado, C.A.D.E.F., dos Santos, S.P.: Challenges for performance assessment and improvement in primary health care: the case of the Portuguese health centres. Health Policy **91**(1), 43–56 (2009). https://doi.org/10.1016/j.healthpol.2008.11.008
14. Santos, S.P., Amado, C.A., Rosado, J.R.: Formative evaluation of electricity distribution utilities using data envelopment analysis. J. Oper. Res. Soc. **62**(7), 1298–1319 (2011). https://doi.org/10.1057/jors.2010.66
15. Atici, K.B., Podinovski, V.V.: Using data envelopment analysis for the assessment of technical efficiency of units with different specialisations: an application to agriculture. Omega **54**, 72–83 (2015). https://doi.org/10.1016/j.omega.2015.01.015
16. Land, K.C., Lovell, C.K., Thore, S.: Chance-constrained data envelopment analysis. Manag. Decis. Econ. **14**(6), 541–554 (1993). https://doi.org/10.1002/mde.4090140607
17. Olesen, O.B., Petersen, N.C.: Chance constrained efficiency evaluation. Manage. Sci. **41**(3), 442–457 (1995). https://doi.org/10.1287/mnsc.41.3.442
18. Cooper, W.W., Deng, H., Huang, Z., Li, S.X.: Chance constrained programming approaches to technical efficiencies and inefficiencies in stochastic data envelopment analysis. J. Oper. Res. Soc. **53**(12), 1347–1356 (2002). https://doi.org/10.1057/palgrave.jors.2601433
19. Sengupta, J.K.: A fuzzy systems approach in data envelopment analysis. Comput. Math. Appl. **24**(8–9), 259–266 (1992). https://doi.org/10.1016/0898-1221(92)90203-T
20. Cooper, W.W., Park, K.S., Yu, G.: IDEA and AR-IDEA: models for dealing with imprecise data in DEA. Manage. Sci. **45**(4), 597–607 (1999). https://doi.org/10.1287/mnsc.45.4.597
21. Despotis, D.K., Smirlis, Y.G.: Data envelopment analysis with imprecise data. Eur. J. Oper. Res. **140**(1), 24–36 (2002). https://doi.org/10.1016/S0377-2217(01)00200-4

22. Soyster, A.L.: Convex programming with set-inclusive constraints and applications to inexact linear programming. Oper. Res. **21**(5), 1154–1157 (1973). https://doi.org/10.1287/opre.21.5.1154
23. Ben-Tal, A., Nemirovski, A.: Robust convex optimization. Math. Oper. Res. **23**(4), 769–805 (1998). https://doi.org/10.1287/moor.23.4.769
24. Ben-Tal, A., Nemirovski, A.: Robust solutions of uncertain linear programs. Oper. Res. Lett. **25**(1), 1–13 (1999). https://doi.org/10.1016/S0167-6377(99)00016-4
25. Ben-Tal, A., Nemirovski, A.: Robust solutions of linear programming problems contaminated with uncertain data. Math. Programm. **88**(3), 411–424 (2000). https://doi.org/10.1007/PL00011380
26. Bertsimas, D., Sim, M.: Robust discrete optimization and network flows. Math. Program. **98**(1–3), 49–71 (2003). https://doi.org/10.1007/s10107-003-0396-4
27. Bertsimas, D., Sim, M.: The price of robustness. Oper. Res. **52**(1), 35–53 (2004). https://doi.org/10.1287/opre.1030.0065
28. Sadjadi, S.J., Omrani, H: Data envelopment analysis with uncertain data: an application for Iranian electricity distribution companies. Energy Policy **36**(11), 4247–4254 (2008). https://doi.org/10.1016/j.enpol.2008.08.004
29. Salahi, M., Toloo, M., Hesabirad, Z.: Robust Russell and enhanced Russell measures in DEA. J. Oper. Res. Soc. **70**(8), 1275–1283 (2019). https://doi.org/10.1080/01605682.2018.1489353
30. Aghayi, N., Raayatpanah, M.A.: A robust optimization approach to overall profit efficiency with data uncertainty: application on bank industry. J. Chin. Inst. Eng. **42**(2), 160–168 (2019). https://doi.org/10.1080/02533839.2018.1553633
31. Toloo, M., Mensah, E.K.: Robust optimization with nonnegative decision variables: a DEA approach. Comput. Ind. Eng. **127**, 313–325 (2019). https://doi.org/10.1016/j.cie.2018.10.006
32. Salahi, M., Toloo, M., Torabi, N.: A new robust optimization approach to common weights formulation in DEA. J. Oper. Res. Soc. **1–13**, (2020). https://doi.org/10.1080/01605682.2020.1718016
33. Peykani, P., Mohammadi, E., Saen, R.F., Sadjadi, S. J., Rostamy–Malkhalifeh, M.: Data envelopment analysis and robust optimization: a review. Expert Syst. e12534 (2020). https://doi.org/10.1111/exsy.12534
34. Smirlis, Y.G., Maragos, E.K., Despotis, D.K.: Data envelopment analysis with missing values: an interval DEA approach. Appl. Math. Comput. **177**(1), 1–10 (2006). https://doi.org/10.1016/j.amc.2005.10.028

An EPQ Model for Delayed Deteriorating Items with Two-Phase Production Period, Variable Demand Rate and Linear Holding Cost

Mustapha Lawal Malumfashi, Mohd Tahir Ismail, Babangida Bature, Dari Sani, and Majid Khan Majahar Ali

Abstract In this paper, an EPQ model for delayed deteriorating items with a two-phase production period variable demand rate and linearly increasing function of time holding cost is developed. The first two stages of this model are two-phase production periods with different production rates and the same demand rate. The third stage is the period after inventory build-up, before deterioration sets in with quadratic time demand rate while the fourth stage is the deterioration period with stock dependent demand rate. It is assumed that back-order is not allowed. At first, this paper shows the impact of change in production rate on cycle length that minimizes the total variable cost per unit time and economic production quantity. Then, a theorem and lemmas are provided to characterize the optimal analytical solutions. A numerical example is given to reveal the applicability of the model developed and sensitivity analysis is carried out to show the effect of change of some system parameter values on the total variable cost and economic production quantity obtained in the numerical example. Suggestions and recommendations are also presented toward reducing the total variable cost while maximizing the profit of an imperfect production cycle with unstable production rate. Finally, it is shown that the developed model improves upon that of the working paper.

Keywords Variable demand rates · Two-phase · EPQ · Quadratic demands · Linear holding cost

M. L. Malumfashi · M. T. Ismail · M. K. M. Ali (✉)
School of Mathematical Sciences, Universiti Sains Malaysia, George Town, Penang, Malaysia
e-mail: majidkhanmajaharali@usm.my

M. L. Malumfashi
e-mail: mustapha.lawal@umyu.edu.ng

D. Sani
Department of Mathematical Sciences, Kaduna State University, Kaduna, Nigeria

M. L. Malumfashi · B. Bature
Department of Mathematics and Statistics, Umaru Musa Yar'adua University, Katsina, Nigeria

M. H. Mohd et al. (eds.), *Modelling, Simulation and Applications of Complex Systems*, Springer Proceedings in Mathematics & Statistics 359,
https://doi.org/10.1007/978-981-16-2629-6_19

1 Introduction

Controlling and maintaining inventory of deteriorating items are some of the must important issues for any business organization. An economic production quantity (EPQ) system is an inventory control system that determines the amount of items to be produced on a single facility by considering appropriate production rate so as to meet a deterministic demand over an infinite planning horizon. Most of the recent economic production quantity modellers consider the ideal cases that the production rate is either constant, time dependent or demand dependent during the inventory build-up periods which is not generally the case. Looking at the state of growing manufacturing industries which are always trying to increase their production to achieve the set objective and the constraints hinder the well-established manufacturing industries such as unexpected machine failures on the EPQ as highlighted in [2], terrorism activities and anti-terrorism policies as explained in [4], epidemics and their precautionary measures detailed in [23] with Covid-19 at the center stage and particular issue of militant activities on oil and gas production industry in Nigeria [17]. This suffices to show the need to consider an economic production quantity EPQ model with a conditional production rate that can solve the production problem caused by the scenarios mentioned earlier.

Holding cost can be defined as the additional cost involved in carrying, storing and maintaining a single of item over a period of time. In most of the inventory models, holding cost assumed to be constant. But this is not generally the case in real business. Items such as palm oil, crude oil, cooking gas, fruits, vegetables, etc. deteriorate over time due to spoilage or other biotic/abiotic constraint. At present, to prevent deterioration, an increase in holding cost is needed to preserve the freshness (quality) of such product. Though, the more the inventory stayed in store, the more the holding cost. As such, the holding cost can be assumed to be linearly time dependent which is very important factor need to be consider in modelling delayed deteriorating inventory.

In the existing literature it is observed that there is almost a huge vacuum in the inventory models for delayed deteriorating items which is based on two phase production periods with different production rates and same demand rate while considering stock dependent demand after deterioration sets in. Few researchers have considered nearly similar but they have taken constant production rate or the production rate as demand dependent only, but when manufacturers predicted that production period may be terminated before the proposed time, production rate that is demand dependent or constant can not be realistic. In the present study considering this realistic approach, the work in [30] is extended by considering two phase production period with the same demand rate but different rates of production and stock dependent demand within deterioration period, which are completely different augment of their work that considered a very rear case (no demand during inventory buildup period).

See Appendix D for details on basic model of [30].

2 Review of Related Literature

In the modern day business, most, if not every business, exhibits holding cost that is a linearly increasing function of time which is the main reason for most of the recent researches on EPQ and EOQ considered linear holding cost as seen in [30] who presented an EPQ model for delayed deteriorating items with quadratic demand and linear holding cost but assumed that there is no demand during inventory build-up which is a very rear case and the demand during deterioration period is constant which is also not always the case and [14] obtained the optimal order quantity and the total present profits value of an inventory model for non-instantaneous deteriorating items with linear function of time holding cost and stock dependent demand. It is assumed that shortages are not allowed. Also, [5] derived an optimal cycle time of an integrated production inventory model for delayed deteriorating item with exponential demand rate, a production rate that is a function of demand rate and linear time varying holding cost. Authors assumed that partially deteriorated items are allowed to sale with a discount rate from original one while units which are completely deteriorated are superfluous and shortages are not allowed. [19] developed an economic production quantity (EPQ) model with stock-dependent demand and shortage were not allowed while the holding cost was assumed to be time varying. [24] provided a fuzzy inventory model for deteriorating items with three parameter Weibull distribution pattern. The demand was assumed to be price, stock, reliability and advertisement dependent and time varying holding cost is considered. Two-parameter Weibull distribution deterioration rate was also considered. The optimum solutions obtained were evaluated under fuzzy environment and optimises the solution by considering production rate and time as decision variables in separate cases. [11] developed an EPQ inventory model with linear holding cost and shortages that are partially backlogged. Holding cost is treated as a linear function of time and production rate was considered as a linear combination of on-hand inventory and demand. In contrast, [6] shows the applicability of non-linear holding cost in a multi-item EOQ with partial backorder. In the same vein, [9] studied a production inventory model considering imperfect production and deterioration of item, simultaneously. Both the serviceable and reworkable items are assumed to deteriorate with time. The deterioration rate is considered to be a type-2 fuzzy number. Such a situation arises when the vendor assigns, with similar priority, a number of experts to determine the rate of deterioration and the decision given by each expert is in linguistic term, which may be replaced by a fuzzy number. Shortages are allowed that are completely backlogged. All the screened items are reworked at the end of the production process. [20] Presented an economic production quantity model for deteriorating items manufactured in a single batch, ramp type of demand rate and none time varying holding cost were considered under the effect of inflation and shortages under fuzziness.

As the linear holding cost discovered un-negligible in modern EPQ/EOQ models, quadratic demand is also on the card as seen in an EPQ model for delayed deteriorating items with quadratic demand before deterioration sets-in and constant demand after deterioration sets-in developed by [15]. It is assumed that shortages are allowed and partially backordered. The authors also assumed that there is no demand during production period. Also, [10] developed a deterministic inventory models for deteriorating items with quadratic demand and variable holding cost. He assumed that salvage value is associated to the deteriorated items and shortages are not allowed. Likewise, [18] presented an inventory model for instantaneous deteriorating items with three-parameter Weibull deterioration. They assumed that the demand is a quadratic function of time and shortages are allowed but fully backlogged. Among the researchers that gave insight on the applicability of quadratic demand are [1, 13, 22, 28] while [8] is one of the most recent work that consider all the factors of modern EPQ models in analysing the relationship of two different EPQ models that is generalized fractional-order economic production quantity (EPQ) model with the uniform demand and production rate and generalized fractional-order EPQ model with the uniform demand, production rate and deterioration. But [21] developed an EPQ model for instantaneous deteriorating items having ramp type demand rate with weibull deterioration under inflation and finite horizon in crisp and fuzzy area. The effect of inflation of none shortage products under finite horizon was examined.

Furthermore, it is also natural that higher inventory on stock attract customers' patronage the most. Therefore, stock dependent demand is always present as shown in [12] which develops an economic production quantity (EPQ) inventory model for deteriorating items with the constant deterioration rate and stock dependent demand. Shortages are not considered while time varying holding cost is considered. [26] presents an inventory model for deteriorating items with stock-dependent demand, time-varying holding cost and shortages where demand is considered constant during stock-out period. [25] developed an EPQ model with stock and selling price dependent demand and variable production rate in interval environment. The production rate is a variable and it varies with the demand rate. Shortages are allowed and it is backlogged fully. The model assumed that there are some defective products which requires reworking in order to make them useful and the possibility of producing some defective items in regular production process and their reworking has been taken into account. The model also assumed that the associated inventory cost parameters are not fixed in reality and may vary time to time depending upon some scenario and considered various cost parameters as interval numbers. [7] developed an EPQ model with linearly time dependent production rate to a certain period and then with constant production rate is in random time horizon under inflation and time value of money. The model assumed that the business period is random and follows exponential distribution with known mean. Demand is linearly stock-dependent. Holding and set up costs are imprecise and the optimistic/pessimistic equivalent of fuzzy objective function is obtained by using possibility/necessity measure of fuzzy event. The model is formulated as a cost minimization problem for a production controlled inventory system. So also [16, 27, 29] developed inventory models with stock dependent demands.

In the present work, we established an EPQ model for delayed deteriorating items with linearly time dependent holding cost and quadratic time dependent demand is considered. The four stages considered in the proposed model are first production period with initial production rate, a production period after change of production rate, period before deterioration begins where inventory depletion depends only on demand rate which is assumed to be quadratic in nature and a period after deterioration sets in where the inventory depletion depends on both deterioration and demand. The holding cost considered is linearly time dependent. To the best of our search, it is found that this type of EPQ model is not been considered by any of the researchers/scientists in inventory literature. As such, this paper aim to determine the best cycle length P that can optimizes the total variable cost the corresponding maximum economic production quantity (EPQ) for maximum profit. Numerical examples are given to show the applicability of the model and sensitivity analysis carried out to see the effect of changes on some system parameters in which managerial insight is given.

3 Notation and Assumptions

Table 1 contains the notation used in this paper.

The model is developed based on the following modified assumptions adopted from [30] with addition of 1, 3, 7 and 9:

1. There are two production rates μ and $\alpha\mu$ where the production rate is assume to stay positively less than 200% of the initial i.e. $0 < \alpha < 2$.
2. It is assumed that the production start instantaneously at zero.
3. It is assumed that the production rate changes Instantaneously with parameter α.
4. All defective items are discarded after inspection.
5. Unconstrained supplies capital.
6. Shortages are not allowed.
7. It is assumed that the demand during the inventory build-up period is constant.
8. Demand rate after production, before deterioration sets in is assumed to be quadratic function of time (i.e. $\rho_1 = a_1 + a_2 t + a_3 t^2$) where $a_3 > 0$.
9. Demand during deterioration period is assumed to be stock dependent defined as $\rho_2 - \beta I(t)$ where $0 \leq \beta < 1$ (as seen in [3] with little modification).
10. The time border of the production system is finite (i.e. $P_1 \leq P_2 \leq P_3 \leq P$).
11. The holding cost is assume to be linearly increasing function of time defined as $h = h_1 + h_2 t$

Table 1 Notation

A	The set up cost per production cycle
μ	The production rate in unit per unit time
ρ	The demand rate during inventory buildup in unit per unit time
ρ_1	The demand rate after production before deterioration starts in unit per unit time
ρ_2	The demand rate at the point deterioration start in unit per unit time
α	Production rate changing parameter
π	Inventory carrying charge
θ	Deterioration rate
C	Unit cost per item
C_θ	The cost of deterioration
C_H	The total holding cost
I_1	The inventory level at the point instantaneous change of production rate occur
I_2	The inventory level at the end of the production period
I_3	The inventory level at the point deterioration starts
$I(t)$	The inventory level during inventory build up before production rate changes, where $0 \leq t \leq P_1$
$I_1(t)$	The inventory level after production rate changes, where $P_1 \leq t \leq P_2$
$I_2(t)$	The inventory level after production stops before deterioration starts, where $P_2 \leq t \leq P_3$
$I_3(t)$	The inventory level within deterioration period, where $P_3 \leq t \leq P$
P_i	Unit time in periods ($i \in \{1, 2, 3\}$)
P	Total cycle length (decision variable)
Z	Total variable cost (decision variable)
EPQ	Economic production quantity per cycle (decision variable)

4 Modelling

Figure 1 shows how the inventory level varies with time due to both change in production rate, demand and deterioration. The production starts at time at $t = 0$ changes at P_1. The production and demand rates during the time interval $[0, P_1]$ are μ and ρ respectively and the inventory rich a level I_1 at time P_1. During the interval $[P_1, P_2]$, the inventories continue accumulating at a rate $\alpha\mu - \rho$ due to change in production rate (increase or decrease with the parameter α) while demand remain the same as initial. The demand within the interval $[P_2, P_3]$ is quadratic function of time ($\rho_1 = a_1 + a_2 t + a_3 t^2$). Deterioration starts when the inventory dropped to I_3 at time P_3. The inventories dropped to zero due to both demand and deterioration time P.

The governing differential equation that describe the situation within $[0, P_1]$ and it's solution are:

$$\frac{dI(t)}{dt} = \mu - \rho; \quad 0 \leq t \leq P_1 \tag{1}$$

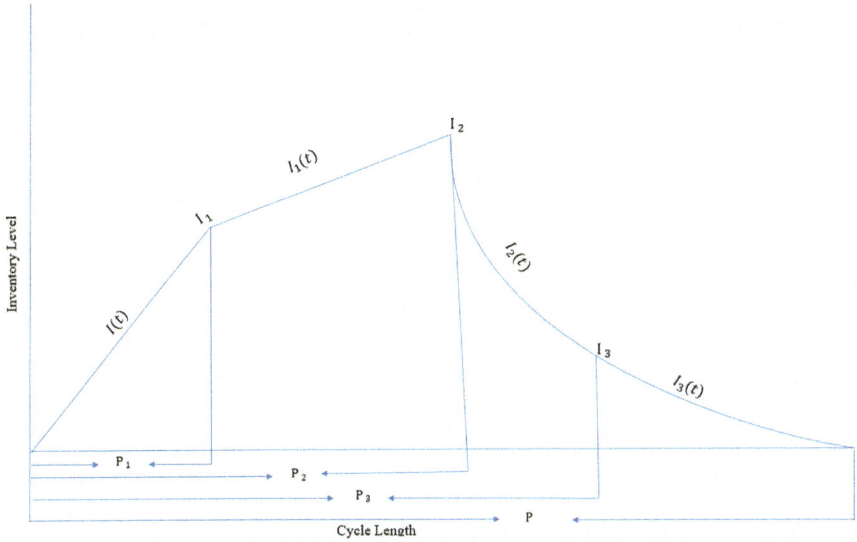

Fig. 1 The inventory system

the solution of Eq. (1) is

$$I(t) = (\mu - \rho)t + k \qquad (where \quad k \quad is \quad constant)$$

applying the boundary conditions $I(0) = 0$ and $I(P_1) = I_1$

$$I(t) = (\mu - \rho)t \tag{2}$$

The differential equation that describe the situation within $[P_1, P_2]$ is given below with its solution.

$$\frac{dI_1(t)}{dt} = \alpha\mu - \rho; \quad P_1 \le t \le P_2 \tag{3}$$

$$I_1(t) = (\alpha\mu - \rho)t + k_1; \qquad (where \quad k_1 \quad is \quad constant)$$

$I_1(P_1) = I_1$ and $I_1(P_2) = I_2$ thus,

$$I_1(t) = (\mu - \rho)P_1 + (\alpha\mu - \rho)(t - P_1); \quad P_1 \le t \le P_2 \tag{4}$$

Below is the differential equation that describe the situation within $[P_2, P_3]$ together with its solution.

$$\frac{dI_2(t)}{dt} = -\rho_1 = -(a_1 + a_2 t + a_3 t^2); \quad P_2 \le t \le P_3 \tag{5}$$

$$I_2(t) = -(a_1 t + a_2\frac{a^2}{2} + a_3\frac{t^3}{3}) + k_2; \quad (where \quad k_2 \quad is \quad constant)$$

applying the boundary conditions $I_2(P_2) = I_2$ and $I_2(P_3) = I_3$

$$I_2(t) = a_1(P_3 - t) + \frac{a_2}{2}(P_3^2 - t^2) + \frac{a_3}{3}(P_3^3 - t^3) + I_3 \tag{6}$$

The differential equation that describe the situation within $[P_3, P]$ and its solution are

$$\frac{dI_3(t)}{dt} + \theta I_3(t) = -(\rho_2 - \beta I_3(t)); \quad P_3 \le t \le P \tag{7}$$

The Integrating Factor (I.F) method of solving ODE is employed to solve (7)

$$I_3(t) = \frac{-\rho_2}{\theta - \beta} + k_3 e^{-(\theta-\beta)t}; \quad (where \quad k_3 \quad is \quad constant)$$

applying the boundary conditions $I_3(P_3) = I_3$ and $I_3(P) = 0$
we have

$$I_3(t) = \frac{-\rho_2}{\theta - \beta} + \left(I_3 + \frac{\rho_2}{\theta - \beta}\right) e^{(\theta-\beta)(P_3-t)} \tag{8}$$

$$I_3 = \frac{\rho_2}{\theta - \beta}\left(e^{(\theta-\beta)(P-P_3)} - 1\right) \tag{9}$$

substitute (9) in (6) and (8) above

$$I_2(t) = a_1(P_3 - t) + \frac{a_2}{2}(P_3^2 - t^2) + \frac{a_3}{3}(P_3^3 - t^3) + \frac{\rho_2}{\theta - \beta}\left(e^{(\theta-\beta)(P-P_3)} - 1\right) \tag{10}$$

$$I_3(t) = \frac{\rho_2}{\theta - \beta}\left(e^{(\theta-\beta)(P-t)} - 1\right) \tag{11}$$

The total cost of deteriorated items is the multiple of cost per unit item C and difference between the total number of items at the time deterioration starts I_3 and the proximate demand within the deterioration period.

$$\begin{aligned} C_\theta &= C(I_3 - \rho_2(P - P_3)) \\ &= C\left[\frac{\rho_2}{\theta - \beta}\left(e^{(\theta-\beta)(P-P_3)} - 1\right) - \rho_2(P - P_3)\right] \end{aligned} \tag{12}$$

The cost associated with storage and carrying charge of the total inventories from production until it is sold or used (i.e. the inventory holding cost) is the multiple of total number of inventories, carrying charge per unit item and cost per unit item, which is given as

$$
\begin{aligned}
C_H &= \pi h\left[\int_0^{P_1} I(t)dt + \int_{P_1}^{P_2} I_1(t)dt + \int_{P_2}^{P_3} I_2(t)dt + \int_{P_3}^{P} I_3(t)dt\right] \\
&= \pi\Bigg[h_1\Bigg[\frac{(\mu-\rho)}{2}P_1^2 + (P_2-P_1)\left(\frac{\alpha\mu}{2}P_2 - \frac{\alpha\mu}{2}P_1 - \frac{\rho}{2}(P_2+P_1)+\mu P_1\right) \\
&\quad + \frac{a_1}{2}(P_3-P_2)^2 + \frac{a_2}{6}(2P_3^3 - 3P_3^2P_2 + P_2^3) + \frac{a_3}{12}(3P_3^4 - 4P_3^3P_2 + P_2^4) \\
&\quad - \frac{\rho_2}{\theta-\beta}\left(P_2 + P_1 - P_3 + P + \frac{1}{\theta-\beta} - (P_2 - P_1 + 1)e^{(\theta-\beta)(P-P_3)}\right)\Bigg] \\
&\quad + h_2\Bigg[\frac{(\mu-\rho)}{3}P_1^3 + \frac{(\alpha\mu-\rho)}{3}(P_2^3 - P_1^3) + \frac{(1-\alpha)}{2}\mu P_1(P_2^2 - P_1^2) \\
&\quad + \frac{a_1}{6}(P_3^3 - 3P_3P_2^2 + 2P_2^3) + \frac{a_2}{8}(P_3^4 - 2P_3^2P_2^2 + P_2^4) \\
&\quad + \frac{a_3}{30}(3P_3^5 - 5P_3^3P_2^2 + 2P_2^5) - \frac{\rho_2}{\theta-\beta}\Bigg(\frac{1}{(\theta-\beta)^2} + \frac{P}{\theta-\beta} + P^2 \\
&\quad - \frac{P_3^2+P_2^2}{2} - \left(P_3^2 - P_2^2 + \frac{1}{(\theta-\beta)^2} + P_3\right)e^{(\theta-\beta)(P-P_3)}\Bigg)\Bigg]\Bigg] \qquad (13)
\end{aligned}
$$

Total Variable Cost. The total variable cost per cycle length is the sum of set-up cost per cycle length, deterioration cost per cycle length and holding cost per cycle length. i.e.

$$
\begin{aligned}
Z(P) &= \frac{A}{P} + \frac{C_\theta}{P} + \frac{C_H}{P} \\
&= \frac{A}{P} + \frac{C}{P}\left[\frac{\rho_2}{\theta-\beta}\left(e^{(\theta-\beta)(P-P_3)} - 1\right) - \rho_2(P-P_3)\right] \\
&\quad + \frac{\pi}{P}\Bigg[h_1\Bigg[\frac{(\mu-\rho)}{2}P_1^2 + (P_2-P_1)\left(\frac{\alpha\mu}{2}P_2 - \frac{\alpha\mu}{2}P_1 - \frac{\rho}{2}(P_2+P_1)+\mu P_1\right) \\
&\quad + \frac{a_1}{2}(P_3-P_2)^2 + \frac{a_2}{6}(2P_3^3 - 3P_3^2P_2 + P_2^3) + \frac{a_3}{12}(3P_3^4 - 4P_3^3P_2 + P_2^4) \\
&\quad - \frac{\rho_2}{\theta-\beta}\left(P_2 + P_1 - P_3 + P + \frac{1}{\theta-\beta} - (P_2 - P_1 + 1)e^{(\theta-\beta)(P-P_3)}\right)\Bigg] \\
&\quad + h_2\Bigg[\frac{(\mu-\rho)}{3}P_1^3 + \frac{(\alpha\mu-\rho)}{3}(P_2^3 - P_1^3) + \frac{(1-\alpha)}{2}\mu P_1(P_2^2 - P_1^2) \\
&\quad + \frac{a_1}{6}(P_3^3 - 3P_3P_2^2 + 2P_2^3) + \frac{a_2}{8}(P_3^4 - 2P_3^2P_2^2 + P_2^4) \\
&\quad + \frac{a_3}{30}(3P_3^5 - 5P_3^3P_2^2 + 2P_2^5) - \frac{\rho_2}{\theta-\beta}\Bigg(\frac{1}{(\theta-\beta)^2} + \frac{P}{\theta-\beta} + P^2 \\
&\quad - \frac{P_3^2+P_2^2}{2} - \left(P_3^2 - P_2^2 + \frac{1}{(\theta-\beta)^2} + P_3\right)e^{(\theta-\beta)(P-P_3)}\Bigg)\Bigg]\Bigg] \qquad (14)
\end{aligned}
$$

Optimality Conditions: The aim of this model is to obtain the optimum value of the cycle length P that minimize the total Variable cost per unit time $Z(P)$. As such, the necessary condition for the optimum value of P is

$$\frac{dZ(P)}{dP} = 0$$

while the sufficient condition is

$$\frac{d^2 Z(P)}{dP^2} > 0.$$

For the necessary condition, we differentiate (14) with respect to P;

$$
\begin{aligned}
\frac{dZ(P)}{dP} = & -\frac{A}{P^2} - \frac{C}{P^2}\bigg[\frac{\rho_2}{\theta-\beta}\Big(e^{(\theta-\beta)(P-P_3)} - 1\Big) - \rho_2(P - P_3)\bigg] \\
& - \frac{\pi}{P^2}\bigg[h_1\bigg[\frac{(\mu-\rho)}{2}P_1^2 + (P_2 - P_1)\Big(\frac{\alpha\mu}{2}P_2 - \frac{\alpha\mu}{2}P_1 - \frac{\rho}{2}(P_2 + P_1) + \mu P_1\Big) \\
& + \frac{a_1}{2}(P_3 - P_2)^2 + \frac{a_2}{6}(2P_3^3 - 3P_3^2 P_2 + P_2^3) + \frac{a_3}{12}(3P_3^4 - 4P_3^3 P_2 + P_2^4) \\
& - \frac{\rho_2}{\theta-\beta}\bigg(P_2 + P_1 - P_3 + P + \frac{1}{\theta-\beta} - (P_2 - P_1 + 1)e^{(\theta-\beta)(P-P_3)}\bigg)\bigg] \\
& + h_2\bigg[\frac{(\mu-\rho)}{3}P_1^3 + \frac{(\alpha\mu-\rho)}{3}(P_2^3 - P_1^3) + \frac{(1-\alpha)}{2}\mu P_1(P_2^2 - P_1^2) \\
& + \frac{a_1}{6}(P_3^3 - 3P_3 P_2^2 + 2P_2^3) + \frac{a_2}{8}(P_3^4 - 2P_3^2 P_2^2 + P_2^4) \\
& + \frac{a_3}{30}(3P_3^5 - 5P_3^3 P_2^2 + 2P_2^5) - \frac{\rho_2}{\theta-\beta}\bigg(\frac{1}{(\theta-\beta)^2} + \frac{P}{\theta-\beta} + P^2 \\
& - \frac{P_3^2 + P_2^2}{2} - \bigg(P_3^2 - P_2^2 + \frac{1}{(\theta-\beta)^2} + P_3\bigg)e^{(\theta-\beta)(P-P_3)}\bigg)\bigg]\bigg] \\
& + \frac{\pi}{P}\bigg[h_1[-\frac{\rho_2}{\theta-\beta} + \rho_2(P_3 - P_2 + 1)e^{(\theta-\beta)(P-P_3)}] + h_2[-\frac{\rho_2}{(\theta-\beta)^2} \\
& - \frac{\rho_2}{\theta-\beta}P + \bigg(-\frac{\rho_2(P_3^2 - P_2^2)}{2} + \frac{\rho_2}{(\theta-\beta)^2} + \rho_2 P_3\bigg)e^{(\theta-\beta)(P-P_3)}]\bigg] \\
& + \frac{C}{P}\Big[\rho_2 e^{(\theta-\beta)(P-P_3)} - \rho_2\Big] \\
= & -\frac{A}{P^2} - \frac{C}{P^2}\bigg[\frac{\rho_2}{\theta-\beta}\Big(e^{(\theta-\beta)(P-P_3)} - 1\Big) + \rho_2 P_3 - \rho_2 P e^{(\theta-\beta)(P-P_3)}\bigg] \\
& - \frac{\pi}{P^2}\bigg[h_1[\frac{(\mu-\rho)}{2}P_1^2 + \frac{(\alpha\mu-\rho)}{2}(P_2^2 - P_1^2) + \frac{a_1}{2}(P_3 - P_2)^2 \\
& + (1-\alpha)\mu P_1(P_2 - P_1) + \frac{a_2}{6}(2P_3^3 - 3P_3^2 P_2 + P_2^3) \\
& + \frac{a_3}{12}(3P_3^4 - 4P_3^3 P_2 + P_2^4) + \frac{\rho_2}{\theta-\beta}(P_3 - P_2 + 1)e^{(\theta-\beta)(P-P_3)} \\
& - \rho_2 P(P_3 - P_2 + 1)e^{(\theta-\beta)(P-P_3)} + \frac{\rho_2}{\theta-\beta}P_2 - \frac{\rho_2}{(\theta-\beta)^2}] \\
& + h_2[\frac{(\mu-\rho)}{3}P_1^3 + \frac{(\alpha\mu-\rho)}{3}(P_2^3 - P_1^3) + \frac{(1-\alpha)}{2}\mu P_1(P_2^2 - P_1^2) \\
& + \frac{a_1}{6}(P_3^3 - 3P_3 P_2^2 + 2P_2^3) + \frac{a_2}{8}(P_3^4 - 2P_3^2 P_2^2 + P_2^4) \\
& + \frac{a_3}{30}(3P_3^5 - 5P_3^3 P_2^2 + 2P_2^5) - \frac{\rho_2}{(\theta-\beta)^3} + \frac{\rho_2 P_2^2}{2(\theta-\beta)} \\
& + \frac{\rho_2}{(\theta-\beta)}\bigg(\frac{P_3^2 - P_2^2}{2} + P_3 + \frac{1}{(\theta-\beta)^2}\bigg)e^{(\theta-\beta)(P-P_3)} + \frac{\rho_2}{2(\theta-\beta)}P^2 \\
& - \rho_2 P\bigg(\frac{P_3^2 - P_2^2}{2} + P_3 + \frac{1}{(\theta-\beta)^2}\bigg)e^{(\theta-\beta)(P-P_3)}]\bigg]
\end{aligned} \tag{15}
$$

Using Maclaurin series expansion of exponential power, the approximate value of $e^{(\theta-\beta)(P-P_3)}$ is $(1+(\theta-\beta)(P-P_3))$, substituting in (15) we have

$$\begin{aligned}
\frac{dZ(P)}{dP} =& -\frac{A}{P^2}-\frac{C}{P^2}[\rho_2(\theta-\beta)P(P_3-P)]-\frac{\pi}{P^2}\Big[h_1[\frac{(\mu-\rho)}{2}P_1^2\\
&+\frac{(\alpha\mu-\rho)}{2}(P_2^2-P_1^2)+(1-\alpha)\mu P_1(P_2-P_1)+\frac{a_1}{2}(P_3-P_2)^2\\
&+\frac{a_2}{6}(2P_3^3-3P_3^2P_2+P_2^3)+\frac{a_3}{12}(3P_3^4-4P_3^3P_2+P_2^4)\\
&+\frac{\rho_2}{\theta-\beta}P_2-\frac{\rho_2}{(\theta-\beta)^2}+\rho_2(P_3-P_2+1)(\frac{1}{\theta-\beta}-P_3)\\
&-\rho_2(P_3-P_2+1)(\theta-\beta)(P-P_3)P]+h_2[\frac{(\mu-\rho)}{3}P_1^3\\
&+\frac{(\alpha\mu-\rho)}{3}(P_2^3-P_1^3)+\frac{(1-\alpha)}{2}\mu P_1(P_2^2-P_1^2)\\
&+\frac{a_1}{6}(P_3^3-3P_3P_2^2+2P_2^3)+\frac{a_2}{8}(P_3^4-2P_3^2P_2^2+P_2^4)\\
&+\frac{a_3}{30}(3P_3^5-5P_3^3P_2^2+2P_2^5)-\frac{\rho_2}{(\theta-\beta)^3}+\frac{\rho_2P_2^2}{2(\theta-\beta)}\\
&+\rho_2\left(\frac{P_3^2-P_2^2}{2}+P_3+\frac{1}{(\theta-\beta)^2}\right)(\frac{1}{\theta-\beta}-P_3)+\frac{\rho_2}{2(\theta-\beta)}P^2\\
&-\rho_2\left(\frac{P_3^2-P_2^2}{2}+P_3+\frac{1}{(\theta-\beta)^2}\right)(\theta-\beta)(P-P_3)P]\Big] \qquad (16)
\end{aligned}$$

Simplifying (16) and equating the resultant to zero, we have

$$MP^2-LP-F=0 \qquad (17)$$

where

$$\begin{aligned}
M =& \Big[C\rho_2h_2(\theta-\beta)+\pi\rho_2h_1(P_3-P_2+1)(\theta-\beta)+\tfrac{\rho_2}{2(\theta-\beta)}\\
&+\pi\rho_2h_2\left(\tfrac{P_3^2-P_2^2}{2}+P_3+\tfrac{1}{(\theta-\beta)^2}\right)(\theta-\beta)\Big],
\end{aligned}$$

$$\begin{aligned}
L =& \Big[C\rho_2h_2(\theta-\beta)P_3+\pi\rho_2h_1(P_3-P_2+1)(\theta-\beta)P_3\\
&+\pi\rho_2h_2\left(\tfrac{P_3^2-P_2^2}{2}+P_3+\tfrac{1}{(\theta-\beta)^2}\right)(\theta-\beta)P_3\Big],
\end{aligned}$$

and

$$
\begin{aligned}
F \;=\; \pi\Bigg[& h_1[\frac{(\mu-\rho)}{2}P_1^2+\frac{(\alpha\mu-\rho)}{2}(P_2^2-P_1^2)+(1-\alpha)\mu P_1(P_2-P_1) \\
& +\frac{a_1}{2}(P_3-P_2)^2+\frac{a_2}{6}(2P_3^3-3P_3^2P_2+P_2^3)+\frac{a_3}{12}(3P_3^4-4P_3^3P_2+P_2^4) \\
& +\frac{\rho_2}{\theta-\beta}P_2-\frac{\rho_2}{(\theta-\beta)^2}+\rho_2(P_3-P_2+1)(\frac{1}{\theta-\beta}-P_3)] \\
& +h_2[\frac{(\mu-\rho)}{3}P_1^3+\frac{(\alpha\mu-\rho)}{3}(P_2^3-P_1^3)+\frac{(1-\alpha)}{2}\mu P_1(P_2^2-P_1^2) \\
& +\frac{a_1}{6}(P_3^3-3P_3P_2^2+2P_2^3)+\frac{a_2}{8}(P_3^4-2P_3^2P_2^2+P_2^4) \\
& +\frac{a_3}{30}(3P_3^5-5P_3^3P_2^2+2P_2^5)+\frac{\rho_2P_2^2}{2(\theta-\beta)}-\frac{\rho_2}{(\theta-\beta)^3} \\
& +\rho_2\left(\frac{P_3^2-P_2^2}{2}+P_3+\frac{1}{(\theta-\beta)^2}\right)(\frac{1}{\theta-\beta}-P_3)]\Bigg]+A
\end{aligned} \tag{18}
$$

Theorem 1. *If* $0<\theta+\beta<1$, $h_1>h_2$, $P_1<P_2<P_3\leq 1$,

$$a_1 \geq \frac{P_3^2P_2(h_1a_3+2h_2a_2)}{h_2(P_3^2-P_2^2)}$$

and

$$A \geq \pi\left[\frac{\rho_2}{(\theta-\beta)^2}\left(h_1+\frac{h_2}{\theta-\beta}\right)+h_2P_2\left(\frac{\alpha\mu-\rho}{2}P_1^2+a_1P_2P_3\right)\right]$$

then $M, L, \& F \in \Re^+$

See the proof in Appendix A

Lemma 1. *The best cycle length that optimizes the total variable cost is*

$$P^*=\frac{1}{2M}\left[L+\left(L^2+4MF\right)^{\frac{1}{2}}\right] \tag{19}$$

See the proof in Appendix B

Lemma 2. *The total cost function* $Z(P)$ *is a convex function of* P *if the hypotheses of Theorem 1 are valid.*

See the proof in Appendix C

The Economic Production Quantity (EPQ) for the corresponding cycle length (P) is computed below.

EPQ^* = total demand during inventory build-up periods + total demand before deterioration start + total demand within deterioration period + number of deteriorated items.

$$
\begin{aligned}
EPQ^* &= \int_0^{P_1}(\mu-\rho)dt+\int_{P_1}^{P_2}(\alpha\mu-\rho)dt+\int_{P_2}^{P_3}(a_1+a_2t+a_3t^2)dt \\
&\quad+\int_{P_3}^{P}(\rho_2-\beta I_3(t))dt+\frac{\rho_2}{\theta-\beta}\left(e^{(\theta-\beta)(P-P_3)}-1\right)-\rho_2(P-P_3) \\
&= (\mu-\rho)P_1+(\alpha\mu-\rho)(P_2-P_1)+a_1(P_3-P_2) \\
&\quad+\frac{a_3}{3}(P_3^3-P_2^3)+\frac{\rho_2}{\theta-\beta}\left(e^{(\theta-\beta)(P-P_3)}-1\right) \\
&\quad+\frac{a_2}{2}(P_3^2-P_2^2)-\rho_2(P-P_3)+\Bigg[\rho_2(P-P_3) \\
&\quad-\frac{\beta\rho_2}{(\theta-\beta)}\left(\frac{1}{(\theta-\beta)}(e^{(\theta-\beta)(P-P_3)}-1)-(P-P_3)\right)\Bigg] \\
&= (\mu-\rho)P_1+(\alpha\mu-\rho)(P_2-P_1)+a_1(P_3-P_2)+\frac{a_2}{2}(P_3^2-P_2^2) \\
&\quad+\frac{a_3}{3}(P_3^3-P_2^3)+\frac{\rho_2}{\theta-\beta}\left(e^{(\theta-\beta)(P-P_3)}-1\right) \\
&\quad-\frac{\beta\rho_2}{(\theta-\beta)}\left(\frac{1}{(\theta-\beta)}(e^{(\theta-\beta)(P-P_3)}-1)-(P-P_3)\right) \\
&= (\mu-\rho)P_1+(\alpha\mu-\rho)(P_2-P_1)+a_1(P_3-P_2)+\frac{a_2}{2}(P_3^2-P_2^2) \\
&\quad+\frac{a_3}{3}(P_3^3-P_2^3)+\frac{\rho_2}{\theta-\beta}\left(e^{(\theta-\beta)(P-P_3)}-1\right) \\
&\quad-\frac{\beta\rho_2}{(\theta-\beta)^2}\left(e^{(\theta-\beta)(P-P_3)}-1\right)+\frac{\beta\rho_2(P-P_3)}{(\theta-\beta)} \qquad (20)
\end{aligned}
$$

5 Numerical Experiments

The proposed model here is illustrated by performing the numerical experiments as follows. The parameter values are adopted from [30] and add the values of parameters absent in their work (μ, α, ρ, β and P_3) as presented in the Table 2.

5.1 Results

The problem described by the parameters previously set is solved and obtain the optimal variable cost per year, the best cycle length and EPQ values by substituting the parameters value above in (14), (19), and (20) respectively. The optimal solutions are presented in Table 3.

Table 2 Experiment parameters

Param.	Value	Param.	Value
A	N3300 per production run	P_1	0.547945
μ	8000 units per production run	P_2	0.821918 years
α	1.6	P_3	0.989123 years
ρ	2300 units per unit time	h_1	0.1
ρ_2	2100 units per unit time	h_2	0.01
π	0.1 units per unit time	a_1	8400.
θ	0.76 units per unit time	a_2	4
C	N140 per unit	a_3	3
β	0.012		

Table 3 Optimal solutions

Decision variables	Value
Z	N2,515.79466 per cycle
P	1.316333415 years
EPQ	8,115.557264 units per cycle

5.2 Comparison Between the Proposed Model and [30]

The aim of both our propose model and [30] is to minimizes the total variable cost within the complete cycle length. The total variable cost $Z = N2,515.79466$ obtained here is minimum to that obtained in their work while the cycle length obtained here $P = 1.316333415$ is minimum. As such, the model we proposed improves upon that proposed in [30] (Table 4).

Table 4 Comparison table

Proposed model	[30]
$Z = N2,515.79466$	Z = N3,545.53
$P = 1.316333415$	P = 1.810959

6 Sensitivity Analysis

Using the results obtained in the above experiment, we perform sensitivity analysis to study the effect of changes in some system's parameters on over the obtained optimal solutions and the results are shown below.

Table 5 shows the effect of change in set-up cost A on the total variable cost Z and EPQ. Both the total variable cost per cycle length ($Z(P)$) and economic production quantity (EPQ) increases as the set-up cost (A) increases and decreases in the reverse case. In a normal circumstance, any change in set-up cost has an impact on both total variable cost and economic production quantity. This indicates that manufacturers should make sure that the production rate is at maximum when the set-up cost is high.

Table 6 shows the effect of change in cost per unit item C on the total variable cost Z and EPQ. As the cost per unit item (C) increases, the total variable cost per cycle length ($Z(P)$) increases while economic production quantity (EPQ) decreases. In real life situation, an increase in cost per unit item lead to the decrease in demand which may reduce economic production required in a cycle. Also, any increase in cost of an item will increase the total variable cost per cycle. As such, it is recommended that Manufacturers make sure that the EPQ is at maximum to lower the cost ratio.

Table 7 shows the effect of change in production rate μ on the total variable cost Z and EPQ. As the production rate (μ) increases, both the total variable cost per cycle ($Z(P)$) and economic production quantity (EPQ) increases. This means that any adjustment in production rate has similar impact on both total variable cost per cycle length and EPQ.

Table 5 The effect of changes in A on Z^* and EPQ^*

% change in A	% change in Z*	% change in EPQ
+50	31.4679	5.1915
+25	16.4289	2.4168
0	0.0000	0.0000
−25	−18.2922	−3.9768
−50	−39.2709	−7.8283

Table 6 The effect of changes in C on Z^* and EPQ^*

% change in C	% change in Z*	% change in EPQ
+50	12.7372	−5.2967
+25	6.9978	−3.3228
0	0.0000	0.0000
−25	−8.8568	3.4155
−50	−20.7672	10.2407

Table 7 The effect of changes in μ on Z^* and EPQ^*

% change in μ	% change in Z*	% change in EPQ
+50	0.3497	42.2292
+25	0.1749	20.8091
0	0.0000	0.0000
−25	−0.1751	−22.0212
−50	−0.3503	−43.4512

Table 8 The effect of changes in π on Z^* and EPQ^*

% change in π	% change in Z*	% change in EPQ
+50	0.0556	10.4439
+25	0.05123	9.4941
0	0.0000	0.0000
−25	−0.0738	−9.4846
−50	−0.2031	−18.9458

Table 9 The effect of changes in θ on Z^* and EPQ^*

% change in θ	% change in Z*	% change in EPQ
+50	13.9100	−5.4808
+25	7.7300	−3.4453
0	0.0000	0.0000
−25	−10.2048	3.6831
−50	−25.1134	11.2154

Table 8 shows the effect of change in inventory carrying charge π on the total variable cost Z and EPQ. As the inventory carrying charge (π) increases, the total variable cost per cycle ($Z(P)$) increases and economic production quantity (EPQ) also increases. It is clear in real life situation that any increase in carrying charge increases the total variable cost. It is recommended that manufacturers should make sure that the EPQ is at maximum whenever the inventory carrying charge increases.

Table 9 shows the effect of change in deterioration rate θ on the total variable cost Z and EPQ. As the deterioration rate (θ) increases, the total variable cost per cycle length ($Z(P)$) increases while the economic production quantity (EPQ) decreases. Naturally, any increase in deterioration rate will increase the total variable cost, also any change in deterioration rate has significant impact on economic production quantity required in a cycle length. The manufacturer should provide adequate row material and excellent storage facilities so that the rate of deterioration should be low.

Table 10 shows the effect of change in first production period P_1 on the total variable cost Z and EPQ. As the 1st phase of production periods (P_1) increases, the total variable cost per cycle length ($Z(P)$) slightly decreases while the economic

Table 10 The effect of changes in P_1 on Z^* and EPQ^*

% change in P_1	% change in Z*	% change in EPQ
+50	−0.0271	−14.2646
+25	−0.01700	−7.1386
0	0.0000	0.0000
−25	0.0324	7.1465
−50	0.0865	14.2974

Table 11 The effect of changes in P_2 on Z^* and EPQ^*

% change in P_2	% change in Z*	% change in EPQ
+50	0.3860	9.5273
+25	0.0965	4.7506
0	0.0000	0.0000
−25	0.0892	−4.7238
−50	0.3561	−9.4205

Table 12 The effect of changes in P_3 on Z^* and EPQ^*

% change in P_3	% change in Z*	% change in EPQ
+50	−15.2301	41.8642
+25	−8.1459	20.8258
0	0.0000	0.0000
−25	9.3210	−20.5893
−50	19.9146	−40.9204

production quantity drastically (EPQ) decreases. In a two phase production periods in cycle length, any increase in a period with smaller production rate will reduce the period with higher production rate. This leads to decrease in economic production quantity which can decreases the total variable cost.

Table 11 shows the effect of change in 2nd phase of production periods P_2 on the total variable cost Z and EPQ. As the 2nd phase of production periods P_2 increases, both total variable cost per cycle length ($Z(P)$) and economic production quantity (EPQ) increases. It is obvious that any increase in a production period with higher production rate will increase the economic production quantity and any increase in EPQ will increase the total variable cost due to the cost per unit item.

Table 12 shows the effect of change in demanding period before deterioration (P_3) on the total variable cost Z and EPQ. As the demanding period before deterioration (P_3) increases, the total variable cost per cycle length ($Z(P)$) decreases while the economic production quantity (EPQ) increases. In any real life business, any increase in life time of items will reduce the deterioration items in a cycle length which clearly reduce cost of production and impacted the economic production quantity.

7 Conclusion

This study investigated an inventory production system for delayed deteriorated goods with two phase production periods and different rates of production in each of the phase in which the demand within the two said periods is the same and constant. The rate of the demand after inventory build-up is quadratic function of time while that of after deterioration set-in is stock dependent. This work also assumed that holding cost as linearly increasing function of time and shortages are not allowed. The best cycle length that optimizes the total variable cost and economic production quantity is proved optimal by theorem and lemmas presented. To show the applicability of the theoretical results, a numerical example is presented and sensitivity analysis also is carried out to show the effect of change of systems' parameter values on total variable cost and economic production quantity obtained in the numerical example. Further, we can develop this model in several ways such as introduction of shortages with partial or full backlogging, exponential demand before deterioration set-in, linearly decreasing function of time demand after deterioration set-in and so on.

8 Appendices

8.1 *Appendix A*

Proof of Theorem 1

It is enough to show that $M > 0$, $L > 0$ and $F > 0$

$$\begin{aligned} M &= C\rho_2 h_2(\theta-\beta) + \pi\rho_2 h_1(P_3 - P_2 + 1)(\theta-\beta) + \frac{\rho_2}{2(\theta-\beta)} \\ &\quad + \pi\rho_2 h_2\left(\frac{P_3^2 - P_2^2}{2} + P_3 + \frac{1}{(\theta-\beta)^2}\right)(\theta-\beta) \\ &= \rho_2(\theta-\beta)\Big[Ch_2 + \pi h_1(P_3 - P_2 + 1) \\ &\quad + \pi h_2\left(\frac{P_3^2 - P_2^2}{2} + P_3 + \frac{1}{(\theta-\beta)^2}\right)\Big] + \frac{\rho_2}{2(\theta-\beta)} \end{aligned}$$

since $0 < \theta - \beta < 1$ and $P_3 > P_2$ then $\pi h_2\left(\frac{P_3^2-P_2^2}{2} + P_3 + \frac{1}{(\theta-\beta)^2}\right) > 0$

clearly, $Ch_2 > 0$, $\frac{\rho_2}{2(\theta-\beta)} > 0$ and $\rho_2(\theta-\beta) > 0$

therefore,

$$\rho_2(\theta-\beta)\left[Ch_2 + \pi h_1(P_3 - P_2 + 1) + \pi h_2\left(\frac{P_3^2 - P_2^2}{2} + P_3 + \frac{1}{(\theta-\beta)^2}\right)\right] > 0 \qquad (21)$$

hence

$$M > 0$$

To show that $L > 0$;

$$\begin{aligned} L &= \Big[C\rho_2 h_2(\theta-\beta)P_3 + \pi\rho_2 h_1(P_3 - P_2 + 1)(\theta-\beta)P_3 \\ &\quad + \pi\rho_2 h_2\left(\frac{P_3^2 - P_2^2}{2} + P_3 + \frac{1}{(\theta-\beta)^2}\right)(\theta-\beta)P_3\Big] \\ &= \Big[\rho_2(\theta-\beta)\Big[Ch_2 + \pi h_1(P_3 - P_2 + 1) \\ &\quad + \pi h_2\left(\frac{P_3^2 - P_2^2}{2} + P_3 + \frac{1}{(\theta-\beta)^2}\right)\Big]\Big]P_3 \end{aligned} \tag{22}$$

since $P_3 > 0$ and the inequality (21) is true, then

$$L > 0$$

To show that $F > 0$,
suppose that

$$A \geq \pi\left[\frac{\rho_2}{(\theta-\beta)^2}\left(h_1 + \frac{h_2}{\theta-\beta}\right) + h_2 P_2\left(\frac{\alpha\mu-\rho}{2}P_1^2 + a_1 P_2 P_3\right)\right] > 0 \tag{23}$$

and

$$a_1 \geq \frac{P_3^2 P_2(h_1 a_3 + 2h_2 a_2)}{h_2(P_3^2 - P_2^2)} > 0 \tag{24}$$

since

$$\begin{aligned} F &= \pi\Big[h_1[\frac{(\mu-\rho)}{2}P_1^2 + \frac{(\alpha\mu-\rho)}{2}(P_2^2 - P_1^2) + (1-\alpha)\mu P_1(P_2 - P_1) \\ &\quad + \frac{a_1}{2}(P_3 - P_2)^2 + \frac{a_2}{6}(2P_3^3 - 3P_3^2 P_2 + P_2^3) + \frac{a_3}{12}(3P_3^4 - 4P_3^3 P_2 + P_2^4) \\ &\quad + \frac{\rho_2}{\theta-\beta}P_2 - \frac{\rho_2}{(\theta-\beta)^2} + \rho_2(P_3 - P_2 + 1)(\frac{1}{\theta-\beta} - P_3)] \\ &\quad + h_2[\frac{(\mu-\rho)}{3}P_1^3 + \frac{(\alpha\mu-\rho)}{3}(P_2^3 - P_1^3) + \frac{(1-\alpha)}{2}\mu P_1(P_2^2 - P_1^2) \\ &\quad + \frac{a_1}{6}(P_3^3 - 3P_3 P_2^2 + 2P_2^3) + \frac{a_2}{8}(P_3^4 - 2P_3^2 P_2^2 + P_2^4) \\ &\quad + \frac{a_3}{30}(3P_3^5 - 5P_3^3 P_2^2 + 2P_2^5) + \frac{\rho_2 P_2^2}{2(\theta-\beta)} - \frac{\rho_2}{(\theta-\beta)^3} \\ &\quad + \rho_2\left(\frac{P_3^2 - P_2^2}{2} + P_3 + \frac{1}{(\theta-\beta)^2}\right)(\frac{1}{\theta-\beta} - P_3)]\Big] + A \end{aligned} \tag{25}$$

substitute

$$\frac{(\alpha\mu-\rho)}{2}(P_2^2 - P_1^2) + (1-\alpha)\mu P_1(P_2 - P_1) = (\mu-\rho)P_1(P_2 - P_1) + \frac{(\alpha\mu-\rho)}{2}(P_2 - P_1)^2$$

and

$$\frac{(\alpha\mu-\rho)}{3}(P_2^3-P_1^3)+\frac{(1-\alpha)}{2}\mu P_1(P_2^2-P_1^2) = \frac{(\mu-\rho)}{2}P_1(P_2^2-P_1^2) + \frac{(\alpha\pi-\rho)}{6}(2P_2^3-3P_2^2P_1+P_1^3)$$

in (25)

$$\begin{aligned} F = \pi\Big[& h_1[\frac{(\mu-\rho)}{2}P_1^2+(\mu-\rho)P_1(P_2-P_1)+\frac{(\alpha\mu-\rho)}{2}(P_2-P_1)^2 \\ & + \frac{a_1}{2}(P_3-P_2)^2+\frac{a_2}{6}(2P_3^3-3P_3^2P_2+P_2^3)+\frac{a_3}{12}(3P_3^4-4P_3^3P_2+P_2^4) \\ & + \frac{\rho_2}{\theta-\beta}P_2-\frac{\rho_2}{(\theta-\beta)^2}+\rho_2(P_3-P_2+1)(\frac{1}{\theta-\beta}-P_3)] \\ & + h_2[\frac{(\mu-\rho)}{2}P_1(P_2^2-P_1^2)+\frac{(\alpha\pi-\rho)}{6}(2P_2^3-3P_2^2P_1+P_1^3) \\ & + \frac{(\mu-\rho)}{3}P_1^3+\frac{a_1}{6}(P_3^3-3P_3P_2^2+2P_2^3)+\frac{a_2}{8}(P_3^4-2P_3^2P_2^2+P_2^4) \\ & + \frac{a_3}{30}(3P_3^5-5P_3^3P_2^2+2P_2^5)+\frac{\rho_2P_2^2}{2(\theta-\beta)}-\frac{\rho_2}{(\theta-\beta)^3} \\ & + \rho_2\left(\frac{P_3^2-P_2^2}{2}+P_3+\frac{1}{(\theta-\beta)^2}\right)(\frac{1}{\theta-\beta}-P_3)]\Big]+A \end{aligned} \tag{26}$$

Using Pascal Triangle, the expansion of $(P_2-P_1)^3$ is

$$(P_2-P_1)^3=P_2^3-3P_2^2P_1+3P_2P_1^2-P_1^3$$

therefore,

$$2P_2^3-3P_2^2P_1+P_1^3=[(P_2-P_1)^3+2P_1^3+P_2^3]-3P_2P_1^2, \tag{27}$$

$$2P_3^3-3P_3^2P_2+P_3^3=[(P_3-P_3)^3+2P_2^3+P_3^3]-3P_3P_2^2 \tag{28}$$

and

$$P_3^3-3P_3P_2^2+2P_2^3=(P_3-P_2)^3+3P_2(P_3^2P_2^2)-6P_3P_2^2 \tag{29}$$

the expansion of $(P_3-P_2)^4$ is

$$(P_3-P_2)^4=P_3^4-4P_3^3P_2+6P_3^2P_2^2-4P_3P_2^3+P_2^3$$

therefore,

$$3P_3^4-4P_3^3P_2+P_2^4=(P_3-P_2)^4+2P_3(P_3^3+2P_2^3)-6P_3^2P_2^2 \tag{30}$$

and

$$P_3^4-2P_3^2P_2^2+P_2^4=(P_3-P_2)^4+4P_3P_2(P_3^2+P_2^2)-8P_3^2P_2^2 \tag{31}$$

while the expansion of $(P_3-P_2)^5$ is

$$(P_3-P_2)^5=P_3^5-5P_3^4P_2+10P_3^3P_2^2-10P_3^2P_2^3+5P_3P_2^4-P_2^5$$

therefore,

$$\begin{aligned} 3P_3^5-5P_3^3P_2^2+2P_2^5 = & (P_3-P_2)^5+5P_3^2P_2(P_3^2+2P_2^2) \\ & +3P_2^5-5P_3P_2^2(3P_3^2+P_2^2) \end{aligned} \tag{32}$$

substitute (27, 28, 29, 30, 31 and 32) in (26)

$$
\begin{aligned}
F &= \pi\Bigg[h_1[\frac{(\mu-\rho)}{2}P_1^2 + (\mu-\rho)P_1(P_2-P_1) + \frac{(\alpha\mu-\rho)}{2}(P_2-P_1)^2 \\
&\quad + \frac{a_1}{2}(P_3-P_2)^2 + \frac{a_2}{6}\left((P_3-P_2)^3 + 2P_2^3 + P_3^3 - 3P_3P_2^2\right) \\
&\quad + \frac{a_3}{12}\left((P_3-P_2)^4 + 2P_3(P_3^3+2P_2^3)\right) - \frac{a_3}{2}P_3^2P_2^2 \\
&\quad + \frac{\rho_2}{\theta-\beta}P_2 - \frac{\rho_2}{(\theta-\beta)^2} + \rho_2(P_3-P_2+1)(\frac{1}{\theta-\beta} - P_3)] \\
&\quad + h_2[\frac{(\mu-\rho)}{3}P_1^3 + \frac{(\mu-\rho)}{2}P_1(P_2^2-P_1^2) \\
&\quad + \frac{(\alpha\pi-\rho)}{6}\left((P_2-P_1)^3 + 2P_1^3 + P_2^3\right) - \frac{(\alpha\pi-\rho)}{2}P_2P_1^2 \\
&\quad + \frac{a_1}{6}\left((P_3-P_2)^3 + 3P_2(P_3^2P_2^2)\right) - a_1P_3P_2^2 \\
&\quad + \frac{a_2}{8}\left((P_3-P_2)^4 + 4P_3P_2(P_3^2+P_2^2)\right) - a_2P_3^2P_2^2 \\
&\quad + \frac{a_3}{30}\left((P_3-P_2)^5 + 5P_3^2P_2(P_3^2+2P_2^2) + 3P_2^5\right) \\
&\quad - \frac{a_3}{6}P_3P_2^2(3P_3^2+P_2^2) - \frac{\rho_2}{(\theta-\beta)^3} + \frac{\rho_2P_2^2}{2(\theta-\beta)} \\
&\quad + \rho_2\left(\frac{P_3^2-P_2^2}{2} + P_3 + \frac{1}{(\theta-\beta)^2}\right)(\frac{1}{\theta-\beta} - P_3)]\Bigg]
\end{aligned}
$$

$$
\begin{aligned}
&= \pi\Bigg[h_1[\frac{(\mu-\rho)}{2}P_1^2 + (\mu-\rho)P_1(P_2-P_1) + \frac{(\alpha\mu-\rho)}{2}(P_2-P_1)^2 \\
&\quad + \frac{a_2}{6}(P_3-P_2)^3 + \frac{a_3}{12}\left((P_3-P_2)^4 + 2P_3(P_3^3+2P_2^3)\right) \\
&\quad + \frac{a_1}{2}(P_3-P_2)^2 + \rho_2(P_3-P_2+1)(\frac{1}{\theta-\beta} - P_3)] \\
&\quad + h_2[\frac{(\mu-\rho)}{3}P_1^3 + \frac{(\mu-\rho)}{2}P_1(P_2^2-P_1^2) + \frac{a_1}{6}(P_3-P_2)^3 \\
&\quad + \frac{a_2}{8}(P_3-P_2)^4 + \frac{(\alpha\pi-\rho)}{6}\left((P_2-P_1)^3 + 2P_1^3 + P_2^3\right) \\
&\quad + \frac{a_3}{30}\left((P_3-P_2)^5 + 5P_3^2P_2(P_3^2+2P_2^2) + 3P_2^5\right) + \frac{\rho_2P_2^2}{2(\theta-\beta)} \\
&\quad + \rho_2\left(\frac{P_3^2-P_2^2}{2} + P_3 + \frac{1}{(\theta-\beta)^2}\right)(\frac{1}{\theta-\beta} - P_3)]\Bigg] \\
&\quad + \frac{\pi h_1a_2}{6}(P_3^3 + 2P_2^3 - 3P_3P_2^2) + \frac{\pi h_1a_3}{6}P_3(P_3^3+2P_2^3) - \frac{\pi h_1\rho_2}{(\theta-\beta)^2} \\
&\quad - \frac{\pi h_1a_3}{2}P_3^2P_2^2 - \frac{\pi h_2(\alpha\mu-\rho)}{2}P_2P_1^2 + \frac{\pi h_2a_1}{2}P_2(P_3^2+P_2^2) \\
&\quad - \pi h_2a_1P_3P_2^2 + \frac{\pi h_2a_2}{2}P_3P_2(P_3^2+P_2^2) - \pi h_2a_2P_3^2P_2^2 \\
&\quad - \frac{\pi h_2a_3}{6}P_3P_2^2(3P_3^2+P_2^2) - \frac{\rho_2\pi h_2}{(\theta-\beta)^3} + A
\end{aligned} \tag{33}
$$

$$
\begin{aligned}
= \pi\Big[& h_1[\frac{(\mu-\rho)}{2}P_1^2 + (\mu-\rho)P_1(P_2-P_1) + \frac{(\alpha\mu-\rho)}{2}(P_2-P_1)^2 \\
& + \frac{a_2}{6}(P_3-P_2)^3 + \frac{a_3}{12}\left((P_3-P_2)^4 + 2P_3(P_3^3+2P_2^3)\right) \\
& + \frac{a_1}{2}(P_3-P_2)^2 + \frac{\rho_2}{\theta-\beta}P_2 + \rho_2(P_3-P_2+1)(\frac{1}{\theta-\beta} - P_3)] \\
& + h_2[\frac{(\mu-\rho)}{3}P_1^3 + \frac{(\mu-\rho)}{2}P_1(P_2^2-P_1^2) + \frac{\rho_2 P_2^2}{2(\theta-\beta)} \\
& + \frac{(\alpha\pi-\rho)}{6}\left((P_2-P_1)^3 + 2P_1^3 + P_2^3\right) + \frac{a_1}{6}(P_3-P_2)^3 \\
& + \frac{a_2}{8}(P_3-P_2)^4 + \frac{a_3}{30}\left((P_3-P_2)^5 + 5P_3^2P_2(P_3^2+2P_2^2) + 3P_2^5\right) \\
& + \rho_2\left(\frac{P_3^2-P_2^2}{2} + P_3 + \frac{1}{(\theta-\beta)^2}\right)(\frac{1}{\theta-\beta} - P_3)]\Big] \\
& + \frac{\pi h_2 a_1}{2}P_2(P_3^2+P_2^2) - \frac{\pi}{2}P_3^2P_2^2(h_1a_3+2h_2a_2) \\
& + \frac{\pi a_2}{6}\left(h_1(P_3^3+2P_2^3-3P_3P_2^2) + 3h_2P_3P_2(P_3^2+P_2^2)\right) \\
& + \frac{\pi a_3}{6}P_3\left(h_1(P_3^3+2P_2^3) - h_2P_2^2(3P_3^2+P_2^2)\right) + A \\
& - \pi\left[\frac{\rho_2}{(\theta-\beta)^2}\left(h_1 + \frac{h_2}{\theta-\beta}\right) + h_2P_2\left(\frac{\alpha\mu-\rho}{2}P_1^2 + a_1P_3P_2\right)\right]
\end{aligned} \tag{34}
$$

substitute (23) in (34)

$$
\begin{aligned}
F = \pi\Big[& h_1[\frac{(\mu-\rho)}{2}P_1^2 + (\mu-\rho)P_1(P_2-P_1) + \frac{(\alpha\mu-\rho)}{2}(P_2-P_1)^2 \\
& + \frac{a_2}{6}(P_3-P_2)^3 + \frac{a_3}{12}\left((P_3-P_2)^4 + 2P_3(P_3^3+2P_2^3)\right) \\
& + \frac{a_1}{2}(P_3-P_2)^2 + \frac{\rho_2}{\theta-\beta}P_2 + \rho_2(P_3-P_2+1)(\frac{1}{\theta-\beta} - P_3)] \\
& + h_2[\frac{(\mu-\rho)}{3}P_1^3 + \frac{(\mu-\rho)}{2}P_1(P_2^2-P_1^2) + \frac{\rho_2 P_2^2}{2(\theta-\beta)} \\
& + \frac{(\alpha\pi-\rho)}{6}\left((P_2-P_1)^3 + 2P_1^3 + P_2^3\right) + \frac{a_1}{6}(P_3-P_2)^3 \\
& + \frac{a_2}{8}(P_3-P_2)^4 + \frac{a_3}{30}\left((P_3-P_2)^5 + 5P_3^2P_2(P_3^2+2P_2^2) + 3P_2^5\right) \\
& + \rho_2\left(\frac{P_3^2-P_2^2}{2} + P_3 + \frac{1}{(\theta-\beta)^2}\right)(\frac{1}{\theta-\beta} - P_3)]\Big] \\
& + \frac{\pi a_3}{6}P_3\left(h_1(P_3^3+2P_2^3) - h_2P_2^2(3P_3^2+P_2^2)\right) \\
& + \frac{\pi a_2}{6}\left(h_1(P_3^3+2P_2^3-3P_3P_2^2) + 3h_2P_3P_2(P_3^2+P_2^2)\right) \\
& + \frac{\pi}{2}P_2\left[h_2a_1(P_3^2+P_2^2) - P_3^2P_2(h_1a_3+2h_2a_2)\right]
\end{aligned} \tag{35}
$$

substitute (24) in the last term of (35)

$$
\begin{aligned}
F \quad = \quad & \pi\Bigg[h_1[\frac{(\mu-\rho)}{2}P_1^2+(\mu-\rho)P_1(P_2-P_1)+\frac{(\alpha\mu-\rho)}{2}(P_2-P_1)^2\\
&+\frac{a_2}{6}(P_3-P_2)^3+\frac{a_3}{12}\left((P_3-P_2)^4+2P_3(P_3^3+2P_2^3)\right)\\
&+\frac{a_1}{2}(P_3-P_2)^2+\frac{\rho_2}{\theta-\beta}P_2+\rho_2(P_3-P_2+1)(\frac{1}{\theta-\beta}-P_3)]\\
&+h_2[\frac{(\mu-\rho)}{3}P_1^3+\frac{(\mu-\rho)}{2}P_1(P_2^2-P_1^2)+\frac{\rho_2P_2^2}{2(\theta-\beta)}\\
&+\frac{(\alpha\pi-\rho)}{6}\left((P_2-P_1)^3+2P_1^3+P_2^3\right)+\frac{a_1}{6}(P_3-P_2)^3\\
&+\frac{a_2}{8}(P_3-P_2)^4+\frac{a_3}{30}\left((P_3-P_2)^5+5P_3^2P_2(P_3^2+2P_2^2)+3P_2^5\right)\\
&+\rho_2\left(\frac{P_3^2-P_2^2}{2}+P_3+\frac{1}{(\theta-\beta)^2}\right)(\frac{1}{\theta-\beta}-P_3)]\Bigg]\\
&+\frac{\pi}{6}\Big[a_2[h_1\left(P_3^3+P_2^2(2P_2-3P_3)\right)+h_2P_3P_2(P_3^2+P_2^2)]\\
&+a_3P_3[h_1(P_3^3+2P_2^3)+h_2P_2^2(3P_3^2+P_2^2)]\Big]>0
\end{aligned}
\tag{36}
$$

since for every value of α, $\alpha\mu>\rho$, $0<\theta+\beta<1$ and $P_3\leq 1$ then $\frac{1}{\theta+\beta}>P_3$. Therefore each term of (36) is positive. hence

$$F>0.$$

Thus $M, L, \& F\in\Re^+$.

8.2 Appendix B

Proof of Lemma 1

The quadratic equation (17) has two solutions (P_* and P_{**}) found as

$$P_*=\frac{1}{2M}\left[L-\left(L^2+4MF\right)^{\frac{1}{2}}\right]$$

and

$$P_{**}=\frac{1}{2M}\left[L+\left(L^2+4MF\right)^{\frac{1}{2}}\right]$$

from the hypothesis of Theorem 1. $M > 0$, $L > 0$ and $F > 0$ then $4MF > 0$

$$\Rightarrow L^2 < \left(L^2 + 4MF\right)$$

$$\Rightarrow L < \left(L^2 + 4MF\right)^{\frac{1}{2}}$$

$$\Rightarrow L - \left(L^2 + 4MF\right)^{\frac{1}{2}} < 0$$

thus

$$P_* = \frac{1}{2M}\left[L - \left(L^2 + 4MF\right)^{\frac{1}{2}}\right] < 0$$

the cycle length is always positive, therefore $P_* \neq P^*$ since

$$L - \left(L^2 + 4MF\right)^{\frac{1}{2}} < 0$$

then

$$L + \left(L^2 + 4MF\right)^{\frac{1}{2}} > 0$$

therefore

$$P_{**} = \frac{1}{2M}\left[L + \left(L^2 + 4MF\right)^{\frac{1}{2}}\right] > 0$$

hence $P_{**} = P^*$

8.3 Appendix C

Proof of Lemma 2

It is enough to show that $\frac{d^2Z(P)}{dP^2} > 0$
Re-differentiating (16)

$$
\begin{aligned}
\frac{d^2Z(P)}{dP^2} &= \frac{2A}{P^3} + \frac{2C}{P^3}[\rho_2(\theta-\beta)P(P_3-P)] + \frac{2\pi}{P\Psi 3}\Big[h_1[\frac{(\mu-\rho)}{2}P_1^2 \\
&+ \frac{(\alpha\mu-\rho)}{2}(P_2^2-P_1^2) + (1-\alpha)\mu P_1(P_2-P_1) + \frac{a_1}{2}(P_3-P_2)^2 \\
&+ \frac{a_2}{6}(2P_3^3-3P_3^2P_2+P_2^3) + \frac{a_3}{12}(3P_3^4-4P_3^3P_2+P_2^4) \\
&+ \frac{\rho_2}{\theta-\beta}P_2 - \frac{\rho_2}{(\theta-\beta)^2} + \rho_2(P_3-P_2+1)(\frac{1}{\theta-\beta}-P_3) \\
&- \rho_2(P_3-P_2+1)(\theta-\beta)(P-P_3)P] + h_2[\frac{(\mu-\rho)}{3}P_1^3 \\
&+ \frac{(\alpha\mu-\rho)}{3}(P_2^3-P_1^3) + \frac{(1-\alpha)}{2}\mu P_1(P_2^2-P_1^2) \\
&+ \frac{a_1}{6}(P_3^3-3P_3P_2^2+2P_2^3) + \frac{a_2}{8}(P_3^4-2P_3^2P_2^2+P_2^4) \\
&+ \frac{a_3}{30}(3P_3^5-5P_3^3P_2^2+2P_2^5) - \frac{\rho_2}{(\theta-\beta)^3} + \frac{\rho_2P_2^2}{2(\theta-\beta)} \\
&+ \rho_2\left(\frac{P_3^2-P_2^2}{2}+P_3+\frac{1}{(\theta-\beta)^2}\right)(\frac{1}{\theta-\beta}-P_3) + \frac{\rho_2}{2(\theta-\beta)}P^2 \\
&- \rho_2\left(\frac{P_3^2-P_2^2}{2}+P_3+\frac{1}{(\theta-\beta)^2}\right)(\theta-\beta)(P-P_3)P]\Big] \\
&- \frac{C}{P^2}\left[2\rho_2(\theta-\beta)P+\rho_2(\theta-\beta)P_3\right] \\
&- \frac{\pi}{P^2}\Big[h_1[-\rho_2(P_3-P_2+1)(\theta-\beta)(2P-P_3)] + h_2[\frac{\rho_2}{(\theta-\beta)}P \\
&- \rho_2\left(\frac{P_3^2-P_2^2}{2}+P_3+\frac{1}{(\theta-\beta)^2}\right)(\theta-\beta)(2P-P_3)]\Big] \\
&= \frac{2A}{P^3} + \frac{C}{P^3}[\rho_2(\theta-\beta)P_3P] + \frac{2\pi}{P^3}\Big[h_1[\frac{(\mu-\rho)}{2}P_1^2 \\
&+ \frac{(\alpha\mu-\rho)}{2}(P_2^2-P_1^2) + (1-\alpha)\mu P_1(P_2-P_1) \\
&+ \frac{a_1}{2}(P_3-P_2)^2 + \frac{a_2}{6}(2P_3^3-3P_3^2P_2+P_2^3) + \frac{\rho_2}{\theta-\beta}P_2 \\
&+ \frac{a_3}{12}(3P_3^4-4P_3^3P_2+P_2^4) + \rho_2(P_3-P_2+1)(\frac{1}{\theta-\beta}-P_3) \\
&- \frac{\rho_2}{(\theta-\beta)^2} - \rho_2(P_3-P_2+1)(\theta-\beta)(P-P_3)P \\
&+ \rho_2(P_3-P_2+1)(\theta-\beta)P^2 - \frac{1}{2}\rho_2(P_3-P_2+1)(\theta-\beta)P_3P] \\
&+ h_2[\frac{(\mu-\rho)}{3}P_1^3 + \frac{(\alpha\mu-\rho)}{3}(P_2^3-P_1^3) + \frac{(1-\alpha)}{2}\mu P_1(P_2^2-P_1^2) \\
&+ \frac{a_1}{6}(P_3^3-3P_3P_2^2+2P_2^3) + \frac{a_2}{8}(P_3^4-2P_3^2P_2^2+P_2^4) \\
&+ \frac{a_3}{30}(3P_3^5-5P_3^3P_2^2+2P_2^5) - \frac{\rho_2}{(\theta-\beta)^3} + \frac{\rho_2P_2^2}{2(\theta-\beta)} \\
&+ \rho_2\left(\frac{P_3^2-P_2^2}{2}+P_3+\frac{1}{(\theta-\beta)^2}\right)(\frac{1}{\theta-\beta}-P_3) \\
&- \rho_2\left(\frac{P_3^2-P_2^2}{2}+P_3+\frac{1}{(\theta-\beta)^2}\right)(\theta-\beta)(P-P_3)P \\
&+ \rho_2\left(\frac{P_3^2-P_2^2}{2}+P_3+\frac{1}{(\theta-\beta)^2}\right)(\theta-\beta)P^2 \\
&- \frac{1}{2}\rho_2\left(\frac{P_3^2-P_2^2}{2}+P_3+\frac{1}{(\theta-\beta)^2}\right)(\theta-\beta)P_3P]\Big]
\end{aligned}
$$

$$
\begin{aligned}
= \; & \frac{2A}{P^3} + \frac{C}{P^3}\,[\rho_2(\theta-\beta)P_3P] + \frac{2\pi}{P^3}\Big[h_1[\frac{(\mu-\rho)}{2}P_1^2 + \frac{a_1}{2}(P_3-P_2)^2 \\
& + \frac{(\alpha\mu-\rho)}{2}(P_2^2-P_1^2) + (1-\alpha)\mu P_1(P_2-P_1) - \frac{\rho_2}{(\theta-\beta)^2} \\
& + \frac{a_2}{6}(2P_3^3 - 3P_3^2P_2 + P_2^3) + \frac{a_3}{12}(3P_3^4 - 4P_3^3P_2 + P_2^4) + \frac{\rho_2}{\theta-\beta}P_2 \\
& + \rho_2(P_3-P_2+1)(\frac{1}{\theta-\beta} - P_3) + \frac{1}{2}\rho_2(P_3-P_2+1)(\theta-\beta)P_3P] \\
& + h_2[\frac{(\mu-\rho)}{3}P_1^3 + \frac{(\alpha\mu-\rho)}{3}(P_2^3-P_1^3) + \frac{(1-\alpha)}{2}\mu P_1(P_2^2-P_1^2) \\
& + \frac{a_1}{6}(P_3^3 - 3P_3P_2^2 + 2P_2^3) + \frac{a_2}{8}(P_3^4 - 2P_3^2P_2^2 + P_2^4) \\
& + \frac{a_3}{30}(3P_3^5 - 5P_3^3P_2^2 + 2P_2^5) - \frac{\rho_2}{(\theta-\beta)^3} + \frac{\rho_2P_2^2}{2(\theta-\beta)} \\
& + \rho_2\left(\frac{P_3^2-P_2^2}{2} + P_3 + \frac{1}{(\theta-\beta)^2}\right)(\frac{1}{\theta-\beta} - P_3) \\
& + \frac{1}{2}\rho_2\left(\frac{P_3^2-P_2^2}{2} + P_3 + \frac{1}{(\theta-\beta)^2}\right)(\theta-\beta)P_3P]\Big]
\end{aligned}
$$

$$
\begin{aligned}
= \; & \frac{1}{P^2}\Big[C\rho_2(\theta-\beta)P_3 + \pi h_1\rho_2(P_3-P_2+1)(\theta-\beta)P_3 \\
& + \pi h_2\left(\frac{P_3^2-P_2^2}{2} + P_3 + \frac{1}{(\theta-\beta)^2}\right)(\theta-\beta)P_3\Big] \\
& + \frac{2}{P^3}\Big[\pi\Big[h_1[\frac{(\mu-\rho)}{2}P_1^2 + \frac{(\alpha\mu-\rho)}{2}(P_2^2-P_1^2) \\
& + (1-\alpha)\mu P_1(P_2-P_1) + \frac{a_1}{2}(P_3-P_2)^2 + \frac{\rho_2}{\theta-\beta}P_2 \\
& + \frac{a_2}{6}(2P_3^3 - 3P_3^2P_2 + P_2^3) + \frac{a_3}{12}(3P_3^4 - 4P_3^3P_2 + P_2^4) \\
& + \rho_2(P_3-P_2+1)(\frac{1}{\theta-\beta} - P_3) - \frac{\rho_2}{(\theta-\beta)^2}] \\
& + h_2[\frac{(\mu-\rho)}{3}P_1^3 + \frac{(\alpha\mu-\rho)}{3}(P_2^3-P_1^3) + \frac{(1-\alpha)}{2}\mu P_1(P_2^2-P_1^2) \\
& + \frac{a_1}{6}(P_3^3 - 3P_3P_2^2 + 2P_2^3) + \frac{a_2}{8}(P_3^4 - 2P_3^2P_2^2 + P_2^4) \\
& + \frac{a_3}{30}(3P_3^5 - 5P_3^3P_2^2 + 2P_2^5) - \frac{\rho_2}{(\theta-\beta)^3} + \frac{\rho_2P_2^2}{2(\theta-\beta)} \\
& + \rho_2\left(\frac{P_3^2-P_2^2}{2} + P_3 + \frac{1}{(\theta-\beta)^2}\right)(\frac{1}{\theta-\beta} - P_3)]\Big] + A\Big]
\end{aligned}
$$

$$
= \frac{1}{P^2}L + \frac{2}{P^3}F
$$

from the hypothesis of Theorem 1, $L > 0$ and $F > 0$

thus, $\frac{1}{P^2}L > 0$ and $\frac{2}{P^3}F > 0$

Hence

$$\frac{d^2Z(P)}{dP^2} > 0$$

8.4 Appendix D

Basic Model of [30]

The model in [30] is developed using the notations in Table 13 below

The equations that described the situation of the model are presented below

$$\frac{dY(t)}{dt} = p; \quad 0 \leq t \leq L_1. \tag{37}$$

$$\frac{dY_1(t)}{dt} = -d_1 = -(c_1 + c_2t + c_3t^2); \quad L_1 \leq t \leq L_2. \tag{38}$$

$$\frac{dY_2(t)}{dt} + \lambda Y_2(t) = -d_2; \quad L_2 \leq t \leq L. \tag{39}$$

the solution of Eqs. (37), (38) and (39) are given below

Table 13 Notation

A	The set up cost per production cycle
p	The production rate in unit per unit time
d_1	The demand rate after production before deterioration starts in unit per unit time
d_2	The demand rate during deterioration period in unit per unit time
i	Inventory carrying charge
λ	Deterioration rate
C	Unit cost per item
$h_1 + h_2t$	linear holding cost
Y_1	Maximum inventory level
Y_2	The inventory level at the point deterioration starts
$Y(t)$	The inventory level during inventory build up
$Y_1(t)$	The inventory level after production before deterioration starts
$Y_2(t)$	The inventory level during deterioration period
L_i	Unit time in periods ($i \in \{1, 2\}$)
L	Total cycle length
Z	Total variable cost
EPQ	Economic production quantity per cycle

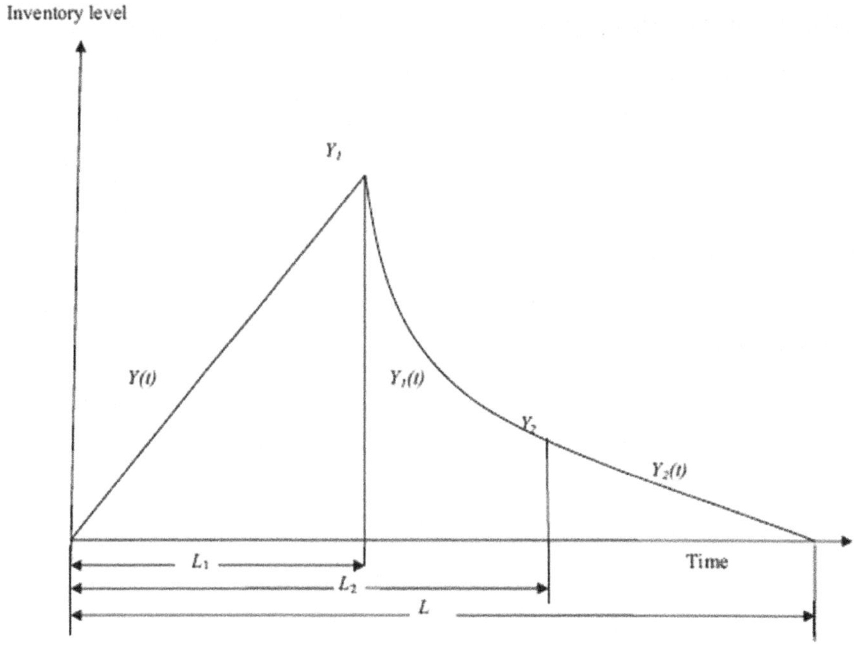

Fig. 2 The inventory system of the basic model

$$Y(t) = \frac{Y_1}{L_1}t; \qquad 0 \le t \le L_1. \tag{40}$$

$$Y_1(t) = c_1(L_2 - t) + c_2\left(\frac{L_2^2}{2} - \frac{t^2}{2}\right) + c_3\left(\frac{L_2^3}{3} - \frac{t^3}{3}\right) + \frac{d_2}{\lambda}(e^{-\lambda(L_2-L)} - 1);$$
$$L_1 \le t \le L_2. \tag{41}$$

$$Y_2(t) = \frac{d_2}{\lambda}(e^{\lambda(L-t)} - 1); \qquad L_2 \le t \le L. \tag{42}$$

The total variable cost and EPQ obtained in this basic model are given by (43) and (44) respectively (Fig. 2).

$$\begin{aligned} Z(L) = {} & \frac{A}{L} + \frac{C}{L}\left[\frac{d_2}{\lambda}\left(e^{-\lambda(L_2-L)} - 1\right) - d_2(L - L_2)\right] + \frac{i}{L}\Bigg[h_1\Bigg[-\frac{d_2L_1}{2\lambda}\left(e^{-\lambda(L_2-L)} - 1\right) \\ & + \frac{d_2L_2e^{-\lambda(L_2-L)}}{\lambda} - \frac{d_2}{\lambda^2} - \frac{d_2L}{\lambda} + \frac{d_2e^{\lambda(L-L_2)}}{\lambda^2} + \frac{c_1}{2}(L_2^2 - L_1L_2) \\ & + \frac{c_2}{12}(4L_2^3 - 3L_1L_2^2 - L_1^3) + \frac{c_3}{12}(3L_2^4 - 2L_1L_2^3 - L_1^4)\Bigg] \\ & + h_2\Bigg[-\frac{d_2L_1^2}{6\lambda}\left(e^{-\lambda(L_2-L)} - 1\right) + \frac{d_2L_2^2e^{-\lambda(L_2-L)}}{2\lambda} - \frac{d_2L}{\lambda^2} - \frac{d_2}{\lambda^3} - \frac{d_2L^2}{2\lambda} \\ & + \frac{d_2L_2e^{\lambda(L-L_2)}}{\lambda^2} + \frac{d_2e^{\lambda(L-L_2)}}{\lambda^3} + \frac{c_1}{6}(L_2^3 - L_1^2L_2) \end{aligned}$$

$$+ \frac{c_2}{24}(3L_2^4 - 2L_1^2L_2^2 - L_1^4) + \frac{c_3}{90}(9L_2^5 - 5L_1^2L_2^3 - 4L_1^5)\Big]\Big] \tag{43}$$

$$\begin{aligned} EPQ^* &= d_1(L_2 - L_1) + d_2(L - L_2) + d_2(L_2) \\ &= c_1(L_2 - L_1) + c_2\left(\frac{L_2^2}{2} - \frac{L_1^2}{2}\right) + c_3\left(\frac{L_2^3}{3} - \frac{L_1^3}{3}\right) + \frac{d_2}{\lambda}\left(e^{-\lambda(L_2 - L^*)} - 1\right) \end{aligned} \tag{44}$$

References

1. Opritescu, D., Hartmann, C., Riedl, W., Ritter, M., Volk, W.: Low-risk bypassing of machine failure scenarios in automotive industry press shops by releasing overall capacity of the production networks. J. Manuf. Syst. **1**(52), 121–130 (2019)
2. Akwen, G.T., Moorthy, R., Daud, S.: The impact of fourth industrial revolution era on Nigeria's counter-terrorism strategy towards Boko haram movement. Malays. J. Hist. Polit. Strateg. Stud. **47**(1), 176–197 (2020)
3. Villarreyna, R., Barrios, M., Vílchez, S., Cerda, R., Vignola, R., Avelino, J.: An economic production quantity model for three levels of production with Weibull distribution deterioration and shortage. Economic constraints as drivers of coffee rust epidemics in Nicaragua. Crop Prot. **127**, 1–12 (2019). 104980
4. Albert, I.O., Danjibo, N., Albert, O.: ZIB back to the past: evolution of kidnapping and hostage taking in the Niger Delta, Nigeria. Beijing Law Rev. **11**, 211–226 (2020)
5. Dari, S., Sani, B.: An EPQ model for delayed deteriorating items with quadratic demand and linear holding cost. OPSEARCH **57**(1), 46–72 (2020)
6. Kumar, S., Malik, A.K., Abhishek, S., Yadav, S.K., Yashveer, S.: An inventory model with linear holding cost and stock- dependent demand for non-instantaneous deteriorating items. In: Advancement in Science and Technology AIP Conference Proceedings, no. 171 (2016)
7. Shilpy, T., Singh, S.R., Rajendra, S., Anubhav, P.S.: An EPQ model for non-instantaneous deteriorating item with time dependent holding cost and exponential demand rate. Int. J. Oper. Res. **23**(2), 145–162 (2015)
8. Pandey, A.: A study of EPQ inventory model for deteriorating items considering stock dependent demand and time varying holding cost. J. Comput. Math. Sci. **8**(7), 323–331 (2017)
9. Mahapatra, G.S., Adak, S., Kaladhar, K.: A fuzzy inventory model with three parameter Weibull deterioration with reliant holding cost and demand incorporating reliability. J. Intell. Fuzzy Syst. **36**(6), 5731–5744 (2019)
10. Monika, V., Singh, S.R.: An EPQ inventory model with variable holding cost and shortages under the effect of learning on setup cost for two warehouses. Int. J. Eng. Technol. Sci. Res. **4**(8), 1222–1235 (2017)
11. Khalilpourazari, S., Pasandideh, S.H.R.: Multi-item EOQ model with nonlinear unit holding cost and partial backordering: moth-flame optimization algorithm. J. Ind. Prod. Eng. **34**(1), 42–51 (2016)
12. Rout, C., Kumar, R.S., Chakraborty, D., Goswami, A.: An EPQ model for deteriorating items with imperfect production, inspection errors, rework and shortages: a type-2 fuzzy approach. OPSEARCH (2019). https://doi.org/10.1007/s12597-019-00390-3
13. Pal, S., Mahapatra, G.S., Samanta, G.P.: A production inventory model for deteriorating item with ramp type demand allowing inflation and shortages under fuzziness. Econ. Model. **46**, 334–345 (2015)
14. Dari, S., Sani, B.: An EPQ model for delayed deteriorating items with quadratic demand and shortages. Asian J. Math. Comput. Res. **22**(2), 87–103 (2017)
15. Mohan, R.: Quadratic demand, variable holding cost with time dependent deterioration without shortages and salvage value. IOSR J. Math. (IOSR-JM) **13**(2), 59–66 (2017)

16. Smaila, S.S., Walford, I.E.C.: An inventory model with three-parameter Weibull deterioration, quadratic demand rate and shortages. Am. J. Math. Manag. Sci. **35**(2), 159–170 (2016)
17. Azharuddin, S., Poonam, M.: Optimal policies for price sensitive quadratic demand with preservation technology investment under inflationary environment. J. Adv. Manuf. Syst. **18**(2), 325–337 (2019)
18. Nita, H.S., Mrudul, Y.J., Urmila, C.: Impact of future price increase on ordering policies for deteriorating items under quadratic demand. Int. J. Ind. Eng. Comput. **7**(3), 423–436 (2016)
19. Dash, B., Pattnaik, M., Pattnaik, H.: Deteriorated economic production quantity (EPQ) model for declined quadratic demand with time value of money and shortages. Appl. Math. Sci. **8**(73), 3607–3618 (2014)
20. Meenakshi, S., Ranjana, G.: An EPQ model for deteriorating items with time and price dependent demand under markdown policy. OPSEARCH **51**(1), 148–158 (2014)
21. Rahaman, M., Mondal, S.P., Shaikh, A.A., Ahmadian, A., Senu, N., Salahshour, S.: Arbitrary-order economic production quantity model with and without deterioration: generalized point of view. Adv. Differ. Equ. (2020). https://doi.org/10.1007/s00500-020-04867-y. Article number: 16
22. Pal, S., Mahapatra, G.S., Samanta, G.P.: An EPQ model of ramp type demand with weibull deterioration under inflation and finite horizon in crisp and fuzzy environment. Int. J. Prod. Econ. **156**, 159–166 (2014)
23. Ashutosh, P.: A study of EPQ inventory model for deteriorating items considering stock dependent demand and time varying holding cost. J. Comput. Math. Sci. **8**(7), 323–331 (2017)
24. Karabi, D.C., Biplab, K., Mantu, D., Tapan, K.D.: An inventory model for deteriorating items with stock-dependent demand, time-varying holding cost and shortages. OPSEARCH **52**(1), 55–79 (2015)
25. Ruidas, S., Seikh, M.S., Nayak, P.K.: An EPQ model with stock and selling price dependent demand and variable production rate in interval environment. Int. J. Syst. Assur. Eng. Manag. (2019). https://doi.org/10.1007/s13198-019-00867-w
26. Roul, J.N., Maity, K., Kar, S., Maiti, M.: Optimal time-dependent production policy under random time horizon. OPSEARCH (2019). https://doi.org/10.1007/s12597-019-00407-x
27. Thinakaran, N., Jayaprakas, J., Elanchezhian, C.: Survey on inventory model of EOQ & EPQ with partial backorder problems. Mater. Today Proc. **16**, 629–635 (2019)
28. Pal, S., Mahapatra, G.S., Samanta, G.P.: A three-layer supply chain EPQ model for price- and stock-dependent stochastic demand with imperfect item under rework. J. Uncertainty Anal. Appl. **4**(1), 1–21 (2016)
29. Singha, S.R., Deepa, K., Shilpy, T.: An economic order quantity model for deteriorating products having stock dependent demand with trade credit period and preservation technology. Uncertain Supply Chain Manag. **4**, 29–42 (2016)
30. Baba, A.A., Baraya, Y.M.: An EPQ model for non-instantaneous deteriorating items with stock dependent demand rate under two-phase production rate and shortages. ABACUS: J. Math. Assoc. Nigeria **46**(1), 225–238 (2019)

An EPQ Model for Delayed Deteriorating Items with Variable Production Rate, Two-Phase Demand Rates and Shortages

Mustapha Lawal Malumfashi, Mohd Tahir Ismail, Amirah Rahman, Dari Sani, and Majid Khan Majahar Ali

Abstract This study proposes an Economic Production Quantity (EPQ) model for delayed deteriorating items, where the production rates vary and the demands are of two phases, constant and quadratic. The model is classified into five stages as follows: (1) Production build up period with initial production rate, (2) Production build up period after change of production rate, (3) Period before deterioration starts, (4) Period after deterioration sets in, and (5) Shortages and backlogs periods. The demand during production build up periods and the demand after deterioration starts are assumed to be constant, while demand before deterioration begins is assumed to be quadratic time dependent. It is also assumed that shortages are allowed and fully backlogged. A theorem and lemmas are framed to characterize the optimal solutions. The necessary and sufficient conditions for optimality are also provided and the best cycle length that optimizes the total variable cost and the EPQ was obtained. A numerical example is given to illustrate the applicability of the model and sensitivity analysis is carried out on the example to discern the effect of changes on some system parameters. R GUI statistical package is employed to demonstrated these effects graphically.

Keywords Variable production rates · EPQ · Quadratic demands · Delayed deterioration · Shortages

M. L. Malumfashi · M. T. Ismail · A. Rahman · M. K. M. Ali (✉)
School of Mathematical Sciences, Universiti Sains Malaysia, George Town, Penang, Malaysia
e-mail: majidkhanmajaharali@usm.my

M. L. Malumfashi
e-mail: mustapha.lawal@umyu.edu.ng

D. Sani
Department of Mathematical Sciences, Kaduna State University, Kaduna, Nigeria

M. L. Malumfashi
Department of Mathematics and Statistics, Umaru Musa Yar'adua University, Katsina, Nigeria

M. H. Mohd et al. (eds.), *Modelling, Simulation and Applications of Complex Systems*, Springer Proceedings in Mathematics & Statistics 359,
https://doi.org/10.1007/978-981-16-2629-6_20

1 Introduction

The goal of any inventory system is to achieve and maintain satisfactory levels of customer service while maintaining the inventory cost within set time intervals. A look at recent inventory control literature reveals that an inventory control or management refers to the policies that improve manufacturing efficiencies using the inventory, and decrease its related costs. In some inventory model, such as [1, 2], the attention was paid on production build-up period without demand which is not usually the case looking at the definition of demand in line with a growing business.

Demand can be defined as the size or quantity of goods or services that consumers are willing to purchase at a particular price and time. Naturally, when the product is introduced to the market, the demand is expected to be very low but may increase with time as a result of suitable marketing strategies applied and popularity of the product in market up to its highest level. Demand is expected to later start declining due to many factors such as the introduction of similar products to the market as in the case of competitive mobile phones across the globe as discussed in [3]. Such demand can be described as quadratic time demand. As the demand approaches it maximum level, the expectation of shortages is inevitable especially for production centres with less safety-stock at the initial production period.

A such, many researchers have developed inventory models with quadratic time dependent demand. The study by [4] obtained an optimal cost of the inventory system with quadratic demand, constant deterioration and salvage value while [1] developed an inventory model for delayed deteriorating items with quadratic time dependent demand. The demand before deterioration starts was assumed to be quadratic and the demand after deterioration starts is assumed to be constant. It was also assumed that shortages were allowed with partial backorder and that there was no demand during the production period. In [2], an EPQ model for delayed deteriorating items was presented, where the demand before deterioration sets in is assumed to be time dependent quadratic demand and the holding (carrying) cost was assumed to be linearly time dependent. It was also assumed that shortages are not allowed and there is no demand during production. Other inventory models with quadratic time demand include [5–9], while [10] modelled an inventory system for deteriorated items with credit period dependent demand. Considering the nature of quadratic demand rate, as the demand started declining after reaching its highest level, the deterioration of the products available is highly expected.

Given the present state of inventory models of deteriorating items, deterioration can be defined as reduction in the inventory quality, or change of original characteristics of items such as colour, physical view and odour. Recent studies [11–16] assumed that deterioration of an inventory starts immediately when the inventories are held in stock. However, this is not generally the case in the real world. Items such as juice, biscuits, and soap normally have a lifespan of maintaining originality, and deterioration starts after that lifespan. Therefore, it is important to reconsider modelling of such delayed deteriorating inventories especially at the present situation of the well-established production centres that the demand of their product is stable and constant.

Looking at the present state of well-established production centres, it is abnormal to neglect constant demand during the inventory build-up period in the study of inventory control. In the type-2 fuzzy approach of an EPQ model for deteriorating items with imperfect production established in [17], demand during inventory build-up was considered constant. The most recent EPQ model with constant demand during inventory build-up period was presented in [18]. The study [19] developed an EPQ model with imperfect production process for time varying demand with shortages and assumed that the production rate is time dependent. Similarly, the study [20] presented an EPQ Model with time dependent demand rate where the production rate is demand dependent as such production rate is also time dependent. Also, an EPQ Model for deteriorating items with quadratic time dependent demand and two-phased (time and demand) dependent production rates was presented in [21], while [22] proposed a model with three levels of production with Weibull distribution deterioration and shortage EPQ. It was assumed that the total cost of production depends on production rate, demand rate and deterioration rate while production rate depends on existing demand. Among the recent work with time dependent production rate is [23] who established an EPQ model for deteriorating items with variable demand rate and allowable shortages The demand during inventory build-up was also time dependent.

Also, many studies established inventory policies on the assumption that production rate depends on time or demand, but this is not always the case. It is natural that most newly established companies with limited resources, along with small and medium-sized industries are always appraising the production planning system to ensure the increase in production rate so as to properly respond to their clients as stated in [24, 25]. On the other hand, some well-established production centres are facing some constraints which normally hinders the production process. This may lead to the changes in production rate in a negative perspective. Examples of this include militant issues affecting oil companies in Nigeria as discussed in [26], uncertainty caused by terrorism and anti-terrorism policies in affected countries as discussed in [27, 28], and the the global issue of coronavirus and anti-coronavirus policies as discussed in [29]. The production rate in this situation is conditional and depends on unexpected factors that may lead to the termination of production before the end of the proposed cycle length, or increase of production rate due to planned upgrade of the industry. This kind of production process is rarely visible in the literature. Our proposed study extends the work of [1] by implementing conditional production rate and demand during the inventory build-up in a production inventory model for delayed deteriorating products which were note visible in the literature. Also, shortages are allowed and completely backlogged at a constant rate which are generally not the argument of [1].

2 Notation and Assumptions

Table 1 contains the notation used in this paper.

The proposed model is developed based on the following modified assumptions adopted from [1] with addition of 1 and 9:

1. There are two production rates p and ap (production rate increases or decreases with parameter a), where the production level is assumed to stay positive and be less than 200% of the initial rate. That is, $a \in \{(0, 1) \cup (1, 2)\}$.
2. Demand rates λ and λ_2 (during production and after deterioration sets in respectively) are assumed to be constant.
3. Demand rate λ_1 (after production but before deterioration sets in) is assumed to be a quadratic function of time. That is, $\lambda_1 = b_1 + b_2 t + b_3 t^2$, where $b_3 > 0$.

Table 1 Notation

λ	Demand rate during inventory buildup in unit per unit time
λ_1	Demand rate after production, before deterioration starts in unit per unit time
λ_2	Demand rate after deterioration start in unit per unit time
p	Production rate in unit per unit time
a	Production rate changing parameter
α	Inventory carrying charge
β	Deterioration rate
A	Set up cost per production cycle
C	Unit cost per item
C_h	Holding cost
C_D	Deterioration cost
c_s	Shortage cost of the unit item
S_c	Cost of shortages
S	Maximum shortage
Q_1	The inventory level at the point where instantaneous change of production rate occurs
Q_2	The inventory level at the end of the production period
Q_3	The inventory level at the point where deterioration sets in
$Q(t)$	The inventory level during production build-up, before the production rate changes, at any time t
$Q_1(t)$	The inventory level after the production rate changes, at any time t
$Q_2(t)$	The inventory level after the production stops, before deterioration sets in, at any time t
$Q_3(t)$	The inventory level after deterioration sets in, at any time t
$N(t)$	The backorder level during shortage period at any time t and rate λ
T	Total cycle length
T_i	Unit time in periods ($i \in \{1, 2, 3, 4, 5\}$)

4. The time border of the production system is finite. That is, $T_1 \le T_2 \le T_3 \le T_4 \le T_5 \le T$.
5. All defective items are discarded after inspection.
6. Production is assumed to be instantaneous.
7. Unconstrained supplies capital.
8. Shortage is allowed and fully backlogged.
9. The items produced during the periods $t \in [0, T_1]$ and $t \in [T_5, T]$ are the same and produced at the same rate.

3 Modelling of Variable Production with Two-Phase Demand and Shortages

As shown in Fig. 1, the inventory level varies with time due to both production phases, demand, deterioration and shortages. Production is assumed to start at time at $t = 0$ with constant rate p and changes at time T_1, the demand rate λ within this interval is also assumed to be constant and the inventory attains the level Q_1 at time T_1. During the interval $[T_1, T_2]$ production rate changes (increases or decreases with the parameter a) to ap where $a \in \{(0, 1) \cup (1, 2)\}$ while demand rate remains constant and the inventory level attains it maximum level Q_2 at time T_2 when the production stopped. Within the interval $[T_2, T_3]$, the inventory level decreases due to demand rate only which assumes to be a quadratic function of time ($\lambda_1 = b_1 + b_2 t + b_3 t^2$) and the inventory drops to level Q_3 at time T_3. Deterioration starts at time T_3 and the inventory level continue dropping due to both constant demand and deterioration up to the time T_4 when the inventory level completely depleted. At time T_4 shortage occurs and attains its maximum level S at time T_5 when reproduction starts and the shortages completely backlogs at the end of the cycle length T.

The governing Differential Equations (DE) of the model are as follows:

$$\frac{dQ(t)}{dt} = p - \lambda; \qquad 0 \le t \le T_1. \quad (1)$$

$$\frac{dQ_1(t)}{dt} = ap - \lambda; \qquad T_1 \le t \le T_2. \quad (2)$$

$$\frac{dQ_2(t)}{dt} = -\lambda_1 = -(b_1 + b_2 t + b_3 t^2); \qquad T_2 \le t \le T_3. \quad (3)$$

$$\frac{dQ_3(t)}{dt} + \beta Q_3(t) = -\lambda_2; \qquad T_3 \le t \le T_4. \quad (4)$$

$$\frac{dN(t)}{dt} = -\lambda; \qquad T_4 \le t \le T_5. \quad (5)$$

$$\frac{dQ(t)}{dt} = p - \lambda; \qquad T_5 \le t \le T. \quad (6)$$

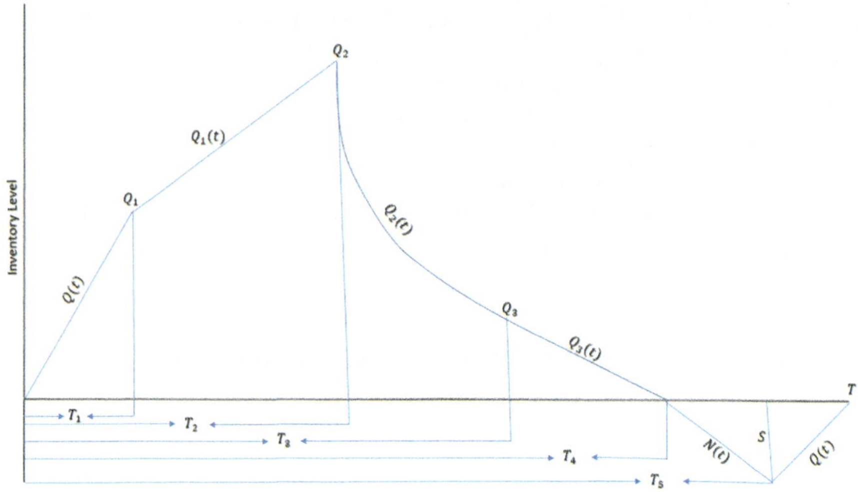

Fig. 1 The inventory system with shortages and full backlog

To obtain the solution of the Differential Equation (1) which describes the situation within the interval $[0, T_1]$

$$Q(t) = (p - \lambda)t + k \tag{7}$$

where k is constant.

Apply the boundary conditions $Q(0) = 0$ and $Q(T_1) = Q_1$ on Eq. (7) we get

$$\begin{aligned} k &= 0 \\ Q_1 &= (p - \lambda)T_1. \end{aligned}$$

thus, the solution of Eq. (1) is

$$Q(t) = (p - \lambda)t. \tag{8}$$

The solution of Differential Equation (2) within $[T_1, T_2]$ is

$$Q_1(t) = (ap - \lambda)t + k_1, \tag{9}$$

where k_1 is constant.

Applying the boundary conditions $Q_1(T_1) = Q_1$ and $Q_1(T_2) = Q_2$ on Eq. (9),

$$Q_1(t) = (p - \lambda)T_1 + (ap - \lambda)(t - T_1). \tag{10}$$

The solution for the Differential Equation (3) which describes the situation within $[T_2, T_3]$ is

$$Q_2(t) = -\left(b_1 t + b_2\frac{t^2}{2} + b_3\frac{t^3}{3}\right) + k_2, \tag{11}$$

where k_2 is constant.

Apply the boundary conditions $Q_2(T_2) = Q_2$ and $Q_2(T_3) = Q_3$ on Eq. (11) we get

$$Q_2(t) = b_1(T_3 - t) + \frac{b_2}{2}(T_3^2 - t^2) + \frac{b_3}{3}(T_3^3 - t^3) + Q_3. \tag{12}$$

The Integrating Factor method of solving ordinary DEs is employed to solve the Differential Equation (4) for the situation within $[T_3, T_4]$.

$$Q_3(t) = \frac{-\lambda_2}{\beta} + k_3 e^{-\beta t}, \tag{13}$$

where k_3 is constant.

Apply the boundary conditions $Q_3(T_3) = Q_3$ and $Q_3(T_4) = 0$ on Eq. (13) we get

$$Q_3(t) = \frac{-\lambda_2}{\beta}\left(Q_3 + \frac{\lambda_2}{\beta}\right)e^{\beta(T_3-t)}. \tag{14}$$

$$Q_3 = \frac{\lambda_2}{\beta}(e^{\beta(T_4-T_3)} - 1). \tag{15}$$

Substitute Eq. (15) into Eq. (12) and Eq. (14) above to get

$$Q_2(t) = b_1(T_3 - t) + \frac{b_2}{2}(T_3^2 - t^2) + \frac{b_3}{3}(T_3^3 - t^3) + \frac{\lambda_2}{\beta}(e^{\beta(T_4-T_3)} - 1). \tag{16}$$

$$Q_3(t) = \frac{\lambda_2}{\beta}(e^{\beta(T_4-t)} - 1). \tag{17}$$

Solving and applying the boundary conditions $N(T_4) = 0$, $Q(T_5) = S$ and $Q(T) = 0$, the solutions of the Differential Equations (5, 6) that describe the situation within shortages and backlogged periods are given below.

$$N(t) = -\lambda(t - T_4); \qquad T_4 \leq t \leq T_5. \tag{18}$$

$$Q(t) = -(p - \lambda)(T - t); \qquad T_5 \leq t \leq T.$$

$$S = \lambda(T_4 - T_5); \tag{19}$$

The cost associated with storage, carrying charge and storage of the total inventory from beginning of production until it is sold or used (i.e. the inventory holding cost) C_h in a production run is the total amount of inventory multiplied by the carrying charge per unit item and cost per unit item, which is given as

$$C_h = \alpha C\left[\int_0^{T_1} Q(t)dt + \int_{T_1}^{T_2} Q_1(t)dt + \int_{T_2}^{T_3} Q_2(t)dt + \int_{T_3}^{T_4} Q_3(t)dt\right] \quad (20)$$

substitute Eqs. (8), (10), (16) and (17) in Eq. (20)

$$\begin{aligned}
C_h &= \alpha C\Bigg[\int_0^{T_1} (p-\lambda)tdt + \int_{T_2}^{T_2} ((p-\lambda)T_1 + (ap-\lambda)(t-T_1))\,dt \\
&+ \int_{T_2}^{T_3}\Bigg(b_1(T_3-t) + \frac{b_2}{2}(T_3^2-t^2) + \frac{b_3}{3}(T_3^3-t^3) \\
&+ \frac{\lambda_2}{\beta}(e^{\beta(T_4-T_3)}-1)\Bigg)dt + \int_{T_3}^{T_4}\frac{\lambda_2}{\beta}(e^{\beta(T_4-t)}-1)dt\Bigg] \\
&= \alpha C\Bigg[\frac{(p-\lambda)T_1}{2} + (p-\lambda)T_1(T_2-T_1) \\
&+ (ap-\lambda)\left[\left(\frac{T_2^2}{2}-T_1T_2\right)-\left(\frac{T_1^2}{2}-T_1^2\right)\right] \\
&+ b_1\left(\frac{T_3^2}{2}-T_3T_2+\frac{T_2^2}{2}\right) + \frac{b_2}{2}\left(\frac{2T_3^3}{3}-T_3^2T_2+\frac{T_2^3}{3}\right) \\
&+ \frac{b_3}{3}\left(\frac{3T_3^4}{4}-T_3^3T_2+\frac{T_2^4}{4}\right) + \frac{\lambda_2}{\beta}\left(e^{\beta(T_4-T_3)}-1\right)(T_3-T_2) \\
&+ \frac{\lambda_2}{\beta}\left(e^{\beta(T_4-T_3)}-1\right) + \frac{\lambda_2}{\beta}(T_3-T_4)\Bigg] \\
C_h &= \alpha C\Bigg[\frac{(p-\lambda)}{2}T_1 + (p-\lambda)T_1(T_2-T_1) + \frac{1}{2}(ap-\lambda)(T_2-T_1)^2 \\
&+ \frac{b_1}{2}(T_3-T_2)^2 + \frac{b_2}{6}(2T_3^3-3T_3^2T_2+T_2^3) \\
&+ \frac{b_3}{12}(3T_3^4-4T_3^3T_2+T_2^4) + \frac{\lambda_2}{\beta}\left(e^{\beta(T_4-T_3)}-1\right)(T_3-T_2) \\
&+ \frac{\lambda_2}{\beta}\left(e^{\beta(T_4-T_3)}-1\right) + \frac{\lambda_2}{\beta}(T_3-T_4)\Bigg]. \qquad (21)
\end{aligned}$$

The number of deteriorated items is the difference between the total number of items at the point deterioration sets in and the demanded items within the period of deterioration give as $Q_3 - \lambda_2(T_4 - T_3)$.

The total cost of deterioration is the multiple of cost per unit item and the total number of deteriorated items given by

$$C_D = C\left[\frac{\lambda_2}{\beta}(e^{\beta(T_4-T_3)}) - \lambda_2(T_4-T_3)\right]. \quad (22)$$

The total cost of shortages is the multiple of cost of shortage per unit item and the total number of items within shortage period and is obtained as

$$\begin{aligned} S_c &= c_s\left[\int_{T_4}^{T_5} -N(t)dt + \int_{T_5}^{T} -Q(t)dt\right] \\ &= \frac{c_s}{2}\left[\lambda(T_5 - T_4)^2 + (p-\lambda)(T - T_5)^2\right]. \end{aligned} \tag{23}$$

The total variable cost is the sum of setup cost, holding cost, deterioration cost and cost of shortages and is given by

$$Z = A + C_D + C_h + S_c.$$

The total variable cost per unit time is given by

$$\begin{aligned} Z(T) &= \frac{\text{Total variable cost}}{\text{Production cycle length}} \\ &= \frac{A}{T} + \frac{C_D}{T} + \frac{C_h}{T} + \frac{S_c}{T} \end{aligned} \tag{24}$$

substitute Eqs. (21), (22) and (23) in Eq. (24)

$$\begin{aligned} Z(T) = {} & \frac{A}{T} + \frac{C}{T}\left[\frac{\lambda_2}{\beta}(e^{\beta(T_4-T_3)} - 1) - \lambda_2(T_4 - T_3)\right] \\ & + \frac{\alpha C}{T}\Bigg[\frac{(p-\lambda)}{2}T_1 + (p-\lambda)T_1(T_2 - T_1) + \frac{1}{2}(ap-\lambda)(T_2 - T_1)^2 \\ & + \frac{b_1}{2}(T_3 - T_2)^2 + \frac{b_2}{6}(2T_3^3 - 3T_3^2T_2 + T_2^3) \\ & + \frac{b_3}{12}(3T_3^4 - 4T_3^3T_2 + T_2^4 + \frac{\lambda_2}{\beta}\left(e^{\beta(T_4-T_3)} - 1\right)(T_3 - T_2) \\ & + \frac{\lambda_2}{\beta^2}\left(e^{\beta(T_4-T_3)} - 1\right) + \frac{\lambda_2}{\beta}(T_3 - T_4)\Bigg] \\ & + \frac{c_s}{2T}\left[\lambda(T_5 - T_4)^2 + (p-\lambda)(T - T_5)^2\right]. \end{aligned} \tag{25}$$

The objective of this model is to determine the optimal value of the cycle length T that provides the minimum total variable cost $Z(T)$. The necessary condition is $\frac{dZ(T)}{dT} = 0$, while the sufficient condition is $\frac{d^2Z(T)}{dT^2} > 0$.

For the necessary condition of optimality, we differentiate Eq. (25)

$$\begin{aligned}\frac{dZ(T)}{dT} &= \frac{-A}{T^2} - \frac{C}{T^2}\left[\frac{\lambda_2}{\beta}(e^{\beta(T_4-T_3)} - 1) - \lambda_2(T_4 - T_3)\right]\\
&- \frac{\alpha C}{T^2}\left[\frac{(p-\lambda)}{2}T_1 + (p-\lambda)T_1(T_2 - T_1)\right.\\
&+ \frac{1}{2}(ap-\lambda)(T_2 - T_1)^2 + \frac{b_1}{2}(T_3 - T_2)^2\\
&+ \frac{b_2}{6}(2T_3^3 - 3T_3^2T_2 + T_2^3) + \frac{b_3}{12}(3T_3^4 - 4T_3^3T_2 + T_2^4)\\
&+ \frac{\lambda_2}{\beta}\left(e^{\beta(T_4-T_3)} - 1\right)(T_3 - T_2) + \frac{\lambda_2}{\beta^2}\left(e^{\beta(T_4-T_3)} - 1\right)\\
&\left.+ \frac{\lambda_2}{\beta}(T_3 - T_4)\right] - \frac{c_s}{2T^2}\left[\lambda(T_5 - T_4)^2 - (p-\lambda)(T^2 - T_5^2)\right]\end{aligned} \tag{26}$$

At $\frac{dZ(T)}{dT} = 0$,

$$\begin{aligned}\frac{c_s(p-\lambda)}{2}T^2 &= A + C\left[\left(\frac{\lambda_2}{\beta}(e^{\beta(T_4-T_3)} - 1) - \lambda_2(T_4 - T_3)\right)\right.\\
&+ \alpha\left(\frac{\lambda_2}{\beta^2}\left(e^{\beta(T_4-T_3)} - 1\right) + \frac{b_1}{2}(T_3 - T_2)^2\right.\\
&\left.\left.+ \frac{b_2}{6}(2T_3^3 - 3T_3^2T_2 + T_2^3) + \frac{b_3}{12}(3T_3^4 - 4T_3^3T_2 + T_2^4)\right)\right]\\
&+ \alpha C\left[\frac{\lambda_2}{\beta}\left((T_3 - T_2)e^{\beta(T_4-T_3)} + T_2 - T_4\right) + \frac{(p-\lambda)}{2}T_1\right.\\
&\left.+ (p-\lambda)T_1(T_2 - T_1) + \frac{1}{2}(ap-\lambda)(T_2 - T_1)^2\right]\\
&+ \frac{c_s}{2}\left(\lambda(T_5 - T_4)^2 + (p-\lambda)T_5^2\right)\\
T^2 &= \frac{2}{c_s(p-\lambda)}\left(A + CK + L\right)\end{aligned} \tag{27}$$

where

$$K = \left[\frac{\lambda_2}{\beta}\left(e^{\beta(T_4-T_3)}-1\right) - \lambda_2(T_4-T_3) + \alpha\left(\frac{\lambda_2}{\beta^2}\left(e^{\beta(T_4-T_3)}-1\right)\right.\right.$$
$$+\frac{b_1}{2}(T_3-T_2)^2 + \frac{b_2}{6}(2T_3^3 - 3T_3^2T_2 + T_2^3)$$
$$\left.\left.+\ \frac{b_3}{12}(3T_3^4 - 4T_3^3T_2 + T_2^4)\right)\right].$$

$$L = \alpha C\left[\frac{\lambda_2}{\beta}\left((T_3-T_2)e^{\beta(T_4-T_3)} + T_2 - T_4\right) + \frac{(p-\lambda)}{2}T_1\right.$$
$$\left.+(p-\lambda)T_1(T_2-T_1) + \frac{1}{2}(ap-\lambda)(T_2-T_1)^2\right]$$
$$+\frac{c_s}{2}\left(\lambda(T_5-T_4)^2 + (p-\lambda)T_5^2\right).$$

Theorem 1. *If* $(p-\lambda) > 0, 0 \le T_1 \le T_2 \le T_3 \le T_4 \le T_5 \le T, e^{\beta(T_4-T_3)} \approx 1+\beta(T_4-T_3)$ *and* $0 < \frac{\frac{\lambda_2}{\beta}(T_4-T_3)}{(ap-\lambda)(T_2-T_1)^2} < \frac{1}{2}$
Then

$$L > 0.$$

Proof. Suppose

$(p-\lambda) > 0, 0 \le T_1 \le T_2 \le T_3 \le T_4 \le T_5 \le T, e^{\beta(T_4-T_3)} \approx 1+\beta(T_4-T_3)$, and

$$then 0 < \frac{\frac{\lambda_2}{\beta}(T_4-T_3)}{(ap-\lambda)(T_2-T_1)^2} < \frac{1}{2} \tag{28}$$

$$L = \alpha C\left[\frac{\lambda_2}{\beta}\left((T_3-T_2)e^{\beta(T_4-T_3)} + T_2 - T_4\right) + \frac{(p-\lambda)}{2}T_1\right.$$
$$\left.+(p-\lambda)T_1(T_2-T_1) + \frac{1}{2}(ap-\lambda)(T_2-T_1)^2\right]$$
$$+\frac{c_s}{2}\left(\lambda(T_5-T_4)^2 + (p-\lambda)T_5^2\right) \tag{29}$$

By Linear Approximation of the Maclaurin series expansion of exponential power,

$$e^{\beta(T_4-T_3)} \approx 1+\beta(T_4-T_3) \tag{30}$$

Substitute Eq. (30) in Eq. (29) to get

$$
\begin{aligned}
L &= \alpha C\left[\frac{\lambda_2}{\beta}\left((T_3 - T_2)(1+\beta(T_4 - T_3)) + T_2 - T_4\right) + \frac{(p-\lambda)}{2}T_1 \right.\\
&\left. +(p-\lambda)T_1(T_2 - T_1) + \frac{1}{2}(ap-\lambda)(T_2 - T_1)^2\right]\\
&+\frac{c_s}{2}\left(\lambda(T_5 - T_4)^2 + (p-\lambda)T_5^2\right)\\
&= \alpha C\left[\frac{\lambda_2}{\beta}\left(\beta(T_4 - T_3) + T_3 - T_4\right) + \frac{(p-\lambda)}{2}T_1\,(1+2(T_2 - T_1)) \right.\\
&\left. +\frac{1}{2}(ap-\lambda)(T_2 - T_1)^2\right] + \frac{c_s}{2}\left(\lambda(T_5 - T_4)^2 + (p-\lambda)T_5^2\right)\\
&= \alpha C\left[\lambda_2(T_4 - T_3) - \frac{\lambda_2}{\beta}(T_4 - T_3) + \frac{(p-\lambda)}{2}T_1\,(1+2(T_2 - T_1)) \right.\\
&\left. +\frac{1}{2}(ap-\lambda)(T_2 - T_1)^2\right] + \frac{c_s}{2}\left(\lambda(T_5 - T_4)^2 + (p-\lambda)T_5^2\right)
\end{aligned} \tag{31}
$$

Substitute Eq. (28) in Eq. (31) to get

$$
\begin{aligned}
L &= \alpha C\left[\lambda_2(T_4 - T_3) + \frac{(p-\lambda)}{2}T_1\,(1+2(T_2 - T_1)) \right.\\
&\left. +\left(\frac{\frac{\lambda_2}{\beta}(T_4 - T_3)}{(ap-\lambda)(T_2 - T_1)^2}\right)(ap-\lambda)(T_2 - T_1)^2 - \frac{\lambda_2}{\beta}(T_4 - T_3)\right]\\
&+\frac{c_s}{2}\left(\lambda(T_5 - T_4)^2 + (p-\lambda)T_5^2\right)\\
&= \alpha C\left[\lambda_2(T_4 - T_3) + \frac{(p-\lambda)}{2}T_1\,(1+2(T_2 - T_1))\right]\\
&+\frac{c_s}{2}\left(\lambda(T_5 - T_4)^2 + (p-\lambda)T_5^2\right) \geq 0
\end{aligned} \tag{32}
$$

since $(p-\lambda) > 0$, and $0 \leq T_1 \leq T_2 \leq T_3 \leq T_4 \leq T_5 \leq T$ Eq. (32) holds. Hence $L > 0$

Remark 1. It follows from [1] that $K > 0$.

Lemma 1. *With the hypotheses of Theorem 1 and Remark 1, the best cycle length is*

$$T^* = \sqrt{\frac{2}{c_s(p-\lambda)}(A + CK + L)}. \tag{33}$$

Proof. The quadratic equation (27) has two solutions ($T^*_{(+)}$ and $T^*_{(-)}$)

$$T^*_{(+)} = \sqrt{\frac{2}{c_s(p-\lambda)}(A + CK + L)}$$

and

$$T^*_{(-)} = -\sqrt{\frac{2}{c_s(p-\lambda)}(A + CK + L)}$$

From Theorem 1 and Remark 1 above,

$$\frac{2}{c_s(p-\lambda)}(A + CK + L) \geq 0$$

since

$$T^*_{(-)} = -\sqrt{\frac{2}{c_s(p-\lambda)}(A + CK + L)} < 0$$

As the cycle length is always positive,

$$T^* = T^*_{(+)} \tag{34}$$

is the optimal cycle length provided that the sufficient condition is satisfied.

Lemma 2. *The total cost function $Z(T)$ is a convex function of T iff the hypotheses of Theorem 1 and Remark 1 are valid.*

Proof. Differentiating Eq. (26)

$$\begin{aligned}
\frac{d^2Z(T)}{dT^2} &= \frac{2A}{T^3} + \frac{2C}{T^3}\left[\frac{\lambda_2}{\beta}(e^{\beta(T_4-T_3)}-1) - \lambda_2(T_4-T_3)\right] \\
&+ \frac{2\alpha C}{T^3}\left[\frac{(p-\lambda)}{2}T_1 + (p-\lambda)T_1(T_2-T_1)\right. \\
&+ \frac{1}{2}(ap-\lambda)(T_2-T_1)^2 + \frac{b_1}{2}(T_3-T_2)^2 \\
&+ \frac{b_2}{6}(2T_3^3 - 3T_3^2T_2 + T_2^3) + \frac{b_3}{12}(3T_3^4 - 4T_3^3T_2 + T_2^4) \\
&+ \frac{\lambda_2}{\beta}\left(e^{\beta(T_4-T_3)}-1\right)(T_3-T_2) + \frac{\lambda_2}{\beta^2}\left(e^{\beta(T_4-T_3)}-1\right) \\
&\left.+ \frac{\lambda_2}{\beta}(T_3-T_4)\right] + \frac{c_s}{T^3}\left[\lambda(T_5-T_4)^2 + (p-\lambda)T_5^2\right] \\
&= \frac{2A}{T^3} + \frac{2C}{T^3}\left[\left(\frac{\lambda_2}{\beta}\left(e^{\beta(T_4-T_3)}-1\right) - \lambda_2(T_4-T_3)\right)\right. \\
&+ \alpha\left[\frac{\lambda_2}{\beta^2}\left(e^{\beta(T_4-T_3)}-1\right) + \frac{b_1}{2}(T_3-T_2)^2\right. \\
&\left.\left.+ \frac{b_2}{6}(2T_3^3 - 3T_3^2T_2 + T_2^3) + \frac{b_3}{12}(3T_3^4 - 4T_3^3T_2 + T_2^4)\right]\right] \\
&+ \frac{2}{T^3}\left[\alpha C\left[\frac{\lambda_2}{\beta}\left((T_3-T_2)e^{\beta(T_4-T_3)} + T_2 - T_4\right)\frac{(p-\lambda)}{2}T_1\right.\right. \\
&\left. + (p-\lambda)T_1(T_2-T_1) + \frac{(ap-\lambda)}{2}(T_2-T_1)^2\right] \\
&\left. + \frac{c_s}{2}\left(\lambda(T_5-T_4)^2 + (p-\lambda)T_5^2\right)\right] \\
&= \frac{2A}{T^3} + \frac{2CK}{T^3} + \frac{2L}{T^3} > 0 \qquad (35)
\end{aligned}$$

Since $L > 0$ and $K > 0$ by the hypothesis of Theorem 1 and Remark 1 respectively, the Eq. (35) is valid.

Hence,

$$\frac{d^2Z(T)}{dT^2} > 0. \qquad (36)$$

The sufficient condition is satisfied, then we concludes that the total variable cost $Z(T)$ is a convex function of T.

The EPQ is the sum of total demand during inventory build-up period, total demand before deterioration sent in, total demand after deteriorated items, total number of deteriorated items and total demand during shortage period.

$$
\begin{aligned}
EPQ^* &= \lambda(T_1) + \lambda(T_2 - T_1) + \lambda_1(T_3 - T_2) + \lambda_2(T_4 - T_3) + d(T_3) \\
&\quad + \lambda(T_5 - T_4) + \lambda(T - T_5) \\
&= \frac{(p-\lambda)}{2}T^2 + \frac{(ap-\lambda)}{2}(T_2 - T_1)^2 + (p-\lambda)T_1(T_2 - T_1) \\
&\quad + \int_{T_2}^{T_3} (b_1 + b_2 t + b_3 t^2)dt + (Q_3 - \lambda_2(T_4 - T_3)) \\
&\quad + \frac{1}{2}\lambda(T_5 - T_4)^2 + \frac{1}{2}(p-\lambda)(T - T_5)^2 \\
&= \frac{(p-\lambda)}{2}T^2 + \frac{(ap-\lambda)}{2}(T_2 - T_1)^2 + (p-\lambda)T_1(T_2 - T_1) \\
&\quad + b_1(T_3 - T_2) + \frac{b_2}{2}(T_3^2 - T_2^2) + \frac{b_3}{3}(T_3^3 - T_2^3) \\
&\quad + \left[\frac{\lambda_2}{\beta}(e^{\beta(T_4 - T_3)} - 1) - \lambda_2(T_4 - T_3)\right] + \frac{1}{2}\lambda(T_5 - T_4)^2 \\
&\quad + \frac{1}{2}(p-\lambda)(T - T_5)^2.
\end{aligned} \tag{37}
$$

4 Numerical Experiments

We illustrate our proposed model by performing the following numerical experiments. We use the parameter values in Table 2 adopted from [1], and add the values of parameters absent in their work (p, a, λ, c_s, T_4 and T_5).

Table 2 Experiment parameters

Param.	Value	Param.	Value
A	N3300 per production run	T_1	0.19178 years
p	6000 units per production run	T_2	0.82192 years
a	1.2	T_3	0.90411 years
λ	2000 units per unit time	T_4	0.92550 years
λ_2	2100 units per unit time	T_5	0.94210 years
α	0.3 units per unit time	b_1	8400.
β	0.2 units per unit time	b_2	4
c_s	N80 per unit	b_3	3
C	N1000 per unit		

4.1 Results

Upon Substituting the parameter values in Table 2 in to Eqs. (23), (25), (33), and (37), we obtained the values of shortages cost per production run, optimal total variable cost per cycle length, the best cycle length and EPQ values respectively. The optimal solutions obtained are shown in Table 3.

Table 3 Optimal solution

Variable	Value
Z	N150,955.7128 per cycle
T	1.283142945 years
EPQ	5,664.4253 units per cycle
S_c	N 58.8448 per cycle

Note that the values obtained for both $EPQ = 5,664.4253$ and $Z = N150,955.7128$ from our proposed model are optimum to that obtained in [1]. As such, our proposed model improves upon that proposed in [1].

4.2 Sensitivity Analysis

We perform sensitivity analysis on the solution obtained to the effect of making changes in some model parameters over the obtained solutions.

i. Figure 2 shows the effect of change in set-up cost A on the best cycle length T, total variable cost Z and EPQ. As the set-up cost A increases, the best cycle length T^* , the total variable cost per unit time Z^* and EPQ increase. Thus, any change in set-up cost affects cycle length, the total variable cost and EPQ. This means that a minimum set-up cost will minimise the total variable cost of the inventory system.
ii. Figure 3 shows the effect of change in production rate p on the best cycle length T, total variable cost Z and EPQ. As the production rate p increases, the best cycle length T^* decreases while the total variable cost Z^* and EPQ increases and the reverse scenario is the case when the production rate decreases. It is clear that any adjustment of production rate has an impact on the best cycle length, the total variable cost and EPQ.
iii. Figure 4 shows the effect of change in deterioration rate β on the best cycle length T, total variable cost Z and EPQ. As the deterioration rate β increases, the total variable cost Z^* increases while both the best cycle length T^* and EPQ decrease. Likewise, the situation reverses when β decreases. This means that the total variable cost can be minimised by reducing the rate of deterioration.

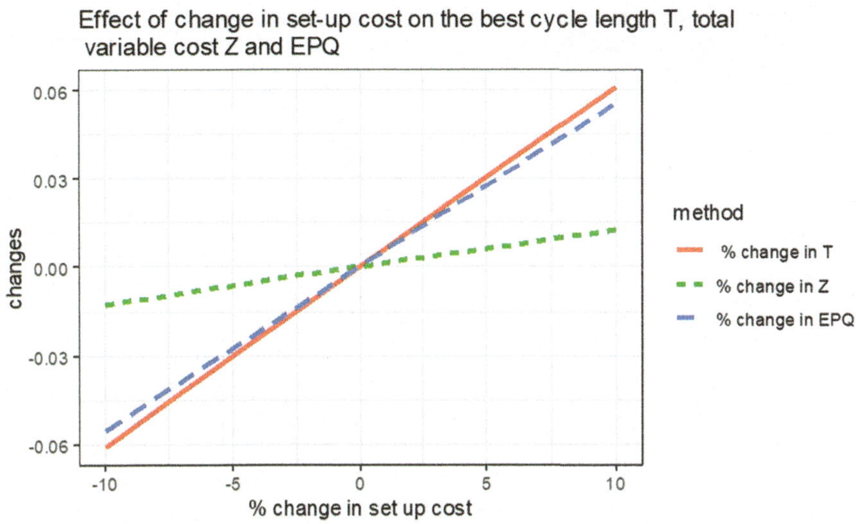

Fig. 2 Effect of change in set-up cost *A* on the optimal solutions

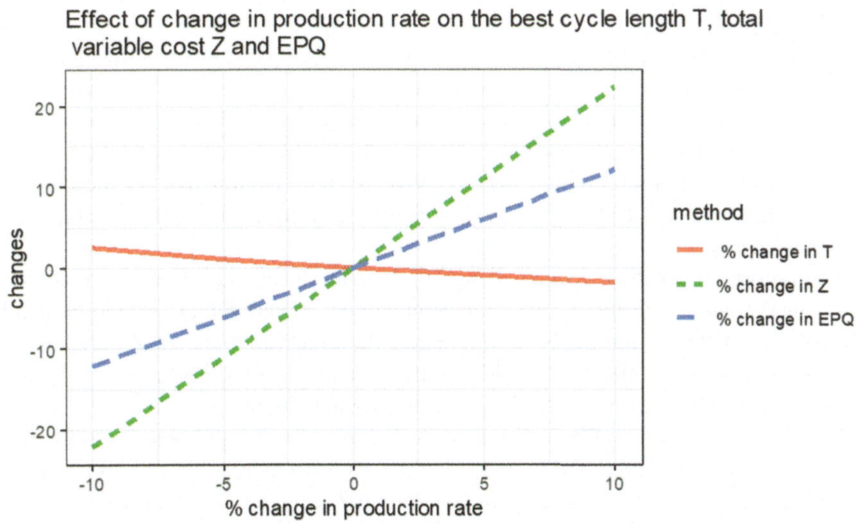

Fig. 3 Effect of change in production rate *p* on the optimal solutions

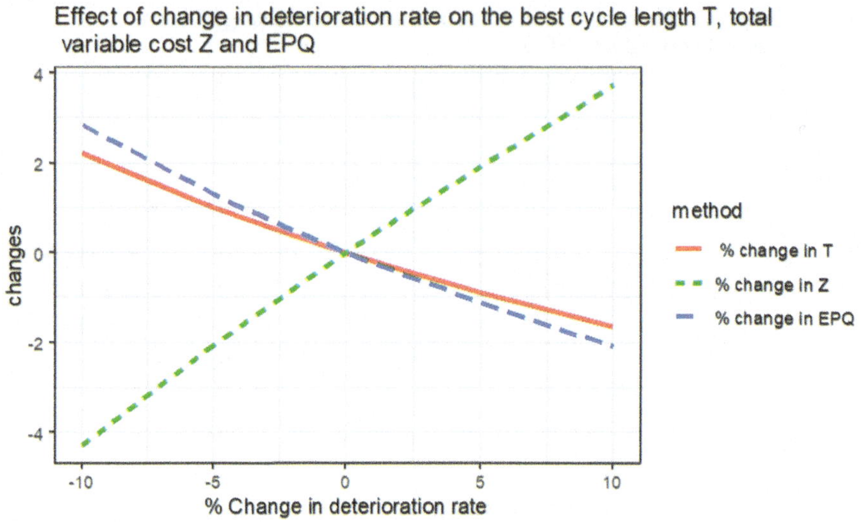

Fig. 4 Effect of change in deterioration rate β on the optimal solutions

iv. Figure 5 shows the effect of change in demand during inventory build-up period λ on the best cycle length T, total variable cost Z and EPQ. As the demand during inventory build-up period λ increases, the best cycle length T^* increases while both the total variable cost Z^* and EPQ decreases and vice versa when the situation change negatively on λ. It is natural that an increase in demand during production has an impact on cycle length, total variable cost and EPQ.

v. Figure 6 shows the effect of change in demand during deterioration period λ_2 on the best cycle length T, total variable cost Z and EPQ. As the demand during deterioration period λ_2 increases, the total variable cost Z^* drastically decreases while the best cycle length T^* and EPQ slightly increases and slightly decreases respectively when the demand decreases.

vi. Figure 7 shows the effect of change in inventory carrying charge α on the best cycle length T, total variable cost Z and EPQ. An increase in inventory carrying charge α increases the best cycle length T^*, the total variable cost Z^* and EPQ. This is expected as the total variable cost always increases if carrying charge increases. Hence, the model needs long cycle length and EPQ in order to reduce the total variable cost.

vii. Figure 8 shows the effect of change in shortage cost of the unit item c_s on the best cycle length T, total variable cost Z and EPQ. As the shortage cost of the unit item c_s increases, the total variable cost Z^* increases, while the best cycle length T^* and EPQ decrease. Also, the case reverses when the shortage cost of unit item decreases. Therefore, to reduce the variable cost, the model needs long cycle length and increase in EPQ.

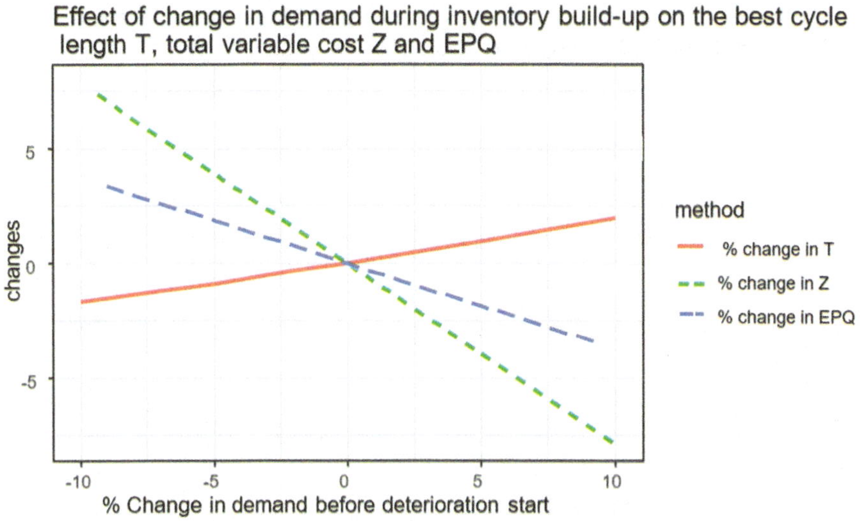

Fig. 5 Effect of change in demand during inventory build-up period λ on the optimal solutions

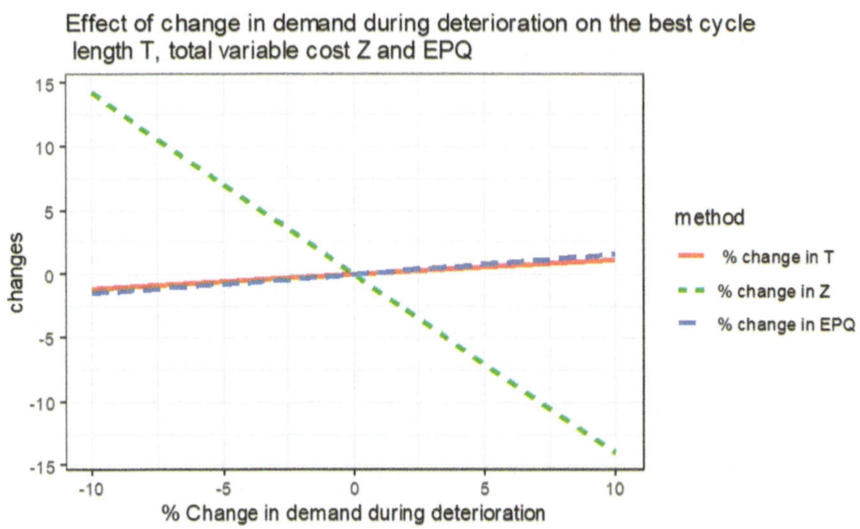

Fig. 6 Effect of change in demand during deterioration period λ_2 on the optimal solutions

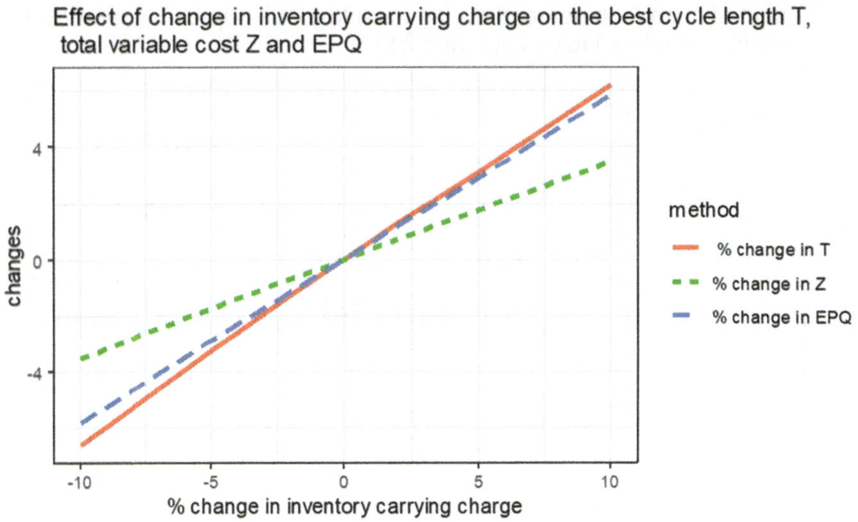

Fig. 7 Effect of change in inventory carrying charge α on the optimal solutions

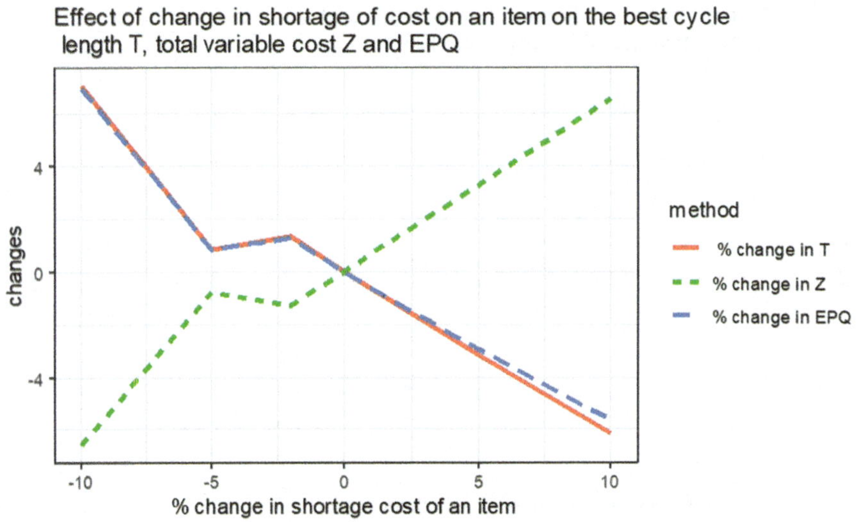

Fig. 8 Effect of change in shortage cost of the unit item c_s on the optimal solutions

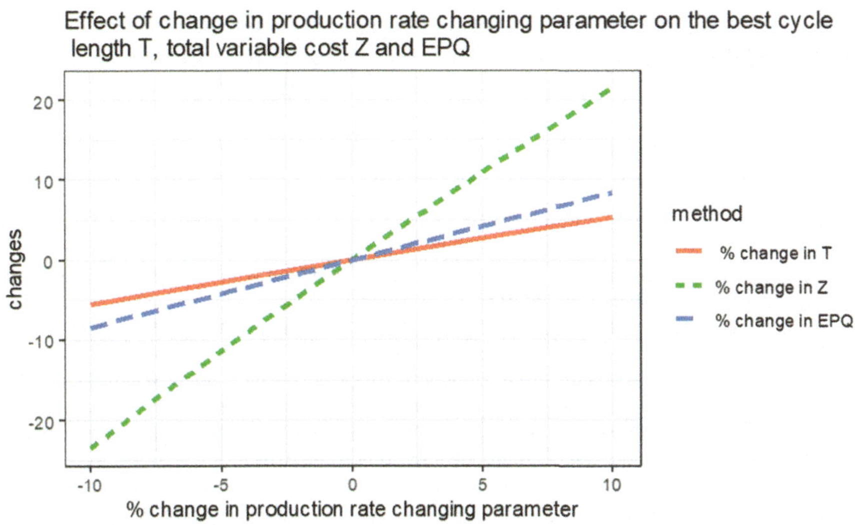

Fig. 9 Effect of change in production rate changing parameter a on the optimal solutions

viii. Figure 9 shows the effect of change in production rate changing parameter a on the best cycle length T, total variable cost Z and EPQ. As the production rate changing parameter a increases, the best cycle length T^*, the total variable cost Z^* and EPQ increase and vice versa. Based on the assumption on a, production rate increases due to the increase in a, consequently EPQ and cycle length increase.

5 Conclusion

In this paper, a practical production inventory model is developed in which conditional production rate is implemented in a model adapted from the study [1]. The present work deals with the problem of determining the optimal cycle length that minimises the total variable cost and maximum EPQ for maximum profit of an imperfect production model for delayed deteriorating items in which the inspection processes are not perfect. The system considered here allows shortages that are fully backlogged. A theorem and two lemmas are presented to ensure the validity of our assumptions and the sufficient condition for optimality. A numerical example is presented for better illustration of the obtained theoretical results. Besides solving the problem optimally, deep insights are also provided through sensitivity analysis of the various parameters on the cycle length, total variable cost function and economic production quantity. The approach in the present work can be further extended to include even more complicated scenarios in inventory management by implementing same model with exponential deterioration rate and linearly decreasing demand during deterioration period or considering similar model of two or more variables.

References

1. Dari, S., Sani, B.: An EPQ model for delayed deteriorating items with quadratic demand and shortages. Asian J. Math. Comput. Res. **22**(2), 87–103 (2017)
2. Dari, S., Sani, B.: An EPQ model for delayed deteriorating items with quadratic demand and linear holding cost. Opsearch **57**(1), 46–72 (2020)
3. Oh, C.G., Park, J.: Insights for sustainability of smartphone business: understanding customer switching behavior in smartphone services. Sustainability **12**(3), 1082, 1–15 (2020)
4. Venkateswarlu, R., Mohan, R.: An inventory model with quadratic demand, constant deterioration and salvage value. Res. J. Math. Stat. Sci. **2**(1), 1–5 (2014)
5. Chowdhury, R.R., Ghosh, S.K., Chaudhuri, K.S.: An order-level inventory model for a deteriorating item with time-quadratic demand and time-dependent partial backlogging with shortages in all cycles. Am. J. Math. Manag. Sci. **33**(2), 75–97 (2014)
6. Khanra, S., Ghosh, S.K., Chaudhuri, K.S.: An EOQ model for a deteriorating item with time dependent quadratic demand under permissible delay in payment. Appl. Math. Comput. **218**(1), 1–9 (2011)
7. Khanra, S., Mandal, B., Sarkar, B.: An inventory model with time dependent demand and shortages under trade credit policy. Econ. Model. **35**(1), 349–355 (2013)
8. Sarkar, T., Ghosh, S.K., Chaudhuri, K.: An optimal inventory replenishment policy for a deteriorating item with time-quadratic demand and time-dependent partial backlogging with shortages in all cycles. Appl. Math. Comput. **218**(18), 9147–9155 (2012)
9. Singh, T., Pattnayak, H.: An EOQ model for a deteriorating item with time dependent quadratic demand and variable deterioration under permissible delay in payment. Appl. Math. Sci. **7**(59), 2939–2951 (2013)
10. Rani, S., Ali, R., Agarwal, A.: Inventory model for deteriorating items in green supply chain with credit period dependent demand. Int. J. Appl. Eng. Res. **15**(2), 157–172 (2020)
11. Dash, B., Pattnaik, M., Pattnaik, H.: Deteriorated economic production quantity (EPQ) model for declined quadratic demand with time value of money and shortages. Appl. Math. Sci. **8**(73), 3607–3618 (2014)
12. Hui-Ling, Y.: Two-warehouse partial backlogging inventory models with three-parameter Weibull distribution deterioration under inflation. Int. J. Prod. Econ. **138**(1), 107–116 (2012)
13. Kumar, R.: A deterministic inventory model with demand & holding cost for Weibull deteriorating items. Int. J. Acad. Res. Dev. **2**(5), 44–48 (2017)
14. Partha, G., Manas, K.M., Manoranjan, M.: Production-inventory models for a damageable item with variable demands and inventory costs in an imperfect production process. Int. J. Prod. Econ. **144**(1), 180–188 (2013)
15. Sunil, T., Leopoldo, E.C., Mark, G., Ali, A.S.: Joint pricing and inventory model for deteriorating items with expiration dates and partial backlogging under two-level partial trade credits in supply chain. Int. J. Prod. Econ. **200**(1), 16–36 (2018)
16. Zaid, T.B.: Optimal economic ordering policy with deteriorating items under different supplier trade credits for finite horizon case. Int. J. Prod. Econ. **133**(1), 216–223 (2011)
17. Rout, C., Kumar, R.S., Chakraborty, D., Goswami, A.: An EPQ model for deteriorating items with imperfect production, inspection errors, rework and shortages: a type-2 fuzzy approach. Opsearch **56**(3), 657–688 (2019)
18. Rahaman, M., Mondal, S.P., Shaikh, A.A., Ahmadian, A., Senu, N., Salahshour, S.: Arbitrary-order economic production quantity model with and without deterioration: generalized point of view. Adv. Difference Equ. **16**(1), 1–30 (2020)

19. Majumder, P., Bera, U.K., Maiti, M.: A multi-item EPQ model with imperfect production process for time varying demand with shortages. Hacet. J. Math. Stat. **45**(3), 929–955 (2016)
20. Agarwal, N.: Economic production quantity (EPQ) model with time-dependent demand and reduction delivery policy. Int. J. Multidiscip. Approach Stud. **2**(3), 48–53 (2015)
21. Anitha, P., Parvathi, P.: An EPQ model for deteriorating items with time dependent demand with reliability and flexibility in a Fuzzy environment. Int. J. Innov. Res. Comput. Commun. Eng. **4**(2), 2037–2047 (2016)
22. Viji, G., Karthikeyan, K.: An economic production quantity model for three levels of production with Weibull distribution deterioration and shortage. Ain Shams Eng. J. **9**(4), 1481–1487 (2018)
23. Khurana, D., Tayal, S., Singh, S.R.: An EPQ model for deteriorating items with variable demand rate and allowable shortages. IJMOR **12**(1), 117–128 (2018)
24. Bakator, M., Ćoćkalo, D., Nikolić, M.: A model for manufacturing optimization and achieving higher productivity in small and medium-sized enterprises. In: Proceedings of VIII International Conference Industrial Engineering and Environmental Protection, Zrenjanin, Serbia, pp. 45–52 (2018)
25. Chuka, C.E., Oguejiofor, N.J., Sunday, A.C.: Evaluation and optimization of inventory control systems in small and medium scale industries. Int. J. Mod. Stud. Mech. Eng. **2**(1), 1–13 (2016)
26. Igbani, S., Okoli, I., Okpu, O.A.: Investigating the economic impact of militant activities on crude oil production in Nigeria. J. Multidiscip. Eng. Sci. Stud. **3**(2), 1317–1322 (2017)
27. Akça, H., Ela, M.: Economic, financial and fiscal effects of terrorism: a literature review. Int. Rev. Manag. Bus. Res. **6**(3), 1051–1071 (2017)
28. Czinkota, M.R., Knight, G., Liesch, P.W., Steen, J.: Terrorism and international business: a research agenda. J. Int. Bus. Stud. **41**(5), 826–843 (2010)
29. Carlsson-Szlezak, P., Reeves, M., Swartz, P.: What coronavirus could mean for the global economy. Harvard Bus. Rev. **3**, 1–10 (2020)

Binary Decision for Discretionary Lane Changing by Time-Effects Factors

Md. Mijanoor Rahman, Mohd. Tahir Ismail, Ahmad Farhan Mohd Sadullah, Noor Saifurina Nana Khurizan, and Majid Khan Majahar Ali

Abstract The vicinity of the on-ramp and off-ramp in the freeway road may experience high collisions in time-varying situations. The binary decision (lane changing and lane keeping) is required in the Discretionary Lane Changing (DLC) by analysing the time-impact factors. In this research, the time-effect factors are used for the binary decision in the DLC, and the cumulative probability distribution functions of these factors are used to demonstrate the influence on the lane changing. The odd ratios of the binary logit model also show that the probability increases or decreases by increasing time of some factors. The accuracy of the decision-making model is 84% by using the Next Generation SIMulation (NGSIM) data. This model shall apply to the decision of the DLC for the next generation of autonomous vehicles on the freeway road.

Keywords Binary-decision · Logit-model · Odd-ratio · Time-effects factor · Next generation simulation

1 Introduction

Lane changing (LC) is a very challenging and risky movement for the vehicle driver, regardless of the human driver or the autonomous vehicle self driver. In order to overcome these problems, the intended DLC driver requires more time to make a binary decision (LC and lane keeping). This binary decision depends on many factors

Md. M. Rahman · Mohd. T. Ismail · N. S. N. Khurizan · M. K. M. Ali (✉)
School of Mathematical Sciences, Universiti Sains Malaysia, George Town, Penang, Malaysia
e-mail: majidkhanmajaharali@usm.my

Md. M. Rahman
e-mail: mijanoor.math@student.usm.my

Md. M. Rahman
Department of Mathematics, Mawlana Bhashani Science and Technology University, Santosh, Tangail 1902, Bangladesh

A. F. M. Sadullah
School of Civil Engineering, Universiti Sains Malaysia, George Town, Penang, Malaysia

M. H. Mohd et al. (eds.), *Modelling, Simulation and Applications of Complex Systems*, Springer Proceedings in Mathematics & Statistics 359,
https://doi.org/10.1007/978-981-16-2629-6_21

that come from microscopic or macroscopic trajectories. When the driver leaves from main-lane to off-ramp, or merges from on-ramp to main-lane, it is called mandatory LC. Due to the forceful and compulsory nature of driving, the DLC driver faces a number of obstructive situations, and the greater disruption is caused in the vicinity of the on-ramp in the congested traffic network [1]. Again, off-ramp vehicle movements may also increase the density of vehicles on the main lane due to the driver's lack of awareness of the times and positions on off-ramp site.

A research study investigated that the on-ramp (merging) vehicle could create a serious problem for the main lane driver [2]. In the previous research studies, the logistic regression model was very commonly used to make the DLC decision by using microscopic and macroscopic factors whereas time was a microscopic factor. In addition, other microscopic factors, such as velocity, acceleration, etc. change values at different times. Only the length of the vehicle (vehicle class) is a macroscopic factor that is the same value for any vehicle. All factors were combined with an equal time interval. The decision of the LC drivers is therefore more influenced by certain time-effects (changing by time) factors. The stop-and-go weaving traffic would result in its acceleration and deceleration trajectories according to [3, 4].

The time-effect analysis to the time-dependent binary logistic regression model was proposed for probabilistic decision-making in [5]. They used microscopic factors such as Subject Vehicle (SV) speed, Rear Vehicle (RV) speed and Front Vehicle (FV) speed on the target lane, the gap between the SV and FV on the current lane, the gap between the SV and FV on the target lane, the FV and the RV types affecting the vehicle merging decisions. However, their model is used for the mandatory LC. In addition, [6] suggested that decisions should be taken by the off-ramp merging driver by analysing microscopic factors as unlike to time-effect analysis. In the literature, however, a few studies have been conducted to discover controlled ramps merging or leaving situations that create a hazard condition for DLC drivers. As a result, there is still a gap in the analysis of time-dependent binary logistic regression with time-varying effects in DLC decision near on-ramp and off-ramp.

2 Review of Related Literature

Most of the LC models were obtained by means of two different forms, such as LC decision and execution. The rule-based LC decision model was invented in [7] for urban roads using some factors such as traffic signals, traffic density and vehicle obstruction. Some researchers then developed and used a rule-based model in some traffic simulation software. The random utility-based LC models was identified in [8], where the model considered three sequential situations, such as LC-decision, LC-choice and LC-execution. In [9], the discretionary-based LC was modeled for congested traffic driver.

An integrated discrete choice and parameter estimation of LC models was established in [10]. In the case of congested traffic, cooperation-based LC movement between SV and RV on the target lane was developed to overcome the risk situation. Based on driver characteristics such as aggressive or non-aggressive behaviour, A cooperative LC model concept was proposed in [11]. The cooperative LC model concept was also developed by using minimising overall braking induced LC, see in [12]. Recently, some research studies have developed the risk-trajectory of LC models assessment using the fault tree analysis [13], the partial binary tree structure classifier [14] and the artificial neural networks [15].

LC models are categorised by the LC decision groups and traffic impact [16]. A binary decision-based DLC model was proposed by using the gap and speed factors in [17]. That model is used in TRANSMODELER software for LC prediction where NGSIM dataset was employed to testing the model accuracy. Their main research factors are SV speed, front gaps at left-lane and right-lane and off-ramp distance. They included a time factor in this model, although time-effect factors are avoided. They suggested that their model could be applied to an autonomous vehicle system. The probabilistic merging behaviour without a time-dependent binary logistic model was explored by using factors [18]. That model used factors such as the remaining distance at the merging point, the merging-vehicle speed, the RV speed and the FV speed at the target lane, the gap between the SV and the FV at the current lane, and the gap between the SV and FV at the target lane.

The LC probability model was proposed in [19], based on relative velocity between FV and SV, and a gap between FV and SV. From Fig. 1, the front-gap, $SF_g = FV_p - SV_p - l_{SV}$, where l_{SV} is length of SV. The other surrounding gaps are SR_g, TR_g and TF_g shown in this figure. The average speed of the NGSIM data set was 38.16 (f/s), and the traffic characteristics in the NGSIM data set were almost homogeneous structure, where vehicle flows were smooth most of the time. The average front gap in the current lane is 30 ft, and the average front gap in the target lane is 50 ft [19].

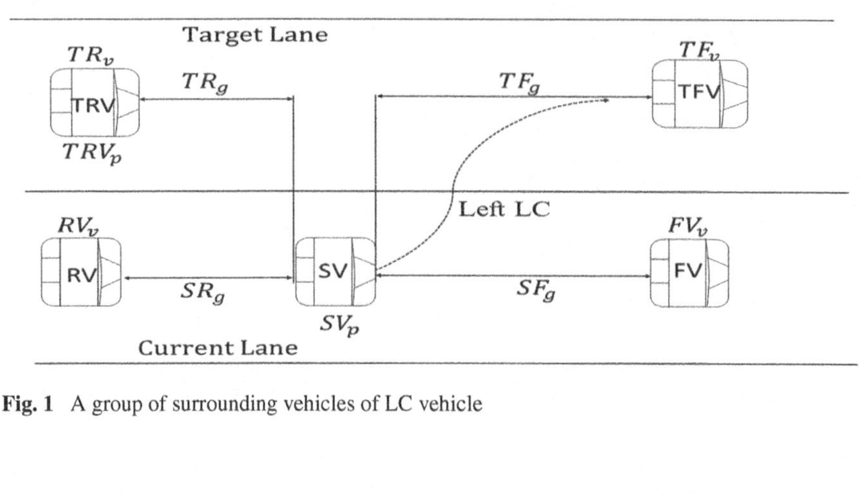

Fig. 1 A group of surrounding vehicles of LC vehicle

A new time-dependent logistic regression model was explored for merging vehicles using time-effect factors in [5], and added a number of other LC factors, such as driver-behaviour, SV-speed, FV-speed and RV-speed at the target lane, front-gap at the target lane and current lane, FV-types and RV-types at the target lane at this model. However, the rear-gap of the target lane and the acceleration of the LC vehicle were not shown.

LC binary decision for freeway off-ramp merging behaviour was explored by using logistic-regression-model and likelihood-function (see in [6]). They used the microscopic dynamical factors which are relative-gaps and relative-speeds, although they did not include time-factor and time-effect factors. The prediction accuracy of any model may show the importance of applications. Two probabilities distributions (speed and density differences between adjacent lanes) as significant factors of the binary logistic regression model was suggested for DLC drivers in the congested traffic area [20]. These factors are only macroscopic data taken from a microscopic NGSIM data set. The binary logistic model to accept or reject the gaps on the neighboring lanes was employed in [21]. The microscopic exploration factors used were very important for decision-making without a time interval.

In [22], the vehicle driver in the waved traffic area either accepts a gap or rejects a gap that is a binary decision for LC. The binary logit decision model was adopted for LC merging behavior from on-ramp using microscopic factors [23]. In [24], an another binary logit model was used for an LC merging vehicle in a congested area of traffic using a microscopic factor such as gap and macroscopic factors such as density and flow rate. Some of the articles mentioned above used a logistic regression model without time-effect factors for LC and lane-merging decisions [20–24]. The binary logistic model was analysed by using the deep learning method for LC decision-making in [25]. The decision on LC was categorised by two steps: the first was the decision on LC and the second was the implementation of LC. In the LC decision, the binary logistic regression model is used to accept and reject gaps by analysing certain microscopic factors such as speed and gap of the surrounding vehicles and speed of the LC vehicle, but time-effect factors have not been included in the logistic regression model. This research work involves thirteen explanatory time-depended factors in the logistic model for LC and lane keeping binary decisions.

3 Notation and Assumptions

Table 1 shows the notation used in this paper.

Table 1 Variable name and description

variable	Description
y_t^j	The decision variables at time t, 0 or 1
SV_v	The speed of SV at time t, (f/s)
FV_v	The speed of FV in the current lane at time t, (f/s)
RV_v	The speed of RV in the current lane at time t, (f/s)
TF_v	The speed of the FV in the target lane at time t, (f/s)
TR_v	The speed of the RV in the target lane at time t, (f/s)
SF_g	The longitudinal gap between the SV and FV in the current lane at time t, (f)
SR_g	The longitudinal gap between the SV and RV in the current lane at time t, (f)
TR_g	The longitudinal gap between the FV and RV in the target lane at time t, (f)
SV_p	The position of SV at time t, (f)
TR_p	The position of RV in target lane at time t, (f)
SV_a	The acceleration of SV at time t, (f/s^2)
TR_a	The acceleration of RV in target lane at time t, (f/s^2)
SV_{cl}	The length (class) of SV (f)

4 Research Approach

In order to avoid crashes on the main lane, the driver sometimes passes through a fixed location called an auxiliary lane. In some cases, the DLC process occurs: (a) changing the desired trajectory; (b) passing the possible jam situations; (c) granting permission to merge or leave the vehicles. The intended driver of the DLC takes the decision by calculating certain factors. This study analyses the decision factors for this DLC action. These are microscopic and macroscopic factors. Microscopic factors observed are more significant for DLC, as time-varying effects are included in these factors.

Figure 2 shows the LC vehicle, entering vehicle and leaving vehicle in motion. The SV counting times before LC are t_i, where, $i = 1, 2, 3, ...$ After starting LC, the counting times are lct_j, where, $j = 1, 2, 3, ...$ When the vehicle completely changes the lane, the time-counts are $t_i + lct_j$ from the time t_i and LC-time lct_j. In this work, it is considered that the drivers LC decision from main-lane-5 to main-lane-4 process at every point measured by time. The driver can take the decision ten times every second. Every factor, therefore, depends on ten times a second named deci-second. At any time, t, the driver of the LC vehicle may make binary decisions: (i) start the LC action from lane-5 to lane-4 (ii) continue driving on lane-5. This model consists of explanatory variables. These variables named time-dependent explanatory factors can change the decision through repeated observations. These time-effects factors give rise to a different LC decision from previous literature.

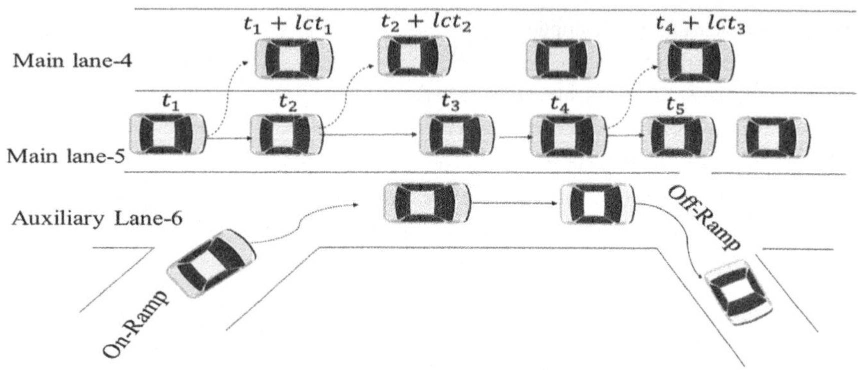

Fig. 2 LC behavior flowchart

4.1 Binary Logistic Model

The binary logistic model is used to investigate the relationship between explanatory and binary response variables. This model depends on the odds ratio of the probability of success and failure. The probability of success at the time t is the binary response value 1 (after crossing the middle line of two lanes for LC action) and the probability of failure at any time t is to continue driving at the main road as the binary value 0. The given explanatory factors are used to determine the probability of the decision of the response variable. Explanatory variables for the logistic regression model are vehicle speed, FV and RV gap, and travel time during LC. The following equation of binary logit model can be described by the probability of LC driver behaviours that are time-dependent decisions. Let y_t^j denotes the decision of the LC driver j at time t. The probability of lane merging starting without time-effects variables is expressed by the logit model:

$$P_i(y_t^j = 1) = 1/(1 + e^{\varphi_0 + \varphi_k x_{jk}}); \quad j = 1, 2, 3, 4, \ldots.N; k = 1, 2, 3, 4, \ldots.K. \quad (1)$$

The probability of reject LC without time variable is expressed by the same model:

$$P_i(y_t^j = 0) = e^{\varphi_0 + \varphi_k x_{jk}}/(1 + e^{\varphi_0 + \varphi_k x_{jk}}); \quad j = 1, 2, 3, 4, \ldots.N; k = 1, 2, 3, 4, \ldots.K. \quad (2)$$

The driver's binary decision may change at any time, when time is variable, then the probability of start LC is expressed by the logit model shown in Eqs. (3) and (4), namely,

$$P_i(y_t^j = 1) = 1/(1 + e^{\varphi_0+\varphi_k x_{jk}+\varphi_{k+1}t});$$
$$j = 1, 2, 3, 4, \ldots.N;\ k = 1, 2, 3, 4, \ldots.K. \quad (3)$$

Again, when, time is variable, then the probability of reject LC is expressed by the same model:

$$P_i(y_t^j = 0) = e^{\varphi_0+\varphi_k x_{jk}+\varphi_{k+1}t}/(1 + e^{\varphi_0+\varphi_k x_{jk}+\varphi_{k+1}t});$$
$$j = 1, 2, 3, 4, \ldots.N;\ k = 1, 2, 3, 4, \ldots.K. \quad (4)$$

and where, φ_k $(k = 1, 2, 3, ...K)$ are the coefficient of $K - th$ influencing factors and x_{jk} is the influencing factors. And $P_i\left(y_t^j = 1\right)$ is the probability that leaving vehicle j starts the action for LC to off-ramp at time t. The previous time can affect the decision choice of the j−th leaving vehicle.

The important point is that the extra variable to the current logit model, the time factor shown in the Eq. (3) and the Eq. (4) could exceed the overall time-adjustment decision of the response with other influence factors. The time-dependent model cannot explain precisely whether the time can be insignificant, or whether the model can be over-fitted to include the time variable to influence factors. This model, therefore, is unpredictable with reality.

4.2 Time-Varying Effect on Logistic Regression Model

The logistic regression model uses time-dependent explanatory variables to classify time-varying effects on the response variable shown in Eqs. (3) and (4). The important point is that the value of the response variable depends on the time interval of the relevant factors. If we use time variables such as Eqs. (3) and (4), the time variable coefficient depends on the previous times. Then, the logistic model is over-fitted to the decision factor, and since the current decision depends on the other variables (current values) and time variable (previous fixed value), it must be unrealistic to make a decision using this model. In order to avoid over-fitting and include impractical coefficients, this work uses another method, the time-dependent logistic regression model, where the linear function of the time-effects of the explanatory variables is presented. This model can therefore be expressed by

$$P_i(y_t^j = 1) = 1/(1 + e^{(\varphi_0+\varphi_{01}t)+(\varphi_k+\varphi_{k1}t)x_{jk}});$$
$$j = 1, 2, 3, 4, \ldots.N;\ k = 1, 2, 3, 4, \ldots.K. \quad (5)$$

The probability of reject LC without time variable is expressed by the same model:

$$P_i(y_t^j = 0) = e^{(\varphi_0+\varphi_{01}t)+(\varphi_k+\varphi_{k1}t)x_{jk}}/(1 + e^{(\varphi_0+\varphi_{01}t)+(\varphi_k+\varphi_{k1}t)x_{jk}});$$
$$j = 1, 2, 3, 4, \ldots.N;\ k = 1, 2, 3, 4, \ldots.K. \quad (6)$$

where, N is the number of observations, and $(\varphi_k + \varphi_{k1}t)$; $(k = 1, 2, ..., K)$ are the time effects adjusted coefficients of the explanatory variables. If the influenced factor is not time-effect then the value, $\varphi_{k1} = 0$. The parameters, φ_k and φ_{k1} are determined by using the maximum log likelihood technique.

4.3 Maximum Log-Likelihood

The SV driver, j should either choose "$y_t^j = 1$" or "$y_t^j = 0$" at any time t, for DLC. Once the SV, j chooses "$y_t^j = 1$", that is, the completion of the LC maneuver (i.e., fully entering the left lane), the decision of the LC driver ends at the time $t + 1$ and then "$y_t^j = 1$" at the next time interval. For LC vehicle, the observation total time interval is $t = 1$ to $t = T_i$. The SV is supposed to complete the LC maneuver at the t−th time. Then, the SV selects the choice of "$y_t^j = 0$" during the time interval 1 to $(t - 1)$, and then "$y_t^j = 1$" in next time interval (i.e., in the target lane) in the time interval $t = 0$ to $t = T_i$, namely,

$$P(y_{t'}^j = 1 \mid y_t^j = 1, \forall t' = t+1, t+2, ..., T_i, y_{t''}^j = 0, \forall t" = 1, 2, ..., t, \sigma_n) = 1$$

The LC vehicle j has choice sequence expressed by the following statements: the first $(t - 1)th$ time decisions are zero, tth time decision is one, and $(T_i - t)th$ time decisions are one.

The probability of any driver's LC decision depends on all observation of a particular decision of sequence, and the product of probabilities is expressed as the product of all choice in the sequence. The probability of a particular decision of the LC driver is as follows: $P_j \left\{choice\ sequence_t \middle| \sigma_n\right\} =$

$$\begin{aligned}
&P(y_{T_i}^j = 1 \mid y_t^j = 1, \forall t' = t+1, t+2, ..., T_i, y_{t''}^j = 0, \forall t" = 1, 2, ..., t, \sigma_n) \times \\
&P(y_{T_i-1}^j = 1 \mid y_t^j = 1, \forall t' = t+1, t+2, ..., T_i, y_{t''}^j = 0, \forall t" = 1, 2, ..., t, \sigma_n) \times \\
&...P(y_t^j = 1 \mid y_{t''}^j = 0, \forall t" = 1, 2, ..., t, \sigma_n) \times \\
&...P(y_2^j = 1 \mid y_1^j = 0, \sigma_n). \\
&= 1 \times 1 \times 1 \times 1 \times 1 \times P(y_t^j = 1 \mid y_{t''}^j = 0, \forall t" = 1, 2, ..., t, \sigma_n) \times \\
&\prod_{t'=1}^{t} P(y_{t'}^j = 0 \mid y_{t''}^j = 0, \forall t" = 1, 2, ..., t' - 1, \sigma_n) \\
&= P(y_t^j = 1 \mid y_{t''}^j = 0, \forall t" = 1, 2, ..., t, \sigma_n) \times \\
&\prod_{t'=1}^{t} P(y_{t'}^j = 0 \mid y_{t''}^j = 0, \forall t" = 1, 2, ..., t' - 1, \sigma_n)
\end{aligned}$$

The LC maneuver of SV can be expressed by the probability of the choice sequence, and the likelihood function for the LC vehicle is the sum of the all probabilities of all sequences. So, the likelihood is given by

$$LK_j = \sum_{t=1}^{T_i} P_j\{choice\ sequence_t\}$$
$$= P_j\{choice\ sequence_1\} + P_j\{choice\ sequence_2\} + P_j\{choice\ sequence_3\} + \ldots$$
$$+ P_j\{choice\ sequence_{T_i}\}$$

The log-likelihood function of all LC driver is

$$LLK = \sum_{j=1}^{K} \log(LK_j) = \sum_{j=1}^{K} \log \sum_{t=1}^{T_i} P_j\{choice\ sequence_t\}$$

where, K is the total number of LC vehicles.

5 Data Collection and Processing

The US 101 (Hollywood Freeway), Los Angeles, California has a multi-lane road on both sides as shown in Fig. 3. On one side, the off-ramp and on-ramp traffic system is connected to the auxiliary lane and the auxiliary lane number at 6. Without this lane, the other five lanes are used for highway vehicles, and the DLC scenarios occur in those five lanes. Lane seven and lane eight are used for both on-ramp and off-ramp. From on-ramp, the driver enters lane five using lane six and from off-ramp, the driver exits lane five using lane six. When the driver goes from lane five to lane four, the LC is called the DLC.

For NGSIM data, the eight cameras are arranged at the top of the high-raised buildings, and the camera captured the vehicle trajectory by microscopy of all lanes at a distance of more than six hundred meters road. These microscopic factors such as vehicle plate number, global longitudinal and lateral positions, local longitudinal position, global time, vehicle velocity and acceleration, vehicle lane position, FV and RV numbers whereas, vehicle length is a macroscopic factor. These are all factors were collected by using these cameras at 7.50 am to 8.05 am in the morning of 15 June 2005 [19, 26]. The time-intervals indicate the congestion of the road at that time. This time is more congested than the other time. Each second, the vehicle trajectory has ten data of all microscopic factors.

This open source data adopted more than one million rows and eighteen columns, focusing on 2169 vehicle trajectories in 15 min. Moreover, it is a big dataset wherein the adopted trajectories are microscopic. This data captures a total of 103 DLC scenarios from lane five to lane four at this time. Influenced factors are front gap and rear gap at the current lane, front gap and rear gap at the target lane, SV velocity,

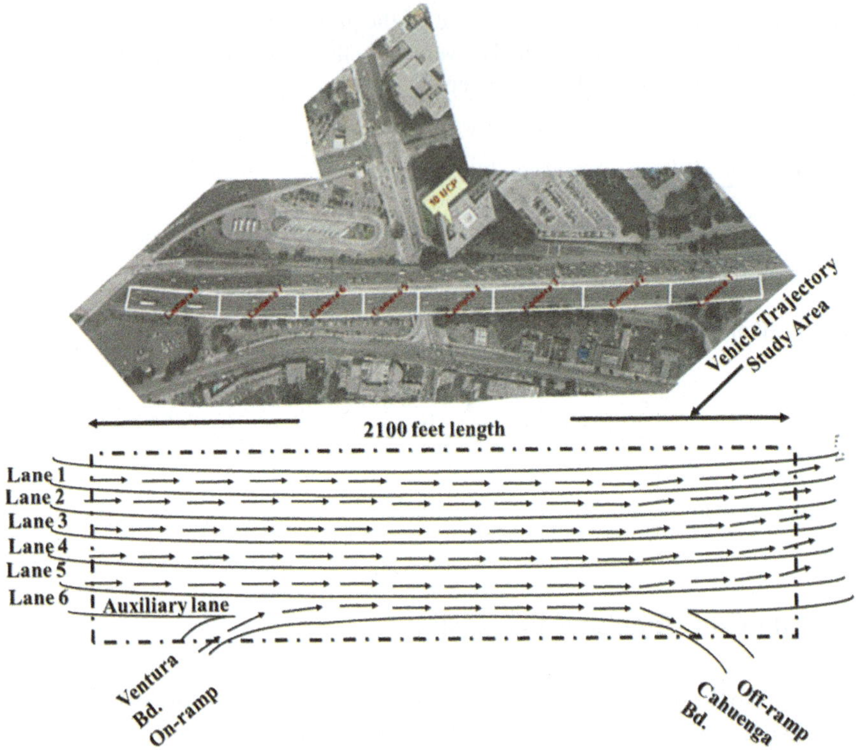

Fig. 3 The data collection area, US 101 (Hollywood Freeway), Los Angeles, California [26] and [19].

FV velocity at the current lane and RV velocity at the target lane. These factors are separated by the time-dependent influenced factors respective current decision.

The LC decision of the driver from lane five to lane four was made before the LC action, and the separate data of the driver's decision depends on the trajectory for all vehicles on lane five and this time starts from four seconds before the vehicle's middle-line crossing time. In four seconds, the driver wants to make LC and lane-continuing binary decisions using LC current trajectories. So, the decision of the driver depends on time. In this work, the maximum likelihood is applied to the decision of the LC driver by using the time-dependent trajectory data for all LC vehicles from lane five to lane four in those 15 min. Since NGSIM data set completely adopts many microscopic information on vehicle trajectories, very recent studies such as [27–33] have used this data set for driver behaviour analysis. Again, this data set is the most popular and open source data in traffic research, because it is a large data set that includes many microscopic information [34]. Other data from the same source (NGSIM) data that I-80 east-bound (north-bound), Emeryville, California data are also frequently used in current research.

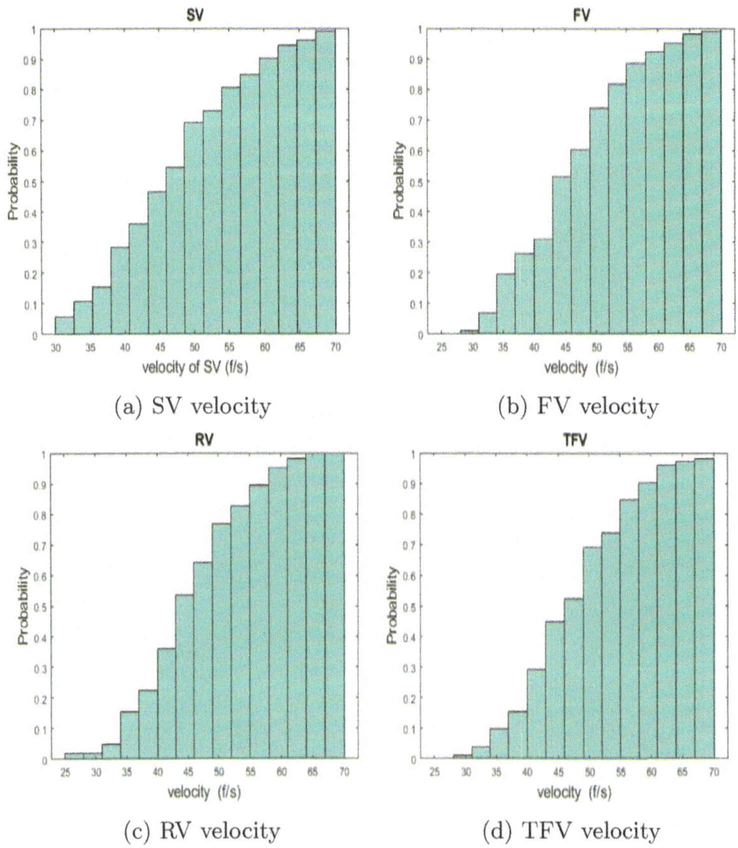

(a) SV velocity (b) FV velocity (c) RV velocity (d) TFV velocity

Fig. 4 The CPDF of current lane explanatory variables

5.1 Data Analysis

The binary response variable depends on some explanatory time-effects factors explained in the Figs. 4 and 5 by the histograms of Cumulative Probability Distribution Functions (CPDF). Figure 4a has shown that the LC vehicle changes the lane at a starting velocity of 30 (f/s) to a final velocity of 70 (f/s). The cumulative probability of the lane merging increases smoothly when the velocity increases smoothly. From the velocity 40 (f/s) to 50 (f/s), the cumulative probability increased more than other intervals. As a result, more vehicles are changing the lane at this interval for SVs velocities. Similarly, Figs. 4b, 4c, 5a and 5b, demonstrate an increased on the cumulative provability from 35 (f/s) to 50 (f/s), 30 (f/s) to 50 (f/s), 30 (f/s) to 55 (f/s) and 35 (f/s) to 50 (f/s) respectively. In these intervals, therefore, more vehicles change the lane due to their respective factors. Again, the other two Figs. 4d and 5d are shown for the gap between FV and SV on the current lane and FV and

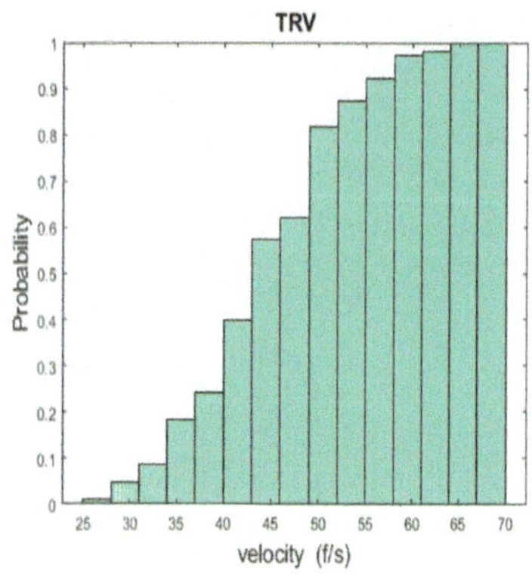

(a) TRV velocity

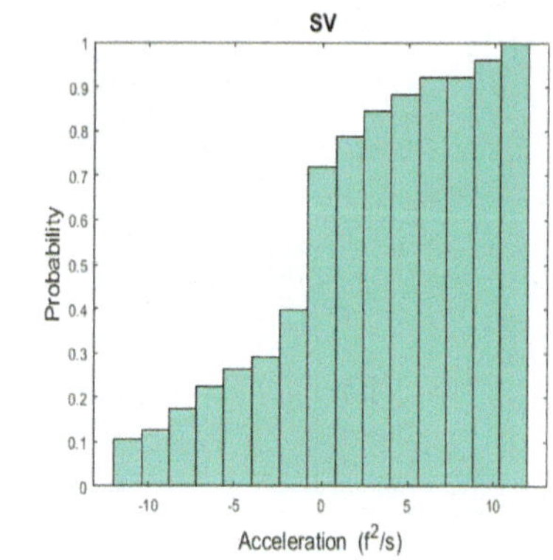

(b) The acceleration of SV

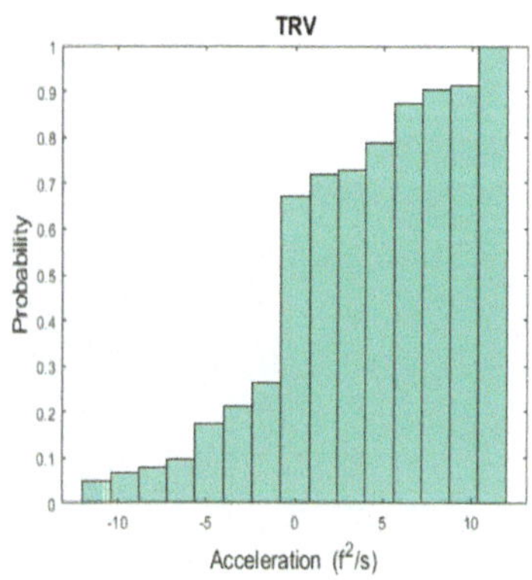

(c) The acceleration of TRV

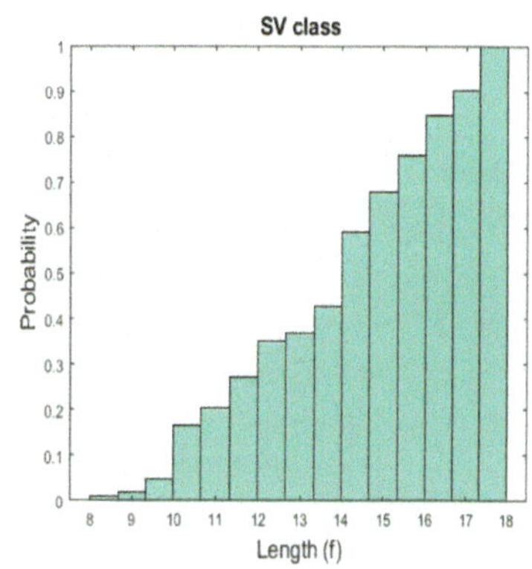

(d) The length of SV

Fig. 5 The CPDF of target lane explanatory variables

SV on the target vehicle. The cumulative probability of a gap increased further by 20 (f) to 100 (f) from Fig. 4d, and 0 (f) to 75 (f) from Fig. 5d. Therefore, in these intervals, more vehicles also change the lane.

6 Results and Discussion

Time-effects explanatory factors with linear function are involved in obtaining a log odds of lane merging success. This study uses the MATLAB package "mnrfit" to determine the probability of a lane merging that 29 data sets (2.9 s data) are used separately in the logit model, each data set comprised 103 vehicle trajectories. Sometimes the p-values are lower than the significant level (considered <0.2) and again few times the p-values are higher than the significant level (considered >0.2) for each factor as the significant level 0.2 in [35]. Some important factors could not be

Table 2 Used factors influence the LC decision

Variable	Influenced without time	Influenced by time	Response of LC probability by increasing time
Constant	2.406	–0.4085	Decreasing
SVv	–0.1045	0.0083	Increasing
FVv	–0.0421	–0.0012	Decreasing
RVv	0.0185	–0.0005	Increasing
TFv	–0.0064	0.0016	Decreasing
TRv	0.0419	–0.0009	Increasing
SFg	0.0057	–0.0003	Increasing
SRg	0.0077	0.0004	Increasing
TRg	0.0005	–	No-effects
SVp	0.0010	–	No-effects
TRp	0.0009	–0.0001	Decreasing
SVa	–0.0302	0.0015	Increasing
TRa	0.0517	–0.0089	Decreasing
SVcl	–0.0082	0.0043	Increasing

avoided in the model at which their significant level crossed the border according to [35]. In addition, a very recent software package used the existing inferential statistics were sometimes unable to predict the significant factor in which 110 row data were tested in [36], although the significance of high-level data could be predicted. Therefore, the identification of significant influence factors and some other statistical model (t-statistics and confidence interval) for all factors at all times seem to be some difficulties in this study. This limitation of this model may be overcome in future research by the use of a further developed software package.

Table 2 shows that the probability of lane merging is reduced over a constant term by increasing the time. Some factors, such as SV velocity, SV front gap at current lane and target lane, are not time-effect factors because these factors multiply by the constant as shown in Table 2. Other factors, such as the FV and RV velocities of the current lane and the same vehicle velocities of the target lane, are time-impact factors. Whereas, the factors of FV velocity and RV velocity at the current lane are multiplied by negative time-effect coefficient values. For the negative coefficient of the factors, the probabilities decrease with increasing time, and for the positive coefficient of the factors the probabilities increase with increasing time. As a result, the probability of FV velocity decreases at the current lane, but the probability of RV velocity increases at the same lane. RV acceleration at the target lane reduces the probability of merging, and the SV class increases the probability of merging. And by increasing the time, the probability of FV velocity on the target lane increases and the probability of RV velocity on the target lane decreases.

6.1 Time Varying Effects at Lane Merging

Figure 6a shows that the odd ratio for the SV velocity increases from below 1 to above 1. Before 1.1 s, for this velocity, the probability of a lane merging is influenced for unsuccess. In addition, after 1.1 s in the same figure, the probability of a lane merging is influenced for success. So, by increasing the SV velocity, odd ratio is increasing with increasing time. This velocity creates heterogeneous effects before 1.1 s and after 1.1 s. By increasing the SV velocity, the front gap decreases, and the probability of a lane merging fails [37, 38]. The similar time-effect is shown in Fig. 6d. Another time-effect is shown in Figs. 6b, 6c, 6e and 6f. In Fig. 6b, the odd ratio for the FV velocity is decreasing from below 1 that indicates the probability of merging is decreasing due to an increase in the front gap [38]. Respectively, in Figs. 6c, 6e and 6f, the RV velocity, TRV velocity and front gap at current lane are slightly time-effect as the odd ratios are slightly decreasing.

Afterwards, the another odd ratio from another factor, such as FV velocity at the target lane shown in Fig. 6d, where it shows heterogeneous effects like as [5]. In this figure, the odd ratio increases from less than 1 to more than 1. So that, the probabilities of lane merging success increase by increasing time. The Figs. 6b and 6c shown respectively that the probability of lane merging reduces if the FV velocity increases, and the probability of lane merging increases if the RV velocity increases. Like as [9, 39] studies, Figs. 6e and 6f show similar scenarios. The odd ratio slowly decreases at Fig. 6f from upper 1 to lower 1 so that the probability of successful lane merging decreases slowly by increasing time.

As above discussion, in Fig. 7a, the gap of SV and RV at current lane have few positive time-effect as odd ratios are slowly increasing. Similarly, Fig. 7e shows that the corresponding factor (SV acceleration) has few positive time-effect for LC success. In addition, Figs. 7b and 7c show fixed effects that the odd ratios do not increase or decrease by increasing time. In these cases, these factors do not influence the probability of a lane merging success or failure by increasing time. So, these are not time-effect factors. Furthermore, The Figs. 7d and 7f have negative time-effect that they influence to LC unsuccess. The SV class also highly time-effect factor shown in Fig. 8a.

6.2 Model Accuracy and Spatial Transferability

A DLC model based on binary decision was proposed in [17], where the accuracy of LC prediction was 82.2% by using gaps and surrounding vehicle speeds. They employed the NGSIM (US 101) data even the model was better than the TRANS-MODELER software used gap acceptance model. They suggested that their model might be applied to the autonomous vehicle system. Again, a time-effects logit model was proposed in [5] for successful lane merging at a working zone with 83.7% accuracy. That study further compared their model to a standard logistic model with 81.6%

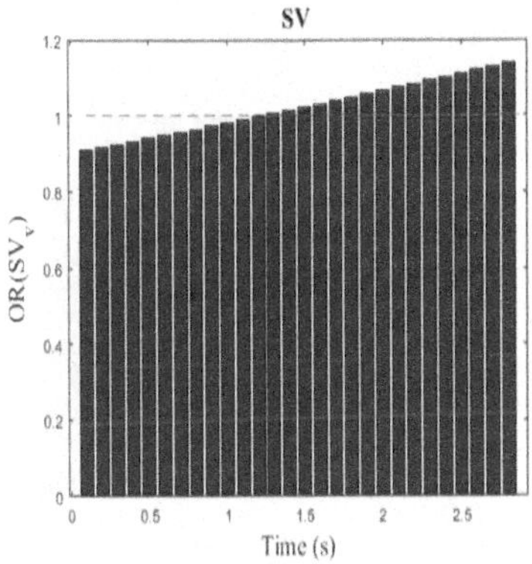

(a) SV velocity

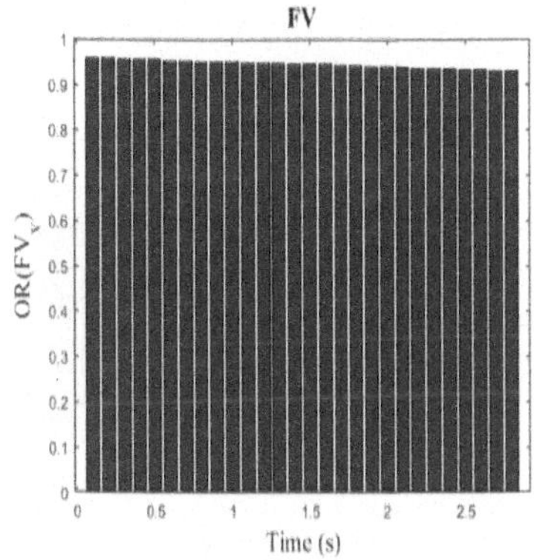

(b) FV velocity

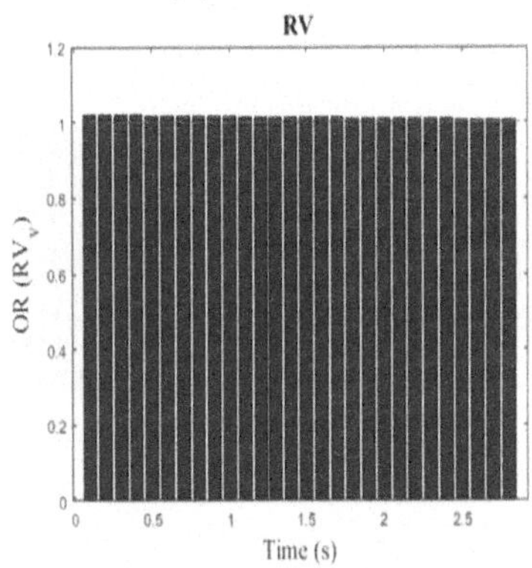

(c) RV velocity

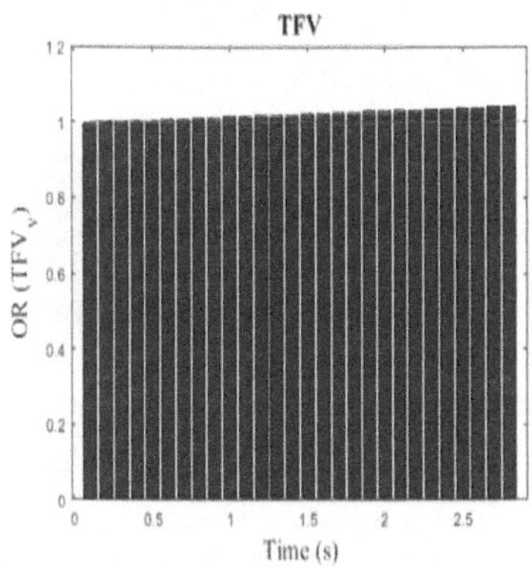

(d) TFV velocity

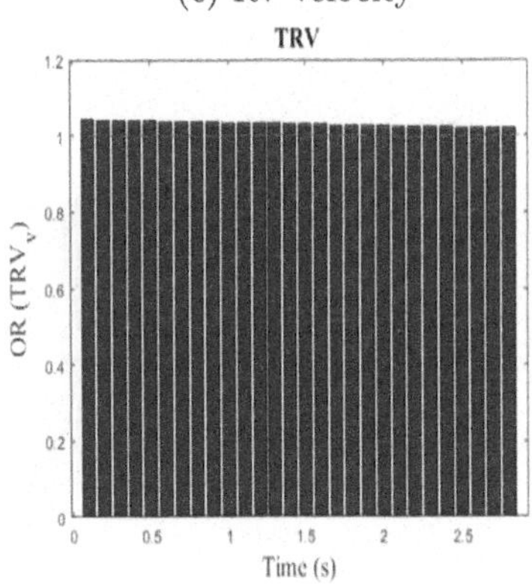

(e) TRV velocity

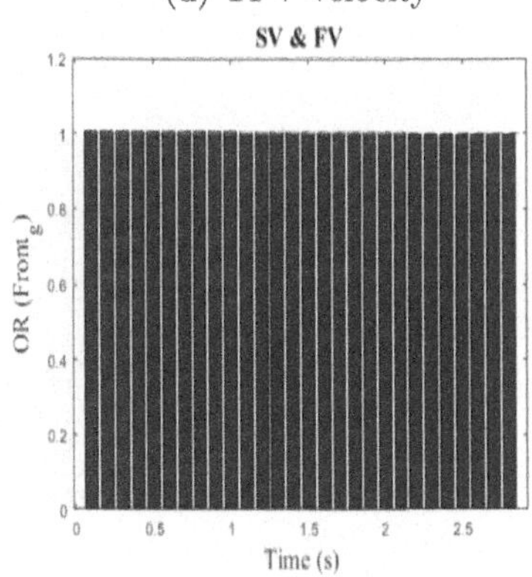

(f) SV and FV gap

Fig. 6 The odd ratios of all explanatory factors with constant by time

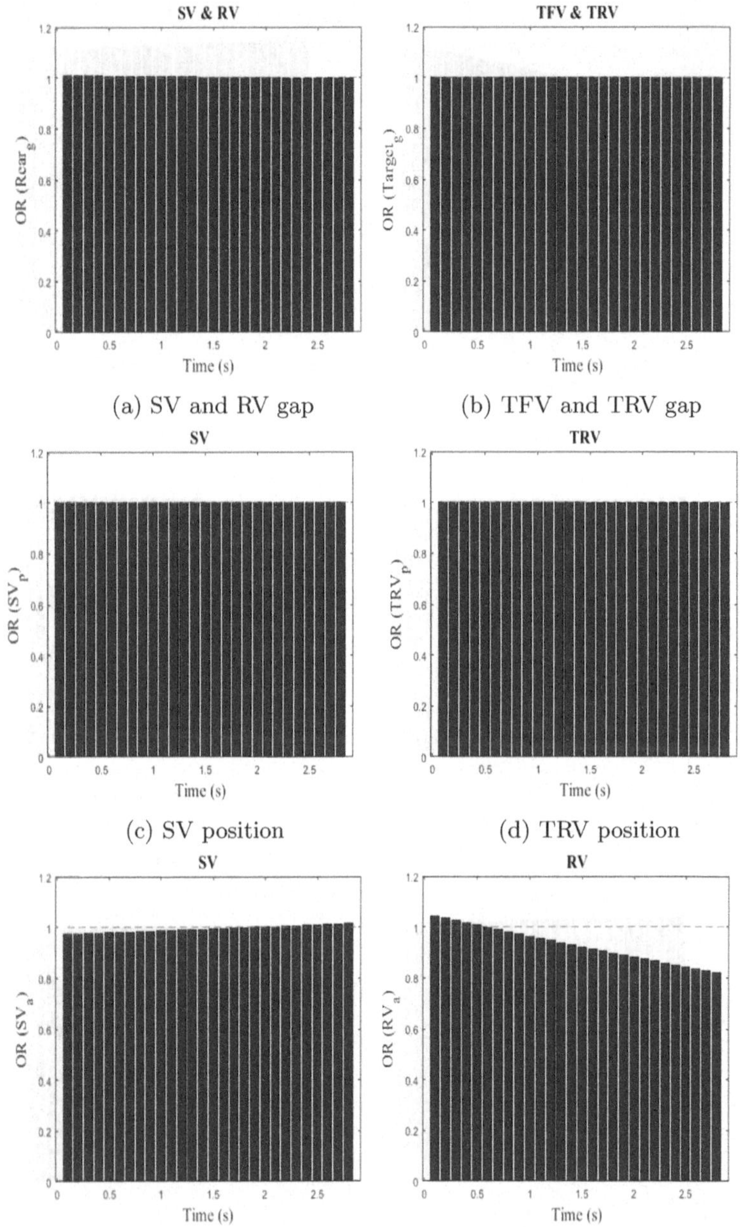

Fig. 7 The odd ratios of all explanatory factors with constant by time

Fig. 8 The odd ratios of all explanatory factors with constant by time

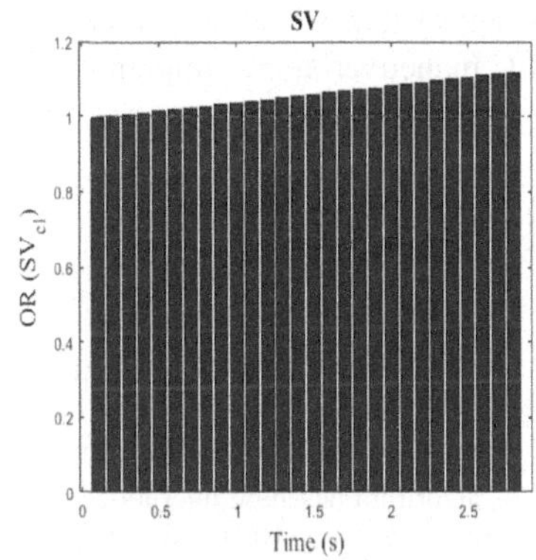

SV class

accuracy for successful lane merging. The 100 training vehicles and the 10 testing vehicles extracted from NGSIM (US 101) data are used in this work. The findings are that in 3 s, the accuracy of this model is 80% for LC decision-making. But after 1 to 3 s, the accuracy for the same decision is 87.5%. The driver can make decisions before 2 s of the LC with an accuracy of 87.5% by using time-effects explanatory factors. This accuracy may be improved by using other time-effect factors or driver characters or weather conditions.

7 Conclusion

The autonomous vehicle on main lane is aware for merging vehicle from the on-ramp and leaving vehicle to the off-ramp. Every time, some vehicles come to merge on main lane, and some vehicles exit from main lane. These circumstances created a risk for all types of vehicles moving on lane 5 in NGSIM (US-101) dataset. Using some influential variables, vehicles on lane five always try to avoid the collision point by DLC. The gaps and speeds of surrounding vehicles are more important explanatory factors for several previous studies to make decisions in these situations. But the time factor as an influencing factor for LC decisions was not included in numerous studies. Although, all of the factors used in previous studies were time-dependent, only the time factor could not be responsible for other factors predicting LC decisions [5, 35]. From the cumulative probability distribution functions of all influencing factors, it is decided that the time-impact factors included in the model can identify and predict LC decisions. The time-effects of all influencing factors are more reliable and logical factors for predicting LC maneuver. Furthermore, the odd-ratios of all time-effects influencing factors can explain the probabilities of lane merging decisions at different times. The accuracy of this model included time-effect factors is higher than the model without time-effect factors. The higher accuracy (84%)

suggests that the autonomous vehicle can apply this model to predict and decide the LC maneuver at on-ramp and off-ramp places.

References

1. Alia, Y., Zuduo, Z., Haque, M.M.: Connectivity's impact on mandatory lane-changing behaviour: evidences from a driving simulator study. Transp. Res. Part C Emerg. Technol. **93C**, 292–309 (2018)
2. Wang, Y., Wenjuan, E., Tang, W., Tian, D., Lu, G., Yu, G.: Automated on-ramp merging control algorithm based on internet-connected vehicles. IET Intell. Transp. Syst. **7**, 371–379 (2013). https://doi.org/10.1049/iet-its.2011.0228
3. Oh, S., Yeo, H.: Impact of stop-and-go waves and lane changes on discharge rate in recovery flow. Transp. Res. Part B **77**, 88–102 (2015). https://doi.org/10.1016/j.trb.2015.03.017
4. Suh, J., Yeo, H.: An empirical study on the traffic state evolution and stop-and-go traffic development on freeways. Transportmetrica A Transp. Sci. **12**(1), 80–97 (2016). https://doi.org/10.1080/23249935.2015.1101508
5. Weng, J., Li, G., Yu, Y.: Time-dependent drivers' merging behavior model in work zone merging areas. Transp. Res. Part C Emerg. Technol. **80**, 409–422 (2017). https://doi.org/10.1016/j.trc.2017.05.007
6. Zhang, L., Chen, C., Zhang, J., Fang, S., You, J., Guo, J.: Modeling lane-changing behavior in freeway off-ramp areas from the Shanghai naturalistic driving study. J. Adv. Transp. (2018). https://doi.org/10.1155/2018/8645709
7. Gipps, P.G.: A model for the structure of lane-changing decisions. Transp. Res. Part B **20**, 403–414 (1986). https://doi.org/10.1016/0191-2615(86)90012-3
8. Ahmed, K.I.: Modeling Freeway Lane Changing Behavior. MIT, Cambridge (1996)
9. Ahmed, K.I.: Modelling Drivers' Acceleration and Lane Changing behavior. MIT, Cambridge (1999)
10. Toledo, T., Zohar, D.: Modeling duration of lane changes. Transp. Res. Rec. J. Transp. Res. Board **1999**, 71–78 (2007). https://doi.org/10.3141/1999-08
11. Hidas, P.: Modelling vehicle interactions in microscopic simulation of merging and weaving. Transp. Res. Part C Emerg. Technol. **13**, 37–62 (2005). https://doi.org/10.1016/j.trc.2004.12.003
12. Kesting, A., Treiber, M., Helbing, D.: General lane-changing model MOBIL for car-following models. Transp. Res. Rec. J. Transp. Res. Board **1999**, 86–94 (2007). https://doi.org/10.3141/1999-10
13. Park, H., Oh, C., Moon, J., Kim, S.: Development of a lane change risk index using vehicle trajectory data. Accid. Anal. Prev. **110**, 1–8 (2018). https://doi.org/10.1016/j.aap.2017.10.015
14. Zhang, L., Wang, S., Chen, C., Yang, M., She, X.: Modeling lane-change risk in urban expressway off-ramp area based on naturalistic driving data. J. Test Eval., 48 (2020)
15. Wang, Y.: Prediction of Mandatory Lane Changing Behavior Using Artificial Neural Network Model. McMaster University, Hamilton (2017)
16. Zheng, Z.: Recent developments and research needs in modeling lane changing. Transp. Res. part B Methodol. **60**, 561–565 (2014)
17. Balal, E., Long, R., Sarkodie-gyan, T.: A binary decision model for discretionary lane changing move based on fuzzy inference system. Transp. Res. Part C **67**, 47–61 (2016). https://doi.org/10.1016/j.trc.2016.02.009
18. Weng, J., Meng, Q.: Modeling speed-flow relationship and merging behavior in work zone merging areas. Transp. Res. Part C Emerg. Technol. **19**, 985–996 (2011). https://doi.org/10.1016/j.trc.2011.05.001
19. Lee, J., Park, M., Yeo, H.: A probability model for discretionary lane changes in highways. KSCE J. Civ. Eng. **20**, 2938–2946 (2016). https://doi.org/10.1007/s12205-016-0382-z

20. Park, M., Jang, K., Lee, J., Yeo, H.: Logistic regression model for discretionary lane changing under congested traffic. Transp. A Transp. Sci. **11**, 333–344 (2015). https://doi.org/10.1080/23249935.2014.994686
21. Li, G., Cheng, J.: Exploring the effects of traffic density on merging behavior. IEEE Access **7**, 51608–51619 (2019). https://doi.org/10.1109/ACCESS.2019.2911302
22. Marczak, F., Daamen, W., Buisson, C.: Empirical analysis of lane changing behavior at a freeway weaving section. Traffic Manag., 139–151 (2016). https://doi.org/10.1002/9781119307822.ch10
23. Chu, T.D., Miwa, T., Morikawa, T.: Discrete choice models for gap acceptance at urban expressway merge sections considering safety, road geometry, and traffic conditions. J. Transp. Eng. **143**, 04017025 (2017). https://doi.org/10.1061/JTEPBS.0000053
24. Chu, K., Lee, M., Sunwoo, M.: Local path planning for off-road autonomous driving with avoidance of static obstacles. IEEE Trans. Intell. Transp. Syst. **13**, 1599–1616 (2012). https://doi.org/10.1109/TITS.2012.2198214
25. Xie, D.F., Fang, Z.Z., Jia, B., He, Z.: A data-driven lane-changing model based on deep learning. Transp. Res. Part C Emerg. Technol. **106**, 41–60 (2019). https://doi.org/10.1016/j.trc.2019.07.002
26. Huang, L., Guo, H., Zhang, R., Wang, H., Wu, J.: Capturing drivers' lane changing behaviors on operational level by data driven methods. IEEE Access **6**, 57497–57506 (2018)
27. Jia, B., Yang, D., Zhang, X., Wu, Y., Guo, Q.: Car-following model considering the lane-changing prevention effect and its stability analysis. Eur.Phys. J. B **93**(8), 1–9 (2020)
28. Khakzar, M., Rakotonirainy, A., Bond, A., Dehkordi, S.G.: A dual learning model for vehicle trajectory prediction. IEEE Access **8**, 21897–21908 (2020)
29. Li, G., Fang, S., Ma, J., Cheng, J.: Modeling merging acceleration and deceleration behavior based on gradient-boosting decision tree. J. Transp. Eng. Part A Syst. **146**(7), 05020005 (2020)
30. Tang, L., Wang, H., Zhang, W., Mei, Z., Li, L.: Driver lane change intention recognition of intelligent vehicle based on long short-term memory network. IEEE Access **8**, 136898–136905 (2020)
31. Li, J., Huang, X., Wang, J., Mu, T.: Lane change behavior research based on NGSIM vehicle trajectory data. In: 2020 Chinese Control And Decision Conference (CCDC), pp. 1865–1870. IEEE (2020)
32. Zhang, S., Zhi, Y., He, R., Li, J.: Research on traffic vehicle behavior prediction method based on game theory and HMM. IEEE Access **8**, 30210–30222 (2020)
33. Zhu, M., Wang, Y., Pu, Z., Hu, J., Wang, X., Ke, R.: Safe, efficient, and comfortable velocity control based on reinforcement learning for autonomous driving. Transp. Res. Part C Emerg. Technol. **117**, 102662 (2020)
34. Wan, Q., Peng, G., Li, Z., Inomata, F.H.T.: Spatiotemporal trajectory characteristic analysis for traffic state transition prediction near expressway merge bottleneck. Transp. Res. Part C Emerg. Technol. **117**, 102682 (2020)
35. Weng, J., Du, G., Li, D., Yu, Y.: Time-varying mixed logit model for vehicle merging behavior in work zone merging areas. Accid. Anal. Prev. **117**, 328–339 (2018). https://doi.org/10.1016/j.aap.2018.05.005
36. Vakhitova, Z.I., Alston-Knox, C.L.: Non-significant p-values? strategies to understand and better determine the importance of effects and interactions in logistic regression. PloS one **13**(11), e0205076 (2018)
37. Mathew, T.V., Velaga, N.R.: Lecture Notes in Civil Engineering. Springer (2017)
38. Patil, G.R., Sangole, J.P.: Behavior of two-wheelers at limited priority uncontrolled T-intersections. IATSS Res. **40**, 7–18 (2016). https://doi.org/10.1016/j.iatssr.2015.12.002
39. Lee, G.: Modeling Gap Acceptance at Freeway Merges. MIT, Cambridge (2006)

Lane Changing Effects on Surrounding Vehicles by Incorporating the Risk Factors

Md. Mijanoor Rahman, Mohd. Tahir Ismail, and Majid Khan Majahar Ali

Abstract The traffic simulation model includes Lane Changing (LC) trajectory that should be able to provide traffic safety by informing the unknown driving trajectory. Difficult types of hindrances such as deceleration, confusion for collision, expected velocity, effects of LC action, etc. will impact LC driver behavior. This study extracts the risk factors from real data and determines the influenced factors of risk perception-based LC decision using a logistic model. Then the effects on these influenced factors are investigated by comparing the before and after LC maneuvers for LC execution. The logistic model provides the influenced factors by 81.51% accuracy of LC decision using the testing dataset (30% of full dataset). The research has found that 20.07% subject vehicle velocities increases for LC execution, and other risk factors such as gaps likely decrease that suggested they are more risk to the collision after LC execution. Therefore, this study is important for decreasing the traffic crash in LC situation.

Keywords Lane changing effect · Logistic model · Surrounding vehicles trajectories

Md. M. Rahman · Mohd. T. Ismail · M. K. M. Ali (✉)
School of Mathematical Sciences, Universiti Sains Malaysia, Penang, Malaysia
e-mail: majidkhanmajaharali@usm.my

Md. M. Rahman
e-mail: mijanoor.math@student.usm.my

Md. M. Rahman
Department of Mathematics, Mawlana Bhashani Science and Technology University, Santosh, Tangail 1902, Bangladesh

M. H. Mohd et al. (eds.), *Modelling, Simulation and Applications of Complex Systems*, Springer Proceedings in Mathematics & Statistics 359,
https://doi.org/10.1007/978-981-16-2629-6_22

1 Introduction

The unplanned Lane Changing (LC) behavior of vehicle drivers in urban roadways influences traffic bottlenecks and increases traffic crashes. Generally, a driver's decision in various hazard situations is the main cause of congestion [1], and merging roadway is estimated the most crucial to high congestion [2]. The time cost of people produced in congested urban traffic areas is more than 6.9 billion hours on the road beside the purchasing cost of fuel is additional 3.1 billion gallons [3–5]. Besides, the driver's distraction, discomfort, and frustration are twisted from traffic congestion and may result in fierce driving behavior [6].

Planned driver behavior is anticipated to resolve many transport problems and provide comfort, safety, productivity, and flexibility during travel. These planned vehicle movements are adopted either in macroscopic or microscopic factor analysis, wherein the former addresses traffic flow, density, and average traffic speed and the latter differentiates the vehicle's trajectory based movements, such as position, velocity, acceleration, gap and time headway [7]. The analysis of the microscopic factor suggests that the driving system is flexible and safe for drivers.

In recent years, researchers developed a gap acceptance model included safety factor, and proposed the distribution of trajectories [8]. That type of LC action provides more safety in congested traffic areas, thereby bringing more comfort when the driver needs increased speed in the freeway road [9]. In that case, the Subject Vehicle (SV) driver tends to identify the gap in the target lane. When the driver identifies the gap at the target lane, the binary decision is applied (e.g. LC and non-LC decisions). If the current gap is less than the minimum safety gap, the Target Rear Vehicle (TRV), the driver had to decelerate to create a more safety gap, and then the SV driver may change his current lane without the crash. In addition, if TRV does not decelerate in the decision time, then the LC is a risk for SV. Therefore, the LC decision depends on TRV's deceleration to avoid the crash. The LC decision time of SV and gap at target lane are shown in Fig. 1.

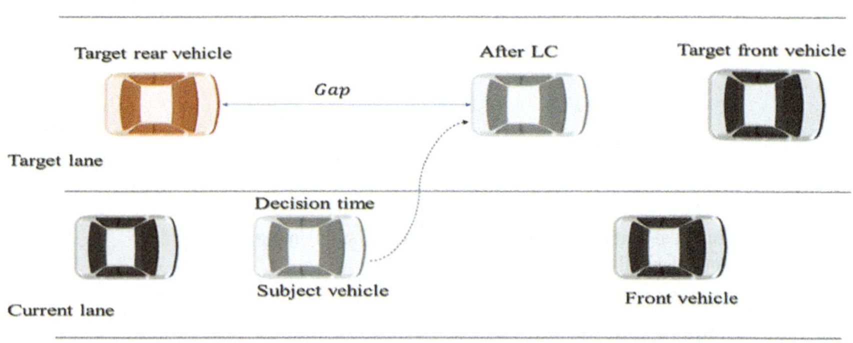

Fig. 1 Lane changing decision time and gap at target lane

The LC decision and execution are continuous procedures for SV. After execution of the LC maneuver, SV arrives at target lane where Target Front Vehicle (TFV) and TRV are new front vehicle and rear vehicle respectively shown in the Fig. 1, and new front and rear trajectories may be threatened for SV. Therefore, the comparative trajectories information of before and after LC are crucially needed to understand the effects of LC action for vehicle controlling system.

A few macroscopic-based analyses were developed to understand the LC effects [10]. In other recent studies, the deceleration rate of TRV is used to find out the LC safety decision using logistic regression model [11–13], whereas they avoided the comparison of before and after LC maneuvers effects. The LC decision model was developed in [14] using safety perception as a threshold value from the deceleration of TRV. However, the effects of LC on before and after trajectories are not focused in that work, although without comparison study, the negative effects of this maneuver may not be decreased, whereas the LC maneuver occurs more traffic accident (see in [14]).

2 Literature Review

The longitudinal driving behavior is a car following driving action, and the lateral driving behavior is an LC driving action. The LC is a crucial matter in any traffic system. The macroscopic LC model was pioneer research in 1961 to develop the driver decision [15]. The first LC decision for vehicle drivers who face up the conflicting situation was proposed by Gipps [16]. That model protected various driving risk situations in an urban roadway wherein the driver behavior was ruled by actions such as keeping the lane or changing the lane. The rule-based LC decision using microscopic factors was developed by Ahmed [17].

The velocity difference and gap factors were used to model the gap acceptance decision where these factors are collected to analyze by microscopic and macroscopic data [18, 19]. Before LC data were analyzed to validate that model. A macroscopic factor such as the vehicle intensity-based LC model was addressed in [20]. That model explored the effects on traffic flows wherein traffic bottleneck was increased before LC. The LC action directly effects traffic flows and impact other surrounding groups of vehicles at macroscopic effect [10, 22–25]. The limited real trajectory data for traffic simulation is one of the limitations as the model is unable to use the before and after LC data. Another data-based comparison method [10] used risk factors such as velocity difference and gap duration. They argued existing LC analysis is unable to explain the risk factor, and found that high-speed vehicle is affected and hindered for LC.

Using statistical significance test, some research found the LC maneuver can affect on immediate TRV as the aggressive driver moves more aggressively [26, 27]. They suggested that further research is needed for understanding the LC effects on the surrounding vehicle dynamics. The effects of LC on surrounding vehicles were addressed in [26] where it suggested to study more that is important for the accuracy of LC models. That study agreed that to explain the LC analysis (such as a statistically fitting of the considered factors of after LC maneuver with collected data) requires further research. The gap acceptance model using gap factors (lag and lead gap) was proposed by [8, 28] wherein both factors are microscopic for before and after LC. However, these factors effect decision-making when the driver tries to change the lane as before LC action. Therefore, their models help predict the gaps as well as risk factors according to vehicle trajectories, and it unable to compare these risk factors before and after LC maneuver. In addition, the influenced factor used in the gap acceptance model was unable to explain the risk after LC [29].

The risk factors were used to determine the LC decision when factors are collected before LC using logistic model [12, 30]. Recently, the microscopic risk factors such as surrounding gaps were analyzed in [31] where it suggested to make LC decision. The microscopic risk factors were used in [32] that these factors are collected before LC. The safety level was discovered by analyzing risk factors such as velocity difference, gap, and agreed that LC is a risk driving behavior [33]. In the LC event, the subject vehicle and surrounding vehicles have associated velocities. The research suggested to develop the risk factor-based prediction model should implement using risk factor that collected after LC maneuver. The decision model was developed in [34] where microscopic safety gap as the risk factor is collected before LC. Therefore, the above-mentioned microscopic LC decision model employed the risk factors only collected before the LC decision, wherein this model could not be able to determine the factors analysis those are collected after LC maneuver. The above-mentioned all literature are summarized in Table 1 wherein the gap of the literature is clearly focused.

In this study, the risk factors such as the velocity of SV and other vehicles such as Rear vehicle (RV), front vehicle (FV), TRV and TFV, surrounding gaps are extracted from real data. This study also determines the influenced factors of risk perception-based LC decision using the logistic model where TRV deceleration trajectory is a response variable. Then the effects of LC on these influenced factors are investigated by comparing the trajectories wherein the descriptive statistical analysis, Levene's-test and T-test are employed to compare before and after LC maneuvers of influenced factors of LC decision.

3 Methodology

Lane changing decision depends on some influenced factors such as surrounding gaps and relative velocities of surrounding vehicles' and LC vehicle's velocities.

Table 1 Similar studies in the literature

Author	BLC	ALC	Analysis		Risk factor		LCD	Mac. E.
			Mac.	Mic.	Velo.	Gap		
[15]	✓		✓		✓			
[16]	✓		✓		✓	✓	✓	
[17]	✓			✓	✓	✓	✓	
[18]	✓		✓	✓	✓	✓	✓	
[19]	✓		✓	✓	✓	✓	✓	
[20]	✓	✓	✓	✓				✓
[23]	✓		✓	✓	✓		✓	
[22]	✓		✓	✓	✓		✓	✓
[21]	✓	✓		✓		✓		
[25]	✓		✓			✓	✓	✓
[27]	✓	✓	✓		✓		✓	✓
[26]	✓		✓		✓	✓	✓	
[28]	✓		✓	✓	✓	✓	✓	
[24]		✓	✓		✓		✓	✓
[8]	✓			✓	✓	✓	✓	
[30]	✓			✓	✓	✓	✓	
[13]	✓				✓	✓	✓	
[10]	✓				✓	✓	✓	
[29]	✓	✓	✓	✓	✓		✓	✓
[31]	✓	✓	✓	✓		✓	✓	
[32]	✓				✓	✓	✓	
[33]	✓					✓		
[34]	✓		✓		✓	✓		
[14]	✓		✓			✓	✓	

BLC-Before LC; ALC-After LC; Mac-Macroscopic; Mic-Microscopic; Velo.-Velocity; LCD-LC decision; Mac. E.-Macroscopic effects

3.1 Definition of Risk Factors

There are some risk factors such as surrounding gaps, time-headway and relative velocities used in some previous studies. The Eq. (1) represents gap between two vehicles, g.

$$g = p_{n-1} - p_n - \frac{1}{2}(l_{n-1} + l_n) \quad (1)$$

where, p_{n-1} and p_n are positions and l_{n-1} and l_n are lengths of FV and SV respectively where the vehicle lengths are excluded motorcycle because the value of vehicle length is more 3 m (according to data). These variables are shown in the Fig. 2. These similar approaches are applied for other surrounding vehicles.

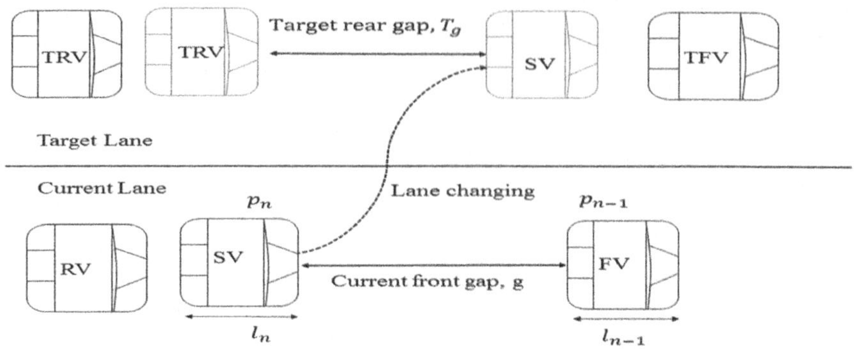

Fig. 2 Lane changing trajectories with surrounding vehicles

3.2 Binary Logistic Model

The binary logistic model is very popular to categorize the factors that influence decision and execution. In that model, there are two parts of factors such as response and influence factors, where the response factor generates from the influence factors. Before LC, the driver continues the car-following trajectories, and after LC he/she again maintains the car-following trajectory. Therefore, the vehicle trajectory (before LC or after LC) is a non-LC trajectory. The influenced factors are collected before LC as LC decision factors and after LC factors are LC execution factors. Further, in the logistic model, the risk factors and other factors are employed to determine the effects for LC execution where the factors have significance in the LC decision and execution. The Eqs. (2) and (3) are used to search the significant factors in which they influenced the LC decision and execution by incorporating the risk factors. The probability of LC decision (risk-free), $P_i\ (y = 0)$ for vehicle i is used by the Eq. (2).

$$P_i\ (y = 0) = \frac{1}{1 + e^{\sum \varphi_k x_k}} \tag{2}$$

The probability of LC decision (risk), $P_i\ (y = 1)$ for vehicle i is expressed by the Eq. (2).

$$P_i\ (y = 1) = \frac{e^{\sum \varphi_k x_k}}{1 + e^{\sum \varphi_k x_k}} \tag{3}$$

where, $y = 0$ represents the risk-free LC decision, and $y = 1$ represents the risky decision of LC execution. Again, where, $\varphi_k\ (k = 0,\ 1,\ 2,\ 3,\ \ldots K)$ is coefficient of $k-$th influencing factor and x_k are influence factors. All collected factors are used in the Eqs. (2) and (3) as a logistic model to determine the probability of either risk-free or risky decision. Further, this model provides the significantly influenced factors are either before LC decision or after LC execution.

4 Data Processing

A search algorithm reproducing-kernel-hilbert-space and non-parametric regression fitting system was discovered for analyzing real trajectory data sets [35]. The vehicle lateral and longitudinal positions during LC were nicely presented in [36]. They have worked on US 101 Next Generation Simulation (NGSIM) real trajectory data for finding the optimization starting and ending points during LC. At LC time, the starting and ending points are very important for the driver because every driver wants to avoid collision with other drivers but worked only on real trajectory data, but not on any specific model or maneuver. A human driver can maintain the LC maneuver without any wireless connection among the drivers by only using optimization starting and ending points.

NGSIM originally collected from US-101 (Hollywood freeway) and I-80 (Berkeley highway) roads [22]. Both roads are straight roads near the off-ramp and on-ramp. There are five lanes on the main roads and the lane width is 3.5 m in every road [9]. This dataset provides good information to explore the LC situations, and it is a more prominent research dataset that could be utilized in any traffic research microscopically. The NGSIM data set is the most popular and open-source data in traffic research [37], because it is a large dataset that included much microscopic information. The dataset is also frequently used in current research. Again, this dataset is perfect data to analyze the congested traffic area. The LC rate and duration can also be determined by using the same dataset. The LC history from the NGSIM dataset can be collected, and the simulation model data and the real trajectory data can be compared by using the same dataset [38]. The NGSIM data dominate in the traffic research area [39], and proposed the data-driven LC model and the gap acceptance model by using this dataset.

The LC starting and ending points from NGSIM (US 101 and I-80) dataset were determined by using reproducing-kernel-hilbert-space and non-parametric regression model with high accuracy [36]. There are 543 LC scenarios in the big data implementation (a total of 543 groups, with each group have 5 vehicles and every vehicle has 5 trajectories with 200 data per trajectory). So, overall, this study manages to implement our proposed model towards 2.715 million data. The five vehicles such as SV, FV, RV, TFV and TRV belong in a group wherein every vehicle has velocity and acceleration profiles, longitudinal and lateral positions, and vehicle length trajectories. This dataset includes the LC starting and ending positions for SV where before starting point represents before LC date and after ending point represents after LC date. The LC duration depends on the starting and ending positions of the vehicle trajectory. The average longitudinal velocity is to use for vehicle longitudinal velocity.

5 Result and Discussions

This study categorizes factors such as general factors (velocities of SV and RV) and risk factors (gaps with surrounding vehicles) as mentioned in introduction. The velocity as a General factor is directly extracted from data whereas risk factor is calculated by using Eq. (1).

5.1 Descriptive Statistics and Variables' Meaning

LC gap acceptance depends on some vehicle trajectory such as velocity and surrounding gaps of SV used in some literature. Although, some previous studies used trajectories as influenced factors and threshold value of critical gap size for LC decision, however these trajectories are being overlooked by driver before LC decision. Table 2 shows all decision variable are taken before LC, and used in binary logistic model. Table 2 also depicts the after LC trajectories as other variables are used to determine the effects for LC decision by comparing the before LC decision and after

Table 2 Descriptive statistics and meaning of all used factors

Variable	Definition	Time	Avg.	Min.	Max.	STD
SV_v	SV average velocity, (m/s)	Before	8.32	0.23	16.80	3.33
		After	9.24	0.64	24.62	4.27
SLV_g	Rear gap between SV and RV, (m)	Before	14.23	0.00	50.00	9.35
		After	10.54	0.00	50.00	10.81
$STFV_g$	Front gap between SV and TFV, (m)	Before	16.26	1.19	80.00	13.77
		After	17.33	0.00	80.00	17.75
$STRV_g$	Rear gap between SV and TRV, (m)	Before	13.66	0.00	50.00	9.30
		After	16.95	0.00	50.00	11.16
$TFRV_g$	Gap between TFV and TRV, (m)	Before	33.09	0.00	100.00	18.52
		After	37.78	0.00	100.00	23.83

Avg.-Average value; Min.-Minimum value; Max.-Maximum value; STD-Standard deviation

Table 3 Correlation coefficient of all factors used in logistic model

	SV_v	SRV_g	$STFV_g$	$STRV_g$	$TFRV_g$
SV_v	1	0.376	0.295	0.138	0.264
SRV_g	0.376	1	0.087	0.161	0.259
$STFV_g$	0.295	0.087	1	0.086	0.594
$STRV_g$	0.138	0.161	0.086	1	0.732
$TFRV_g$	0.264	0.259	0.594	0.732	1

LC execution. All variables as factors adopted descriptive statistical values (average, minimum and maximum values and standard division) shown in the Table 2.

5.2 Correlation Analysis

A correlation analysis is performed for all factors in the dataset. The objective of the correlation coefficient analysis is to study that the high correlation between every paired factor may be excluded from the LC analysis, because high correlated factors may create bias at the significance level in the logistic model. All factors used in the correlation analysis have the same sample size and are paired before and after LC. All correlation coefficients are determined by using the MATLAB package "corr" are depicted in Table 3. Most of the correlation coefficient values are significantly different from 0, and most of the p-values of paired factors are less than 0.05. The maximum correlation coefficient of each paired factors used in LC decision model is 0.732 for $STRV_g$ and $TFRV_g$ in Table 3. This value indicated that all factors used in this study are not highly correlated and they do not create bias in the model. Besides, the lowest value of the correlation coefficients in Table 3 is 0.086. This small positive value indicated that no factor has a high negative correlation and the used factors do not create bias in the model like a highly correlated value.

5.3 Significance Testing of Factors by Logistic Model

Logistic model is using eqs. (2) and (3) to category the data according to response variable. The used influenced variables as factors are either significant (p-value < 0.153) or insignificant (p-value > 0.153). Insignificant factor may provide a bias in the parameter evaluations and statistical inferences [13]. In that case, the study uses the MATLAB package "mnrfit" and "mnrval" to employ this model, and removes the insignificant factors to increase the model accuracy in the testing dataset, although for removing some insignificant factors (time-to-collision and time headway) from the model, the accuracy is decreased in the training dataset. However, the accuracy

Table 4 Result of logistic model

Factor	Coefficient	T-test	p-value
Constant	–0.548	–1.435	0.151
SV_v	–0.061	–1.43	0.153
SRV_g	0.026	1.738	0.082
$STFV_g$	–0.155	–2.891	0.004
$STRV_g$	–0.142	–2.436	0.015
$TFRV_g$	0.127	2.436	0.015

increasing is not an object of this study, because the significant factors influence more in the response variable.

The significant factors are shown in Table 4 whereas the risk factors are the velocity of SV, gaps between SV and RV, SV and TFV, SV and TRV, and TFV and TRV. These five factors improve the model performance, although some other removed factors which are insignificant improved the model performance as mentioned before. In Table 4, the higher positive value indicates the increased factor likely influences to TRV decelerate trajectory as safety LC maneuver whereas a higher negative value indicates the increased factor likely influences to TRV accelerate trajectory as risky LC maneuver. Therefore, some factors associated with parameters adopt negative values. However, in the training dataset, the LC situations are likely risky because of negative parameter values.

This logistic model is used on 543 SVs for LC decision with surrounding groups of vehicles, where 70% of these vehicles are training vehicles and the other 30% vehicles are testing vehicles. The accuracy (81.51%) of this model obtain from testing vehicles. Therefore, the used factors are highly influenced by the LC decision of SV. The five factors significantly influenced the LC decision shown in Table 4. Although this study used many other factors in the logistic model but the other factors are statistically insignificant where the p-values of the cut-in variable are more than 0.20. The threshold p-value is 0.20 (see in [13]) because the p-value of some variables was might not be deleted for accuracy of the model.

5.4 *Velocity Effects for LC*

There are some surrounding vehicles that adopted velocities directly effect for LC maneuver [27], here the study analyses the velocity of SV because it is influenced factor of LC decision as risk-avoiding LC maneuver. Therefore, the comparative analysis of SV velocity before and after LC depicts in some figures by drawing a middle line. The down star sign in Fig. 3 generates from a vehicle by before and after LC represents that the before LC velocity is more than after LC velocity. Similarly,

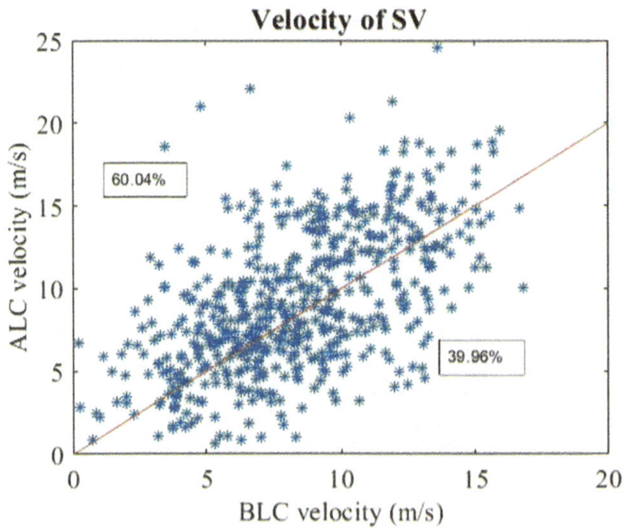

Fig. 3 A group of surrounding vehicles of LC vehicle

the upper point also generated from a vehicle represents that after LC value is more than before LC. This similar procedure was applied by [22, 23].

Figure 3 represents the comparative analysis for the SV velocity, where the before LC velocities are put vertically, and the after LC velocities are put horizontally, and the middle red line separates the maximum velocities of before and after LC action. From Fig. 3, 60.04% SVs increase the velocity after LC action, and 39.976% SV decreases the velocity after LC action. Therefore, the SV velocity increasing is the positive effect of LC, where the effect of LC may have positive or negative [40].

5.5 Surrounding Gaps for Subject Vehicle

The four gaps such as the rear gap in the current lane and front gap, rear gap, and another gap of TFV and TRV at target lane are considered as surrounding gaps for SV. The rear gap between SV and RV before and after LC in current lane shown in Fig. 4 where after LC, gap increases (62.98–28.91)% = 34.07%. Since, after LC, SV goes to the target lane, so the increased gap gives RV more opportunity to increase speed, and SV faces up the front gap in the target lane. In addition, Fig. 5 shows the gaps between SV and TFV before and after LC maneuver wherein after LC 49.36% gaps increase and 49.91% gaps decrease for LC maneuver. Since the difference between these two gaps is very small so the effect is less too for LC.

Figure 6 depicts the gaps of before and after LC execution between SV and TRV, wherein after LC, (56.54–37.75)% = 18.89% gaps decrease for LC maneuver. Since

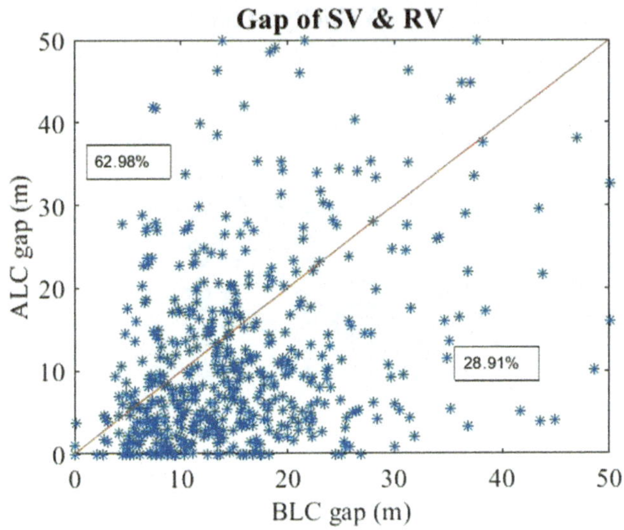

Fig. 4 Rear gap at current lane

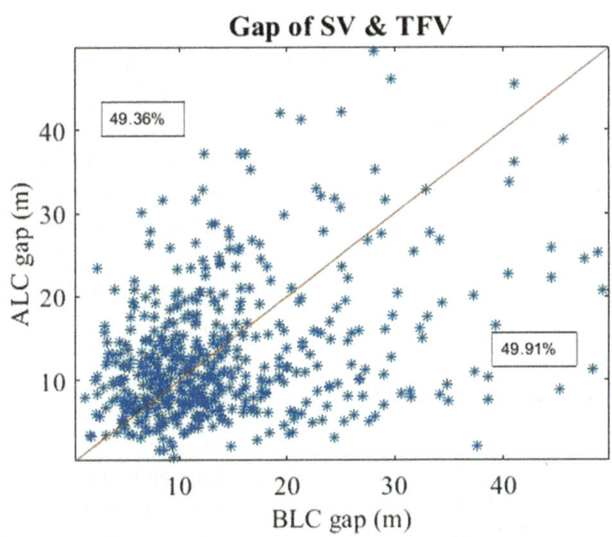

Fig. 5 Front gap at target lane

the 18.89 % gaps decrease after LC. So, the LC effect on the rear gap at the target lane is very high and the rear crash likely increases after LC because of target rear gap decreased. Another fact is that the gap of TFV and TRV at the target lane is equal to the sum of the other two gaps (gap of SV and TFV and gap of SV and TRV) and SV length. This gap is another significant factor in the logistic model used for LC decisions,

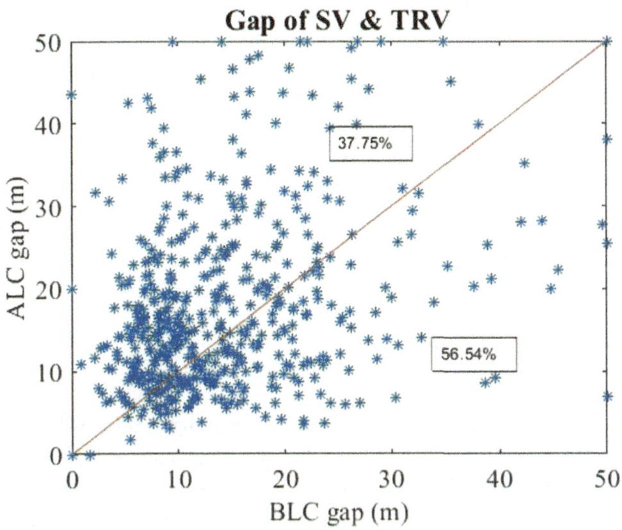

Fig. 6 Rear gap at target lane

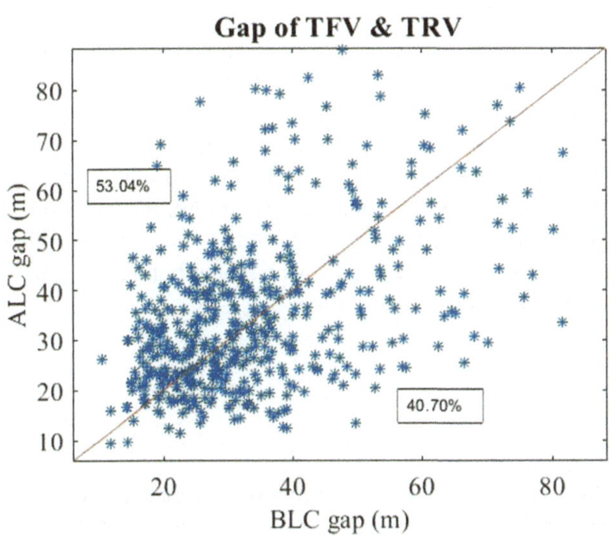

Fig. 7 Gap between TFV and TRV

and this factor is a less correlated factor with other factors. Interestingly, in Fig. 7, the gap increases (53.04–40.70)% = 12.34% after LC. Therefore, the increasing gap is a positive effect on LC action.

Table 5 F-test result

	SV_v	SRV_g	$STFV_g$	$STRV_g$	$TFRV_g$
h	1	1	1	1	1
p-value	0.000	0.001	0.000	0.000	0.000
F-Stats	0.608	0.749	0.602	0.695	0.604
$df1$	542	542	542	542	542
$df2$	542	542	542	542	542

5.6 *F-test: Result Analysis*

The above section adopted comparisons between before and after LC trajectories should be tested statistically where means and variances are either equal or unequal. If variances of any two trajectories are unequal, then the means of these datasets have to be tested to determine the comparisons of the significant differences between the two trajectories. Therefore, Table 5 shows the result of Levene's-test used MATLAB package "vartest2" to compare the variance every two samples of before and after LC trajectories.

In this section, the effects of LC action are discussed according to before and after LC trajectories. From Levene's-test in Table 5, The value of F-Stats for every factor is neither small nor bigger that represents the test is efficient, the all degree of freedom ($df1$ and $df2$) is 542, and p-value is not more than 0.001. Therefore, the standard deviations of the two datasets are not equal, and it is a significant difference between the standard deviations of the two datasets. Therefore, this test concludes that the average trajectories used in the decision model after LC action is significantly unequal to the average trajectories used in the decision model before LC action whereas the gap of SV and TFV are equal at these time.

5.7 *T-test:*

The Levene's-test is for explaining the equality of variances, and the T-test is for explaining the mean differences for unequal variances. Both tests are discussed before and after LC action in this section. Levene's-test provides the conditions of equality of variances, which means that this statistical test is used for checking, whether the standard deviations of these two datasets are significantly equal or not. The hypothesis is that the used trajectories changed for LC action. The p value is 0.000. So, the null hypothesis is rejected. The T-test used MATLAB package "testt2" shown result in Table 6 where $Sd1$ and $Sd2$ represent the unequal standard divisions of before and after LC respectively, and the p-value is more than 0.05 for all factors except one factor (gap of SV and TFV, $STFV_g$). The value of T-Stats for every factor is neither small nor bigger that represents the test is efficient. So, the null hypothesis

Table 6 T-test result

	SV_v	SRV_g	$STFV_g$	$STRV_g$	$TFRV_g$
h	1	1	0	1	1
p-value	0.000	0.000	0.266	0.000	0.000
$Sd1$	3.333	9.35	13.77	9.304	18.515
$Sd2$	4.275	10.806	17.745	11.164	23.832
T-Stats	–3.945	6.028	–1.113	–5.287	–3.620

is accepted, and the averages of two datasets are significantly equal except for one factor. That means, the group of all populations has a different mean and there is a significant difference between these averages. So, lane changes give the significantly improved velocities for SVs in this data.

6 Conclusion

Some researchers developed the simulation model for increasing traffic efficiency. To increase traffic efficiency, this study determines the risk factors by using the logistic model with an accuracy of 81.51% for the testing dataset. The explored risk factors are SV velocity and some surrounding gaps that significantly influenced the LC decision. Some surrounding vehicles of SV are directly affected for LC action, and some vehicles could increase the velocities by this action [25]. Further, this study has compared the SV trajectories and surrounding trajectories of each group. The dataset has 543 groups of vehicles, and each group has 5 vehicles. We have also associated the group vehicles' velocities before and after LC action. And the SV trajectories are compared with surrounding trajectories before and after LC action. Figure 3 gives the important information that 20% SV (LC vehicle) increases the velocities by comparing with before and after LC action, that the results support by some other research (see in [26, 41]). The hypothesis T-test has given important information that the SV average velocity after LC action is significantly greater than the SV average velocity before LC action. But, when we test the hypothesis of surrounding gaps before LC with after LC action, wherein most of the gaps decrease because of increasing the velocity of SV vehicle. The decreasing gaps may create a crash between SV and other vehicles. Eventually, after LC action, only the average velocities of SV have increased a very small value. The finding results from this study depicts that, LC has effects to increase the LC vehicle velocity, and decrease some surrounding gaps as risk factors. Future research may include further risk factors to increase the accuracy of the LC decision model wherein the influenced factors adopted simulation model that will provide a safe and efficient journey.

References

1. Malikopoulos, A., Aguilar, J.P.: An optimization framework for driver feedback systems. IEEE Trans. Intell. Transp. Syst. **14**(2), 955–964 (2013)
2. Margiotta, R.A., Snyder, D.: An agency guide on how to establish localized congestion mitigation programs (Tech. Rep.). United States. Federal Highway Administration. Office of Operations (2011)
3. Afrin, T., Yodo, N.: A survey of road traffic congestion measures towards a sustainable and resilient transportation system. Sustainability **12**(11), 4660 (2020)
4. Rahaman, I., Reza, M.F., Hasib, M.H.H., Hossain, M.I., Hossain, S.A., Sarkar, P.K.: A low cost intelligent multi wireless sensor network perspective on real time traffic surveillance. In: 2019 International Conference on Computer, Communication, Chemical, Materials and Electronic Engineering (ic4me2), pp. 1–4 (2019)
5. Schrank, D., Eisele, B., Lomax, T., Bak, J.: 2015 urban mobility scorecard (2015). https://trid.trb.org/view/1367337
6. Malikopoulos, A.A., Aguilar, J.P.: Optimization of driving styles for fuel economy improvement. In: 2012 15th international IEEE conference on intelligent transportation systems, pp. 194–199 (2012)
7. Rahman, M.M., Ismail, M.T., Ali, M.K.M.: Comparing the calibration methods for intelligent driver model using beijing data. Int. J. Vehicle Syst. Modell. Testing, Inderscince Publishers
8. Balal, E., Cheu, R.L., Sarkodie-Gyan, T.: A binary decision model for discretionary lane changing move based on fuzzy inference system. Trans. Res. Part C: Emerging Technol. **67**, 47–61 (2016)
9. Yang, M., Wang, X., Quddus, M.: Examining lane change gap acceptance, duration and impact using naturalistic driving data. Transp. Res. Part C: Emerg. Technol. **104**, 317–331 (2019)
10. Guo, M., Wu, Z., Zhu, H.: Empirical study of lane-changing behavior on three Chinese freeways. PLoS ONE **13**(1), 1–22 (2018). https://doi.org/10.1371/journal.pone.0191466
11. Cunto, F., Saccomanno, F.: Calibration and validation of simulated vehicle safety performance at signalized intersections. Accid. Anal. Prev. **40**(3), 1171–1179 (2008)
12. Weng, J., Du, G., Li, D., Yu, Y.: Time-varying mixed logit model for vehicle merging behavior in work zone merging areas. Accid. Anal. Prev. **117**, 328–339 (2018)
13. Weng, J., Xue, S., Yan, X.: Modeling vehicle merging behavior in work zone merging areas during the merging implementation period. pp. 1–9 (2015)
14. Wang, C., Sun, Q., Li, Z., Zhang, H.: Human-like lane change decision model for autonomous vehicles that considers the risk perception of drivers in mixed traffic. Sensors **2259**(20) (2020)
15. Gazis, D.C., Herman, R., Rothery, R.W.: Nonlinear follow-the-leader of traffic flow. JSTOR **9**(4), 545–567 (1961)
16. Gipps, P.G.: A model for the structure of lane-changing decisions. Transp. Res. Part B **20**(5), 403–414 (1986). https://doi.org/10.1016/0191-2615(86)90012-3
17. Ahmed, K.I.: Modeling drivers' acceleration and lane changing behavior. MIT (1999)
18. Toledo, T.: Modeling integrated lane-changing behavior. TRB 2003 Annual Meeting (2003)
19. Toledo, T., Choudhury, C.F., Ben-Akiva, M.E.: Lane-changing model with explicit target lane choice. Transp. Res. Record **1934**, 157–165 (2005). https://doi.org/10.3141/1934-17
20. Jin, W.L.: A kinematic wave theory of lane-changing traffic flow. Transp. Res. Part B: Methodol. **44**(8–9), 1001–1021 (2010). https://doi.org/10.1016/j.trb.2009.12.014
21. Rahman, M., et al.: Review of microscopic lane-changing models and future research opportunities. IEEE Trans. Intell. Transp.Syst. **14**(4), 1942–1956 (2013). https://doi.org/10.1109/TITS.2013.2272074
22. Moridpour, S., Sarvi, M., Rose, G., Mazloumi, E.: Lane-changing decision model for heavy vehicle drivers. J. Intell. Transp. Syst. Technol. Planning, Oper. **16**, 24–35 (2012). https://doi.org/10.1080/15472450.2012.639640
23. Moridpour, S., Rose, G., Sarvi, M.: Effect of surrounding traffic characteristics on lane changing behavior. J. Transp. Eng. **136**(11), 973–985 (2010). https://doi.org/10.1061/(asce)te.1943-5436.0000165

24. Ko, H.T., Liu, X.H., Guo, M.M., Wu, Z.: A new traffic model with a lane-changing viscosity term. Chinese Phys. B **24**, 098901 (2015). https://doi.org/10.1088/1674-1056/24/9/098901
25. Zheng, Z., Ahn, S., Chen, D., Laval, J.: The effects of lane-changing on the immediate follower: Anticipation, relaxation, and change in driver characteristics. Transp. Res. Part C Emerg. Technol. **26**, 367–379 (2013)
26. Zheng, Z.: Recent developments and research needs in modeling lane changing. Transp. Res. Part B: Methodol. **60**, 561–565 (2014)
27. Lv, W., Song, W.G., Liu, X.D., Ma, J.: A microscopic lane changing process model for multilane traffic. Phys. A Stat. Mech. its Appl. **392**, 1142–1152 (2013). https://doi.org/10.1016/j.physa.2012.11.012
28. Balal, E., Cheu, R.L., Gyan-sarkodie, T.: Analysis of discretionary lane changing parameters on freeways. Int. J. Transp. Sci. Technol. **3**, 277–296 (2014). https://doi.org/10.1260/2046-0430.3.3.277
29. Dutta, M., Ahmed, M.A.: Gap acceptance behavior of drivers at uncontrolled T-intersections under mixed traffic conditions. J. Mod. Transp. **26**(2), 119–132 (2018). https://doi.org/10.1007/s40534-017-0151-9
30. Weng, J., Li, G., Yu, Y.: Time-dependent drivers' merging behavior model in work zone merging areas. Transp. Res. Part C Emerg. Technol. **80**, 409–422 (2017)
31. Zhou, H., Sun, Y., Qin, X., Xu, X., Yao, R.: Modeling discretionary lane-changing behavior on urban streets considering drivers' heterogeneity. Transp. Lett. **12**(3), 1–10 (2019). https://doi.org/10.1080/19427867.2019.1581465
32. Zhou, J., Tkachenko, P., Re, L.: Gap acceptance based safety assessment of autonomous overtaking function. In: 2019 IEEE Intelligent Vehicles Symposium (IV), (Iv), pp. 2113–2118 (2019)
33. Chen, T., Shi, X., Wong, Y.D.: Key feature selection and risk prediction for lane-changing behaviors based on vehicles' trajectory data. Accid. Anal. Prev. **129**, 156–169 (2019). https://doi.org/10.1016/j.aap.2019.05.017
34. Ni, J., Han, J., Liu, Z., Dong, F.: Situation assessment for lane-changing risk based on driver's perception of adjacent rear vehicles. Int. J. Automotive Technol. **21**(2), 427–439 (2020). https://doi.org/10.1007/s12239
35. Marinho, Z., et al.: Functional gradient motion planning in reproducing kernel hilbert spaces. pp. 1–17 (2016). https://homes.cs.washington.edu/~bboots/files/RKHSMP.pdf
36. Dong, C., Zhang, Y., Dolan, J.M.: Lane-change social behavior generator for autonomous driving car by non-parametric regression in Reproducing Kernel Hilbert Space, IEEE International Conference on Intelligent Robots and Systems, 2017-Septe, pp. 4489–4494 (2017). https://doi.org/10.1109/IROS.2017.8206316.
37. Wan, Q., Peng, G., Li, Z., Inomata, F.H.T.: Spatiotemporal trajectory characteristic analysis for traffic state transition prediction near expressway merge bottleneck. Transp. Res. Part C: Emerging Technol. **117**, 102682 (2020)
38. Thiemann, C., Treiber, M., Kesting, A.: Estimating acceleration and lane-changing 42 dynamics based on NGSIM trajectory data. pp. 1–22 (2008). https://doi.org/10.3141/2088-10.
39. Bi, H., Mao, T., Wang, Z., Deng, Z.: A data-driven model for lane-changing in traffic simulation. Ladislav Kavan and Chris Wojtan 1–11 (2016)
40. Pan, T.L., Lam, W.H.K., Sumalee, A., Zhong, R.X.: Modeling the impacts of mandatory and discretionary lane-changing maneuvers. Transp. Res. Part C Emerg. Technol. **68**, 403–424 (2016). https://doi.org/10.1016/j.trc.2016.05.002
41. Gaddam, H.K., Rao, K.R.: Speed-density functional relationship for heterogeneous traffic data: a statistical and theoretical investigation. J. Mod. Transp. **27**, 61–74 (2019). https://doi.org/10.1007/s40534-018-0177-7

The manufacturer's authorised representative in the EU is Springer Nature Customer Service Centre GmbH, Europaplatz 3, 69115 Heidelberg, Germany. If you have any concerns regarding our products, please contact ProductSafety@springernature.com

Printed and bound by CPI Group (UK) Ltd, Croydon, CR0 4YY
30/09/2026
02225295-0014